Design of Controlled Release Drug Delivery Systems

Design of Controlled Release Drug Delivery Systems

Xiaoling Li, Ph.D.

Bhaskara R. Jasti, Ph.D.

Department of Pharmaceutics and Medicinal Chemistry
Thomas J. Long School of Pharmacy and Health Sciences
University of the Pacific
Stockton, California

McGraw-Hill

New York Chicago San Francisco Lisbon London Madrid Mexico City Milan New Delhi San Juan Seoul Singapore Sydney Toronto

The McGraw-Hill Companies

Cataloging-in-Publication Data is on file with the Library of Congress.

1 2 3 4 5 6 7 8 9 0 DOC/DOC 0 1 0 9 8 7 6 5

ISBN 0-07-141759-1

The sponsoring editor for this book was Kenneth P. McCombs, the editing supervisor was David E. Fogarty, and the production supervisor was Richard C. Ruzycka. It was set in Century Schoolbook by International Typesetting and Composition. The art director for the cover was Anthony Landi.

Printed and bound by RR Donnelley.

This book was printed on recycled, acid-free paper containing a minimum of 50% recycled, de-inked fiber.

This book is dedicated to our beloved wives, Xinghang Ma and Hymavathy Jasti, and to our children, Richard Li, Louis Li, Sowmya Jasti, and Sravya Jasti. The perseverance and tolerance of our spouses over the years when our eyes were glued on computer screen, and the play-time sacrifice of our children are highly appreciated.

XIAOLING AND BHASKARA

Contents

Contributors

Brian Barclay, PE (MSChE). *Engineering Fellow, ALZA Corporation, a Johnson & Johnson Company, Mountain View, Calif.* (CHAP. 7)

Bret Berner, Ph.D. *Vice President, Depomed, Inc., Menlo Park, Calif.* (CHAP. 6)

Shiladitya Bhattacharya, M. Pharm. *Ph.D. Candidate, Department of Pharmaceutics and Medicinal Chemistry, Thomas J. Long School of Pharmacy and Health Sciences, University of the Pacific, Stockton, Calif.* (CHAP. 13)

Ramesh R. Boinpally, Ph.D. *Research Investigator, OSI Pharmaceuticals, Boulder, Colo.* (CHAP. 13)

Rajasekhar Chiruvella, M. Pharm. *College of Pharmaceutical Sciences, Kakatiya University, Warangal, India* (CHAP. 13)

Xin Guo, Ph.D. *Assistant Professor, Department of Pharmaceutics and Medicinal Chemistry, Thomas J. Long School of Pharmacy and Health Sciences, University of the Pacific, Stockton, Calif.* (CHAP. 11)

Manjusha Gutta, M.S. *Department of Pharmaceutics and Medicinal Chemistry, Thomas J. Long School of Pharmacy and Health Sciences, University of the Pacific, Stockton, Calif.* (CHAP. 13)

Bhaskara R. Jasti, Ph.D. *Associate Professor, Department of Pharmaceutics and Medicinal Chemistry, Thomas J. Long School of Pharmacy and Health Sciences, University of the Pacific, Stockton, Calif.* (EDITOR, CHAPS. 2, 3, 9)

Tzuchi "Rob" Ju, Ph.D. *Group Leader, Abbott Laboratories, North Chicago, Ill.* (CHAP. 4)

Jin-Seok Kim, Ph.D *Associate Professor, College of Pharmacy, Sookmyung Women's University, Seoul, South Korea* (CHAP. 10)

Amit Kokate, M.S. *Ph.D. Candidate, Department of Pharmaceutics and Medicinal Chemistry, Thomas J. Long School of Pharmacy and Health Sciences, University of the Pacific, Stockton, Calif.* (CHAP. 2)

Phanidhar Kotamraj, M. Pharm. *Ph.D. Candidate, Department of Pharmaceutics and Medicinal Chemistry, Thomas J. Long School of Pharmacy and Health Sciences, University of the Pacific, Stockton, Calif.* (CHAP. 7)

Ick Chan Kwon, Ph.D. *Principal Research Scientist, Biomedical Research Center, Korea Institute of Science and Technology, Seoul, South Korea* (CHAP. 10)

Xiaoling Li, Ph.D. *Professor and Chair, Department of Pharmaceutics and Medicinal Chemistry, Thomas J. Long School of Pharmacy and Health Sciences, University of the Pacific, Stockton, Calif.* (EDITOR, CHAP. 2)

Puchun Liu, Ph.D. *Sr. Director, Emisphere Technologies, Inc., Tarrytown, N.Y.* (CHAP. 4)

Ravichandran Mahalingam, Ph.D. *Post Doctoral Research Fellow, Department of Pharmaceutics and Medicinal Chemistry, Thomas J. Long School of Pharmacy and Health Sciences, University of the Pacific, Stockton, Calif.* (CHAP. 9)

Venugopal P. Marasanapalle, M.S. *Ph.D. Candidate, Department of Pharmaceutics and Medicinal Chemistry, Thomas J. Long School of Pharmacy and Health Sciences, University of the Pacific, Stockton, Calif.* (CHAP. 2)

Rudi Mueller-Walz, Ph.D. *Head, SkyePharma AG, Muttenz, Switzerland* (CHAP. 8)

Jae Hyung Park, Ph.D. *Full Time Lecturer, College of Environment and Applied Chemistry, Kyung Hee University, Gyeonggi-do, South Korea* (CHAP. 10)

Yihong Qiu, Ph.D. *Research Fellow, Abbott Laboratories, North Chicago, Ill.* (CHAP. 4)

Harish Ravivarapu, Ph.D. *Sr. Manager, SuperGen, Inc., Pleasanton, Calif.* (CHAP. 9)

Appala Raju Sagi, M.S. *Scientist, Corium International, Inc., Redwood City, Calif.* (CHAP. 13)

Anant Shanbhag, M.S. *Chemist II, ALZA Corporation, a Johnson & Johnson Company, Mountain View, Calif.* (CHAP. 3)

Sastry Srikonda, Ph.D. *Director, Xenoport Inc., Santa Clara, Calif.* (CHAP. 7)

Rama A. Shmeis, Ph.D. *Principal Scientist, Boehringer-Ingelheim Pharmaceuticals, Inc., Ridgefield, Conn.* (CHAP. 5)

Amir H. Shojaei, Ph.D. *Director, Shire Pharmaceuticals, Inc., Wayne, Pennsylvania.* (CHAP. 6)

James A. Uchizono, Pharm.D., Ph.D. *Assistant Professor, Department of Pharmaceutics and Medicinal Chemistry, Thomas J. Long School of Pharmacy and Health Sciences, University of the Pacific, Stockton, Calif.* (CHAP. 1)

Andrea Wamsley, Ph.D. *Department of Pharmaceutics and Medicinal Chemistry, Thomas J. Long School of Pharmacy and Health Sciences, University of the Pacific, Stockton, Calif.* (CHAP. 12)

Zeren Wang, Ph.D. *Associate Director, Boehringer-Ingelheim Pharmaceuticals, Inc., Ridgefield, Conn.* (CHAP. 5)

Noymi Yam, M.S. *Senior Research Engineer, ALZA Corporation, a Johnson & Johnson Company, Mountain View, Calif.* (CHAP. 3)

Preface

Discovery of a new chemical entity that exerts pharmacological effects for curing or treating diseases or relieving symptoms is only the first step in the drug developmental process. In the developmental cycle of a new drug, the delivery of a desired amount of a therapeutic agent to the target at a specific time or duration is as important as its discovery. In order to realize the optimal therapeutic outcomes, a delivery system should be designed to achieve the optimal drug concentration at a predetermined rate and at the desired location. Currently, many drug delivery systems are available for delivering drugs with either time or spatial controls, and numerous others are under investigation. Many books and reviews on drug delivery systems based on drug release mechanism(s) have been published. As the technology evolves, it is crucial to introduce these new drug delivery concepts in a logical way with successful examples, so that the pharmaceutical scientists and engineers working in the fields of drug discovery, development, and bioengineering can grasp and apply them easily.

In this book, drug delivery systems are presented with emphases on the design principles and their physiological/pathological basis. The content in each chapter is organized with the following sections:

- Introduction
- Rationale for the system design
- Mechanism or kinetics of controlled release
- Key parameters that can be used to modulate the drug delivery rate or spatial targeting
- Current status of the system/technology
- Future potential of the delivery system

Prior to discussing individual drug delivery system/technology based on the design principles, the basic concepts of pharmacokinetics and biological barriers to drug delivery are outlined in the first two chapters.

For each specific design principle, the contributors also briefly introduce the relevant pharmacokinetics (where necessary) and include the challenges of different biological barriers that need to be overcome.

It is our belief that this book provides distinctive knowledge to pharmaceutical scientists, bioengineers, and graduate students in the related fields and can serve as a comprehensive guide and reference to their research and study.

We would like to thank all the authors for their contributions to this book project. Especially, we would like to thank Mr. Kenneth McCombs at McGraw-Hill for his patience, understanding, and support in editing this book.

XIAOLING LI, PH.D.
BHASKARA R. JASTI, PH.D.
Department of Pharmaceutics and Medicinal Chemistry
Thomas J. Long School of Pharmacy and Health Sciences
University of the Pacific
Stockton, California

Design of Controlled Release Drug Delivery Systems

Chapter

1

Application of Pharmacokinetics and Pharmacodynamics in the Design of Controlled Delivery Systems

James A. Uchizono

Thomas J. Long School of Pharmacy and Health Sciences
University of the Pacific
Stockton, California

1.1 Introduction

In biopharmaceutics, more specifically drug delivery, pharmaceutical scientists generally are faced with an engineering problem: develop drug delivery systems that hit a desired target. The target in pharmacokinetics is generally a plasma/blood drug concentration that lies between the minimum effect concentration (MEC) and minimum toxic concentration (MTC) (Fig. 1.1).

In 1937, Teorell's two articles,[1a,1b] "Kinetics of Distribution of Substances Administered to the Body," spawned the birth of pharmacokinetics. Thus his work launched an entire area of science that deals

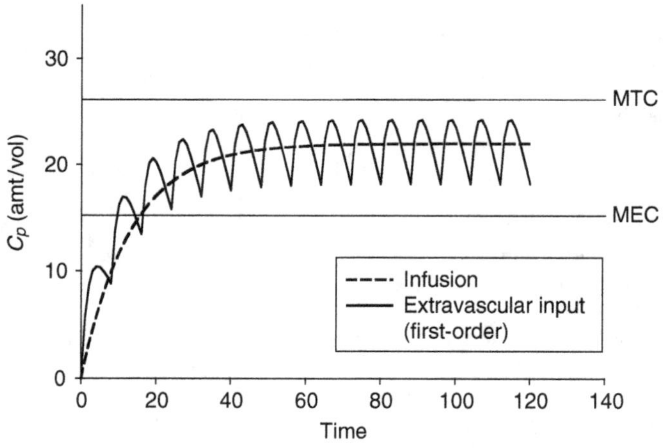

Figure 1.1 Therapeutic window.

with the quantitative aspects that undergird the kinetic foundation of controlled release delivery systems: designing a delivery device or system that achieves a desired drug plasma concentration C_p or a desired concentration profile. To be effective clinically but not toxic, the desired steady-state C_p must be greater than the MEC and less than the MTC. This desired or target steady-state C_p may be achieved by using a variety of dosage forms and delivery/dosage strategies.

1.2 Pharmacokinetics and Pharmacodynamics

Pharmacokinetics and pharmacodynamics provide the time-course dynamics between drug concentration and desired target effect/outcome necessary in the development of optimal drug delivery strategies. The basic premise is that if one is able to model the dynamics governing the translation of drug input into drug concentration in the plasma C_p or drug effect accurately, one potentially can design input drug delivery devices or strategies that maximize the effectiveness of drug therapy while simultaneously minimizing adverse effects. Figure 1.2 shows the relationship between the three main processes that convert the dose into an effect. The pharmacokinetic model translates the dose into a plasma concentration C_p; the link model maps C_p into the drug concentration at the effect site C_e; finally, the pharmacodynamic model converts C_e into the measured effect. For most drugs, C_p is in one-to-one correspondence with the corresponding effect; therefore, most delivery devices can focus primarily on achieving a desired steady-state drug plasma concentration $C_{p,ss}$. Therefore, in this chapter the focus will be on the use of pharmacokinetics to guide the design of controlled release delivery systems that achieve their intended concentration. Some issues arising owing to C_p versus effect nonstationarity (either time- or state-varying pharmacokinetics or pharmacodynamics) will be discussed in the section entitled, "Limitations of Using Pharmacokinetics Only to Design Controlled Release Delivery Systems."

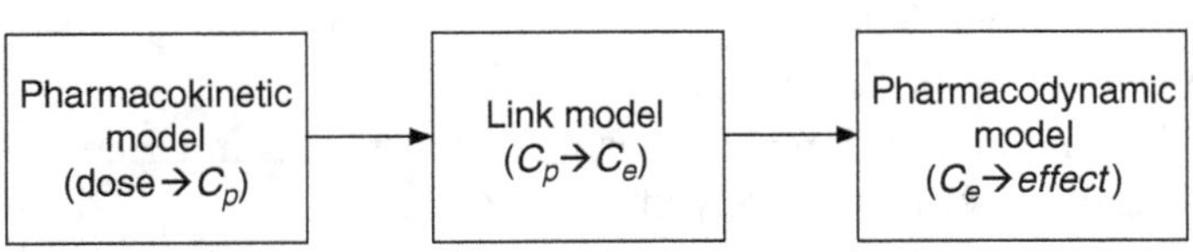

Figure 1.2 Relationship between the pharmacokinetic, link, and pharmacodynamic models.

1.3 LADME Scheme and Meaning of Pharmacokinetic Parameters

The frequently used acronym *LADME*, which stands for liberation, absorption, distribution, metabolism, and excretion, broadly describes the various biopharmaceutical processes influencing the pharmacokinetics of a drug. Since each of aspect of LADME can influence the pharmacokinetics of a drug and ultimately the design of controlled release delivery devices, this section will review and explain the relationship between LADME processes and eight common pharmacokinetic parameters (F, K, V_d, $t_{1/2}$, Cl, k_a, $t_{\max}$, $C_{p,\max}$).

Each of the LADME processes can have an impact on a drug's pharmacokinetics profile, some more than others depending on the physicochemical properties of the drug, dosage formulation, route of administration, rates of distribution, patient's specific anatomy/physiology, biotransformation/metabolism, and excretion. From a pharmacokinetics perspective, liberation encompasses all kinetic aspects related to the liberation of drug from its dosage form into its active or desired form. For example, free drug released from a tablet or polymeric matrix in the gut would be liberation. Although liberation is first in the LADME scheme, it does not need to occur first. For example, ester prodrug formulations can be designed to improve gut absorption by increasing lipophilicity. These ester formulations deliver the prodrug into the systemic circulation, where blood esterases or even chemical decomposition cleaves the ester into two fragments, a carboxcylic acid and an alcohol; the desired free drug can be liberated as either the carboxcylic acid or the alcohol depending on the chemical design. Liberation kinetics can be altered by other physicochemical properties, such as drug solubility, melting point of vehicle (suppository), drug dissolution, gastrointestinal pH, etc. Overall liberation kinetics are fairly well known because they generally can be estimated from in vitro experiments. The foundational principles governing the liberation of drug from delivery systems were laid by many, who rigorously applied the laws and principles of physics and physical chemistry to drug delivery systems.[2–12]

1.3.1 Maximum concentration, time to maximum concentration, and first-order absorption rate constant $C_{p,\max}$, $t_{\max}$, k_a

Although liberation and absorption can overlap, absorption is much more difficult to model accurately and precisely in pharmacokinetics. A great deal of work in this area by Wagner-Nelson[13–15] and Loo-Riegelman[16,17] reflects the complexities of using pharmacokinetics and diffusion models to describe the rate of drug absorption. Since most drugs are delivered via the oral route, the gastrointestinal (GI) tract is described briefly. In the GI tract, the source of these complexities lies in the changing environmental conditions surrounding the drug and delivery modality as it moves along the GI tract. Most drugs experience a mix of zero- and first-order kinetic absorption; this mixing of zero- and first-order input results in nonlinearities between dose and C_p (see "Linear versus Nonlinear Pharmacokinetics"). A widely used simplification assumes that extravascular absorption (including the gut) is a first-order process with a rate constant k_a or $k_{e.v}$ or k_{abs}; practically, $C_{p,max}$ and t_{max} are also used to characterize the kinetics of absorption. $C_{p,max}$ (i.e., the maximal C_p) can be determined directly from a plot of C_p versus time; it is the maximum concentration achieved during the absorption phase. t_{max} is amount of time it takes for $C_{p,max}$ to be reached for a given dose [see Fig. 1.14; the equations for $C_{p,max}$ and t_{max} are given in Eqs. (1.28) and (1.29)].

1.3.2 Bioavailability *F*

While pharmacokinetics describing the rate of absorption are quite complex owing to simultaneous kinetic mixing of passive diffusion and multiple active transporters (e.g., P-glycoprotein,[18] amino acid[19]) and enzymes (cytochrome P450s[20–23]) pharmacokinetics describing the *extent* of absorption are well characterized and generally accepted, with area under the C_p curve (AUC) (Eq. 1.1) being the most widely used pharmacokinetics parameter to define extent of absorption. AUC is closely and sometimes incorrectly associated with bioavailability. AUC is a measure of extent of absorption, not rate of absorption; true bioavailability is made up of both extent and rate of absorption. The rate of absorption tends to be more important in acute-use medications (e.g., pain management), and the extent of absorption is a more important factor in chronic-use medications.[24] Frequently, the unitless ratio pharmacokinetics parameter F will be used to represent absolute bioavailability under steady-state conditions or for medications of chronic use.

$$\mathrm{AUC} = \int C_p(t)\,dt \tag{1.1}$$

$$F = \frac{\mathrm{AUC}_{\mathrm{e.v.}} / \mathrm{dose}_{\mathrm{e.v.}}}{\mathrm{AUC}_{\mathrm{i.v.}} / \mathrm{dose}_{\mathrm{i.v.}}} \tag{1.2}$$

In Eq. (1.2), the e.v. and i.v. subscripts stand for extravascular and intravenous, respectively. F is a unitless ratio, $0 < F \leq 1$, that compares the drug's availability given in a nonintravenous route compared with the availability obtained when the drug is given by the intravenous route. F is also known as the fraction of dose that reaches the systemic circulation (i.e., posthepatic circulation).

1.3.3 Volume of distribution V_d

Volume of distribution V_d has units of volume but is not an actual physiologically identifiable volume. The first common definition of V_d is that "it is the volume that it appears the drug is dissolved in." The second definition for V_d is that "it is the proportionality constant that links the amount of drug in the body to the concentration of drug measured in the blood." Clinically, in general, the larger V_d is, the greater is the extent of drug partitioning and the greater is the amount of drug being removed from the site of measurement. Most drugs have a V_d of between 3.5 and 1000 L; there are cases where V_d is greater than 20,000 L (as in some antimalarial drugs).

1.3.4 Clearance Cl

Systemic clearance Cl can be defined as the volume of blood/plasma completely cleared of drug per unit time. Systemic clearance is calculated by dividing the amount of drug reaching the systemic circulation by the resulting AUC (Eq. 1.3). At any given C_p, the amount of drug lost per unit time can be determined easily by multiplying Cl × C_p.

$$\mathrm{Cl} = \frac{(F)(S)\,\mathrm{dose}}{\mathrm{AUC}} \tag{1.3}$$

1.3.5 First-order elimination rate constant *K* and half-life $t_{1/2}$

The first-order elimination rate constant K can be determined as shown in Eq. (1.4) and has units of 1/time. The larger the value of K, the more rapidly elimination occurs. Once K has been determined, then calculating the half-life $t_{1/2}$ is straightforward (Eq. 1.5).

$$K = \frac{\mathrm{Cl}}{V_d} \tag{1.4}$$

$$t_{1/2} = \frac{\ln(2)}{K} = \frac{(V_d)\ln(2)}{\mathrm{Cl}} \tag{1.5}$$

Equations (1.4) and (1.5) were written intentionally in these two forms to indicate that K and $t_{1/2}$ are functions of Cl and V_d, and not vice versa. The anatomy and physiology of the body, along with the physicochemical properties of the drug, combine to form the biopharmaceutical properties, such as Cl and V_d, which can be found in many reference books.[25]

Clinically, the two pharmacokinetics parameters $t_{1/2}$ and systemic clearance Cl are very important when determining patient-specific dosing regimen. A patient's drug concentration is at steady state clinically when the drug concentration is greater than 90 percent of the true steady-state level (some clinicians use 96 percent, but nearly all use at least 90 percent). According to the preceding definition of $t_{1/2}$, it will take a patient approximately 3.3 half-lives to reach 90 percent of the true steady state (this assumes no loading dose and that each dose is the same size); at 5 half-lives, the patient will be approximately 96 percent to the true steady-state level. While $t_{1/2}$ is an important pharmacokinetics parameter when determining the dosing interval, the size of the dose is not based on $t_{1/2}$. Two other pharmacokinetics parameters, V_d (volume of distribution) and Cl (systemic clearance), help to determine the size of the dose.

1.4 Pharmacokinetics and Classes of Models

Many books and review articles have been written about pharmacokinetics.[24,26–28] And as one would suspect, there are multiple ways to model the kinetic behavior of a drug in the body. The three most common classes of pharmacokinetic models are *compartmental, noncompartmental,* and *physiological modeling*. Although physiologic modeling[24,26,29,30] gives the most accurate view of underlying mechanistic kinetic behavior, it requires fairly elaborate experimental and clinical setups. Noncompartmental modeling[31–37] is based on statistical moment theory and requires fewer a priori assumptions regarding physiological drug distribution and mechanisms of drug elimination. Over the last 10 years, increased computational capabilities and sophisticated nonlinear parameter-estimation software packages have encouraged the reintroduction of noncompartmental modeling strategies. In compartmental modeling, the underlying idea is to bunch tissues and organs that similarly affect the kinetic behavior of a drug of interest together to form compartments. While the compartmentalization of tissues and organs leads to a loss of information (e.g., mechanistic), the plasma kinetic behavior of most drugs can be approximated with tractable models with as few compartments as one, two, or three. In addition to these three general model classifications, the issue of linearity versus nonlinearity has an impact on all three general classifications. These terms describe the relationship between dose and C_p. Regardless of modeling paradigm, the clinical goal of pharmacokinetics is to determine an optimal dosing strategy based on patient-specific parameters, measurements, and/or disease state(s), where *optimal* is defined by the clinician. The development of many new controlled release delivery devices over the last 20 or so years has given the clinician many alternative dosing inputs.

1.4.1 Linear versus nonlinear pharmacokinetics

A general understanding of the definitions of *linear* and *nonlinear* will be helpful when discussing drug input into the body, whether that dose or input is delivered by classic delivery means or by novel controlled release delivery systems. Linear and nonlinear pharmacokinetics are differentiated by the relationship between the dose and the resulting drug concentration. A linear pharmacokinetics system exhibits a proportional relationship between dose and C_p for all doses, whereas nonlinear pharmacokinetics systems do not.

Linear pharmacokinetics. For a simple linear pharmacokinetics case, the body can be modeled as a single drug compartment with first-order kinetic elimination—where the dose is administered and drug concentrations are drawn from the same compartment. For an intravenous bolus dose, the expected drug plasma concentration C_p versus time curves are shown in Fig. 1.10. The kinetics for this system are described by Eq. (1.6). The well-known solution to this equation is given by Eq. (1.7), and a linearized version of this solution is given in Eq. (1.8) and shown graphically in Fig. 1.13.

$$\frac{d(C_p)}{dt} = -K(C_p) \tag{1.6}$$

$$C_p = \frac{\text{dose}}{V_d} e^{-Kt} = C_p^0 e^{-Kt} \tag{1.7}$$

$$\log_e C_p = \log_e C_p^0 - Kt \quad \text{or} \quad \log_{10} C_p = \log_{10} C_p^0 - \frac{K}{2.303} t \tag{1.8}$$

where V_d is volume of distribution, and K is the first-order kinetic rate constant of elimination. According to Eq. (1.7) the linear relationship between dose and C_p holds for all sized doses.

If for the same one-compartment model the input is changed from an intravenous bolus to first-order kinetic input (e.g., gut absorption), the expected C_p versus time curves are shown in Fig. 1.14. The kinetics for this system are described by

$$\frac{d(C_p)}{dt} = k_a C_a - KC_p \tag{1.9}$$

where k_a is the first-order kinetic rate input constant, and C_a is the driving force concentration or concentration of drug at the site of administration. The integrated solution for Eq. (1.9) is given by Eq. (1.10):

$$C_p = \frac{(F)(S)(\text{dose})k_a}{V_d(k_a - K)}\left(e^{-Kt} - e^{-k_a t}\right) \tag{1.10}$$

Although Eq. (1.10) is linear with respect to dose, it is not linear with respect to its parameters (k_a and K). The definition of linear and nonlinear pharmacokinetic models is based on the relationship between C_p and dose, not with respect to the parameters.

Nonlinear pharmacokinetics. *Nonlinear pharmacokinetics* simply means that the relationship between dose and C_p is not directly proportional for all doses. In nonlinear pharmacokinetics, drug concentration does not scale in direct proportion to dose (also known as *dose-dependent kinetics*). One classic drug example of nonlinear pharmacokinetics is the anticonvulsant drug phenytoin.[38] Clinicians have learned to dose phenytoin carefully in amounts greater than 300 mg/day; above this point, most individuals will have dramatically increased phenytoin plasma levels in response to small changes in the input dose.

Many time-dependent processes appear to be nonlinear, yet when the drug concentration is measured carefully relative to the time of dose, the underlying dose-to-drug-concentration relationship is directly proportional to the dose and therefore is linear (see "Time- and State-Varying Pharmacokinetics and Pharmacodynamics").

1.4.2 Time- and state-varying pharmacokinetics and pharmacodynamics

Time- and state-varying pharmacokinetics or pharmacodynamics refer to the dynamic or static behavior of the parameters used in the model. Time-varying would encompass phenomena such as the circadian variation of C_p owing to underlying circadian changes in systemic clearance. While time-varying can be considered a subset of the more general state-varying models, state-varying parameters can change as an explicit function of time and/or as an explicit function of another pharmacokinetic or pharmacodynamic state variable (e.g., metabolite concentration, AUC, etc.).

Time-varying. Figure 1.3 shows two possible C_p versus time plots that could arise from a pharmacokinetic/pharmacodynamic system where Cl (bottom panel) or receptor density (top panel) varies sinusoidally

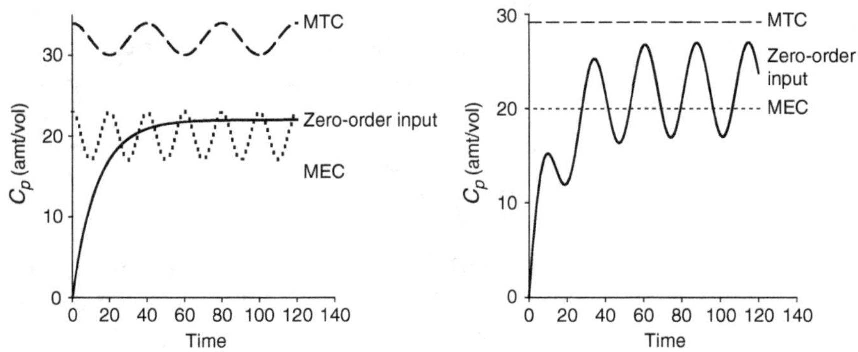

Figure 1.3 Plots showing two different scenarios caused by time- or state-varying pharmacokinetic or pharmacodynamic parameters.

with time. The solid line is the drug-concentration-versus-time profile in response to a zero-order input in both plots. The top panel shows the MEC and MTC (dotted and dashed lines, respectively) changing as a function of time—indicating that one or more pharmacodynamic parameters is changing (e.g., receptor density). The bottom panel shows stationary MEC and MTC, but the concentration-time profile is oscillating as a function of time—indicating that one or more pharmacokinetic parameters is changing (e.g., Cl). In either case, the C_p curve periodically drops below the MEC—thus rendering the drug ineffective during the periods where C_p is less than the MEC.

State-varying. Figure 1.4 shows two plots of concentration-time profiles and MEC/MTC behavior for pharmacokinetic/pharmacodynamic systems with stationary parameters (top panel) and nonstationary

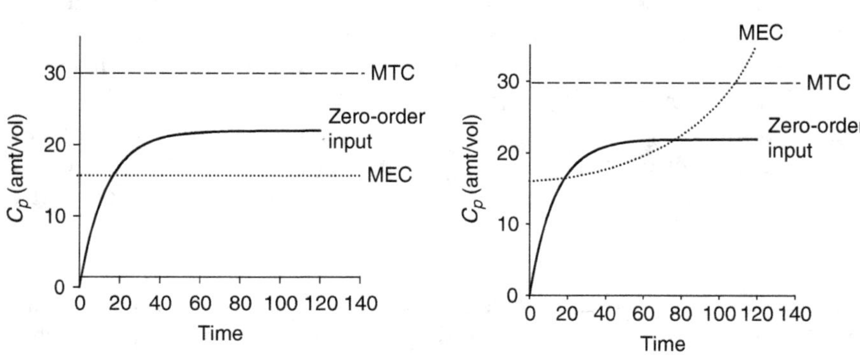

Figure 1.4 Alteration of MEC in a state-varying system.

parameters (bottom panel). In both plots, the solid line is the drug-concentration-versus-time profile in response to a zero-order input, and the dotted and dashed lines represent the MEC and MTC, respectively. The bottom panel could represent the presence of pharmacodynamic drug tolerance (e.g., receptor desensitization). In the bottom panel, the drug starts out effective, and then, as drug tolerance develops, C_p is no longer greater than MEC, resulting in drug ineffectiveness.

1.5 Pharmacokinetics: Input, Disposition, and Convolution

In linear pharmacokinetics, the drug concentration-versus-time-course profile is the result of two distinct kinetic components—*input* and *disposition*. Nearly all dosage forms, both old and new, can be classified into one of three kinetic categories—instantaneous, zero order, or first order. Since the physiology, anatomy, and drug physicochemical characteristics largely determine the disposition component, if we want to have any control over the drug's concentration profile, we must modulate the input. The next subsection identifies the kinetic order of the most commonly used dosing inputs, followed by a subsection on the kinetic order of different disposition models and a concluding subsection describing the mathematical combination of an input function and disposition function to give a complete drug concentration kinetic profile.

1.5.1 Input

The regulation of drug input into the body is the core tenet of controlled release drug delivery systems. With advances in engineering and material sciences, controlled release delivery systems are able to mimic multiple kinetic types of input, ranging from instantaneous to complex kinetic order. In this section three of the most common input functions found in controlled release drug delivery systems will be discussed—*instantaneous, zero order*, and *first order*.

Instantaneous input (IB). Truly *instantaneous input* (IB) does not physically exist; in fact, the kinetic order is mathematically undefined. However, when the input kinetics are exceedingly fast compared with distribution and elimination kinetics, the dose provides a relatively "instantaneous" input. The best example of an "instantaneous" input is an intravenous bolus dose—where the drug is administered over a short period (<5 minutes) and directly into the systemic circulation. Figure 1.5 (left panel) shows this type of input being given at time = t'. When the intravenous bolus is given repetitively at a fixed interval

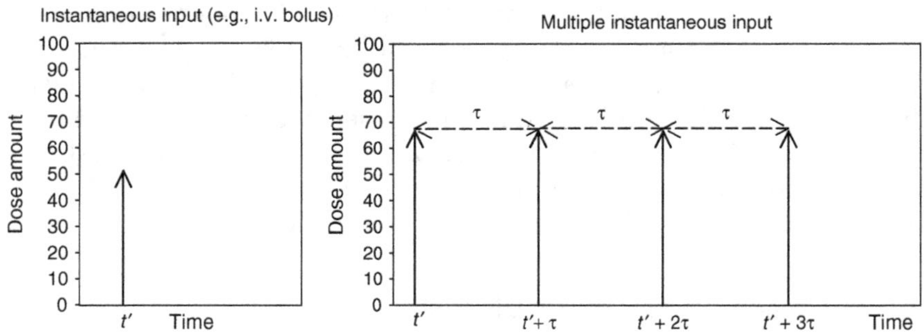

Figure 1.5 Figure showing single and multiple instantaneous input.

τ, as shown in the right panel of the figure, the resulting plot is similar to the desired results for pulsatile controlled release delivery systems.

For instantaneous input (intravenous bolus), the derivative for dose amount with respect to time is undefined because in the limit $dt = 0$, thus resulting in a zero for the denominator of d (dose amount)/dt or an undefined derivative (Eq. 1.11):

$$\text{Intravenous bolus} = \frac{d\,(\text{dose amount})}{dt} \tag{1.11}$$

Zero-order input (I0). *Zero-order kinetic input* (I0) refers to an input system that delivers drug at a constant rate. The best example of zero-order input devices are intravenous infusions (>30 minutes' duration). Historically, pharmaceutical scientists have focused on zero-order delivery systems because these systems achieve relatively stable C_p levels—thereby helping to minimize side effects owing to peak drug concentrations and lack of efficacy owing to subtherapeutic trough drug concentrations (see "Convolution of Input and Disposition"). A zero-order system delivers the same amount of drug per unit time from its initiation to termination, as shown in Fig. 1.6.

The differential equation describing this kinetic behavior from t_{start} to t_{end} in Fig. 1.6 is

$$\frac{d(\text{drug amount})}{dt} = k_0 \tag{1.12}$$

where k_0 is a zero-order kinetic rate constant and is equal to 1 in Fig. 1.6. Equation (1.12) describes an input rate process that is independent of

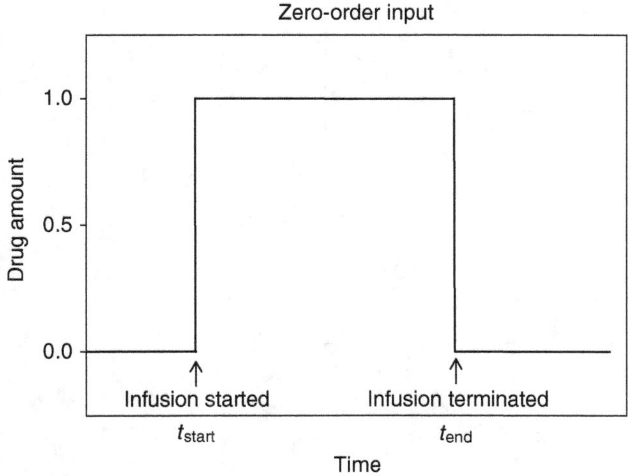

Figure 1.6 Zero-order release kinetics.

the amount of drug in the reservoir holding the drug (i.e., drug amount does not explicitly appear on the right-hand side of the equation). The units of k_0 are mass/time.

First-order input (I1). *First-order kinetic input* (I1) delivers drug at a rate proportional to the concentration gradient driving the transfer of drug movement. A classic example of a first-order kinetic process is the passive diffusion of drug across a homogeneous barrier. The differential equation describing first-order kinetic behavior is shown in Eq. (1.13):

$$\frac{dC_p}{dt} = k_a C_{\text{site of absorption}} \qquad (1.13)$$

The rate of appearance of drug in the plasma C_p is directly proportional to the concentration of drug at the site of absorption ($C_{\text{site of absorption}}$). Like all first-order rate constants, the units of the absorption rate constant k_a are 1/time. A plot of drug amount versus time is shown in Fig. 1.7.

1.5.2 Disposition

The kinetic order of a drug disposition is determined primarily by the relationship between the patient's physiology/anatomy and the physicochemical properties of drug. Disposition is made up of three major components: (1) distribution, (2) metabolism, and (3) excretion. These

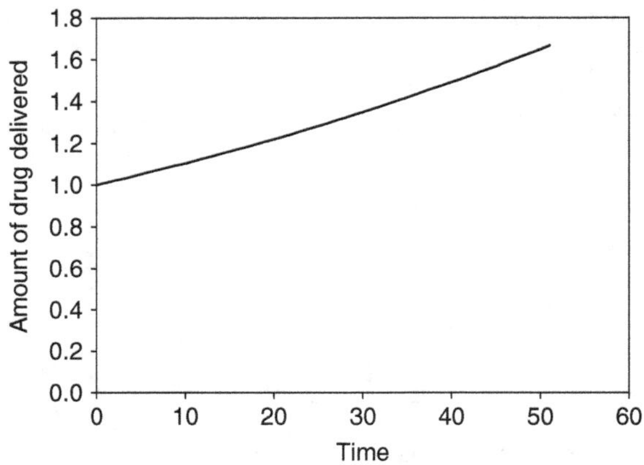

Figure 1.7 Plot showing amount of drug delivered versus time for a first-order delivery process.

three processes occur simultaneously in the body. As time passes from initiation of therapy to its end, any one of these three components can dominantly shape the drug concentration profile. *Distribution* is the movement of drug between tissues (e.g., blood to adipose tissue) and generally is considered to be bidirectional first-order kinetic processes (e.g., k_{12}, k_{21}). *Metabolism* is the biotransformation of the drug by enzymes or chemical reactions into its metabolites. As long as the $C_p << K_m$ (where K_m is the enzyme's Michaelis constant), drug will be metabolized under pseudo-first-order kinetics k_m (different from the Michaelis constant K_m). Although physiologically most metabolized drugs are excreted in the urine or feces, the metabolite does not contribute to the first-order rate constant k_e used to describe excretion of the parent compound; the loss of drug to metabolite biotransformation has been accounted for already by the metabolism first-order rate constant k_m. *Excretion* has three components: (1) filtration, (2) passive and active secretion, and (3) passive and active reabsorption. As long as $C_p << T_{50,1}$ for secretion and $C_{\text{urine}} << T_{50,2}$ for reabsorption, both transport processes will be approximately first order (Eqs. 1.14 and 1.15). Filtration is directly proportional to C_p and does not saturate (or exhibit dose-dependent nonlinear kinetics) at therapeutic drug concentrations. If all three excretion processes are exhibiting first-order behavior, then a single first-order rate constant k_e can be used to describe excretion (Eq. 1.16):

$$T_{\text{secrection}} = \frac{T_{\max,1} C_p}{T_{50,1} + C_p} \qquad (1.14)$$

$$T_{\text{reabsorption}} = \frac{T_{\text{max},2} C_{\text{urine}}}{T_{50,2} + C_{\text{urine}}} \tag{1.15}$$

$$k_e = k_{\text{filtration}} + k_{\text{secretion}} - k'_{\text{reabsorption}} \tag{1.16}$$

The minus sign and prime on $k_{\text{reabsorption}}$ indicate a different driving force concentration than for filtration and secretion and drug transport in the opposite direction. *Elimination* is the generic term given to the first-order rate constant K, or sometimes β, describing the parent drug lost by both metabolism k_m and excretion k_e (Eq. 1.17):

$$K = k_m + k_e \tag{1.17}$$

Factors analogous to those affecting gut absorption also can affect drug distribution and excretion. Any transporters or metabolizing enzymes can be taxed to capacity—which clearly would make the kinetic process nonlinear (see "Linear versus Nonlinear Pharmacokinetics"). In order to have linear pharmacokinetics, all components (distribution, metabolism, filtration, active secretion, and active reabsorption) must be reasonably approximated by first-order kinetics for the valid design of controlled release delivery systems.

1.5.3 Convolution of input and disposition

To obtain a complete drug concentration profile, both the input and disposition kinetics must be known or assumed. If the input is an intravenous bolus, zero or first order, and disposition is first order, then the input and disposition can be combined mathematically through the convolution operation, represented by the $*$ symbol. Mathematically, this is represented as

$$C_p(t) = \text{input}\ (t) * \text{disposition}\ (t) = \int_0^t \text{input}(\tau) \times \text{disposition}\ (t - \tau)\ d\tau \tag{1.18}$$

If we know input(t) and $C_p(t)$, we can extract disposition(t). The easiest way to accomplish this deconvolution (extraction) is to give an intravenous bolus dose and measure $C_p(t)$, which will exactly mirror the underlying disposition(t). Once disposition(t) is known (and assumed not to change for the same drug), $C_p(t)$ can be predicted for any input(t), or more important for controlled release of delivery systems, the input(t) needed to produce a specific $C_p(t)$ profile can be determined easily. Pharmacokinetics is simply the convolution of an input(t) with the drug/patient's disposition(t). Putting the whole package together, pharmacokinetics includes all kinetic aspects of input (liberation, absorption) and disposition (distribution, metabolism, and excretion).

1.6 Compartmental Pharmacokinetic Modeling

1.6.1 Single-dose input systems

Compartmental modeling is by far the most commonly used pharmacokinetics modeling technique. In compartmental modeling, tissues having similar kinetic drug concentration profiles are lumped together into a compartment. For example, a common three-compartment model (Fig. 1.8) may have one compartment representing the blood/plasma and other tissues that reach their steady-state concentration very rapidly (< 3 hours) for a given dose (i.e., kinetically similar to the blood). This compartment is commonly called the *central compartment* and usually contains such organs as the blood/plasma, kidney, lungs, liver, and most other large internal organs. The second compartment in this three-compartment model could be called *shallow tissues*; these tissues do not reach their steady-state concentration as rapidly as the central compartment but still reach steady-state somewhat quickly (3 to 8 hours). Examples of the shallow compartment might be organs such as muscle, eyes, and other smaller internal organs, as well as sometimes the skin. The third compartment consists of tissues that reach their steady-state concentration slowly; examples of the *deep compartment* are adipose tissue, brain, and sometimes skin tissues (particularly when the drug

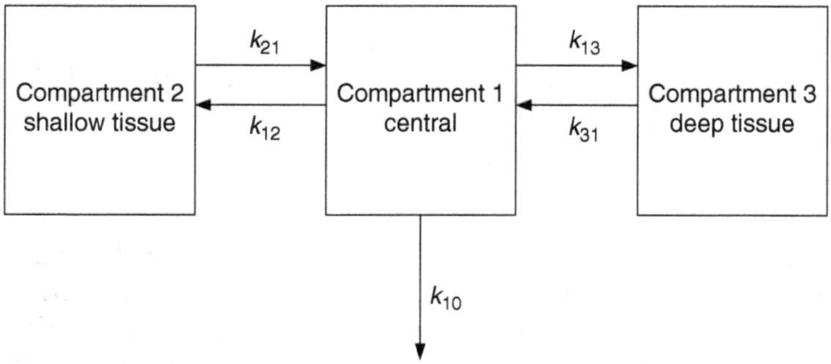

Figure 1.8 Classic three-compartment model with elimination from only the central compartment. All microrate constants are first order.

is sequestered in the statrum corneum). While three or more compartment models frequently can be justified to describe the kinetic concentration profile of a drug accurately, researchers prefer to deal with less cumbersome yet pragmatically useful compartmental models. Below are four of the simplest and most useful compartmental models. The four models—designated *IBD1, IBD2, I0D1*, and *I1D1*, can be linked easily back to one of the three input types described earlier convolved with either a one-compartment first-order elimination disposition or a two-compartment first-order elimination disposition. Although there would be six combinations (three different inputs convolved with two different dispositions), two of these six combinations have been removed for simplicity sake: (1) zero-order input with two-compartment disposition and (2) first-order input with two-compartment disposition. The abbreviations are defined as IB = intravenous bolus input; I0 = zero-order input; I1 = first-order input; D1 = one-compartment disposition, first-order elimination, and instantaneous distribution between the blood and all organs/tissues; and D2 = two-compartment disposition, first-order elimination, and first-order transfer of drug between the blood and some organs/tissues.

Instantaneous input and one-compartment disposition (IBD1). In this model, all tissues/organs are considered kinetically similar to the blood/plasma [i.e., (drug) changes in the blood are communicated to all tissues/organs, and all tissues/organs instantaneously respond to the blood (drug) change]. This one-compartmental model can be represented by Fig. 1.9.

The differential equation and its solution that describe this model are given by

$$\frac{d(C_p)}{dt} = -K(C_p) \tag{1.19}$$

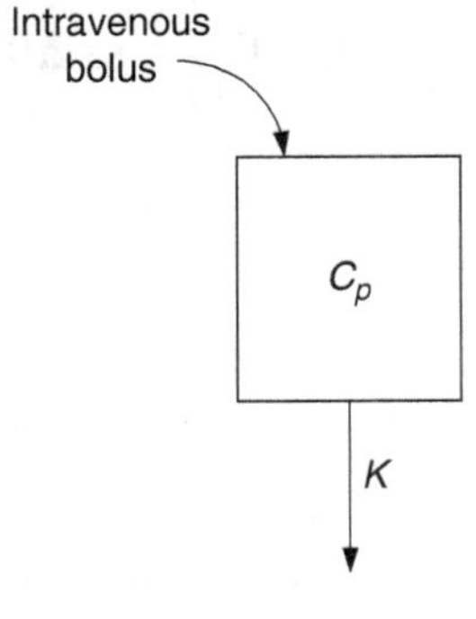

Figure 1.9 One compartment box diagram.

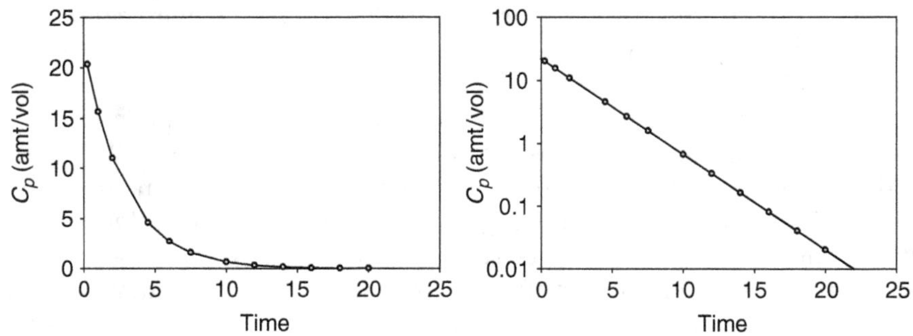

Figure 1.10 C_p versus time for intravenous bolus input and one-compartment disposition.

$$C_p = C_p{}^0 e^{-Kt} = \frac{\text{dose}}{V_d} e^{-Kt} \tag{1.20}$$

where K is the first-order rate constant of elimination, and C_p^0 is C_p when time is extrapolated back to zero (see Fig. 1.10 for the IBD1 plot). The other related parameters can be determined [shown in Eq. (1.7)]. If one has C_p versus time data, K and V_d can be determined by a software nonlinear regression or through graphic means; the former is more accurate and is preferred. For an intravenous bolus dose, the parameter F equals 1. $t_{1/2}$ is simply calculated as $t_{1/2} = \ln(2)/K$. The systemic clearance can be calculated by the following equation:

$$\text{Cl} = \frac{\text{dose}}{\text{AUC}_{0\to\infty}} = \frac{\text{dose}}{C_p^0/K} \tag{1.21}$$

Zero-order input and one-compartment disposition (I0D1). This model and the I0D1 model only differ from the first model (IBD1) by the kinetic order of the input; the disposition component remains the same. The compartmental box diagram is shown in Fig. 1.11. The differential

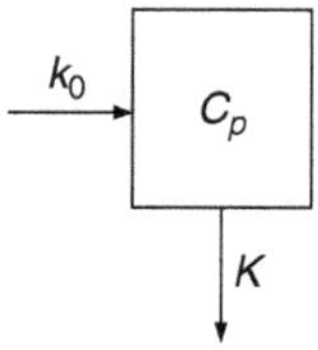

Figure 1.11 Zero-order input and one-compartment disposition box diagram.

equation (Eq. 1.22) describes an I0D1 model, and its integrated form is shown in Eqs. (1.23) and (1.24):

$$\frac{d(C_p)}{dt} = \frac{k_0}{V_d} - K(C_p) \tag{1.22}$$

$$C_p = \underbrace{\left[\frac{k_0}{\underbrace{(K)(V_d)}_{\text{Cl}}}\right]}_{C_{p,\text{ss}}}\left(1 - e^{-(k)(t)}\right) = \frac{k_0}{\text{Cl}}\left(1 - e^{-(k)(t)}\right) \tag{1.23}$$

$$C_p = C_{p,\text{ss}}\left(1 - e^{-(k)(t)}\right) \tag{1.24}$$

where k_0 is the zero-order input rate constant (units of amount per time), K is the first-order elimination rate constant (units of 1/time), V_d is the volume of distribution, and C_p is the drug plasma concentration.

Figure 1.12 shows the concentration profile plotted on Cartesian coordinates and semilog scales respectively. When a zero-order input has been left on for an amount of time equal to 3.3 to 5 half-lives, C_p is considered to be at steady state (90 and 96 percent of true steady state, respectively) and is designated as $C_{p,\text{ss}}$. There is only one volume term, V_d, because the disposition is only one compartment.

First-order input and one-compartment disposition (I1D1). In this model, the input is first order, and the disposition is one compartment. The

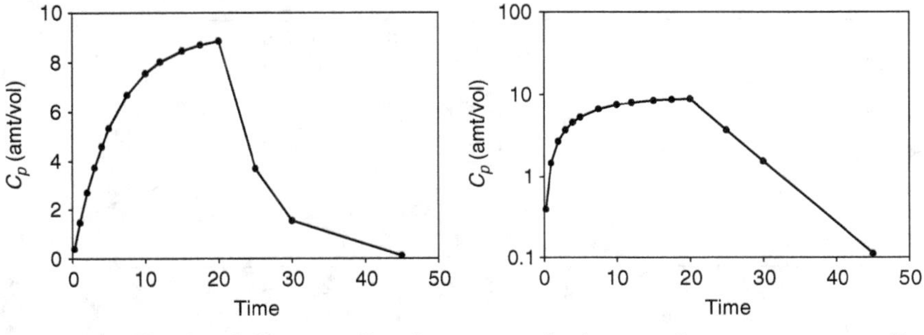

Figure 1.12 Simulated C_p versus time for a zero-order input and one-compartment disposition.

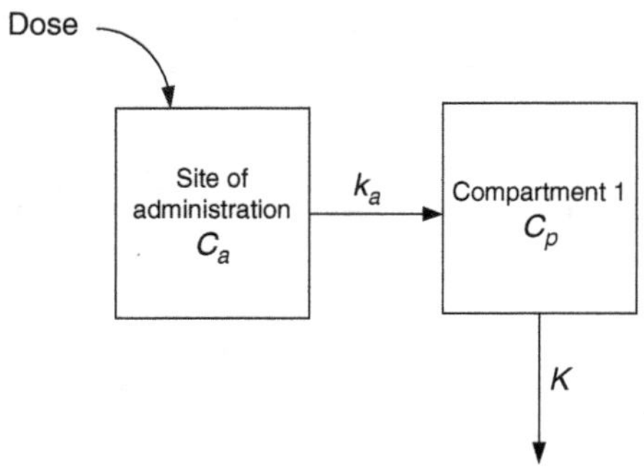

Figure 1.13 First-order input and one-compartment disposition box diagram.

compartmental box diagram (shown in Fig. 1.13) and C_p versus time plot are shown in Fig. 1.14.

This model generally describes a situation where the drug is administered into a depot tissue (e.g., tablet in the gut or injection into muscle), and the drug transports across a biomembrane in a first-order manner. In this case, the drug concentration at the depot site provides the driving force to move the drug out of the depot and into the systemic circulation. The differential equation and its integrated form describing I1D1 are given by Eqs. (1.25) and (1.26):

$$\frac{d(C_p)}{dt} = k_a\left(C_{\text{depot}}\right) - K(C_p) \tag{1.25}$$

$$C_p = \frac{(F)(S)(\text{dose})(k_a)}{V_d(k_a - K)}\left(e^{-K(t)} - e^{-k_a(t)}\right) \tag{1.26}$$

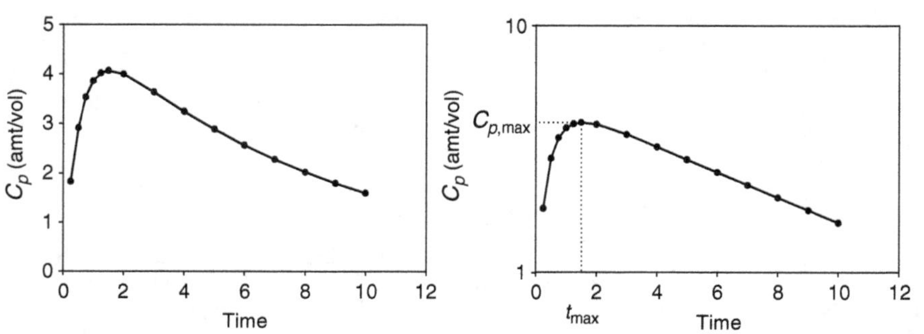

Figure 1.14 Simulated data of C_p versus time for first-order input and one-compartment disposition.

where k_a is the first-order absorption rate constant (units of 1/time), K is the first-order rate constant of elimination (units of 1/time), V_d is the volume of distribution (units of volume), F is the fraction of the administered dose that is delivered to the systemic circulation as parent compound (no units, varies from 0 to 1), and S is the formulation salt factor (no units, varies from 0 to 1). V_d can be determined by Eq. (1.27):

$$V_d = \frac{(F)(S)(\text{dose})}{(\text{AUC}_{0\to\infty})(K)} \tag{1.27}$$

which is the same equation for $V_{d(\text{area})}$ in two-compartment disposition models. In one-compartment disposition models, $V_{d(\text{area})}$ degenerates into V_d. Two other parameters, $C_{p,\max}$ and $t_{\max}$, which identify the maximum drug concentration reached and the time at which that maximum is reached, respectively, are useful in the design of controlled release delivery systems. These values are calculated by Eqs. (1.28) and (1.29), and the graphic representation of these values can be seen on Fig. 1.14.

$$t_{\max} = \frac{\ln(k_a/K)}{k_a - K} \tag{1.28}$$

$$C_{p,\max} = \frac{(F)(S)(\text{dose})(k_a)}{V_d(k_a - K)}\left(e^{-K(t_{\max})} - e^{-k_a(t_{\max})}\right) \tag{1.29}$$

Instantaneous input and two-compartment disposition (IBD2). The IBD2 model adds another level of reality and complexity to the IBD1 model yet still remains relevant and computationally accessible to most scientists. The two-compartment model is a nice compromise that allows for distribution and elimination kinetics, whereas one-compartment models only have elimination kinetics. In the IBD2 model, the input is an intravenous bolus dose, and the disposition consists of two compartments—a "central" compartment and a "tissue" compartment (Fig. 1.15).

The central compartment represents the blood/plasma and any other tissue that rapidly equilibrates, relative to the distribution rate, with the blood/plasma (e.g., liver or heart tissue). The tissue compartment represents all other tissues that keep the drug and reach steady-state concentrations more slowly than the tissues of the central compartment. Since the two-compartment model is fairly robust in describing a bulk of all drugs, we will limit our discussion to two compartments with elimination

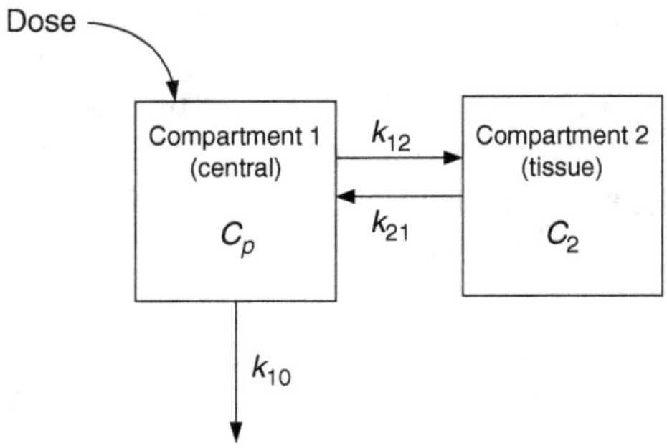

Figure 1.15 Intravenous bolus input and two-compartment disposition box diagram.

only from the central compartments; elimination can occur from either compartment or both. The differential equations and integrated forms of this model with elimination from the central compartment only are shown in Eqs. (1.30) through (1.32):

$$\frac{d(C_p)}{dt} = k_{21}(C_2) - k_{10}(C_p) - k_{12}(C_p) \tag{1.30}$$

$$\frac{d(C_2)}{dt} = k_{12}(C_p) - k_{21}(C_2) \tag{1.31}$$

with a general solution of

$$C_p = Ae^{-\alpha t} + Be^{-\beta t} \tag{1.32}$$

where A, B, α, and β depend on site(s) of elimination in the model. α and β are called *macro rate constants* and contain the model micro rate constants (k_{10}, k_{12}, k_{21}).

For a two-compartment model with only central compartment elimination, A, B, α, and β and the micro rate constants k_{10}, k_{12}, and k_{21} can be determined from Eqs. (1.33) and (1.34):

$$A = \frac{\text{dose}(\alpha - k_{21})}{V_1(\alpha - \beta)} \qquad B = \frac{\text{dose}(k_{21} - \beta)}{V_1(\alpha - \beta)} \tag{1.33}$$

$$\alpha + \beta = k_{10} + k_{12} + k_{21} \qquad \alpha\beta = k_{10}k_{21}$$

$$k_{21} = \frac{A\beta + B\alpha}{A + B} \qquad k_{10} = \frac{\alpha\beta}{k_{21}} \qquad k_{12} = (\alpha + \beta) - (k_{21} + k_{10}) \tag{1.34}$$

Combining Eqs. (1.32) and (1.33) produces Eq. (1.35), which contains both the micro and macro rate constants:

$$C_p = \overbrace{\frac{\text{dose}(\alpha - k_{21})}{V_1(\alpha - \beta)}}^{A} e^{-\alpha t} + \overbrace{\frac{\text{dose}(k_{21} - \beta)}{V_1(\alpha - \beta)}}^{B} e^{-\beta t} \tag{1.35}$$

When an intravenous bolus is administered to a two-compartment model (Fig. 1.16, *left panel*), it can be difficult to discern two significant and distinct kinetic behaviors in the curve on the Cartesian coordinate plot, but two distinct linear regions are seen easily when the data are plotted on semilog paper (Fig. 1.16, *right panel*). In the "distributive phase," the decrease in the C_p kinetic profile is not due only to distribution but also to both distribution and elimination—hence the sharp slope. In the "elimination" or "postdistributive" phases, the central compartment is at steady state with the tissue compartment, and the C_p kinetic profile is primarily due to elimination of drug (which includes drug being transferred from the tissues into the central compartment).

There are multiple volume terms associated with this model: V_1 and V_2 (volumes of compartments 1 and 2), $V_{d,ss}$ (volume at steady state), $V_{d,\text{area}}$ or $V_{d\cdot\beta}$, and $V_{d,\text{extrap}}$. Each is useful, but under specific conditions. These volume terms do *not* represent a specific physiological space; their utility is primarily the conversion of amount of drug into a concentration. Of these many volume terms, $V_{d,ss}$ is probably the most relevant in the design of controlled release delivery systems.

V_1 and V_2 (volume of distribution, compartments 1 and 2). V_1 and V_2 are the volume of distribution of compartments one and two, respectively (Eq. 1.36):

$$V_1 = \frac{\text{dose}}{A+B} \qquad V_2 = V_1\left(\frac{k_{12}}{k_{21}}\right) \tag{1.36}$$

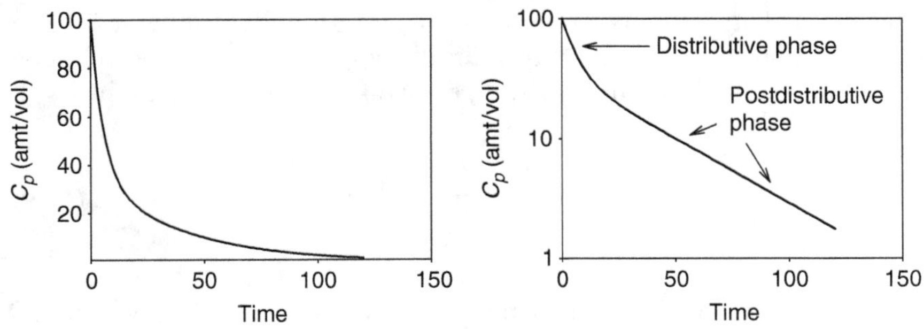

Figure 1.16 C_p versus time for intravenous bolus input and two-compartment disposition.

V_1 most likely would be used when calculating a loading dose for a drug exhibiting two-compartment behavior. V_2 is almost never used in the determination of dosing but sometimes may be used in blood protein or tissue-binding calculations and in the estimation of $V_{d,ss}$.

$V_{d,\mathrm{extrap}}$ (volume of distribution, extrapolated). $V_{d,\mathrm{extrap}}$ is given by

$$V_{d,\mathrm{extrap}} = \frac{\mathrm{dose}_{\mathrm{i.v.\ bolus}}}{B} = \frac{V_1(\alpha - \beta)}{k_{21} - \beta} \tag{1.37}$$

Although $V_{d,\mathrm{extrap}}$ is the same as V_d from the one-compartment disposition model, one should apply this volume term cautiously to systems greater than one compartment. As Eq. (1.37) shows, $V_{d,\mathrm{extrap}}$ is dependent on the elimination rate from the central compartment (k_{10}) in a complex interaction between α and β. Of all the volume terms, $V_{d,\mathrm{extrap}}$ overestimates the volume to the greatest degree and is probably the least useful in the design of controlled release delivery systems.

$V_{d,\mathrm{area}}$ or $V_{d,\beta}$ (volume of distribution, area or β). $V_{d,\mathrm{area}}$ is given by Eq. (1.38):

$$V_{d,\mathrm{area}} = V_{d,\beta} = \frac{\mathrm{dose}_{\mathrm{i.v.\ bolus}}}{(\mathrm{AUC})(\beta)} \tag{1.38}$$

This volume term also depends on β and/or k_{10} and overestimates the volume. However, when terminal concentration-time data are used (i.e., distribution is at steady state and elimination is the process significantly altering C_p), this volume term will produce an accurate conversion factor between C_p and the amount of drug in the body. While $V_{d,\mathrm{area}}$ overestimates the volume, it can be useful in the design of controlled release delivery systems, particularly in pulsatile delivery.

$V_{d,ss}$ (volume of distribution, steady state). This volume term and $V_{d,\mathrm{area}}$, are probably the two most useful in appropriate dose determination. $V_{d,ss}$ is used in calculating maintenance doses for an individual whose drug concentration is at steady state. Unlike $V_{d,\mathrm{area}}$, $V_{d,ss}$ does not change with changes in elimination (i.e., α, β, or k_{10} does not show up in $V_{d,ss}$):

$$V_{d,ss} = V_1\left(1 + \frac{k_{12}}{k_{21}}\right) = V_1 + V_2 \tag{1.39}$$

Generally, since the goal of some controlled release delivery systems is to achieve and maintain the drug at a steady-state concentration within the therapeutic window, $V_{d,ss}$ is used frequently to determine the dose that will achieve this therapeutic target.

Cl, *AUC*, $t_{1/2,\alpha}$, $t_{1/2,\beta}$. Assuming that A, B, α, and β have been obtained either graphically or from a nonlinear regression software package, for a two-compartment disposition model, the equations for Cl, *AUC*, $t1/2$, α, and $t1/2$, β are given in Eqs. (1.40) through (1.42):

$$\text{Cl} = \frac{\text{dose}}{\text{AUC}} = (V_{d,\text{area}})(\beta) = (V_1)(k_{10}) \tag{1.40}$$

$$\text{AUC} = \frac{A}{\alpha} + \frac{B}{\beta} \tag{1.41}$$

$$t_{1/2,\alpha} = \frac{\ln(2)}{\alpha} \qquad t_{1/2,\beta} = \frac{\ln(2)}{\beta} \tag{1.42}$$

1.6.2 Multiple-dosing input systems and steady-state kinetics.

Since the goal of most controlled release delivery systems is to maintain the drug concentration within the therapeutic window, the effect of multiple-dosing strategies (used in chronic diseases) on C_p will be discussed. In this section we assume that the blood/plasma drug concentration achieves its steady state rapidly with all involved tissues, especially the concentration at site of effect C_e or biophase concentration. The MEC and MTC are determined by C_e. If C_p and C_e are in a steady-state relationship or rapidly reach a steady-state relationship, then controlling C_p should effectively control C_e and presumably the response generated by C_e. *This relationship between C_p and C_e is one of the foundational assumptions of using pharmacokinetics in the design of most controlled release delivery devices.*

Zero-order input and one-compartment disposition (I0D1). The simplest case for achieving a drug plasma concentration between the MEC and MTC is to use a zero-order input. In Fig. 1.17, the six time points show time as measured in half-lives. At 3.3 half-lives, C_p is at approximately 90 percent of its true steady-state value; at 5 half-lives, C_p is at approximately 96 percent of its true steady-state value.

Multiple instantaneous input and one-compartment disposition (IBD1). In the case of IBD1 single-dose input, the C_p kinetic profile is given by Eq. (1.43) for any time t after the bolus dose has been given:

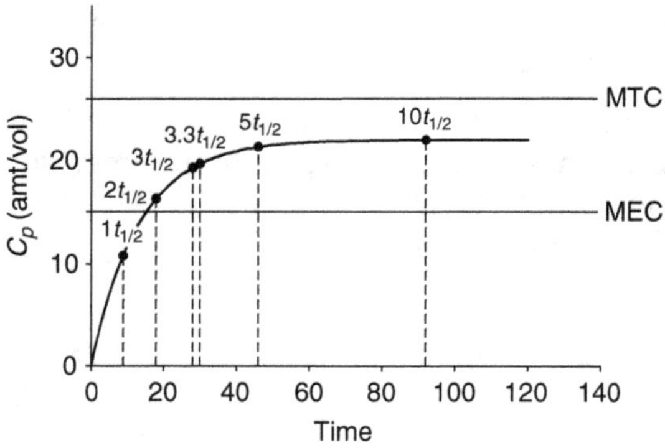

Figure 1.17 C_p versus time profile for zero-order input and one-compartment disposition.

$$C_p = C_p^0 e^{-Kt} = \frac{\text{dose}}{V_d} e^{-Kt} \tag{1.43}$$

If drug had been administered in equally sized multiple intravenous bolus doses at equally spaced τ time intervals (e.g., $\tau = 6$ h), then an accounting of accumulated drug between doses must be instituted. Figure 1.18, similar to the zero-order input, shows that repetitive instantaneous dosing will produce an average C_p profile similar to Fig. 1.17.

For multiple intravenous bolus doses, C_p in the nth dosing period is

$$\Rightarrow C_{p,n} = C_p^0 \frac{\left(1 - e^{-nK\tau}\right) e^{-Kt}}{\left(1 - e^{-K\tau}\right)} = C_p^0(\text{MDF})e^{-Kt} \tag{1.44}$$

MDF is the multiple dosing factor and is defined in Eq. (1.45):

$$\text{MDF} = \frac{\left(1 - e^{-nK\tau}\right)}{\left(1 - e^{-K\tau}\right)} \tag{1.45}$$

As Schoenwald[28] points out, the MDF is quite mobile in its application and can be applied to obtain two important concentrations in the design of controlled release delivery systems—$C_{p,\max}$ and $C_{p,\min}$. Under multiple-dosing intravenous bolus input, applying MDF to $C_{p,\max}$ and $C_{p,\min}$ gives Eqs. (1.46) and (1.47):

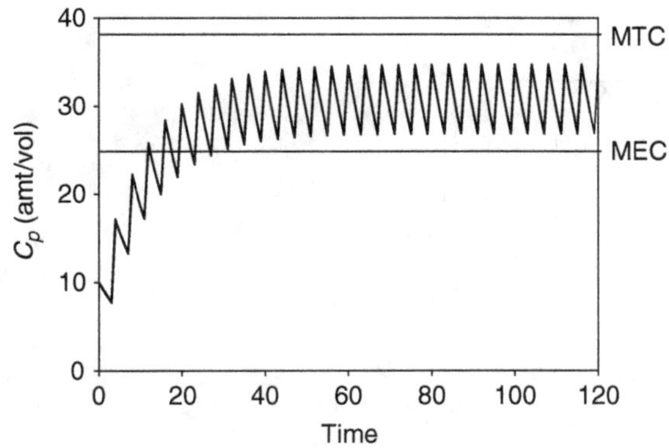

Figure 1.18 C_p versus time profile for multiple equal-sized instantaneous inputs into a one-compartment disposition model.

$$C_{p,\max} = C_p^0 \frac{\left(1-e^{-nK\tau}\right)}{\left(1-e^{-K\tau}\right)} \qquad C_{p,\mathrm{ss},\max} = C_p^0 \frac{1}{\left(1-e^{-K\tau}\right)} \tag{1.46}$$

$$C_{p,\min} = C_{p,\max}\left(e^{-K\tau}\right) \qquad C_{p,\mathrm{ss},\min} = C_{p,\mathrm{ss},\max}\left(e^{-K\tau}\right) \tag{1.47}$$

Multiple first-order input and one-compartment disposition (I1D1). In a similar manner, MDF can be applied to the I1D1 single-dose equations (Eq. 1.26) to get the multiple-dose equation for C_p (Eq. 1.48):

$$C_p = \frac{(F)(S)(\mathrm{dose})(k_a)}{V_d(k_a - K)}\left[\frac{\left(1-e^{-nK\tau}\right)}{\left(1-e^{-K\tau}\right)}e^{-K(t)} - \frac{\left(1-e^{-nk_a\tau}\right)}{\left(1-e^{-k_a\tau}\right)}e^{-k_a(t)}\right] \tag{1.48}$$

A plot of Eq. (1.48) is shown by Fig. 1.19.

The calculation for $t_{\max}$ from Eq. (1.29) needs to be modified to $t_{\max,\mathrm{MD}}$ as follows:

$$t_{\max,\mathrm{MD}} = \frac{\ln\left[\dfrac{k_a(1-e^{-K\tau})}{K(1-e^{-k_a\tau})}\right]}{k_a - K} \tag{1.49}$$

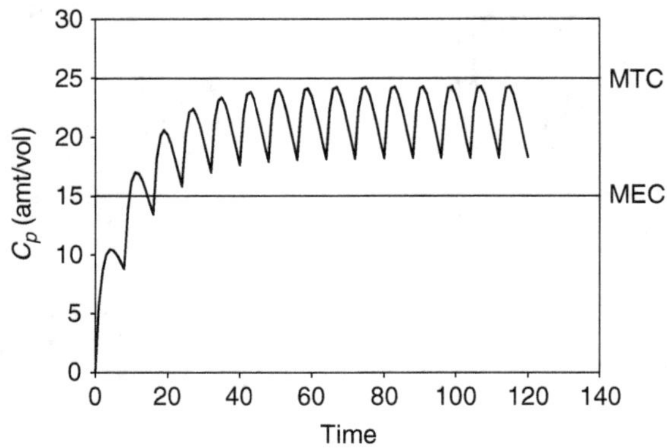

Figure 1.19 C_p versus time profile for multiple equal-sized first-order inputs into a one-compartment disposition model.

Using $t_{\max,\mathrm{MD}}$, $C_{p,\max}$, and $C_{p,\min}$ prior to steady state (Eqs. 1.50 and 1.51) and at steady state (Eqs. 1.51 and 1.52) can be calculated:

$$C_{p,\max} = \frac{(F)(S)(\text{dose})(k_a)}{V_d(k_a - K)} \times \left[\frac{(1-e^{-nK\tau})}{(1-e^{-K\tau})} e^{-Kt_{\max,\mathrm{MD}}} - \frac{(1-e^{-nk_a\tau})}{(1-e^{-k_a\tau})} e^{-k_a t_{\max,\mathrm{MD}}} \right] \tag{1.50}$$

$$C_{p,ss,\max} = \frac{(F)(S)(\text{dose})(k_a)}{V_d(k_a - K)} \times \left[\frac{1}{(1-e^{-K\tau})} e^{-Kt_{\mathrm{Max,MD}}} - \frac{1}{(1-e^{-k_a\tau})} e^{-k_a t_{\max,\mathrm{MD}}} \right] \tag{1.51}$$

$$C_{p,\min} = \frac{(F)(S)(\text{dose})(k_a)}{V_d(k_a - K)} \times \left[\frac{(1-e^{-nK\tau})}{(1-e^{-K\tau})} e^{-Kt_{\max,\mathrm{MD}}} - \frac{(1-e^{-nk_a\tau})}{(1-e^{-k_a\tau})} e^{-k_a t_{\max,\mathrm{MD}}} \right] = C_{p,\max}(e^{-K\tau}) \tag{1.52}$$

$$C_{p,\text{ss},\min} = \frac{(F)(S)(\text{dose})(k_a)}{V_d(k_a - K)} \times \left[\frac{1}{\left(1 - e^{-K\tau}\right)} e^{-Kt_{\max,\text{MD}}} - \frac{1}{\left(1 - e^{-k_a\tau}\right)} e^{-k_a t_{\max,\text{MD}}} \right] = C_{p,\text{ss},\max}\left(e^{-K\tau}\right) \tag{1.53}$$

Lastly, in multiple-dose regimens and controlled release delivery systems, it is important to know the amount of drug accumulating over the dosing interval. The accumulation factor R gives a quantitative indication of the fraction of drug that remains in the body after the first dose compared with the amount of drug that accumulates at steady state. Equation (1.54) gives the expression for R:

$$R = \frac{C_{p,ss,\max}}{C_{p,\max,1st\ dose}} = \frac{1}{1 - e^{-K\tau}} \tag{1.54}$$

1.7 Applications of Pharmacokinetics in the Design of Controlled Release Delivery Systems

1.7.1 Design challenges for controlled release delivery systems

Of the many design goals that need to be achieved in a successful controlled delivery system, two are closely related to pharmacokinetics: (1) the achievement of a sufficient input flux of drug and (2) the achievement of a desired drug concentration-time profile. While both goals require the balancing of design parameters and biopharmaceutical properties of the drug, the first goal is governed primarily by system design (input), and the second is governed primarily by the physiology of the body (disposition). Therefore, it is not surprising that pharmacokinetics, which is the convolution of input and disposition, can be of great help in the design of controlled release delivery systems. In the following subsections, the challenges of achieving these two goals from a pharmacokinetics perspective will be explored.

Achievement of a sufficient input flux of drug. The achievement of sufficient input drug flux is probably the greatest challenge to designing a successful controlled release delivery system. While some controlled release delivery systems face even greater design challenges, such as in pulsatile drug delivery[39,40] or tissue/site-specific drug targeting (improved local bioavailability), sufficient achievable input flux is still the predominant design issue. To calculate the necessary input flux J_{in} to achieve a desired $C_{p,ss}$, only the systemic clearance Cl needs to be known (Eq. (1.55):

$$J_{in} = (\mathrm{Cl})(C_{p,ss}) \tag{1.55}$$

If the input flux cannot closely match the amount of drug being lost per unit time, then the desired $C_{p,ss}$ will not be achieved. The human body superbly insulates the systemic circulation from exogenous input through many anatomical barriers (e.g., lipid bilayers, ciliated epithelia, cornified stratum, cellular tight junctions) and physiological barriers [e.g., wide-ranging gut pH and the presence of gut, hepatic, and renal drug-metabolizing enzymes (CYPs, GST) and efflux transporters (P-glycoprotein, MRP)] (see Chap. 2). These barriers can severely limit both the extent (*AUC*) and the rate of input (k_{abs}) or simply bioavailability. Numerous techniques, many discussed in subsequent chapters, seek to improve the extent (kinetic exposure profile) and/or input rate by achieving a sufficient input flux. Input flux generically can be increased by (1) increasing the absorption site surface area (e.g., larger patch area, absorption in small intestine versus stomach), (2) using permeability enhancers (e.g., ethanol) to selectively modulate barrier properties, (3) creating prodrug entities (e.g., ester conjugates, PEG conjugates, chemical targeting groups) to increase absorption and/or to decrease metabolic/chemical degradation, (4) administering the drug via a route not susceptible to first-pass metabolism, (5) targeting the drug to a specific region or tissue, thus effectively reducing the amount of drug needed systemically, or (6) increasing drug potency. Although increasing drug potency does not really increase input drug flux, if one is able to increase the potency of a drug then, the desired $C_{p,ss}$ would be effectively lowered, thus reducing the necessary input flux— and thus indirectly "increasing" the relative input flux. Of these six general techniques for improving input flux, the first five are highly related to the design of the delivery system and the pharmacokinetics of zero- and first-order inputs.

Achievement of a desired drug concentration-time profile. Although physiological processes govern the disposition of drug in the body, several pharmacokinetic parameters are still useful for evaluating drugs as candidates for controlled release delivery systems. In addition to potency, the pharmacokinetic parameters systemic clearance Cl, volume of distribution V_d, and the elimination rate constant K or half-life $t_{1/2}$ can provide useful information in the design of delivery systems. One important link and design consideration between Cl and V_d and controlled release delivery systems is $t_{1/2}$. The half-life indicates how quickly C_p can be modulated intentionally upward or downward; in other words, half-life indicates how well the shape of C_p can be controlled (Fig. 1.20). The shorter the half-life, the more closely C_p will mimic the shape of the input. The solid line indicates the shape and time course of the administered input of two different hypothetical drugs that variy only in half-life. The dotted and dashed lines show the resulting concentration-time profiles for each drug, the shorter and longer half-life, respectively. From a controllability perspective, the biopharmaceutical properties of the drug with a shorter half-life make it more desirable for controlled release delivery systems.

Although it may appear that drug delivery systems with efficacies requiring only a single steady-state drug concentration may not benefit directly from increased concentration-time controllability, one also should consider the following two pharmacokinetic benefits: the shorter the half-life, the less time is needed to reach steady state, and the less time is need to fully eliminate the drug (e.g., in an overdose). In therapies requiring complex drug concentration-time profiles (e.g., diseases requiring pulsatile delivery), the controllability of C_p is paramount to a successful system.

Ideal drug candidate for controlled release delivery systems. From a pharmacokinetic and pharmacodynamic perspective, the ideal drug candidate for controlled release delivery systems would have high potency and

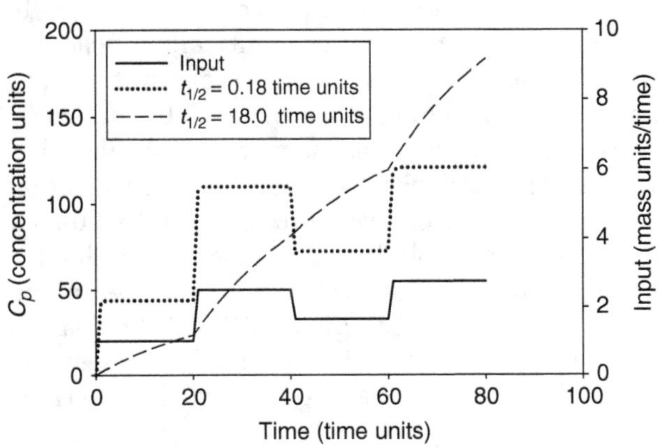

Figure 1.20 Plot showing the relationship between elimination half-life $t_{1/2}$ and the controllability of the concentration-time profile.

minimal adverse effects, have a short half-life, have stationary pharmacokinetics and pharmacodynamics, and have physicochemical properties conducive to achieving sufficient input flux. Since "perfect" drug candidates do not exist, the designer must weigh the benefits against the negatives to optimize the design elements of the controlled release delivery system to achieve success.

When pharmacokinetics/pharmacodynamics of a drug behavior within the therapeutic window are well described by linear or nonlinear pharmacokinetics/pharmacodynamics models with stationary parameters, the dosing "gold standard," or zero-order input (e.g., intravenous infusion), easily achieves and maintains a C_p that falls within the therapeutic index (i.e., greater than the MEC and less than the MTC) (see Fig. 1.1). Although multiple intravenous bolus doses can be designed to achieve steady-state concentrations (see Fig. 1.18) within the therapeutic index, this therapeutic modality generally is reserved for patients who are hospitalized, critically ill, and/or unable to use other extravascular dosage forms. In outpatient settings, oral administration is the most common dosing regimen used to achieve a therapeutic $C_{p,ss}$. Whether the oral dosage form is a zero- or first-order input device, a $C_{p,ss}$ that falls within the therapeutic index is still easily achievable (see Fig. 1.1). Some limitations of using pharmacokinetics alone, along with examples of pharmacokinetics and pharmacodynamics used in the design of controlled release delivery systems, are provided in the following sections.

1.7.2 Limitations of using pharmacokinetics only to design controlled release delivery systems

For a majority of drugs in controlled release delivery systems, pharmacokinetics alone have been sufficient to guide the choice of kinetically based design elements necessary to achieve therapeutically efficacious systems. However, there are cases when just a simple constant steady-state drug concentration is not sufficient or appropriate. The increased depth of research in the areas of physiology, biochemistry, molecular biology, and drug biotransformation and transport has uncovered many complex biological rhythms (menstrual, circadian, etc.) and feedback loops/cascades (drug tolerance or tachyphylaxis), as discussed at length in Chap. 13. Some of these complexities can cause a loss of the unique one-to-one relationship between $C_{p,ss}$ and C_e and/or C_e and the desired effect. In these cases, the design of controlled release delivery systems will need to incorporate specific kinetic behavior(s) in order to be therapeutically effective.

What happens when the MEC or MTC (i.e., therapeutic index) does vary with time or state of the system (see Figs. 1.3 and 1.4)? When the underlying physiological and pharmacological systems possess nonstationary parameters (e.g., due to CYP3A4 induction, P-glycoprotein induction, drug tolerance), the once "gold standard" zero-order input that easily achieved the desired therapeutic outcome frequently fails to maximize drug therapy effectiveness or fails to minimize unwanted or adverse effects. For example, when patients with the coronary malady angina use transdermally delivered (zero-order input) nitroglycerin (NTG) to decrease cardiac oxygen demands and consequently reduce anginal discomfort/pain, studies show that these patients can rapidly develop an NTG tolerance that reduces therapeutic benefits within 24 hours of patch administration.[41–43] From a pharmacokinetics/pharmacodynamics standpoint, the Food and Drug Administration (FDA) only required that the NTG patches achieve and maintain an NTG plasma concentration within the NTG-naïve MEC/MTC therapeutic range for approval (see Fig. 1.4, *left panel*). In hindsight, had the FDA also required pharmacodynamics studies, the patch probably would not have been approved because the NTG tolerance would have stood out immediately as a problem; NTG clearly exhibits state-varying or nonstationary behavior (see Fig. 1.4 *right panel*). Under these conditions, it is foreseeable that a strong dependence on pharmacokinetics/pharmacodynamics modeling will improve the success of therapies using controlled release delivery systems and optimized dosing strategies. A new "gold standard" of therapy, which may be based on pulsatile delivery, may emerge for these complex nonstationary systems.

1.7.3 Examples of pharmacokinetic/pharmacodynamic considerations in controlled release delivery systems design

In this section several articles pertaining to pharmacokinetic/pharmacodynamic considerations in controlled release delivery systems design will be presented. These articles tend to indicate whether pharmacokinetics alone or pharmacokinetics and pharmacodynamics are needed to design specific controlled release delivery systems.

Protein and peptide delivery. The successful delivery of peptides and proteins, as therapeutic modalities, always has been plagued by low extravascular bioavailability,[19,44–47] stability issues,[48,49] and complex measures of effect.[50] In addition to being drugs themselves, proteins and peptides can be used as drug carrier systems.[51] In this case, the proteins/peptides are not themselves the drug but are considered part of the prodrug entity. In other systems, the proteins/peptides can act as a targeting sequence to direct the protein/peptide–attached drug to its proper site of action.[52,53] Since endogenous proteins and peptides are used in nearly every function within the body, a specific quantitative pharmacodynamics marker or outcome must be resolutely identified and linked to the pharmacokinetics before the optimal input profile can be determined.[54–56]

Polymeric controlled release delivery systems. As new polymers become more suitable biomaterials, polymers play an increasingly important role in delivery device construction,[57,58] drug targeting,[59] and gene and protein delivery.[60,61] "Smart" polymers and gels are also being developed to create a closed-loop drug delivery feedback system.[39,62–64] When trying to achieve a desired flux of drug from the polymeric matrix, the release kinetics[65] and drug/matrix solubility and loading[66–68] must be addressed. The desired flux will require just pharmacokinetics in the simplest cases and pharmacokinetics and pharmacodynamics in more complicated cases. The pharmacokinetics and pharmacodynamics of these drug delivery systems tend to be complex as well,[69,70] especially at the cellular level, where polymer-drug conjugates behave differently from unbound free drug.

Other controlled release delivery devices. Several other controlled release delivery devices not in the preceding two categories are liposomal,[71,72] transdermal,[73–75] and transmucosal.[76] Liposomes have pharmacokinetic and pharmacodynamic problems similar to those found in the design of drug carrier systems. Since the goal is to achieve an effective concentration of drug directly at the site of action, the plasma level is not necessarily indicative of an end-therapeutic goal, and therefore, systemic circulation pharmacokinetics alone will not be sufficient. In these cases, either a pharmacodynamic marker needs to be used, or the drug concentration at the site of action must be obtained to titrate the drug carrier system.

In transdermal or transmucosal controlled release delivery systems, the general goal is to deliver a desired flux of drug across the skin or respective mucosal barrier (nasal, buccal, vaginal, rectal). Most current transdermal systems deliver small-molecular-weight drugs that do not have clinically significant time- or state-varying pharmacokinetics/pharmacodynamics (except for nitroglycerin and possibly nicotine). Obviously, drug candidates that do exhibit such pharmacokinetic or pharmacodynamic behavior will need appropriate pharmacokinetic and/or pharmacodynamic consideration in the design of their controlled release delivery systems.

Future controlled release delivery systems for diseases requiring mixed zero-order and instantaneous input. As the pathological and physiological mechanisms of diseases become further elucidated, the design of more effective controlled release delivery systems becomes a greater challenge.

One example of a medical condition requiring complex controlled release delivery is Type 1 diabetes.

In Type 1 diabetes, the external control of systemic insulin levels is critical to therapeutic care. Elaborate models of glucose homeostasis,[77] insulin kinetics,[78] and endogenous insulin secretion[77,79] have been developed and studied. Endogenous insulin is released on at least three different time scales: diurnal (day/night), ultradian (one pulse approximately every 40 minutes), and high frequency (one pulse approximately every 6 minutes).[77] Research[77,80] indicates that Type 1 diabetes is not well controlled by either zero-order or pulsatile insulin delivery alone. Effective therapy[81–83] needs to simultaneously control basal insulin needs (zero-order and pulsatile) *and* bolus insulin needs (pulsatile). The ability to deliver drug (insulin) in both a zero-order and a pulsatile manner within a single controlled release delivery system is the future challenge for these systems.

1.8 Conclusions

As controlled release drug delivery systems have matured, the challenge to develop sophisticated zero-order release products has been handled more easily than 15 to 20 years ago. However, as we learn more about normal physiological control over various bodily functions and disease states, we find that the body does not respond well to zero-order delivery of some drugs (e.g., growth hormone, insulin, nitroglycerin, etc.). In these special cases, both the pharmacokinetics and pharmacodynamics should be considered, which further substantiates the growing need for advanced controlled delivery devices that can deliver drug as impulses (pulsatile delivery[40]) and/or respond to the rate of change of some surrogate marker (e.g., receptor density, pH, glucose concentration, cAMP concentration, neurotransmitter, etc.). Lastly, with regard to regulatory issues, although pharmacokinetic data without pharmacodynamic data may be sufficient in establishing safety and efficacy for some drugs, pharmacokinetic data alone may not be sufficient for other drugs, especially the newer, more sophisticated drugs that may have state-varying or complex pharmacodynamics.

References

1a. T. Teorell. Kinetics of distribution of substances administered to the body. I. The extravascular (*) modes of administration. *Int. Arch. Pharm.* **57**:205–225, 1937.

1b. T. Teorell. Kinetics of distribution of substances administered to the body. II. The intravascular modes of administration. *Int. Arch. Pharm.* **57**:226–240, 1937.

2. G. E. Amidon, W. I. Higuchi, and N. F. Ho. Theoretical and experimental studies of transport of micelle-solubilized solutes. *J. Pharm. Sci.* **71**:77–84, 1982.

3. J. T. Carstensen. Stability of solids and solid dosage forms. *J. Pharm. Sci.* **63**:1–14, 1974.

4. J. T. Carstensen, J. B. Johnson, D. C. Spera, and M. J. Frank. Equilibrium phenomena in solid dosage forms. *J. Pharm. Sci.* **57**:23–27, 1968.
5. G. L. Flynn. Buffers: pH control within pharmaceutical systems. *J. Parent. Sci. Technol.* **34**:139–162, 1980.
6. G. L. Flynn. Solubility concepts and their applications to the formulation of pharmaceutical systerms. I. Theoretical foundations. *J. Parent. Sci. Technol.* **38**:202–209, 1984.
7. G. L. Flynn, S. H. Yalkowsky, and T. J. Roseman. Mass transport phenomena and models: theoretical concepts. *J. Pharm. Sci.* **63**:479–510, 1974.
8. T. Higuchi. Some physical chemical aspects of suspension formulation. *J. Am. Pharm. Assoc.* **47**:657–660, 1958.
9. T. Higuchi, M. Gupta, and L. W. Busse. Influence of electrolyte, pH, and alcohol concentration on the solubilities of acidic drugs. *J. Am. Pharm. Assoc.* **42**:157–161, 1953.
10. W. I. Higuchi, N. A. Mir, and S. J. Desai. Dissolution rates of polyphase mixtures. *J. Pharm. Sci.* **54**:1405–1410, 1965.
11. R. G. Stehele and W. I. Higuchi. Diffusional model for transport rate studies across membranes. *J. Pharm. Sci.* **56**:1367–1368, 1967.
12. T. Yotsuyanagi, W. I. Higuchi, and A. H. Ghanem. Theoretical treatment of diffusional transport into and through an oil-water emulsion with an interfacial barrier at the oil-water interface. *J. Pharm. Sci.* **62**:40–43, 1973.
13. J. G. Wagner. Biopharmaceutics: Absorption aspects. *J. Pharm. Sci.* **50**:359–387, 1961.
14. J. G. Wagner. An overview of the analysis and interpretation of bioavailability studies in man. *Pharmacology* **8**:102–117, 1972.
15. J. G. Wagner. An overview of the analysis and interpretation of bioavailability studies in man. *Arzneimittelforschung* **26**:105–108, 1976.
16. J. Loo and S. Riegelman. New method for calculating the intrinsic absorption rate of drugs. *J. Pharm. Sci.* **57**:918–928, 1968.
17. J. Loo and S. Riegelman. Assessment of pharmacokinetic constants from postinfusion blood curves obtained after IV infusion. *J. Pharm. Sci.* **59**:53–55, 1970.
18. J. Hunter and B. H. Hirst. Intestinal secretion of drugs: The role of P-glycoprotein and related drug efflux systems in limiting oral drug absorption. *Adv. Drug Deliv. Rev.* **25**:129–157, 1997.
19. E. Walter, T. Kissel, and G. E. Amidon. The intestinal peptide carrier: a potential transport system for small peptide–derived drugs. *Adv. Drug Deliv. Rev.* **20**:33–58, 1996.
20. L. Z. Benet and C. L. Cummins. The drug efflux–metabolism alliance: Biochemical aspects. *Adv. Drug Deliv. Rev.* **50**:S3–S11, 2001.
21. V. J. Wacher, L. Salphati, and L. Z. Benet. Active secretion and enterocytic drug metabolism barriers to drug absorption. *Adv. Drug Deliv. Rev.* **46**:89–102, 2001.
22. K. E. Thummel, K. L. Kunze, and D. D. Shen. Enzyme-catalyzed processes of first-pass hepatic and intestinal drug extraction. *Adv. Drug Deliv. Rev.* **27**:99–127, 1997.
23. P. B. Watkins. The barrier function of CYP3A4 and P-glycoprotein in the small bowel. *Adv. Drug Deliv. Rev.* **27**:161–170, 1997.
24. M. Gibaldi and D. Perrier. *Pharmacokinetics*. New York: Marcel Dekker, 1982.
25. J. G. Hardman, L. E. Limbird, and A. Gilman. *Goodman & Gilman's The Pharmacological Basis of Therapeutics*. New York: McGraw-Hill, 2001.
26. M. Rowland and T. N. Tozer. *Clinical Pharmacokinetics: Concepts and Applications*. Philadelphia: Lippincott Williams & Wilkins, 1995.
27. L. Z. Benet, G. Levy, and B. L. Ferraillo. *Pharmacokinetics: A Modern View*. New York: Plenum Press, 1984.
28. R. D. Schoenwald. *Pharmacokinetic Principles of Dosing Adjustments: Understanding the Basics*. Boca Raton, FL: CRC Press, 2001.
29. E. B. M. Ekblad and V. Licko. A model eliciting transient responses. *Am. J. Physiol.* **246**:R-114–R121, 1984.
30. W. J. Jusko and H. C. Ko. Physiologic indirect response models characterize diverse types of pharmacodynamic effects. *Clin. Pharmacol. Ther.* **56**:406–419, 1994.
31. L. Z. Benet. Mean residence time in the body versus mean residence time in the central compartment. *J. Pharm. Biopharm.* **13**:555–558, 1985.

32. W. R. Gillespie and P. Veng-Pedersen. The determination of mean residence time using statistical moments: It is correct. *J. Pharm. Biopharm.* **13**: 549–554, 1985.
33. E. Nakashima and L. Z. Benet. General treatment of mean residence time, clearance, and volume parameters in linear mammillary models with elimination from any compartment. *J. Pharm. Biopharm.* **16**:475–492, 1988.
34. P. Veng-Pedersen. Noncompartmentally based pharmacokinetic modeling. *Adv. Drug Deliv. Rev.* **48**:265–300, 2001.
35. K. Yamaoka, T. Nakagawa, and T. Uno. Statistical moments in pharmacokinetics. *J. Pharm. Biopharm.* **6**:547–558, 1978.
36. N. Watari and L. Z. Benet. Determination of mean input time, mean residence time, and steady-state volume of distribution with multiple drug inputs. *J. Pharm. Biopharm.* **17**:593-599, 1989.
37. H. Cheng, W. R. Gillespie, and W. J. Jusko. Review article: mean residence time concepts for non-linear pharmacokinetic systems. *Biopharm. Drug Dispos.* **15**:627–641, 1994.
38. E. Martin, T. N. Tozer, L. B. Sheiner, and S. Riegelman. The clinical pharmacokinetics of phenytoin. *J. Pharm. Biopharm.* **5**:579–596, 1977.
39. G. P. Misra and R. A. Siegel. Multipulse drug permeation across a membrane driven by a chemical pH-oscillator. *J. Control. Release* **79**:293–297, 2002.
40. T. Bussemer, I. Otto, and R. Bodmeier. Pulsatile drug-delivery systems. *Crit. Rev. Ther. Drug Carrier Syst.* **18**:433–458, 2001.
41. A. Wiegand, K. Bauer, R. Bonn, et al. Pharmacodynamic and pharmacokinetic evaluation of a new transdermal delivery system with a time-dependent release of glyceryl trinitrate. *J. Clin. Pharmacol.* **32**:77–84, 1992.
42. S. Savonitto, M. Motolese, and E. Agabiti-Rosei. Antianginal effect of transdermal nitroglycerin and oral nitrates given for 24 hours a day in 2456 patients with stable angina pectoris. *Int. J. Clin. Pharmacol. Ther.* **33**:194–203, 1995.
43. L. Cloarec-Blanchard, C. Funck-Brentano, A. Carayon, and P. Jaillon. Rapid development of nitrate tolerance in healthy volunteers: assessment using spectral analysis of short-term blood pressure and heart rate variability. *J. Cardiovasc. Pharmacol.* **24**:266–273, 1994.
44. R. D. Egleton and T. P. Davis. Bioavailability and transport of peptides and peptide drugs into the brain. *Peptides* **18**:1431–1439, 1997.
45. B. J. Aungst, H. Saitoh, D. L. Burcham, et al. Enhancement of the intestinal absorption of peptides and non-peptides. *J. Control. Release* **41**:19–31, 1996.
46. U. B. Kompella and V. H. L. Lee. Delivery systems for penetration enhancement of peptides and protein drugs: design considerations. *Adv. Drug Deliv. Rev.* **46**:211–245, 2001.
47. J. E. Talmadge. The pharmaceutics and delivery of therapeutic polypeptides and proteins. *Adv. Drug Deliv. Rev.* **10**:247–299, 1993.
48. W. Wang. Instability, stabilization, and formulation of liquid protein pharmaceuticals. *Int. J. Pharm.* **185**:129–188, 1999.
49. M. C. Manning, K. Patel, and R. T. Borchardt. Stability of protein pharmaceuticals. *Pharm. Res.* **6**:903–918, 1989.
50. N. B. Modi. Pharmacokinetics and pharmacodynamics of recombinant proteins and peptides. *J. Control. Release* **29**:269–281, 1994.
51. D. K. F. Meijer, G. Molema, F. Moolenaar, et al. (Glyco)-protein drug carriers with an intrinsic therapeutic activity: the concept of dual targeting. *J. Control. Release* **39**:163–172, 1996.
52. D. K. F. Meijer, L. Beljaars, G. Molema, and K. Poelstra. Disease-induced drug targeting using novel peptide-ligand albumins. *J. Control. Release* **72**:157–164, 2001.
53. H. Brooks, B. Lebleu, and E. Vives. Tat peptide–mediated cellular delivery: Back to basics. *Adv. Drug Deliv. Rev.* **57**:559–577, 2005.
54. N. A. Mazer. Pharmacokinetic and pharmacodynamic aspects of polypeptide delivery. *J. Control. Release* **11**:343–356, 1990.
55. D. D. Breimer. Pharmacokinetic and pharmacodynamic basis for peptide drug delivery system design. *J. Control. Release* **21**:5–10, 1992.
56. M. Hashida, R. I. Mahato, K. Kawabata, et al. Pharmacokinetics and targeted delivery of proteins and genes. *J. Control. Release* **41**:91–97, 1996.

57. M. M. Goldenberg. An extended-release formulation of oxybutynin chloride for the treatment of overactive urinary bladder. *Clin. Ther.* **21**:634–642, 1999.
58. J. S. Grundy and R. T. Foster. The nifedipine gastrointestinal therapeutic system (GITS). Evaluation of pharmaceutical, pharmacokinetic and pharmacological properties. *Clin. Pharm.* **30**:28–51, 1996.
59. J. Kopecek. Targetable polymeric anticancer drugs: Temporal control of drug activity. *Ann. N. Y. Acad. Sci.* **618**:335–344, 1991.
60. S.-O. Han, R. I. Mahato, Y. K. Sung, and S. W. Kim. Development of biomaterials for gene therapy. *Mol. Ther.* **2**:302–309, 2000.
61. Y.-P. Li, Y.-Y. Pei, X.-Y. Zhang, et al. PEGylated PLGA nanoparticles as protein carriers: Synthesis, preparation and biodistribution in rats. *J. Control. Release* **71**:203–211, 2001.
62. J. Kopecek. Smart and genetically engineered biomaterials and drug delivery systems. *Eur. J. Pharm. Sci.* **20**:1–16, 2003.
63. G. Wang, K. Kuraoda, T. Enoki, et al. Gel catalysts that switch on and off. *Proc. Natl. Acad. Sci. USA* **97**:9861–9864, 2000.
64. J. Kost and R. Langer. Responsive polymeric delivery systems. *Adv. Drug Deliv. Rev.* **46**:125–148, 2001.
65. A. C. Balazs, D. F. Calef, J. M. Deutch, et al. The role of polymer matrix structure and interparticle interactions in diffusion-limited drug release. *Biophys. J.* **47**:97–104, 1985.
66. V. M. Rao, J. L. Haslam, and V. J. Stella. Controlled and complete release of a model poorly water-soluble drug, prednisolone, from hydroxypropyl methylcelluose matrix tablets using $(SBE)_{7m}$-β-cyclodextrin as a solubilizing agent. *J. Pharm. Sci.* **90**:807–816, 2001.
67. B. R. Jasti, B. Berner, S.-L. Zhou, and X. Li. A novel method for determination of drug solubility in polymeric matrices. *J. Pharm. Sci.* **93**:2135–2141, 2004.
68. S. W. Kim, Y. H. Bae, and T. Okano. Hydrogels: Swelling, drug loading, and release. *Pharm. Res.* **9**:282–290, 1992.
69. P. Caliceti and F. M. Veronese. Pharmacokinetic and biodistribution properties of poly(ethylene glycol)-protein conjugates. *Adv. Drug Deliv. Rev.* **55**:1261–1277, 2003.
70. M. Hashida and Y. Takakura. Pharmacokinetics in design of polymeric drug delivery systems. *J. Control. Release* **31**:163–171, 1994.
71. X. Guo and F. C. J. Szoka. Chemical approaches to triggerable lipid vesicles for drug and gene delivery. *Acc.Chem. Res.* **36**:335–341, 2003.
72. A. Gabizon. Emerging role of liposomal drug carrier systems in cancer chemotherapy. *J. Liposome Res.* **13**:17–20, 2003.
73. Y. N. Kalia, A. Naik, J. Garrison, and R. H. Guy. Iontophoretic drug delivery. *Adv. Drug Deliv. Rev.* **56**:619–658, 2004.
74. Y. N. Kalia and R. H. Guy. Modeling trandermal drug release. *Adv. Drug Deliv. Rev.* **48**:159–172, 2001.
75. C. H. Purdon, C. G. Azzi, J. Zhang, et al. Penetration enhancement of transdermal delivery: Current permutations and limitations. *Crit. Rev. Ther. Drug Carrier Syst.* **21**:97–132, 2004.
76. Y. Song, Y. Wang, R. Thakur, et al. Mucosal drug delivery: Membranes, methodologies, and applications. *Crit. Rev. Ther. Drug Carrier Syst.* **21**:195–256, 2004.
77. R. A. Ritzel and P. C. Butler. Physiology of glucose homeostasis and insulin secretion. In J. L. Leahy (ed.), *Insulin Therapy.* New York: Marcel Dekker, 2002, pp. 61–71.
78. A. Pilo, R. Navalesi, and E. Ferrannini. Insulin kinetics after portal and peripheral injection of $[^{125}I]$insulin. I. Data analysis and modeling. *Am. J. Physiol.* **230**:1626–1629, 1976.
79. K. S. Polonsky, B. D. Given, and V. C. E. Twenty-four-hour profiles and pulsatile patterns of insulin secretion in normal and obese subjects. *J. Clin. Invest.* **81**:442–448, 1988.
80. J. L. Leahy. Intensive insulin therapy in type 1 diabetes mellitus. In J. L. Leahy (ed), *Insulin Therapy.* New York: Marcel Dekker, 2002, pp. 87–112.

81. C. B. Cook, J. P. McMichael, R. Lieberman, et al. The intelligent dosing system: application for insulin therapy and diabetes management. *Diabetes Technol. Ther.* **7**:58–71, 2005.
82. B. W. Bequette and J. Desemone. Intelligent dosing system: Need for design and analysis based on control theory. *Diabetes Technol. Ther.* **6**:868–873, 2004.
83. A. Ahmann. Lessons on developing a better basal insulin. *Diabetes Technol. Ther.* **6**:596–600, 2004.

Chapter

2

Physiological and Biochemical Barriers to Drug Delivery

Amit Kokate, Venugopal P. Marasanapalle, Bhaskara R. Jasti, and Xiaoling Li

Thomas J. Long School of Pharmacy and Health Sciences
University of the Pacific
Stockton, California

2.1 Introduction

The human body is a complex system with a unique organization that has undergone advanced development during evolution. The human body has adapted to resist and overcome the invading biological or chemical moieties. A drug, being a chemical/biological agent or a mixture of chemical/biological agents, is recognized as a xenobiotic by the human body. Subsequently, the drug will be prevented from entering the body and/or eliminated after its entry. As a result, the defense mechanisms of the human body become barriers to the delivery of drugs. Conceptually, any obstacle that prevents a drug from reaching its site of action is considered to be a barrier to drug delivery. A drug may encounter physical, physiological, enzymatic, or immunological barriers on its way to the site of action.

The goal of this book is to discuss the different strategies of designing controlled release drug delivery systems. A good understanding of the barriers to drug delivery will provide a sound foundation for the design of controlled release drug delivery systems. In this chapter the barriers to the delivery of drugs are discussed based on the anatomy and physiology of the respective organ system.

2.2 Barriers to Peroral Controlled Release Drug Delivery

Oral drug delivery is the preferred route for drug administration because of its convenience, low cost, and high patient compliance compared with several other routes. About 90 percent of drug products are administered via the oral route.[1] Orally administered drugs travel through different regions of the alimentary canal. The anatomy of the gastrointestinal (GI) tract and the physiological and biochemical barriers that it offers are described in the following sections.

2.2.1 Anatomical organization of the gastrointestinal tract

Anatomically, the alimentary canal can be divided into a conduit region and digestive and absorptive regions. The conduit region includes the mouth, pharynx, esophagus, lower rectum, and anal canal. The digestive and absorptive regions consist of the stomach, small intestine, and all the large intestine except the distal portion. A schematic representation of the human GI tract is shown in Fig. 2.1. The physical characteristics of the GI tract are summarized in Table 2.1. Orally ingested materials are absorbed predominantly through the duodenum and proximal jejunum. The alimentary canal from the mouth to the anus is lined by a mucous membrane (or mucosa) consisting of an epithelium, lamina

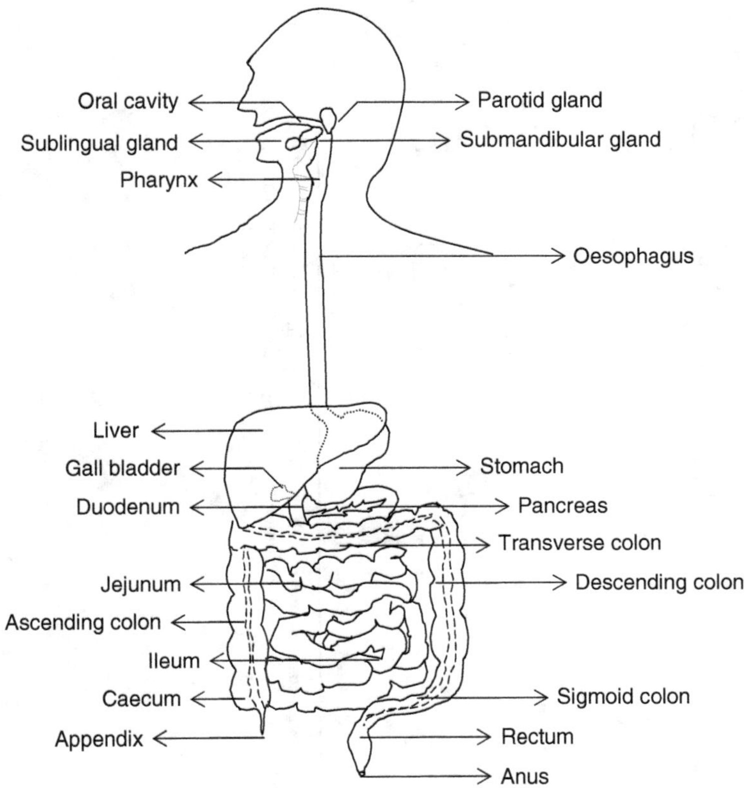

Figure 2.1 Schematic representation of the gastrointestinal tract.

propria (supporting loose connective tissue), and muscularis mucosae (thin layer of smooth muscle). The conduit region of the GI tract possesses a multilayer structure (stratified squamous nature) and is partially keratinized (gums and hard palate in the mouth), making it a formidable barrier to the entry of several substances. The stomach, small intestine, and colon are lined by simple columnar epithelium (single layer), which is usually involved in the absorption and/or secretion of substances.

Stomach. The stomach is the first digestive organ that a drug encounters in the alimentary canal. The human stomach can be divided anatomically into (1) the cardia, the gastroesophageal sphincter, (2) the fundus, the uppermost part of stomach, (3) the body, a reservoir for food and fluids, and (4) the antrum, the lower part of the stomach. The diameter of the stomach in the fed state is quite variable depending on the area. The antrum diameter does not change much in the fed state, whereas the diameter of the body may increase several-fold.

TABLE 2.1 Summary of the Physical Characteristics of GI Tract in Normal Adults[2–5]

Region of the GI tract	Physical characteristics					
	Length, cm	Internal diameter, cm	Volume, mL	Surface area, cm^2	pH	Average residence time
Entire GI tract	530–870	3–9		2×10^6	1.5–7	Up to 38 h
Mouth cavity	15–20	10		700		
Esophagus	20	2–4		200		
Stomach						
Fasted state	25	15	25–50		1.4–2.1	0.5–1.5 h
Fed state			1000–1600		2–5	2–6 h
Small intestine	370–630	3–5		$2.1–5.9 \times 10^6$*	4.4–7.4	3 ± 1 h
Duodenum	20–30	3–5		113,000–283,000*	4.9–6.4	3–10 min
Jejunum	150–260	3–5		270,000–750,000*	4.4–6.4	0.5–2 h
Ileum	200–350	3–5		360,000–1,050,000*	6.5–7.4	0.5–2.5 h
Large intestine	150	3–9		15,000	5.5–7.4	Up to 27 h
Caecum	7	7		500	5.5–7	
Colon	90–150	3–9		15,000	7.4	
Rectum	11–16	2.5		150	7	

* In the small intestine there is an amplification factor of 600 times the mucosal cylinder, with the folds of Kerckring accounting for 3 times, the villi 10 times, and the microvilli 20 times. The ranges of surface area are based on the range of diameters and length of each segment. This is according to the American Gastroenterological Association Teaching Project: Unit VII-A, Physiology of intestinal water and electrolyte absorption.

The mucosa of the stomach is folded into rugae, but they do not increase the absorptive surface area significantly. Consequently, the role of stomach in drug absorption is limited, and it acts primarily as a reception area for oral dosage forms. Nonionic, lipophilic molecules of moderate size can be absorbed through the stomach only to a limited extent owing to the small epithelial surface area and the short duration of contact with the stomach epithelium in comparison with the intestine. The relative residence times of stomach in the fasted and fed states are shown in Table 2.1. After passing through stomach, the next organ that a drug encounters is the small intestine.

Small intestine. The small intestine is the most important part for nutrient and drug absorption in the GI tract. The small intestine consists of the short, fixed duodenum and the relatively long, more mobile jejunum (the proximal half) and ileum (the distal half). The duodenum is the first part of the small intestine, connecting the stomach to the jejunum. The pancreatic and bile ducts open into the duodenum. The jejunum is broader and has a thicker wall than the ileum. The folds in the ileum are sparse and low in comparison with those in the jejunum. But the ileum has distinctive lymphoid tissue patches (Peyer's patches). The volume of the small intestine can vary from as little as 50 to 100 mL to several liters. The colon volume is also quite variable from 100 up to 2000 mL.

Blood is supplied to the jejunum and ileum via the superior mesenteric artery arising from the aorta. Venous blood returns to the superior mesenteric vein, which combines with the splenic vein to form the portal vein. Thus drugs absorbed through these parts pass initially through the liver.

The intestinal epithelium is composed of absorptive cells (enterocytes) interspersed with goblet cells (specialized for secretion of mucus) and a few enteroendocrine cells (that release hormones). The cells between the villi, known as *crypts of Lieberkühn*, are a protected site for stem cells and Paneth cells. The stem cells are involved in replenishment of the enterocytes and goblet cells once every 4 days. The Paneth cells (secretory epithelial cells) located at the ends of intestinal crypts are involved in secretion of antibacterial proteins into the crypt lumen, thereby providing protection to the stem cells lining the crypt walls. The enterocyte is the most important cell type involved in drug absorption in the intestinal epithelium. The epithelial cells are separated from the underlying tissue by a basement membrane composed of collagen and glycoproteins. The lamina propria, lying beneath and supporting the delicate epithelium, is a connective tissue with lymphatic drainage. The lamina propria is heavily infiltrated in the ileum with lymphocytes (Peyer's patches). A thin layer of smooth muscle at the interface of mucosa and submucosa called the *muscularis mucosa* is responsible for local mucosal movements and brings the enterocytes in close proximity with the GI contents.

The small intestinal mucosa possesses a large number of folds (folds of Kerckring). Long finger like projections called *villi* (0.5 to 1 mm long) project out of the mucosa in the small intestine (predominantly duodenum and jejunum). The villi, in turn, are divided into tiny microvilli (about 1000 per enterocyte) of about a micron in length. The microvilli give a brush-border appearance to the intestinal epithelium. These surface modifications of the small intestine enhance the surface area available for absorption by about 600-fold[6] (see Table 2.1; Fig. 2.2). The jejunal villi are long and closely packed in comparison with those in the ileum. This results in a greater surface-area-to-length ratio in the proximal region of the small intestine than in the distal region.[7]

The enterocytic membrane consists of lipids and proteins, apart from carbohydrates, water, and a number of chelated calcium and magnesium

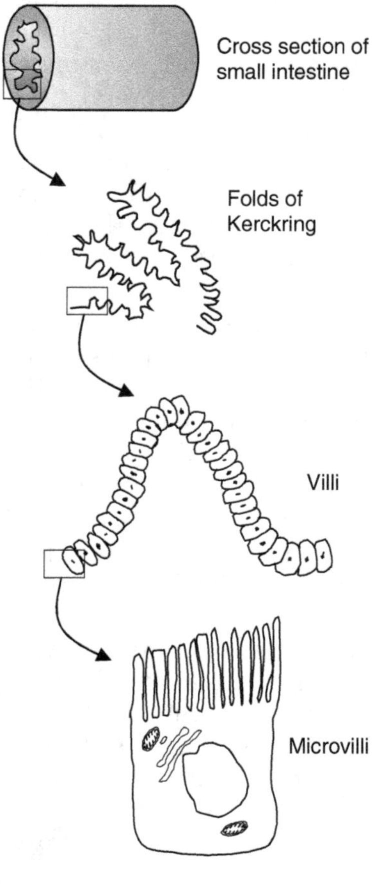

Figure 2.2 Surface characteristics of the small intestine. [*Adapted from D. Weiner in Biological Foundations of Biomedical Engineering (1976) with the permission of Lippincott, Williams and Wilkins.*]

ions. Most of lipids in the enterocytic membrane are glucosylceramide (~33 percent) and cholesterol (20 percent of total lipids), apart from phosphatidylcholine, phosphatidylethanolamine, phosphatidylserine, phosphatidylglycerol, and sphingomyelin.[8] Glucosylceramide is a cerebroside, which, in turn, is a simple glycosphingoid. The presence of an uncommon fatty acid composition (C24:0 and C24:1 type of 2-hydroxy fatty acids) in the cerebrosides results in interlipid hydrogen bonding through the hydroxyl groups of the fatty acids and the glucosyl group and the NH group of the ceramide backbone.[9] This results in a higher order-disorder transition temperature T_M for glucosylceramide, which is well above the body temperature. A higher proportion of glucosylceramide in the lumenal or brush-border membrane tends to stiffen the membrane, reducing the membrane fluidity and resulting in poor permeability for substrates. Cholesterol in the membrane aids in modulating the fluidity by promoting the mixing of lipids possessing a range of transition temperatures.[8] Three different classes of proteins are associated with the membrane: intrinsic (integral) membrane proteins, extrinsic (peripheral) membrane proteins, and lipid-anchored proteins (such as G proteins and kinases). Carbohydrate-rich materials covalently bound to the peripheral proteins with sialic acid at the terminal positions are also present on the cell surface.[10–13]

Large intestine. The colon and rectum form the large intestine. The colon is divided into five regions: cecum, ascending colon, transverse colon, descending colon, and sigmoid colon (see Fig. 2.1). Histologically, colonic mucosa resembles small intestinal mucosa, the absence of villi being the major difference (Fig. 2.3). The microvilli of the large intestine enterocytes are less organized than those of the small intestine.[14] The resulting decrease in the surface area of the colon leads to a low absorption potential in comparison with the small intestine (see Table 2.1). However, the colonic residence time is longer than that for the small intestine, providing extended periods of time for the slow absorption of drugs.

The colon contains about 400 different species of anaerobic and aerobic bacteria.[15] The most common anaerobes in the colon are *Bifidobacterium* spp., *Eubacterium* spp., *Peptostreptococcus* spp., *Fusobacterium* spp., and *Bacteroides* spp. The most common aerobe in the colon is *Escherichia coli*.[16,17] Colonic bacteria are involved in the synthesis of B complex vitamins and a majority of vitamin K.[18]

The distal portion of the large intestine is the rectum. Rectal absorptive capacity is considerably less than that of the upper GI tract owing to a limited surface area, a result of the absence of microvilli. Also, the blood supply to colon and rectum is less than that to the small intestine. The rectal artery branching off the inferior mesenteric artery of the

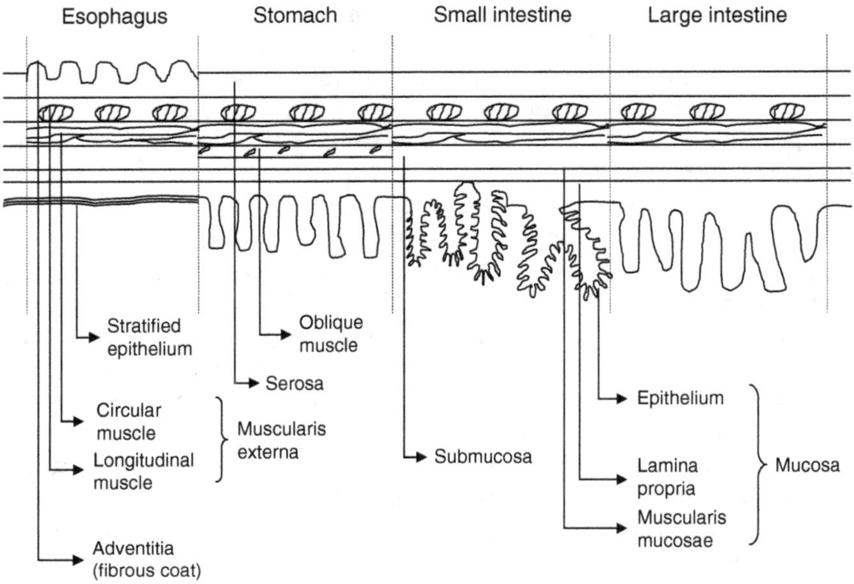

Figure 2.3 Layers of the gastrointestinal tract.

descending aorta supplies blood to rectum. The superior hemorrhoidal vein drains blood from the upper part of rectum into the hepatic portal system. However, the inferior and middle hemorrhoidal veins, which drain the lower part of rectum, connect directly to the systemic circulation through the inferior vena cava. From a drug delivery standpoint, this is of great importance because drugs administered in the lower region of rectum bypass the hepatic portal system, avoid the first-pass effect, and have higher bioavailability. In addition to the blood vessels, lymphatic vessels run along the large intestines and rectum. The lymphatic vessels dispose lymph into the thoracic duct, which subsequently secretes into the left subclavian vein. This provides a second route of delivering drugs to the systemic circulation.

2.2.2 Physiological and biochemical characteristics of the gastrointestinal tract

Parietal (oxyntic) cells located predominantly in the body and fundus of the stomach secrete hydrochloric acid. The gastric pH is not constant owing to variation in acid secretion and gastric content. The pH in the stomach at different states of feeding and various parts of the small intestine is shown in Table 2.1. The mucus layer restricts the diffusion of hydrogen ions secreted by the intestinal epithelial cells. As a result, the pH of this 700-μm-thick microclimate region is on the order 5.8 to

6.3, which is lower than that of the bulk solution in the intestinal tract.[19,20]

Mucus is secreted by the cardiac, neck, and surface mucous cells in the stomach and goblet cells in the crypts of the intestine.[6] Mucus is composed of water, electrolytes, sloughed epithelial cells, fatty acids, phospholipids, and glycoproteins.[21] The primary function of mucus is to protect the epithelium from physical injury and pathogens. The mucus in the stomach and colon has been reported to be relatively more hydrophobic in comparison with other parts in the GI tract.[22] Within the human stomach, the mucosal surfaces of the body and antrum were found to have the highest hydrophobicity.[21] The surface hydrophobicity in the stomach has been attributed to the secretion of phospholipids by surface mucous cells.[23] The positively charged phospholipids might be adsorbed onto the mucus, which is negatively charged owing to the presence of sialic acid and sulfated oligosaccharides.[22] The hydrophobic surface has been proposed to protect the gastric mucosa by forming a nonwettable barrier against the acidic lumenal contents.[24] The major phospholipids in the gastric mucosa were shown to be phosphatidylcholine (PC) and phosphatidylethanolamine (PE).[25] The major PC species present in the stomach lavage of rat and pig were found to be palmitoyloleoyl (PC 16:0/18:1) and palmitoyllinoleoyl (PC 16:0/18:2) derivatives.[26] The thickness of the mucus layer decreases gradually along the GI tract from the stomach (50 to 500 μm) to colon (16 to 150 μm).[27] Mucus content in the duodenum and jejunum is scanty and wispy compared with that in the ileum, which has a thick, relatively solid layer of mucus between the villi and the luminal contents.[7] The chief organic constituent of mucus is a complex mixture of glycoproteins called *mucins* (1 to 2 × 10^6 Da) with a polypeptide core known as *apomucin* covered by *O*-linked carbohydrate chains. Thirteen different mucins (MUC1–13) have been discovered, with MUC2 mucin being predominant in the GI tract.[28] The mucus layer is a potential barrier to the oral absorption of hydrophobic drugs, positively charged drugs such as tetracycline, quaternary amines (owing to binding by the negatively charged mucin), and large molecules. The mucus layer is functionally a part of the unstirred water layer. The barrier properties of the mucus layer might vary based on the drug-induced changes in the intestinal flora as mucins are degraded by colonic bacterial flora. The viscosity of colonic contents and the presence of gas bubbles and dietary fibers increase the thickness of unstirred water layer and contribute to physical barriers of drug absorption. A carbohydrate-rich glycoprotein layer (glycocalyx) covers the apical membrane to a distance of about 0.1 μm from the tips of the microvilli apart from filling the intermicrovillous space. The glycocalyx serves as a host to a variety of enzymes (mainly adsorbed pancreatic enzymes) such as α-amylase, trypsin, lipase, etc.[29]

Various cytochrome P450s (CYPs) capable of carrying out phase I metabolism have been identified in the human GI tract, as shown in Table 2.2. These enzymes pose one of the major biochemical barriers to oral drug bioavailability. The CYP3A subfamily accounts for 30 percent of total hepatic CYP content and 70 percent of total enterocytic CYP content.[30] Of these, CYP3A4 is the most abundant enzyme in the human small intestine, especially in the brush-border membrane. Drugs that are substrates to CYP3A4 generally undergo significant first-pass metabolism via hydroxylation and *N*-dealkylation reactions.[30]

Intestinal monoamine oxidase (MAO) metabolizes drugs such as sumatriptan,[31] phenylephrine,[32] and other amine-containing drugs. A number of hydrolytic and phase II enzymes, such as acetyl transferases, glutathione-*S*-transferases, methyl transferases, sulfo transferases, and UDP-glucuronosyl transferases, are also present in the intestinal mucosa (see Table 2.2). The amounts of metabolic enzymes decrease

TABLE 2.2 Drug-Metabolizing Enzymes in the GI Tract

Enzymes	Substrates
Cytochrome P450 (CYP):	
1A1	Polycyclic aromatic hydrocarbons (PAHs), substituted coumarins, debrisoquine
1B1	17β-Estradiol, PAHs
2C9	Phenytoin, tolbutamide, (*S*)-warfarin
2C19	Phenytoin, tolbutamide, (*S*)- and (*R*)-warfarin, omeprazole
2D6	Dextromethorphan, debrisoquine
2E1	Verapamil, tamoxifen
3A4/3A5	Estrogens, sirolimus, cyclosporine, tamoxifen
2S1	All-*trans* retinoic acid
4F12	Ebastine
2J2	Arachidonic acid, astemizole
Non-CYP phase I enzymes:	
Alcohol dehydrogenase	Ethanol
Xanthine oxidase	Theophylline, 6-mercaptopurine, allopurinol
Monoamine oxidase	Sumatriptan, phenylephrine
Phase II enzymes:	
Acetyl transferases	Aromatic and heterocyclic amines, hydrazines, and sulfonamides
Glutathione-*S*-transferases	Acetaminophen
Methyl transferases	(–)-Epicatechin
Sulfotransferases	Phenols
UDP-glucuronosyl transferases	Aromatic amines
Enzymes from colonic microbial flora:	
Azoreductase	Sulfasalazine
Nitroreductase	Chloramphenicol

SOURCES: From refs. 40 to 43.

along the GI tract. This is paralleled by an increase in the microbial flora capable of carrying out both phase I and phase II metabolism with the aid of β-glycosidase, β-glucuronidase, azoreductase, nitroreductase, pectinase, amylase, xylanase, dextranase and 7α-hydroxysteroid dehydrogenase.[33] Apart from drug metabolism, the enzymes produced by the microflora have the potential to bioactivate prodrugs. The bacteria can carry out the following reactions: (1) hydrolysis of glucuronides, esters, and amides, (2) aromatization of ethereal sulfates, (3) reduction of C—C double bonds, *N*-hydroxy compounds, carboxyl groups, alcohols, and phenols, and (4) deamination, dealkylation, decarboxylation, and dehalogenation.[15] A lower surface area in the colon results in comparatively smaller amounts of enzymes such as CYP. Thus it may be easy to saturate colonic wall enzymes, and degradation may be lesser than in the small intestine. The number of CYP enzymes located in the rectum is minimal.

Active secretion of drugs also might be a significant barrier for drug bioavailability. The product of the *MDR1* gene, P-glycoprotein (P-gp) is a 170-kDa multidrug efflux pump. P-gp is a dimer comprised of 1280 amino acids with 12 transmembrane segments and 2 adenosine 5'-triphosphate (ATP)–binding domains.[34] P-gp was first recognized as an ATP-dependent efflux pump of chemotherapeutic agents in cancer cells. Subsequently, P-gp was found to be expressed in various normal human tissues, including gastrointestinal epithelia (esophagus, stomach, jejunum, and colon), brain, liver, and adrenal gland.[35] P-gp is expressed on the apical membrane of the mature epithelial cells in the small intestine. The P-gp in the GI tract is functionally identical to P-gp present in other epithelial cells and in cancer cells. It is involved in the efflux of a wide range of structurally unrelated molecules. However, the molecules tend to be large and amphipathic with one or more aromatic rings.[36]

CYP3A and P-gp share a large number of substrates and inhibitors, suggesting that they might form a coordinated barrier to drug bioavailability.[30] It is possible that P-gp adjusts the entry rate of a substrate so that it undergoes maximal intestinal metabolism by CYP. The expression of P-gp increases along the length of the intestine, whereas CYP3A4 has the highest expression in the proximal part of the intestine.[37,38]

Membranous epithelial M cells are a part of the organized mucosa-associated lymphoid tissue (O-MALT). These cells are specialized for antigen sampling. Also, they are exploited as a route of invasion by several pathogens.[39] M cells are concentrated in follicle-associated epithelial (FAE) tissue called *Peyer's patches* in the small intestine. As M cells carry out endocytotic transport, they can be potential vehicles for mucosal drug and vaccine delivery.

2.2.3 Barriers to peroral drug delivery

The barriers offered by the GI tract are manifold when compared with nonperoral routes. The solubility, stability, and absorption of several drugs are affected by the acidic nature of the gastric contents, intestinal enzymes, and microflora-associated enzymes in colon. In addition, intestinal efflux transporters might limit a drug's absorption across the gut. Factors such as variable gastric emptying rate and the presence of food result in erratic bioavailability of certain drugs.[3,44] Rectal drug administration can overcome the first-pass metabolism associated with peroral drug delivery. However, this route of drug administration has the disadvantages of a limited surface area and a relatively impermeable membrane in comparison with the small intestine.

2.3 Barriers to Nonperoral Controlled Release Drug Delivery

2.3.1 Skin

Skin forms the body's protection against the entry of foreign substances, pathogens, and radiation and prevents the loss of endogenous contents, including water. The advantages of drug delivery through the skin include easy accessibility, convenience, prolonged therapy, avoidance of liver first-pass metabolism, and a large surface area.

Structure of skin. Skin is composed of two layers, the epidermis and the dermis, separated by a basement membrane zone. Hypodermis, composed of adipose tissue, sweat glands, and pacinian corpuscles, is not part of the skin.[45]

Epidermis. The epidermis is a 50- to 100-μm-thick layer made of keratinocytes that migrate outward from the basal cell into highly differentiated nondividing cells.[46,47] During differentiation, keratinocytes transform from polygonal (cuboidal) cells to spinous (prickly) cells, flattened granular cells, and finally to flattened polyhedral dead corneocytes full of the protein keratin.[45,47]

Epidermis can be divided into stratum basale (SB), stratum spinosum (SS), stratum granulosum (SG), and stratum corneum (SC). The epidermal cell layers are interconnected by desmosomes.[45,48] A cross section of the skin epidermis is shown in Fig. 2.4.

Stratum basale is a single layer composed of stem cells and their derivative cells. It is attached to the basement membrane by hemidesmosomes.[48] The cells in this layer are columnar or cuboidal in shape and characterized by large nuclei (high nuclear-cytoplasmic ratio) and keratin filaments (tonofilaments). The basal layer contains keratins K14 and K15, melanocytes (which are pigment-forming cells), Langerhans

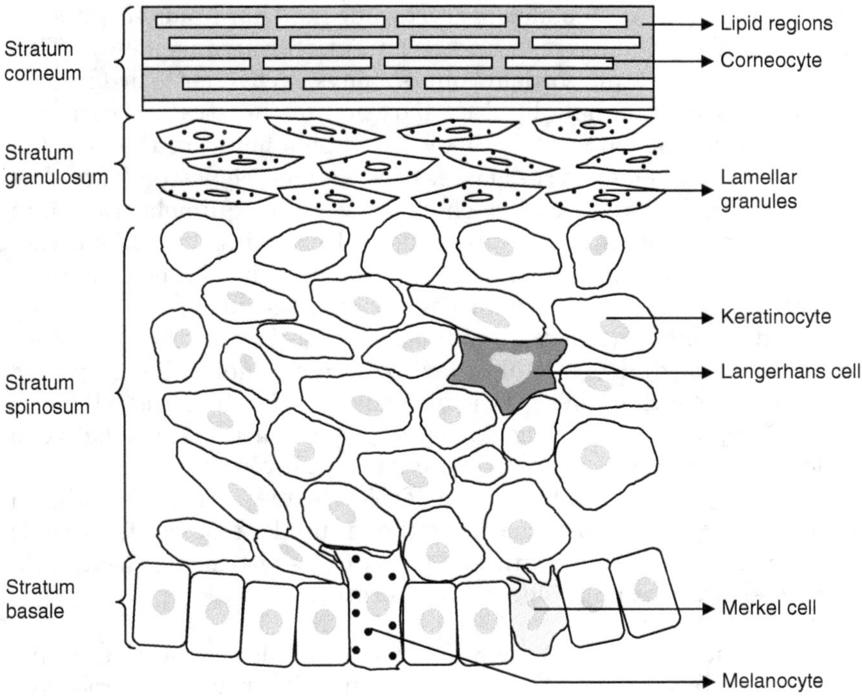

Figure 2.4 Layers of the skin epidermis.

cells (which play a role in the immune response of the body), and Merkel cells (which are associated with the nervous system).[45,48]

The stratum spinosum layer contains abundant desmosomes, lipid lamellar bodies (Odland bodies), keratinosomes, and membrane-coating granules (MCGs). Lipid lamellar bodies are parallel stacks of polar lipid-enriched disks enclosed in a trilaminar membrane.[48] The lamellar bodies also contain hydrolytic enzymes capable of converting polar lipids such as glycolipids and phospholipids to nonpolar products such as ceramides and free fatty acids, respectively.[46,49]

The stratum granulosum layer is characterized by darkly staining keratohyalin granules, which are composed of profillaggrin, loricin (a cysteine-rich protein), and Keratin 1 and Keratin 10. An increase in the number of lamellar bodies, along with protein synthesis, accompanies the differentiation process. Lamellar bodies in this region of skin are composed of glucosyl ceramides, phospholipids, cholesterol, and enzymes such as lipases, sphingomyelinase, β-glucosylcerebrosidase, and phosphodiesterases.[48,50]

The stratum corneum is 10 to 20 μm in thickness (about 18 to 21 cell layers thick).[46,48] It is composed of corneocytes (enucleated, keratin-filled, dead keratinocytes) embedded in crystalline lamellar lipids secreted by lamellar bodies, giving rise to a brick-and-mortar arrangement.[51] This arrangement of stratum corneum offers a physical barrier to drug transport across the skin. The lipid composition of the stratum corneum varies in different parts of the body. The lipids found in the stratum corneum mainly belong to the classes of ceramides, cholesterol, and free fatty acids, with 16 to 33 carbon chains present in equimolar ratios. In addition, small amounts of triglycerides, glycosphingolipids, and cholesterol sulfate are detected in the stratum corneum.[48,52] The ceramides present in the stratum corneum possess small head groups and long fatty acid chains. The numerous functional groups on the ceramides can form intra- and intermolecular hydrogen bonds. The fatty acid chains are longer than those of the phospholipids found in the cell membranes of living cells.[51] Strong van der Waals interactions exist between the long hydrocarbon chains of ceramides and free fatty acids.[46] The presence of hydrogen bonds and van der Waals interactions, in addition to saturated, aliphatic long chains, in the ceramides and free fatty acids results in higher melting points and thus in a less permeable solid crystalline (gel) state at physiological temperature.[53]

Dermis. The dermis is a 2- to 3-mm-thick vascular connective tissue. It contains collagen, elastic fibers, and various cells, including fibroblasts, mast cells, macrophages, and lymphocytes, that present an immunological barrier.[45,47,48] The mast cells originate from the bone marrow stem cells and are present in the GI tract, skin, and pulmonary tract apart from being distributed in the blood.[54] The mast cells secrete histamine and neutral proteinases (tryptase, chymase, and carboxypeptidase) on activation.[55] Some drugs could challenge these mast cells and cause hypersensitivity reactions. Blood vessels and epidermal appendages such as hair follicles, sweat glands, and sebaceous glands are present in the dermis. Collagen, elastin, and glycosaminoglycans in the dermis collectively are called the *extracellular matrix* (ECM).[48] The ECM permits free diffusion of nutrients and acts as a "biological glue" in holding together the dermal cells.

The dermis can be divided into two layers, the papillary layer adjacent to the epidermis and the deeper reticular layer. Blood vessels and sensory nerve endings are present in this layer. The reticular layer is made up of collagen and elastin lattice. Papillary and reticular layers provide structural and nutritional support.[45]

Enzymes and transporters in skin. Skin shows differences in its enzyme activity at different sites of the body. Sebaceous glands were found to be most enzymatically active, followed by differentiated keratinocytes and then basal cells.[47,56] Also, hair follicles in rodent skin were found to exhibit enzyme activity.[47,57,58] It is generally believed that the epidermal layer has higher enzyme activity than the dermal layer.[47] Cytochrome P450 and epoxide hydrolases are the most common enzymes found in the skin. CYP1A1, -1B1, -2B6, -2E1, and -3A5 were reported in the keratinocytes.[59] Glucuronide, sulfate, and glutathione conjugation of xenobiotics has been observed. Transporters such as organic anion transport protein (OATP B, D, E), multidrug resistance protein (MDR1), lung resistance protein, and multidrug resistance–associated proteins (MRP1, 3–6) were reported to be present in the keratinocytes.[60,61]

Barriers to transdermal drug delivery. The outermost layer of skin, the stratum corneum, is a formidable permeability barrier to drugs intended for systemic use. The crystalline lamellar lipids in which the corneocytes are embedded offer a physical barrier to drug transport.[48,51] The arrangement of corneocytes offers tortuosity to the path of hydrophobic molecules, and the lipid organization offers resistance to hydrophilic molecules.[46] Drug diffusion across the stratum corneum depends on the hydrophilicity, size, and hydrogen-bonding capacity of the permeant.[62] The enzymes responsible for keratinocyte differentiation could be capable of degrading (metabolizing) xenobiotics.[47]

2.3.2 Eye

The eye is a specialized sensory organ of photoreception. The eye is an easily accessible organ for local or systemic drug delivery.[63] The anatomical and physiological characteristics of the eye are described, and the different barriers to drug delivery via ocular route are outlined in this section.

Structure of the eye. The eye can be divided into two compartments: the anterior and posterior segments. An internal cross section of an eye is shown in Fig. 2.5.

Anterior segment. Externally, the anterior segment of eye is made up of cornea, conjunctiva, and sclera. Internally, it consists of anterior chamber, iris/pupil, posterior chamber, and ciliary body.[64] The cornea, an optically transparent tissue that aids in refraction of light to the eye for focusing, is 1 mm thick at the periphery and 0.5 to 0.6 mm thick in the center.[45] It is composed of squamous and basal columnar epithelium, Bowman's membrane, substantia propria (stroma), limiting lamina, and the endothelium. The conjunctiva is a thin, transparent, vascularized mucous membrane with an area of 18 cm^2 covering the eye globe and the inner eyelids.[65] It maintains the precorneal tear film and protects the eye. It produces mucus and lubricates the surface of the eye. It is made up of stratified columnar epithelium and lamina propria. The conjunctival epithelium is divided into bulbar (covering the eyeball),

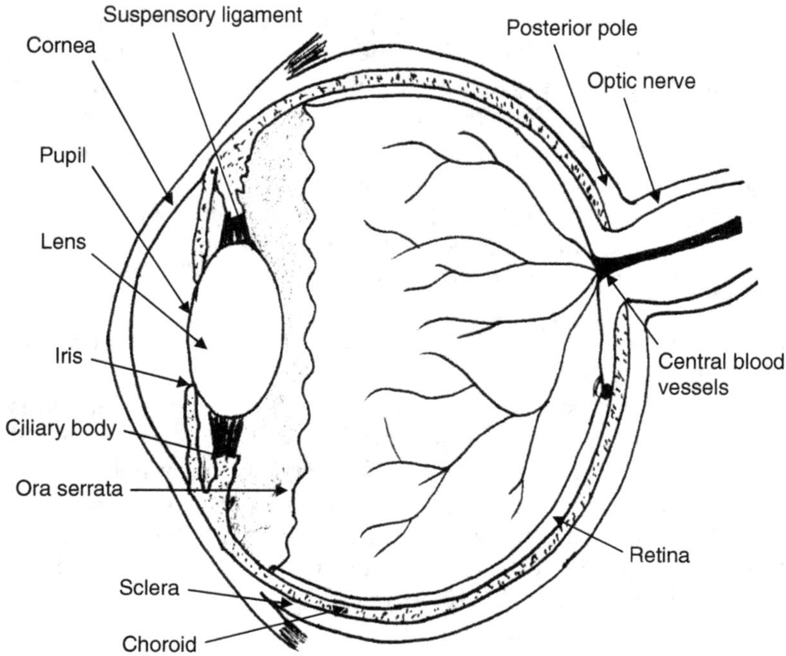

Figure 2.5 Internal structure of the eye.

fornix (covering the cornea), and palpebral (covering the eyelid) conjunctivae. The sclera, the white outer coat of the eyeball, provides structural integrity, size, and shape to the eye. There are three layers in the sclera, the anterior episclera, the middle scleral stroma, and the posterior lamina fusca. The sclera is composed of gellike mucopolysaccharides, elastic fibers, bundles of dense collagen fibrils, and fibroblasts.[45,65] The iris is a diaphragm around the pupil (lens) and controls the amount of light entering the inner eye. The ciliary body is made up of ciliary muscles, which aid in accommodation.

The anterior surface of the eye is constantly rinsed by tear fluid secreted at a flow rate of about 1 μL/min by the main lachrymal gland of the lachrymal apparatus. Tears eventually drain into the nasal cavity through the nasolachrymal ducts. Tear fluid contains mucin, lysozyme, lactoferrin, prealbumin, and serum proteins. It functions as an antibacterial lubricant and aids in draining out foreign substances. The normal volume of tear fluid is 5 to 10 μL.[65]

Posterior segment. Externally, the posterior segment consists of the optic nerve and associated vasculature, and internally, it consists of the lens, vitreous, and rear ocular tissues.[64] Vitreous is a colorless medium consisting of about 99 percent water, dissolved type II collagen, sodium hyaluronate, and proteoglycans.[45,64] The retina is the inner nervous layer of the eye responsible for the sensory function of sight. The choroid is a dark brown vascular layer attached to the sclera and is believed to provide nourishment to the retina.

Barriers to ocular drug delivery. The presence of epithelial tight junctions and rapid drainage of the instilled drug solution by tears from the precorneal area result in the permeation of less than 5 percent of the applied dose of a drug.[66,67] The cellular calcium levels and actin filaments of the cytoskeleton play a major role in the integrity of tight junctions.[68–72]

The different factors that influence the permeability of drugs across the cornea are distribution coefficient (log $D_{O/W}$), ionic equilibrium, and charge. Since the corneal epithelium is negatively charged above its isoelectric point (pI 3.2), cationic molecules with a log $D_{O/W}$ of 2 to 3 at pH 7.4 exhibit good permeation.[73–80] The cornea has significant levels of various esterases, peptidases, proteases, and other enzymes that can limit the availability of drugs such as peptides but can transform prodrugs to active moieties.[65,81–83]

A single layer of flat hexagonal cells of the corneal endothelium covers the posterior corneal surface and hydrates the cornea. The corneal endothelium can allow diffusion of molecules of dimensions up to 20 nm.[65] The stroma has a highly organized hydrophilic tissue structure that comprises 90 percent of cornea.[63] It has an open structure that can allow molecules up to 500,000 Da in size to pass through. However, the stroma may be a diffusion barrier to lipophilic drugs.[65,84] Drug-melanin interactions such as the one with timolol can form a barrier and reduce bioavailability.[85]

In addition to the corneal route, topically applied ocular drugs may be absorbed via a noncorneal absorption route that involves drug transport across the bulbar conjunctiva and underlying sclera into the uveal tract and vitreous humor.[65] The intercellular spaces of the conjunctival epithelium are wider than those of the corneal epithelium. As a result, the conjunctiva has higher permeability than the cornea to agents such as mannitol, inulin, and FITC-dextran.[86] The penetration of peptides, however, is limited by enzymatic degradation.[87] The limit of molecular size for conjunctival penetration is between 20,000 and 40,000 Da. Vitreous can act as an aqueous and unstirred diffusion barrier to drug permeation.[64]

Apart from physical barriers, topical drug delivery to eye is also affected by the volume, viscosity, pH, tonicity of vehicle, and type of drug. Constant drainage by tear fluid minimizes topical drug absorption and increases systemic absorption. As a result, only about 5 percent of total dose is effectively absorbed through the intraocular route, resulting in low ocular bioavailability and possible systemic side effects.[65] Drug permeation also can be influenced by protein binding in the tear fluid (about 0.7 percent).[63]

The disadvantages of this route of administration include irritation, stinging, tear formation, and blurring of vision owing to mydriasis or miosis, which could result in patient noncompliance. Drug absorption enhancement has been achieved by increasing the lipophilicity of drugs through prodrug formation, periocular administration (which includes subconjunctival injection, subtenon's injection, etc.), and coadministration of different drugs.[88]

2.3.3 Oral cavity

Drug delivery via this route is promising owing to ease of administration and a rich supply of blood and lymphatic vessels. In addition, this route offers high permeability to drugs and good reproducibility. Drugs absorbed via the buccal mucosa enter the systemic circulation directly through the jugular vein. This ensures a rapid onset of action and avoids first-pass liver metabolism, gastric acid hydrolysis, and intestinal enzymatic degradation.[89–91]

Buccal tissue is a robust tissue owing to its continuous exposure to a multitude of substances and its high cellular turnover rate.[92,93] The absence of Langerhans cells in the oral mucosal tissues reduces sensitivity to potential allergens.[94,95] Hence irritation and hypersensitivity reactions due to drugs and their formulation excipients may be minimal in the short term as well as chronic treatment by this route.

Anatomy and physiology of the oral cavity. The different anatomical regions of the oral cavity and mucosal tissues (excluding the odontal structures) are shown in Fig. 2.6. The various target sites for drug delivery may include the inner surfaces of the upper and lower lips, gums (gingiva), hard and soft palate, floor of the mouth (sublingual), and tongue and buccal mucosal tissue (cheek).

The structure, thickness, and blood flow of the mucosa vary within the oral cavity.[96] The oral mucosal tissue consists of a *keratinized* epithelium in the masticatory region consisting of the gums (gingivae), palatal mucosa, and inner sides of the lips. The sublingual (floor of mouth) and the buccal mucosa are *nonkeratinized*.[94] The keratinized areas of the hard palate and gingival tissue resist shear forces and abrasion caused by food materials.[97] The masticatory regions have an underlying submucosa in the hard palate. Submucosa is absent in the gingiva. Submucosa contains mucus salivary glands, greater palatine nerves, and blood vessels. It serves to anchor the buccal mucosa to the periosteum of the maxillae and palatine bones.[45]

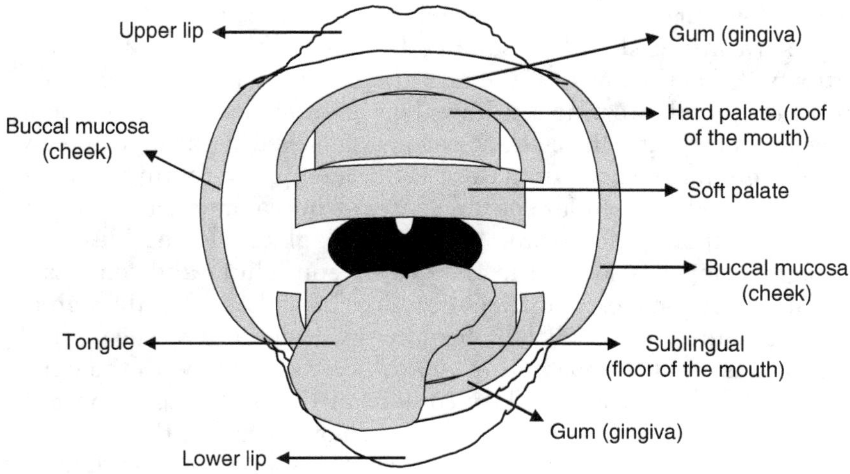

Figure 2.6 The anatomic regions of the oral cavity.

The cross-sectional structure of the buccal mucosa is shown in Fig. 2.7. The buccal mucosa consists of the epithelium and the underlying connective tissue, the lamina propria, separated by a basement membrane.

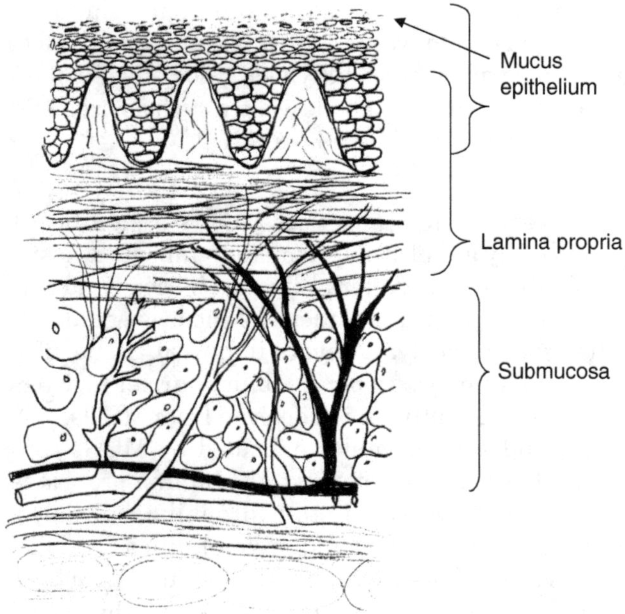

Figure 2.7 Cross-sectional structure of buccal tissue.[99]

Buccal epithelium. The buccal epithelium is composed of 40 to 50 layers of nonkeratinized stratified squamous cells with varying degrees of maturity.[93] The thickness of the buccal epithelium is 500 to 800 μm.[98] Unlike the small intestine, the buccal epithelium lacks tight junctions. The cohesion of epithelial cells, however, is achieved by the lipid and glycolipid contents extruded from the cellular membrane-coating granules (MCGs) in the intercellular space.[96,98,100] The buccal mucosa is made up of phospholipids, cholesterol sulfate, and glycosylceramides. The Malpighian layer is located in the deeper epithelium and consists of cells at various stages of differentiation. The cells in the Malpighian layer are round and loosely held together by desmosomes. The paracellular path is less tortuous in this layer when compared with the upper superficial layer.[96] The papillary contour of the basal region permits efficient vascularization of the cell. The turnover time for the epithelial cells is 5 to 6 days.[94] The epithelium terminates with the basal lamina, a 1- to 2-μm-thick proteinaceous fibrous matrix.[96]

Lamina propria. The structure of the lamina propria is characterized by a hydrated matrix that has a loose structure.[96,101] The lamina propria is made up of collagen fibrils, a supporting layer of connective tissue, blood vessels, and smooth muscle.[96]

Submucosa. The submucosa is a relatively dense connective tissue with a few accessory salivary glands (mucus acinus).[102] The saliva is secreted primarily by parotid, submandibular, and sublingual glands at a rate of 0.5 to 2 L/day.[94] Apart from water, saliva is composed of electrolytes, mucin (forms mucus with water), amylase, lysozyme (a bacteriostatic enzyme), IgA antibodies, and metabolic wastes such as urea and uric acid.[18] The pH of saliva varies between 6.8 and 7.2.[96]

Barriers to buccal drug delivery. The oral mucosa offers a significant barrier to the toxins produced by a multitude of microorganisms it hosts. Buccal mucosa, like the small intestine, offers a lipoidal barrier, and this route is preferable for small, lipophilic molecules.[103] Mucus is not as significant a barrier as the other underlying layers in buccal tissue. The superficial layers of the epithelium pose a permeability barrier to drugs owing to the presence of intercellular material derived from MCGs.[93,100] Lipids such as ceramides and acylceramides extruded by MCGs contribute to the permeability barrier.[104] In contrast, the lamina propria is not a barrier to drug permeation, especially to hydrophilic molecules, and may not hinder the permeability of even large molecules.[96,101] The carrier-mediated transport of a few hydrophilic compounds such as amino acids and monosaccharides has been reported through the buccal mucosa.[96,101,105]

Buccal mucosal permeability also varies with the anatomical site; for example, keratinized sites of the oral cavity hinder the permeation of hydrophilic molecules.[96] Therefore, it is important to consider these factors in selecting the transmucosal route for drug delivery in the oral cavity. A limited surface area (100 to 170 cm^2) and the need to mask taste for bitter drugs are some of the drawbacks associated with this route of drug delivery.

2.3.4 Nose

In the past few years, use of the nasal cavity as an alternative route for drug delivery has attracted great interest in the pharmaceutical industry, especially for systemic delivery of drugs that possibly can be delivered only through injection. The nasal route is favored for drugs that are given in small doses and require rapid onset of action, have extensive GI and hepatic degradation, and are used chronically.[106,107] The nasal route of delivery is advantageous because of its rich blood supply, accessibility, noninvasiveness, low risk of overdose, and possibility of self-medication.[108,109]

Nasal anatomy and physiology. The nose is the first organ of the respiratory tract. The structure of the nasal cavity is shown in Fig. 2.8.

Nasal cavity. The nasal cavity has a volume of 15 to 20 cm^3 and a surface area of 150 to 180 cm^2.[110,111] The human nasal cavity is divided into two halves by the midline septum.[45,107] Each cavity consists of three regions: (1) vestibules, which are the anterior sections of the nasal cavity, (2) respiratory region, consisting of turbinates or chonchae (superior, middle, and inferior), and (3) olfactory region, which constitutes about 10 percent of nasal area.[45,107]

The nasopharynx plays an important role in optimizing the temperature and moisture content of the inhaled air. It also protects the body from particles and microorganisms by filtering the particulate matter in the inhaled air to prevent it from reaching the lower airways.[107] The nasal turbinates divide the nasal passageway into narrow slits, resulting in a turbulent airflow, which helps in warming and humidifying the air, and an increased surface area.[107,110]

Nasal epithelium. The nasal cavity superior to the nostrils (vestibule) is covered by skin containing hair follicles and sebaceous and sweat glands. The skin is continuous with the inner nasal mucosa. Posteriorly, the epithelium is a pseudostratified ciliated columnar epithelium that covers the respiratory regions (formed by the maxilloturbinates).[45,107,110] The superior turbinate (also called the *ethmoturbinate*) is lined by a thicker mucosa consisting of olfactory receptors and supporting cells.[45,110]

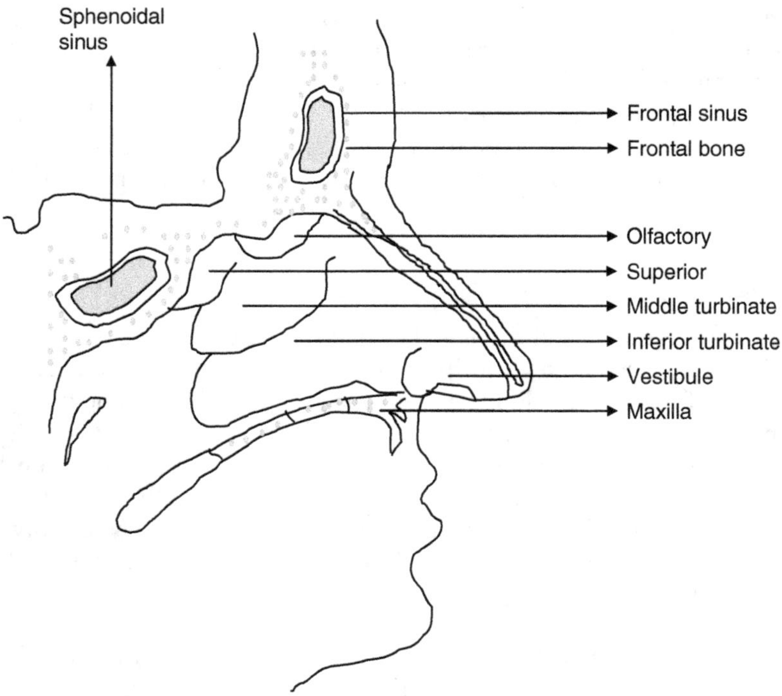

Figure 2.8 Structure of the nasal cavity.

Apart from the columnar cells (ciliated and nonciliated), the nasal epithelium also consists of basal and goblet cells.[45] The ciliated and nonciliated columnar cells are connected by tight junctions and are covered with microvilli (~300 microvilli per cell).[111] The 4- to 6-μm-long cilia beat the overlying mucus layer at a frequency of 1000 strokes per minute and propel the mucus from the anterior to the posterior part of the nasal cavity.[107] The cilia in the nasal vestibule and the mucus layer covering the respiratory area of the nasal cavity trap particulates, which are then carried down to the esophagus by the mucociliary clearance mechanism.[107]

Nasal mucus. The nasal mucus protects the body against airborne substances. Nasal mucus consists of mucopolysaccharides complexed with sialic acid, sloughed epithelial cells, bacteria, water (95 percent), glycoproteins and lipids (0.5 to 5 percent), mineral salts (0.5 to 1 percent), and free proteins (albumin, immunoglobulins, lysozyme, interferon, lactoferin, etc., 1 percent).[13,45,111,112] The surface pH of the nasal mucosa is 5.5 to 6.5.[113]

Barriers affecting nasal absorption. Unlike other biological membranes, it was noted that physicochemical properties such as charge and lipophilicity might not be of significant importance for transnasal drug delivery of drugs with a molecular weight of less than 300 Da. It has been hypothesized that the absorption of small molecules takes place via the aqueous channels of the membrane.[106,114]

Anatomical and physiological factors that influence nasal absorption include membrane transport, deposition, enzyme degradation, and mucociliary clearance.[106,115] The relative bioavailability of small lipophilic drugs delivered by this route approaches 100 percent owing to good absorption.[107] However, the bioavailability of macromolecules such as proteins and polar drugs larger than 1000 Da is low and ranges from 0.5 percent to 5 percent owing to low permeability through the membrane.[107,108]

The nasal route of drug delivery avoids the liver first-pass effect, but the pseudo-first-pass effect owing to nasal metabolism of drugs is still a concern. Many enzymes such as carboxylesterase, aldehyde dehydrogenase, glutathione transferases, UDP-glucoronyl transferase, epoxide hydrolases, CYP-dependent monoxygenases, exo- and endopeptidases and proteases are present in the nasal mucosa.[106–108,110,116] CYP enzymes are present abundantly in the olfactory epithelium.[107,110]

Substances deposited in the nasal cavity that are not readily absorbed are cleared in about 15 to 20 minutes by mucociliary clearance. The mucociliary clearance mechanism affects the absorption of polar drugs that have low permeability across membranes. This defense mechanism transports the particles down the throat.[107,108] Mucociliary clearance can be affected by the drug moiety, formulation, hormones, and disease states.[108]

2.3.5 Lung

The lung as a drug delivery site offers many advantages. Drugs can be delivered noninvasively via lung for systemic effects. A rich blood supply and large effective surface area (140 m^2) contributes to a high bioavailability and a quick onset of action.[117] Delivery of proteins and peptides may be feasible owing to the low metabolic activity of this surface.[117]

Anatomy and physiology. The human respiratory system is divided into upper and lower respiratory tracts. The upper respiratory system consists of the nose, nasal cavities, nasopharynx, and oropharynx. The lower respiratory tract consists of the larynx, trachea, bronchi, and alveoli, which are composed of respiratory tissues.

The left and right lungs are unequal in size. The right lung is composed of three lobes: the superior, middle, and inferior lobes. The smaller left lung has two lobes.[13,45] The nasopharynx is a passageway from the nose to the oral pharynx. The larynx controls the airflow to the lungs and aids in phonation. The larynx leads into the cartilaginous and fibromuscular tube, the trachea, which bifurcates into the right and left bronchi. The bronchi, in turn, divide into bronchioles and finally into alveoli. The respiratory tree can be differentiated into the conducting zone and the respiratory zone. The conducting zone consists of the bronchi, which are lined by ciliated cells secreting mucus and terminal bronchioles. The respiratory zone is composed of respiratory bronchioles, alveolar ducts, atria, and alveoli.[118]

The epithelium in the conducting zone gets thinner as it changes from pseudostratified columnar to columnar epithelium and finally to cuboidal epithelium in the terminal bronchioles. The upper part of the conducting zone (from the trachea to the bronchi) consists of ciliated and goblet cells (secrete mucus). These cells are absent in the bronchioles. Alveoli are covered predominantly with a monolayer of squamous epithelial cells (type I cells) overlying a thin basal lamina.[18] Cuboidal type II cells present at the junctions of alveoli secrete a fluid containing a surfactant (dipalmitoylphosphatidylcholine), apoproteins, and calcium ions.[118]

The lungs are covered extensively by a vast network of blood vessels, and almost all the blood in circulation flows through lungs. Deoxygenated blood is supplied to the lungs by the pulmonary artery. The pulmonary veins are similar to the arteries in branching, and their tissue structure is similar to that of systemic circulation. The total blood volume of the lungs is about 450 mL, which is about 10 percent of total-body blood volume.[118]

Pulmonary permeability and barriers to drug delivery. The most important barrier to the delivery of drugs to the pulmonary epithelium is the path that a drug follows when administered through inhalation. An inhalation formulation of salbutamol and terbutaline with particle size of 3 μm (when compared with particle sizes of 1.5 or 5 μm) was found to provide the best relief from asthma owing to the most optimal deposition.[119] The larger particles settled peripherally in the respiratory system and could not deliver drug in sufficient amounts. The smaller particles were believed to either settle peripherally or were expelled out during expiration, resulting in reduced drug deposition in the lung. Alveolar epithelium is permeable to lipophilic drugs; however, it is relatively less permeable to proteins. The tight junctions of the cuboidal epithelial lining of the lumen limit drug absorption between the respiratory airway and the internal environment. The tight junctions are believed to affect the absorption of hydrophilic drugs.[119] The presence of proteases such as neutral endopeptidase and cathepsin H may result in degradation of peptide drugs.[119] Other enzymes, namely, peroxidases, and inflammatory and immunomodulatory mediators are also present in the respiratory passageways.[120] Aerosols with an optimal particle size (1 to 3 μm diameter) are required to deliver drugs to the deep lung area and yet not be expelled during expiration.[119,121] Current inhalation delivery devices deliver only about 10 to 20 percent of the dose in the correct particle size.

2.3.6 Vaginal mucosa

The potential of vaginal drug delivery is under investigation because of the advantages of prolonged, less frequent administration, low dose requirements, and continuous release. A drug delivered by the vaginal route does not face stability and degradation barrier issues.[122]

Anatomy and physiology. The vagina, also called the *birth canal*, is a thin-walled fibromuscular tubular structure, 8 to 12 cm in length, that lies inferior to the uterus, posterior to the urethra and bladder, and anterior to the rectum. The vaginal orifice is the external opening of the vagina. The vagina of a healthy woman consists of numerous folds called *rugae* that help in distension. The vaginal wall consists of four layers: the epithelial layer, the tunica adventitia, the muscular layer, and connective tissue.[122]

Vaginal epithelium and tunica adventitia. The epithelium is characterized by nonkeratinized stratified squamous epithelial cells. The glycogen levels in the vaginal epithelium increase after puberty and diminish after menopause. Estrogen, progesterone, luteinizing hormone (LH), and follicle-stimulating hormone (FSH) levels affect the vaginal wall.[45] Estrogen increases the thickness of the epithelial layer.[122] The vaginal epithelium is thickest before ovulation and thinnest after menses. Although the vagina lacks mucus glands, it receives mucus from the cervical glands.[45] The tunica adventitia is made up of collagen and elastin and is innervated extensively with blood vessels and lymphatics.[122]

Vaginal muscles and connective tissue. The muscular layers are composed of longitudinal and circular smooth muscles. The two layers are connected by oblique decussating fasciculi.[45] The connective tissue to the exterior of the muscular layer consists of vascular plexuses. The vaginal wall lacks hair follicles, fat cells, and other glands.[122] Blood supply to the vagina is provided by the vaginal artery, a branch of the internal iliac artery. Vaginal veins drain blood from the vagina into the internal iliac veins. The internal iliac veins lead through the common iliac vein into the inferior vena cava, which takes blood to the heart to be redistributed to the whole body, thus bypassing hepatic first-pass metabolism.[45]

The vagina contains a normal flora of bacteria, mostly harmless, that feed on the cervical mucous nutrients. The bacterial flora (e.g.,

Doderlein's bacillus, Lactobacillus acidophilus) present in the vagina metabolize glycogen, resulting in the production of lactic acid and hydrogen peroxide, thereby causing a slightly acidic environment (pH 4 to 5). The acidic environment inhibits the growth of other pathogens.[45,122,123] Conversely, after menopause, the cervical mucous is low in glycogen, resulting in a higher pH in the range of 7.0 to 7.4.[45] The intravaginal pH also can be influenced by the presence of cervical mucus and vaginal transudate.[123]

Enzymes in the vagina. The vaginal mucosa contains β-glucuronidase, acid phosphatase, α-napthylesterase, DPNH diaphorase, phosphoamidase, succinic dehydrogenase, and acid phosphatase, which could affect drug availability when drugs are delivered via this route.[123]

Barriers to vaginal drug delivery. As stated earlier, different hormones affect the makeup of the vaginal wall. The stage of the menstrual cycle thus is important because it affects the epithelial lining of the vagina and thus the absorption of drugs. In chronic drug delivery, the pH differences caused by alteration in vaginal flora as a result of the menstrual cycle could be of importance. The applicability of this delivery route to the female gender only is another shortcoming of this route of drug delivery.

2.4 Physiological and Biochemical Barriers to Controlled Release Drug Delivery

The design of controlled release drug delivery systems should be done with a thorough knowledge of the anatomy and physiology of the route of administration. The different barriers and drug delivery considerations for each route are summarized below.

When designing a dosage form for peroral delivery, the drug stability in a range of pH values covering the entire GI tract needs to be considered. A dosage form should be so designed that the exposure of a drug is minimal at the unstable pH to minimize degradation. A classic example of this would be an enteric-coated dosage form. The oral absorption of poorly absorbable drugs can be enhanced by using permeation enhancers. However, the irreversible loss of cell monolayer integrity and the systemic adverse effects associated with the existing permeation enhancers have limited their clinical use. Polymers such as chitosan and its derivatives are being explored as permeation enhancers chiefly owing to their lack of cytotoxicity and minimal absorption.[124] Variable gastric emptying might lead to nonuniform absorption profiles and incomplete drug release.[125] This is especially true for drugs absorbed within a narrow absorption window. As a means to prolong the gastric retention time and thereby the drug absorption, floating dosage forms and bioadhesive systems have been and are being developed. Oral drug delivery also can be optimized by using prodrugs. Ester prodrugs are being used widely to enhance lipophilicity and thus the oral bioavailability in comparison with the parent drug.[126] Intestinal transporters such as peptide transporters (PEPT1, PEPT2) also can be targeted for oral drug delivery by synthesizing prodrugs. Ganciclovir per se is not recognized as a substrate by the peptide transporters. However, the valyl ester of ganciclovir, valganciclovir, exhibits greater bioavailability in comparison with the parent drug owing to transport by PEPT1 and PEPT2.[127]

The colonic or rectal route can be preferred over the oral route for drugs that cause irritation or emesis when administered orally. The proximal colon can be accessed by the oral route, and the distal colon can be accessed by anal route. To deliver drugs via the proximal colon, drugs have to be protected from digestion in the acidic and basic environments of stomach and small intestine, respectively. Prolonged or sustained delivery of drugs can be achieved in the large intestine owing to the slow peristaltic movement of substances. The rectal route of drug delivery to the distal colon has the limitation of triggering the defecation reflex. Active transporters for drugs have not been reported, thereby limiting drug design for a potential carrier-mediated or targeted drug delivery. The presence of bacterial enzymes such as pectinase, amylase, xylanase, etc. in colon has led to the development of colon-specific drug delivery systems based on polysaccharides such as pectin, chitosan, and guar gum.[128,129] Since the rectum is normally void, it is possible for suppositories to be administered and be absorbed without having to permeate through feces. Second, suppositories can be administered without having the rectum reflexively expel the medication. Finally, the external anal sphincter would firmly retain the suppository within the rectum. Although rectal drug delivery offers a few advantages over the oral route, the combination of limited absorptive area and highly impermeable epithelial lining may necessitate the use of permeation enhancers to improve permeability.

The nonperoral mucosal delivery routes such as buccal, nasal, and vaginal sites offer barriers to drug molecules similar to that of the peroral route. Drugs delivered via these routes have to be small (<300 Da), lipophilic in nature, and with low dosage regimen requirements. The different approaches used to deliver drugs across these mucosae include the use of enzyme inhibitors, penetration enhancers, bioadhesive patches, prodrugs, liposomes, and solubility modifiers.[96,106,130]

The dead cells on the surface of skin offer a formidable barrier to drugs. The use of permeation enhancers in controlled delivery patches and techniques such as iontophoresis, electroporation, and ultrasound are applied to enhance the permeation of drugs across skin.[131]

The eye offers different kinds of barriers, such as rapid drainage of delivered drugs and a combination of lipophobic and hydrophobic barriers. Strategies of drug retention include the use of hydrogels, viscosity-imparting agents, and ointments. Other strategies, such as suspension formulations, inclusion of penetration enhancers, bioadhesive polymers, phase-transition agents, colloidal systems, liposomes, nanoparticles, and prodrugs, have been applied to deliver drugs through this organ. The different delivery devices include inserts, minidisks, and contact lenses.[63] Improving the tolerability of and compliance with ophthalmic drug delivery systems is also a major focus in formulation technology.

Drug delivery to the lung is a pharmaceutical technology issue. Novel devices capable of delivering substantial amounts of a formulation in the appropriate particle range may overcome this barrier.

The advent of biotechnology and combinatorial chemistry has led to the discovery of potent molecules that are either very large and susceptible proteins or lipophilic small drug moieties. These drugs pose great challenges to pharmaceutical scientists working on their successful delivery. Depending on the site of drug administration, the desired pharmacological effects, and pharmacokinetic profiles, different delivery strategies are being investigated to deliver these drugs. Understanding the different barriers posed by the body is critical to the successful design of controlled delivery systems.

Acknowledgments

We would like to thank Verne E. Cowles, Ph.D., DepoMed, Inc., Foster City, CA, for his help on GI contents.

References

1. Gyurosiova, L., Laitinen, L., Raiman, J., et al. Permeability profiles of M-alkoxy-substituted pyrrolidinoethylesters of phenylcarbamic acid across caco-2 monolayers and human skin. *Pharm. Res.* **19**(2):162–168, 2002.
2. Macheras, P., Reppas, C., and Dressman, J. B. *Biopharmaceutics of Orally Administered Drugs*. London: Ellis Horwood, 1995.
3. Fleisher, D., Li, C., Zhou, Y., et al. Drug, meal and formulation interactions influencing drug absorption after oral administration: Clinical implications. *Clin. Pharmacokinet.* **36**(3):233–254, 1999.
4. Guyton, A. C., and Hall, J. E. *Gastrointestinal Physiology*, Vol. 9. Philadelphia: Saunders, 1996.
5. Kararli, T. T. Comparison of the gastrointestinal anatomy, physiology, and biochemistry of humans and commonly used laboratory animals. *Biopharm Drug Dispos* **16**(5):351–380, 1995.
6. Ganong, W. F. *Review of Medical Physiology*, Vol. 20. New York: Mc-Graw Hill, 2001.
7. Smyth, D. H. *Biomembranes*. New York: Plenum Press, 1974.
8. Curatolo, W. The lipoidal permeability barriers of the skin and alimentary tract. *Pharm. Res.* **4**(4):271–277, 1987.
9. Curatolo, W., and Neuringer, L. J. The effects of cerebrosides on model membrane shape. *J. Biol. Chem.* **261**(36):17177–17182, 1986.

10. Rambourg, A., and Leblond, C. P. Electron microscope observations on the carbohydrate-rich cell coat present at the surface of cells in the rat. *J. Cell. Biol.* **32**(1):27–53, 1967.
11. Paulson, J. C., and Colley, K. J. Glycosyltransferases: Structure localization, and control of cell type-specific glycosylation. *J. Biol. Chem.* **264**(30):17615–17618, 1989.
12. Kurosawa, N., Hamamoto, T., Lee, Y. C., et al. Molecular cloning and expression of GalNAc alpha-2,6-sialyltransferase. *J. Biol. Chem.* **269**(2):1402–1409, 1994.
13. Washington, N., Washington, C., and Wilson, G. C. *Biological Pharmaceutics: Barriers to Drug Absorption.* New York: Taylor and Francis, 2001.
14. Carr, K. E., and Toner, P. G. *Morphology of the Intestinal Mucosa.* New York: Springer-Verlag, 1984.
15. Edwards, C. Physiology of the colorectal barrier. *Adv. Drug Deliv. Rev.* **28**(2):173–190, 1997.
16. Evaldson, G., Heimdahl, A., Kager, L., and Nord, C. E. The normal human anaerobic microflora. Scand. *J. Infect. Dis. Suppl.* **35**:9–15, 1982.
17. Huijsdens, X. W., Linskens, R. K., Mak, M., et al. Quantification of bacteria adherent to gastrointestinal mucosa by real-time PCR. *J. Clin. Microbiol.* **40**(12): 4423–4427, 2002.
18. Marieb, E. N. *Human Anatomy and Physiology*, 4th ed. Menlo Park, CA: Benjamin/Cummings Science, Publishing, 1998.
19. Lucas, M. Determination of acid surface pH in vivo in rat proximal jejunum. *Gut* **24**(8):734–739, 1983.
20. Shiau, Y. F., Fernandez, P., Jackson, M. J., and McMonagle, S. Mechanisms maintaining a low-pH microclimate in the intestine. *Am. J. Physiol.* **248**(6 pt 1):G608–617, 1985.
21. Mack, D. R., and Sherman, P. M. Hydrophobicity and the gastrointestinal tract: Methods of determination, its source and implications for bacterial adherence. *Colloids Surfaces B* **15**:355–363, 1999.
22. Lichtenberger, L. M. The hydrophobic barrier properties of gastrointestinal mucus. *Annu. Rev. Physiol.* **57**:565–583, 1995.
23. Anand, B. S., Romero, J. J., Sanduja, S. K., and Lichtenberger, L. M. Phospholipid association reduces the gastric mucosal toxicity of aspirin in human subjects. *Am. J. Gastroenterol.* **94**(7):1818–1822, 1999.
24. Lichtenberger, L. M., Romero, J. J., Kao, Y. C., and Dial, E. J. Gastric protective activity of mixtures of saturated polar and neutral lipids in rats. *Gastroenterology* **99**(2):311–326, 1990.
25. Schmitz, M. G., and Renooij, W. Phospholipids from rat, human, and canine gastric mucosa: Composition and metabolism of molecular classes of phosphatidylcholine. *Gastroenterology* **99**(5):1292–1296, 1990.
26. Bernhard, W., Postle, A. D., Rau, G. A., and Freihorst, J. Pulmonary and gastric surfactants: A comparison of the effect of surface requirements on function and phospholipid composition. *Comp. Biochem. Physiol.* [*A*] **129**(1):173–182, 2001.
27. Norris, D. A., Puri, N., and Sinko, P. J. The effect of physical barriers and properties on the oral absorption of particulates. *Adv. Drug Deliv. Rev.* **34**(2–3):135–154, 1998.
28. Paulsen, F. [Mucins in otorhinolaryngology.] *Hno* **50**(3):209–216, 2002.
29. Ugolev, A. M., Iezuitova, N. N., and Smirnova, L. F. *Role of Digestive Enzymes in the Permeability of the Enterocyte.* New York: Springer-Verlag, 1984.
30. Wacher, V. J., Salphati, L., and Benet, L. Z. Active secretion and enterocytic drug metabolism barriers to drug absorption. *Adv. Drug Deliv. Rev.* **46**(1–3):89–102, 2001.
31. Warner, P. E., Brouwer, K. L., Hussey, E. K., et al. Sumatriptan absorption from different regions of the human gastrointestinal tract. *Pharm. Res.* **12**(1):138–143, 1995.
32. Kanfer, I., Dowse, R., and Vuma, V. Pharmacokinetics of oral decongestants. *Pharmacotherapy* **13**(6 pt 2):116S–128S; discussion 143S–146S, 1993.
33. Ilett, K. F., Tee, L. B., Reeves, P. T., and Minchin, R. F. Metabolism of drugs and other xenobiotics in the gut lumen and wall. *Pharmacol. Ther.* **46**(1):67–93, 1990.
34. Gottesman, M. M., and Pastan, I. Biochemistry of multidrug resistance mediated by the multidrug transporter. *Annu. Rev. Biochem.* **62**:385–427, 1993.
35. Watkins, P. B. The barrier function of CYP3A4 and P-glycoprotein in the small bowel. *Adv. Drug Deliv. Rev.* **27**(2–3):161–170, 1997.

36. Seelig, A., and Landwojtowicz, E. Structure-activity relationship of P-glycoprotein substrates and modifiers. *Eur. J. Pharm. Sci.* **12**(1):31–40, 2000.
37. Fojo, A. T., Ueda, K., Slamon, D. J., et al. Expression of a multidrug-resistance gene in human tumors and tissues. *Proc. Natl. Acad. Sci. USA*, **84**(1):265–269, 1987.
38. Chan, L. M., Lowes, S., and Hirst, B. H. The ABCs of drug transport in intestine and liver: Efflux proteins limiting drug absorption and bioavailability. *Eur. J. Pharm. Sci.* **21**(1):25–51, 2004.
39. Kraehenbuhl, J. P., and Neutra, M. R. Epithelial M cells: Differentiation and function. *Annu. Rev. Cell. Dev. Biol.* **16**:301–332, 2000.
40. Zhang, Q. Y., Dunbar, D., Ostrowska, A., et al. Characterization of human small intestinal cytochromes P450. *Drug Metab. Dispos.* **27**(7): 804–809, 1999.
41. Anzenbacher, P., and Anzenbacherova, E. Cytochromes P450 and metabolism of xenobiotics. *Cell. Mol. Life Sci.* **58**:737–747, 2001.
42. Rylander, T., Neve, E. P., Ingelman-Sundberg, M., and Oscarson, M. Identification and tissue distribution of the novel human cytochrome P450 2S1 (CYP2S1). *Biochem. Biophys. Res. Commun.* **281**(2):529–535, 2001.
43. Lampen, A., Zhang, Y., Hackbarth, I., et al. Metabolism and transport of the macrolide immunosuppressant sirolimus in the small intestine. *J. Pharmacol. Exp. Ther.* **285**(3):1104–1112, 1998.
44. Fleisher, D., Lippert, C. L., Sheth, N., et al. Nutrient effects on intestinal drug absorption. *J. Controlled Release.* **11**(1–3):41–49, 1990.
45. Bannister, L. H., Berry, M. M., Collins, P., et al. *Gray's Anatomy: The Anatomical Basis of Medicine and Surgery*, 38th ed. London: Churchill-Livingstone, 1995.
46. Bouwstra, J. A., Honeywell-Nguyen, P. L., Gooris, G. S., and Ponec, M. Structure of the skin barrier and its modulation by vesicular formulations. *Prog. Lipid Res.* **42**(1):1–36, 2003.
47. Steinstrasser, I., and Merkle, H. P. Dermal metabolism of topically applied drugs: Pathways and models reconsidered. *Pharm. Acta Helv.* **70**(1):3–24, 1995.
48. Menon, G. K. New insights into skin structure: Scratching the surface. *Adv. Drug Deliv. Rev.* **54**(suppl. 1):S3–17, 2002.
49. Wertz, P. W., and van den Bergh, B. The physical, chemical and functional properties of lipids in the skin and other biological barriers. *Chem. Phys. Lipids.* **91**(2):85–96, 1998.
50. Menon, G. K., and Ghadially, R. Morphology of lipid alterations in the epidermis: A review. *Micros. Res. Technol.* **37**:180–192, 1997.
51. Bouwstra, J. A., and Honeywell-Nguyen, P. L. Skin structure and mode of action of vesicles. *Adv. Drug Deliv. Rev.* **54**(suppl. 1):S41–55, 2002.
52. Wertz, N. P., and Downing, L. D. *Epidermal Lipids.* New York: Oxford University Press, 1991.
53. Coderch, L., Lopez, O., de la Maza, A., and Parra, J. L. Ceramides and skin function. *Am. J. Clin. Dermatol.* **4**(2):107–129, 2003.
54. Basset, C., Holton, J., O'Mahony, R., and Roitt, I. Innate immunity and pathogen-host interaction. *Vaccine.* **21**(suppl. 2):S12–23, 2003.
55. He, S. H. Key role of mast cells and their major secretory products in inflammatory bowel disease. *World J. Gastroenterol.* **10**(3):309–318, 2004.
56. Coomes, M. W., Norling, A. H., Pohl, R. J., et al. Foreign compound metabolism by isolated skin cells from the hairless mouse. *J. Pharmacol. Exp. Ther.* **225**(3):770–777, 1983.
57. Merk, H. F., Mukhtar, H., Kaufmann, I., et al. Human hair follicle benzo[*a*]pyrene and benzo[*a*]pyrene 7,8-diol metabolism: Effect of exposure to a coal tar-containing shampoo. *J. Invest. Dermatol.* **88**(1):71–76, 1987.
58. Baron, J., Kawabata, T. T., Redick, J. A., et al. *Localization of Carcinogen-Metabolizing Enzymes in Human and Animal Tissues.* Amsterdam: Elsevier, 1983.
59. Baron, J. M., Holler, D., Schiffer, R., et al. Expression of multiple cytochrome P450 enzymes and multidrug resistance–associated transport proteins in human skin keratinocytes. *J. Invest. Dermatol.* **116**(4):541–548, 2001.
60. Stinchcomb, A. L. Xenobiotic bioconversion in human epidermis models. *Pharm. Res.* **20**(8):1113–1118, 2003.

61. Schiffer, R., Neis, M., Holler, D., et al. Active influx transport is mediated by members of the organic anion transporting polypeptide family in human epidermal keratinocytes. *J. Invest. Dermatol.* **120**(2):285–291, 2003.
62. Cevc, G. Lipid vesicles and other colloids as drug carriers on the skin. *Adv. Drug Deliv. Rev.* **56**(5):675–711, 2004.
63. Kaur, I. P., and Kanwar, M. Ocular preparations: the formulation approach. *Drug Dev. Ind. Pharm.* **28**(5):473–493, 2002.
64. Rittenhouse, K. D., and Pollack, G. M. Microdialysis and drug delivery to the eye. *Adv. Drug Deliv. Rev.* **45**(2–3):229–241, 2000.
65. Jarvinen, K., Jarvinen, T., and Urtti, A. Ocular absorption following topical delivery. *Adv. Drug Deliv. Rev.* **16**:3–19, 1995.
66. Lee, V. H., and Robinson, J. R. Mechanistic and quantitative evaluation of precorneal pilocarpine disposition in albino rabbits. *J. Pharm. Sci.* **68**(6):673–684, 1979.
67. Maurice, D. M., and Mishima, S. Ocular pharmacokinetics. In *Handbook of Experimental Pharmacology: Pharmacology of the Eye*, Vol. 69, Berlin: Springer-Verlag, 1984, pp. 19–116.
68. Gumbiner, B., Lowenkopf, T., and Apatira, D. Identification of a 160-kDa polypeptide that binds to the tight junction protein ZO-1. *Proc. Natl. Acad. Sci. USA.* **88**(8):3460–3464, 1991.
69. Bhat, M., Toledo-Velasquez, D., Wang, L., et al. Regulation of tight junction permeability by calcium mediators and cell cytoskeleton in rabbit tracheal epithelium. *Pharm. Res.* **10**(7):991–997, 1993.
70. Rojanasakul, Y., and Robinson, J. R. The cytoskeleton of the cornea and its role in tight junction permeability. *Int. J. Pharm.* **68**:135–149, 1991.
71. Martinez-Palomo, A., and Erlij, D. Structure of tight junctions in epithelia with different permeability. *Proc. Natl. Acad. Sci. USA.* **72**(11):4487–4491, 1975.
72. Grass, G. M., Wood, R. W., and Robinson, J. R. Effects of calcium chelating agents on corneal permeability. *Invest. Ophthalmol. Vis. Sci.* **26**(1):110–113, 1985.
73. Kaur, I. P., and Smitha, R. Penetration enhancers and ocular bioadhesives: Two new avenues for ophthalmic drug delivery. *Drug. Dev. Ind. Pharm.* **28**(4):353–369, 2002.
74. Schoenwald, R. D., and Ward, R. L. Relationship between steroid permeability across excised rabbit cornea and octanol-water partition coefficients. *J. Pharm. Sci.* **67**(6):786–788, 1978.
75. Schoenwald, R. D., and Huang, H. S. Corneal penetration behavior of beta-blocking agents. I. Physiochemical factors. *J. Pharm. Sci.* **72**(11):1266–1272, 1983.
76. Friedrich, S. W., Cheng, Y. L., and Saville, B. A. Theoretical corneal permeation model for ionizable drugs. *J. Ocul. Pharmacol.* **9**(3):229–249, 1993.
77. Rojanasakul, Y., and Robinson, J. R. Transport mechanisms of the cornea: Characterization of barrier permselectivity. *Int. J. Pharm.* **55**:237–246, 1989.
78. Rojanasakul, Y., Wang, L. Y., Bhat, M., et al. The transport barrier of epithelia: A comparative study on membrane permeability and charge selectivity in the rabbit. *Pharm. Res.* **9**(8):1029–1034, 1992.
79. Liaw, J., Rojanasakul, Y., and Robinson, J. R. The effect of drug charge type and charge density on corneal transport. *Int. J. Pharm.* **88**:111–124, 1992.
80. Sasaki, H., Yamamura, K., Mukai, T., et al. Enhancement of ocular drug penetration. *Crit. Rev. Ther. Drug Carrier Syst.* **16**(1):85–146, 1999.
81. Lee, V. H., Carson, L. W., Kashi, S. D., and Stratford, R. E., Jr. Metabolic and permeation barriers to the ocular absorption of topically applied enkephalins in albino rabbits. *J. Ocul. Pharmacol.* **2**(4):345–352, 1986.
82. Harris, D., Liaw, J., and Robinson, J. R. Routes of delivery: Case studies (7): Ocular delivery of peptide and protein drugs. *Adv. Drug Deliv. Rev.* **8**:331–339, 1992.
83. Stratford, R. E., Jr., and Lee, V. H. L. Aminopeptidase activity in homogenates of various absorptive mucosae in the albino rabbit: Implications in peptide delivery. *Int. J. Pharm.* **30**:73–82, 1986.
84. Kamai, Y., and Ushiki, T. The three-dimensional organization of collagen fibrils in the human cornea and sclera. *Invest. Ophthalmol. Vis. Sci.* **32**:2244–2258, 1991.
85. Salminen, L., and Urtti, A. Disposition of ophthalmic timolol in treated and untreated rabbit eyes. A multiple and single dose study. *Exp. Eye Res.* **38**(2):203–206, 1984.

86. Huang, A. J., Tseng, S. C., and Kenyon, K. R. Paracellular permeability of corneal and conjunctival epithelia. *Invest. Ophthalmol. Vis. Sci.* **30**(4):684–689, 1989.
87. Hayakawa, E., Chien, D. S., Inagaki, K., et al. Conjunctival penetration of insulin and peptide drugs in the albino rabbit. *Pharm. Res.* **9**(6):769–775, 1992.
88. Geroski, D. H., and Edelhauser, H. F. Transscleral drug delivery for posterior segment disease. *Adv. Drug Deliv. Rev.* **52**(1):37–48, 2001.
89. Shojaei, A. Buccal mucosa as a route for systemic drug delivery: A review. *J. Pharm. Pharmaceut. Sci.* **1**:15–30, 1998.
90. de Vries, M. E., Bodde, H. E., Verhoef, J. C., and Junginger, H. E. Developments in buccal drug delivery. *Crit. Rev. Ther. Drug Carrier Syst.* **8**(3):271–303, 1991.
91. Squier, C. A. The permeability of oral mucosa. *Crit. Rev. Oral Biol. Med.* **2**(1):13–32, 1991.
92. Merkle, H. P., and Wolany, G. Buccal delivery for peptide drugs. *J. Control. Release,* **21**:155–164, 1992.
93. Chen, L. L., Chetty, D. J., and Chien, Y. W. A mechanistic analysis to characterize oramucosal permeation properties. *Int. J. Pharm.* **184**(1):63–72, 1999.
94. Shojaei, A. H. Buccal mucosa as a route for systemic drug delivery: A review. *J. Pharm. Pharmaceut. Sci.* **1**(1):15–30, 1998.
95. Bodde, E. H., de Vries, M. E., and Junginger, E. H. Mucoadhesive polymers for the buccal delivery of peptides, structure-adhesiveness relationships. *J. Control. Release.* **13**:225–231, 1990.
96. Veuillez, F., Kalia, Y. N., Jacques, et al. Factors and strategies for improving buccal absorption of peptides. *Eur. J. Pharm. Biopharm.* **51**(2):93–109, 2001.
97. Squier, C. A., and Hill, M. W. *Oral Mucosa.* St. Louis: Mosby, 1989.
98. Rathbone, M. J. *Oral Mucosal Drug Delivery.* New York: Marcel Dekker, 1996.
99. Audus, K. L., and Raub, T. J. *Biological Barriers to Protein Delivery.* New York: Plenum Press, 1993.
100. Gandhi, B. R., and Robinson, R. J. Oral cavity as a site for bioadhesive drug delivery. *Adv. Drug Deliv. Rev.* **13**:43–74, 1994.
101. Harris, D., and Robinson, J. R. Drug delivery via the mucous membranes of the oral cavity. *J. Pharm. Sci.* **81**:1–10, 1992.
102. Burkitt, G. H., Young, B., and Heath, J. W. *Wheater's Functional Histology,* Vol. 3. Edinburgh: Churchill-Livingstone, 1993.
103. Washington, N., Washington, C., and Wilson, G. C. *Drug Delivery to the Oral Cavity or Mouth.* New York: Taylor and Francis, 2001.
104. Chen, H. L., Chetty, J. D., and Chien, W. Y. A mechanistic analysis to characterize oral mucosal permeation properties. *Int. J. Pharm.* **184**:63–72, 1999.
105. Rathbone, J. M., and Hadgraft, J. Absorption of drugs from the human oral cavity. *Int. J. Pharmacol.* **74**:9–24, 1991.
106. Hussain, A. A. Intranasal drug delivery. *Adv. Drug Deliv. Rev.* **29**(1–2):39–49, 1998.
107. Illum, L. Nasal drug delivery: Possibilities, problems and solutions. *J. Control. Release.* **87**(1–3):187–198, 2003.
108. Arora, P., Sharma, S., and Garg, S. Permeability issues in nasal drug delivery. *Drug Discov. Today.* **7**(18):967–975, 2002.
109. Dahlin, M., and Bjork, E. Nasal administration of a physostigmine analogue (NXX-066) for Alzheimer's disease to rats. *Int. J. Pharm.* **212**(2):267–274, 2001.
110. Sarkar, A. M. Drug metabolism in the nasal mucosa. *Pharm. Res.* **9**(1):1–9, 1992.
111. Dahl, R., and Mygind, N. Anatomy, physiology and function of the nasal cavities in health and disease. *Adv. Drug Deliv. Rev.* **29**(1–2):3–12, 1998.
112. Khanvilkar, K., Donovan, M. D., and Flanagan, D. R. Drug transfer through mucus. *Adv. Drug Deliv. Rev.* **48**(2–3):173–193, 2001.
113. Pujara, C. P., Shao, Z., Duncan, M. R., and Mitra, A. K. Effects of formulation variables on nasal epithelial cell integrity: Biochemical evaluations. *Int. J. Pharm.* **114**(2):197–203, 1995.
114. Hussain, A., Hamadi, S., Kagashima, M., et al. Does increasing the lipophilicity of peptides enhance their nasal absorption? *J. Pharm. Sci.* **80**(12):1180–1181, 1991.
115. Harris, A. S., Nilsson, I. M., Wagner, Z. G., and Alkner, U. Intranasal administration of peptides: Nasal deposition, biological response, and absorption of desmopressin. *J. Pharm. Sci.* **75**(11):1085–1088, 1986

116. Mitra, A. K., and Krishnamoorthy, R. Prodrugs for nasal drug delivery. *Adv. Drug Deliv. Rev.* **29**(1–2):135–146, 1998.
117. Forbes, B., Wilson, C. G., and Gumbleton, M. Temporal dependence of ectopeptidase expression in alveolar epithelial cell culture: Implications for study of peptide absorption. *Int. J. Pharm.* **180**(2):225–234, 1999.
118. Guyton, A. C., and Hall, J. E. *Textbook of Medical Physiology*, 9th ed. Philadelphia: Saunders, 1996.
119. Labiris, N. R., and Dolovich, M. B. Pulmonary drug delivery: I. Physiological factors affecting therapeutic effectiveness of aerosolized medications. *Br. J. Clin. Pharmacol.* **56**(6):588–599, 2003.
120. Agu, R. U., Ugwoke, M. I., Armand, M., et al. The lung as a route for systemic delivery of therapeutic proteins and peptides. *Respir. Res.* **2**(4):198–209, 2001.
121. Groneberg, D. A., Witt, C., Wagner, U., et al. Fundamentals of pulmonary drug delivery. *Respir. Med.* **97**(4):382–387, 2003.
122. Alexander, N. J., Baker, E., Kaptein, M., et al. Why consider vaginal drug administration? *Fertil. Steril.* **82**(1):1–12, 2004.
123. Woolfson, A. D., Malcolm, R. K., and Gallagher, R. Drug delivery by the intravaginal route. *Crit. Rev. Ther. Drug Carrier Syst.* **17**(5):509–555, 2000.
124. Thanou, M., Verhoef, J. C., and Junginger, H. E. Oral drug absorption enhancement by chitosan and its derivatives. *Adv. Drug Deliv. Rev.* **52**(2):117–126, 2001.
125. Reddy, L. H., and Murthy, R. S. Floating dosage systems in drug delivery. *Crit. Rev. Ther. Drug Carrier Syst.* **19**(6):553–585, 2002.
126. Taylor, M. Improved passive oral drug delivery via prodrugs. *Adv. Drug Deliv. Rev.* **19**:131–148, 1996.
127. Sugawara, M., Huang, W., Fei, Y. J., et al. Transport of valganciclovir, a ganciclovir prodrug, via peptide transporters PEPT1 and PEPT2. *J. Pharm. Sci.* **89**(6):781–789, 2000.
128. Sinha, V. R., and Kumria, R. Polysaccharides in colon-specific drug delivery. *Int. J. Pharm.* **224**(1–2):19–38, 2001.
129. Macleod, G. S., Collett, J. H., and Fell, J. T. The potential use of mixed films of pectin, chitosan and HPMC for bimodal drug release. *J. Control. Release.* **58**(3):303–310, 1999.
130. Senel, S., and Hincal, A. A. Drug permeation enhancement via buccal route: Possibilities and limitations. *J. Control. Release.* **72**(1–3):133–144, 2001.
131. Hadgraft, J. Skin deep. *Eur. J. Pharm. Biopharm.* **58**:291–299, 2004.

Chapter

3

Prodrugs as Drug Delivery Systems

Anant Shanbhag,* Noymi Yam,* and Bhaskara Jasti

Thomas J. Long School of Pharmacy and Health Sciences
University of the Pacific
Stockton, California

Present affiliation: ALZA Corporation, Mountain View, California.

3.1 Introduction

Innovations in drug discovery and development, fueled by rapid advances in technology, have led to novel therapeutics for the prevention and treatment of diseases, greatly improving the quality of patients' lives. These innovations have been driven by increasing investments in research and development by pharmaceutical companies, which to some extent have contributed to the upward-spiraling costs of health care, especially prescription medications. The number of new molecular entities (NMEs) receiving Food and Drug Administration (FDA) approval has declined steadily over the years amid concerns over safety and efficacy, with a worrisome nadir of 24 new approvals in 2001. The application of high-throughput screening (HTS) has resulted in bulging drug discovery pipelines full of novel therapeutics with improved receptor binding and efficacy but often without adequate physicochemical or pharmacokinetic properties, resulting in costly failures. Despite these developments, new drug applications and approvals did not increase in the last decade. Consequently, development of line extensions seems to be a logical course of action for pharmaceutical companies in order to protect their revenue pool. Such line extensions can be achieved by designing novel drug delivery systems to deliver the existing drugs on the market. An attractive alternative is a chemical delivery system such as a prodrug or soft drug that changes the drug molecule itself to improve the drug's physicochemical properties and safety/tolerability profile.[1]

Historically, the term *prodrug* or *proagent* was coined by Albert[2] in the late 1950s to denote chemical derivatives that could temporarily alter the physicochemical properties of drugs in order to increase their therapeutic utility and reduce associated toxicity. Prodrugs also have been synonymously referred to as *latentiated drugs*, *bioreversible derivatives*, and *congeners*.[3–6] However, the term *prodrug* gained wider acceptance and usually describes compounds that undergo chemical transformation within the body prior to exhibiting pharmacologic activity. Some of the earliest examples of prodrugs are methenamine and aspirin. In the early stages, prodrugs were obtained fortuitously rather than intentionally; an example is prontosil, which was discovered in the 1930s and later identified as a prodrug of the antibiotic sulfanilamide.[7]

A prodrug strategy can be implemented for existing marketed chemical entities (post hoc design). The prodrug strategy also can be implemented in early discovery (ad hoc design) during lead optimization to address the physicochemical aspects of the NMEs and to improve the chances of success.[8]

Prodrugs are pharmacologically inactive compounds that result from transient chemical modifications of a biologically active species and are designed to convert to biologically active species in vivo by a predictable mechanism.[9] Soft drugs are pharmaceutical agents that are converted to active species in the biological system. However, soft drugs are active isosteric or isoelectric analogues of a lead compound that are metabolized or deactivated in a predictable and controllable fashion after achieving their therapeutic role.[10,11] They are usually desired for local activity and administered at or near the site of action. Hence they exhibit pharmacological effect locally and distribute away from the intended site as inactivated metabolites, thus avoiding undesired side effects or toxicities. Therefore, they can be designed to improve the therapeutic index by simplifying the activity/distribution profile, reducing systemic side effects, eliminating drug interactions by avoiding metabolic routes involving saturable enzyme systems, and preventing long-term toxicity owing to accumulation.

Prodrugs and soft drugs can be used strategically to address different problems. Prodrugs and soft drugs are treated by the FDA as new chemical entities, and in most cases they require complete toxicological evaluation prior to submission. The soft-drug approach is gaining acceptance as a way to build a metabolic pathway to a drug in order to achieve predictable metabolism and address the safety and toxicity issues.

This chapter discusses the rationale, design concepts, and application of prodrugs and their current status in drug development and delivery.

3.2 Rationale for Prodrug Design

A large number of the new molecular entities with promising therapeutic profiles are dropped from the screening stage because of their inferior physicochemical and biopharmaceutical properties. These undesired properties result in poor absorption, extensive metabolism, and low bioavailability because of physical, biological, or metabolic barriers. If the chemical structure of the drug or lead compound can be modified to overcome these barriers and then revert to the pharmacologically active form, the drug can be delivered efficiently. The rationale for the design of prodrugs is to achieve favorable physicochemical characteristics (e.g., chemical stability, solubility, taste, or odor), biopharmaceutical properties (e.g., oral absorption, first-pass metabolism, permeability across biological membranes such as the blood-brain barrier, or reduced toxicity), or pharmacodynamic properties (e.g., reduced pain or irritation). The objectives in designing a prodrug are described in Table 3.1.

Meeting a single objective often addresses others; for example, improved solubility may simultaneously address low absorption and bioavailability and provide an improved plasma concentration-time

TABLE 3.1 Objectives in Prodrug Design

Pharmacodynamic objectives	Pharmaceutical objectives	Pharmacokinetic objectives
Mask reactive species to improve its therapeutic index	Improve solubility	Improve oral absorption
Activate cytotoxic agent in situ	Improve chemical stability	Decrease presystemic metabolism
	Improve taste, odor	Improve absorption by nonoral routes
	Decrease irritation and pain	Improve plasma concentration-time profile
		Provide organ/tissue-selective delivery of active agent

SOURCE: Ref. 12.

profile. Thus prodrugs can be designed to achieve a myriad of objectives with significant overlaps.

Multiple benefits associated with prodrug design include increased bioavailability with ester prodrugs, increased permeability with hydroxyl amine prodrugs, enhanced solubility with prodrug salts, enhanced stability with PEGylated prodrugs, enhanced absorption with prodrugs targeted at intestinal transporters, and improved cancer therapy with gene- and receptor-targeted prodrugs.

3.3. Principles of Prodrug Design

Successful design of prodrug-based controlled delivery systems is generally based on efficient transformation of the promoiety into the active drug in vivo at the desired organ or tissue. Despite the diversity in chemical structure, most prodrugs can be classified on the basis of a common chemical linkage (e.g., esters, amides, and salts). In addition to these common types of prodrugs, recent prodrug approaches include conversion of promoieties into primary and secondary amines, imides, hydroxyls, thiols, carbonyls, and carboxyls. Some unconventional prodrug approaches include conjugates of drug with cyclodextrins,[12] microsphere encapsulated prodrugs using biodegradable polyester microspheres,[13] and lysosome-associated hydrophobic prodrugs.[14] Prodrugs can address myriad inherent limitations of a drug: poor stability, negative aesthetic features (taste, odor, or pain on injection), incomplete absorption, excessively rapid absorption, extensive drug metabolism prior to reaching the systemic circulation, high toxicity, and poor specificity.[15–17] Specific applications of prodrugs in controlled delivery most commonly aim to prolong therapeutic action, enhance absorption, or alter tissue distribution.

3.3.1 Ester-based prodrugs

Owing to the properties of carbonyl group, esters generally are more hydrophobic (and consequently more lipophilic) than their parent compounds. Using specifics of their chemical structure, properties of ester prodrugs can be broadly modulated to achieve particular stability and solubility profiles, provide good transcellular absorption, resist hydrolysis during the initial phase of absorption, and transform rapidly and efficiently at the site of action.[18–21] Biotransformation of an ester prodrug to its active form usually involves enzymatic or nonenzymatic hydrolysis; in many cases the initial enzymatic cleavage is followed by nonenzymatic rearrangement. Ester prodrugs can be designed with single or multiple functional groups. Some examples are shown in Table 3.2.

Ester prodrugs often are designed for absorption enhancement by introducing lipophilicity and masking ionization groups of an active compound. For example, valacyclovir, the L-valyl ester prodrug of acyclovir used for treatment of herpes, demonstrates an oral bioavailability that is three to five times greater than its parent compound.[24,25] The prodrug structure, shown in Fig. 3.1, is responsible for the enhanced carrier-mediated intestinal absorption via the hPEPT1 peptide transporter.[26] Rapid and complete conversion of valacyclovir to acyclovir results in higher plasma concentrations, allowing for reduced dosing frequency.

Similarly, oral absorption, as well as transdermal penetration, of the long-acting angiotensin-converting enzyme (ACE) inhibitor enalaprilat is improved considerably by esterification of one of its carboxyl groups. The improved pharmacokinetic properties are attributed to the significantly higher lipophilicity of the ethyl ester prodrug enalapril (Fig. 3.2)[27]

Ester prodrugs are also designed to *reduce side effects*[28] by changing the physicochemical properties of active compounds that cause tissue irritation. For example, piroxicam, a nonsteroidal anti-inflammatory drugs (NSAID), is well absorbed after oral administration but causes gastrointestinal (GI) bleeding, perforation, and ulceration. Ampiroxicam (Fig. 3.3), a nonacidic prodrug, is an ester carbonate prodrug of piroxicam with comparable therapeutic efficacy to piroxicam and reduced ulcerogenic and GI side effects.

Another application of ester prodrugs is related to *stability improvement* of parent compounds by modifying particularly unstable functional groups present in active agents.[29] For example, potassium tricyclo[5.2.1.0(2,6)]-decan-8-yl dithiocarbonate (D609) is a selective antitumor agent, potent antioxidant, and cytoprotectant. D609 has a strong potential to be developed as a unique chemotherapeutic agent

TABLE 3.2 Examples of Ester Prodrugs

General design objective	Goal	Drug	Prodrug
Improvement of physicochemical properties	Improve taste	Chloramphenicol	Chloramphenicol palmitate[22]
	Improve taste	Clindamycin	Clindamycin palmitate[30]
	Decrease pain on injection	Clindamycin	Clindamycin phosphate[30]
	Increase solubility	Palcitaxel	PEG-Palcitaxel[23]
	Increase solubility	Prednisolone	Prednisolone sodium succinate[30]
	Decrease solubility	Erythromycin	Erythromycin[30] Ethyl succinate
Improvement of pharmacokinetic properties	Improve absorption	Adefovir	Adefovir[31] Dipivoxil
	Improve absorption	Ro-64-0802	Oseltamivir[31]
	Target to specific transporters	Dopamine	Levodopa[31]
	Increase duration of action	Doxorubicin	PEG-Doxorubicin[31]
	Increase oral absorption	Ampicillin	Pivampicillin[30]
	Extend duration	Fluphenazine	Fluphenazine decanoate[30]
	Increase site specificity	Dromostanolone	Dromostanolone propionate[30]
	Decrease side effects	Aspirin and *N*-acetyl-*p*-aminophenol	Benorylate[30]

that may provide dual therapeutic benefits against cancer, e.g., accelerating tumor cell death while protecting normal tissues from damage. However, D609 contains a dithiocarbonate (xanthate) group [O—C(=S)S(–)/O—C(=S)SH] that is chemically unstable, being readily

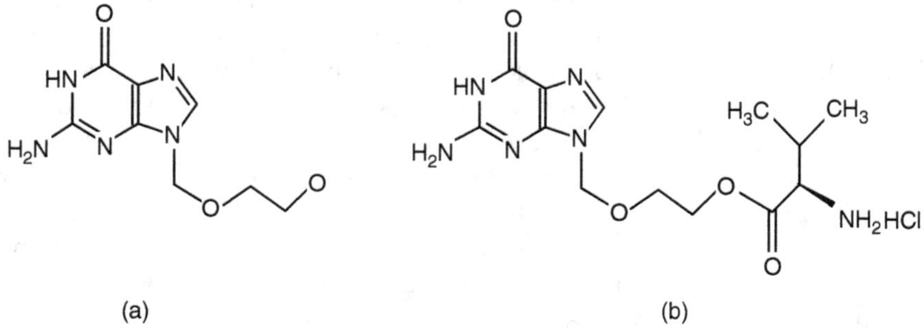

Figure 3.1 Acyclovir (*a*) and valacyclovir (*b*).

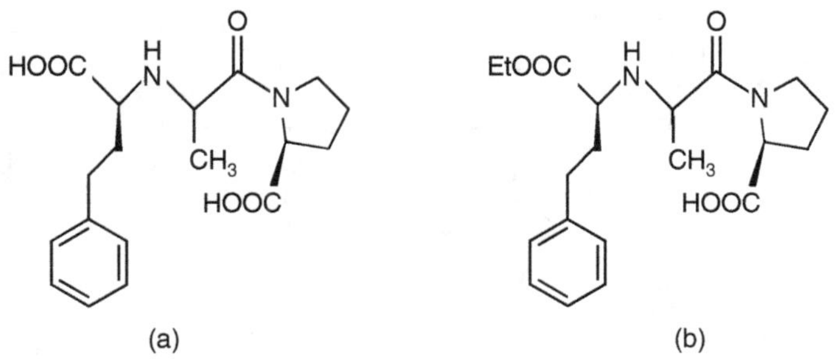

Figure 3.2 Enalaprilat (*a*) and enalapril (*b*).

oxidized to form a disulfide bond with subsequent loss of all biological activities. A series of *S*-(alkoxyacyl)-D609 prodrugs that connect the xanthate group of D609 to an ester via a self-immolative methyleneoxyl group was designed recently. These *S*-(alkoxylacyl)-D609 prodrugs release D609 in two steps: esterase-catalyzed hydrolysis of the acyl ester bond followed by conversion of the resulting hydroxymethyl D609 to formaldehyde and D609. The prodrugs are stable at ambient conditions but are readily hydrolyzed by esterases to liberate D609 in a controlled manner.

An interesting variation of the ester-prodrug approach is creation of a *double prodrug*, where two functional groups are modified simultaneously to achieve the combined physicochemical properties that would maximize permeability enhancement. A double prodrug of the direct platelet and thrombin aggregation inhibitor Melagatran was developed by converting the carboxylic acid to an ester and hydroxylating the imidine moiety to reduce its basicity.[30] Melagatran (Fig. 3.4) originally exhibited a low oral bioavailability of 5 percent, which was attributed

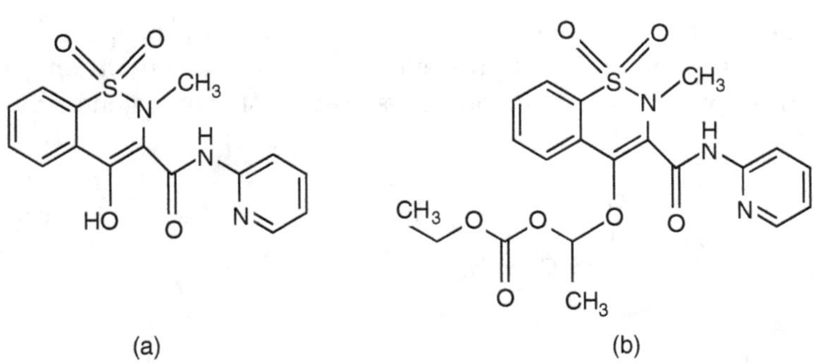

Figure 3.3 Piroxicam (*a*) and ampiroxicam (*b*).

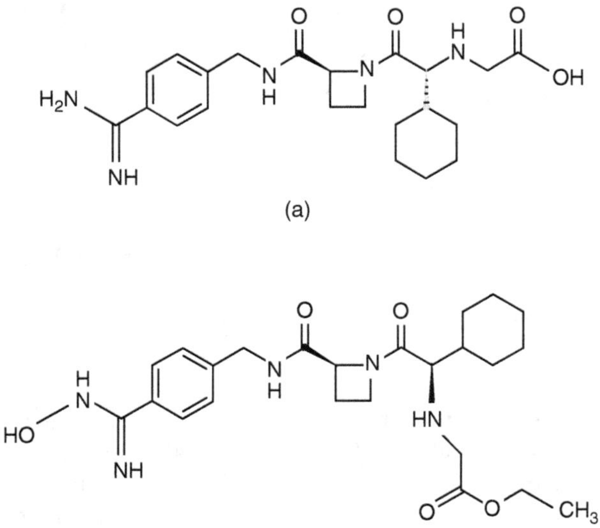

Figure 3.4 Melagatran (*a*) and ximelagatran (*b*).

to the presence of two strongly basic groups and a carboxylic acid group causing the compound to exist as a zwitterion at intestinal pH. The resulting prodrug, ximelagatran, is uncharged at intestinal pH and has 80-fold improved permeability and an oral bioavailability of 20 percent.

3.3.2 Amide-based prodrugs

Amide prodrugs are relatively similar to ester prodrugs in terms of the chemical nature of the intermolecular linkage (Fig. 3.5). In vivo activation of the amide prodrug generally involves enzymatic cleavage, shown schematically in Fig. 3.6. Because of the need to protonate an amide leaving group, amide hydrolysis is much more pH-sensitive than ester hydrolysis.[31,32]

Owing to their particular chemical structure, amide prodrugs can be designed for targeting peptide and nutrient transporters to *enhance permeability*. In this application, amide prodrugs are also shown to generally

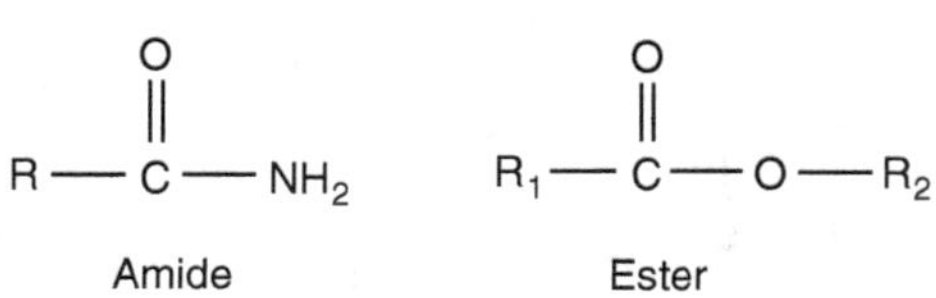

Figure 3.5 Chemical structures of amides and esters.

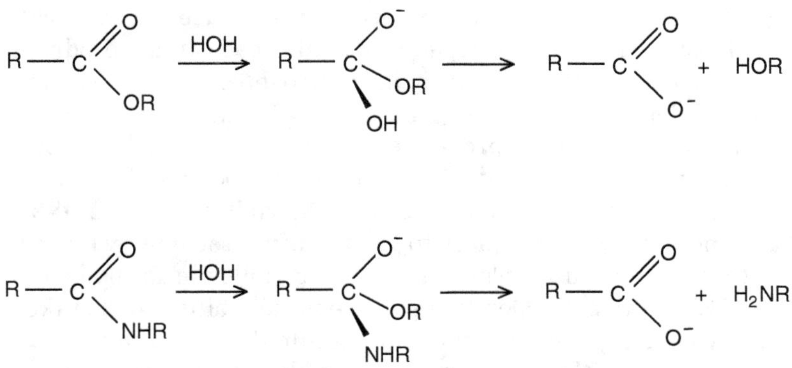

Figure 3.6 Typical path of enzymatic hydrolysis of esters and amides.

provide superior *physicochemical stability* compared with more conventional ester derivatives.

An interesting example of this application is a promising amide prodrug, LY-544344, developed for the metabotropic glutamate receptor 2 agonist (mGluR2) LY-35470.[20] As shown in Fig. 3.7, LY-35470 contains two carboxylic acid groups and a basic amino group. It exists as zwitterions at physiologic pH, resulting in low oral bioavailability (approximately 6 percent) in humans. Simple esterification does not provide a compound with adequate stability; hence the primary amine was derivatized with L-alanine to yield LY-544344, which acted as an excellent substrate for intestinal transporter hPepT1. LY-544344 demonstrated excellent solid and solution stability. In vivo tests in dogs showed rapid absorption and a 17-fold higher exposure to the parent compound. In clinical studies, LY-544344 exhibited an approximately 13-fold increase in systemic exposure of LY-35470 in comparison with LY-35470 administration.

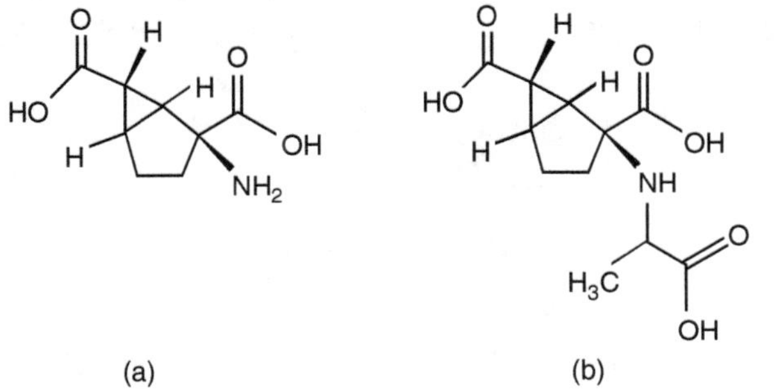

Figure 3.7 LY-354740 (*a*) and LY-544344 (*b*).

The modification of gabapentin, an anticonvulsant used in the treatment of epilepsy and postherpetic neuralgia, also uses amide prodrugs to target active transport pathways in order to overcome enhancement of central nervous system (CNS) penetration. Gabapentin[20] exhibits suboptimal pharmacokinetic properties, including saturable absorption, high interpatient variability, lack of dose proportionality, and a short half-life. Its amide prodrug, XP-13512, shown in Fig. 3.8, is stable chemically and is converted rapidly to gabapentin, presumably by nonspecific esterases present in tissues encountered following oral absorption. XP-13512 shows pH-dependent passive permeability and is taken up in vitro by cells expressing either the sodium-dependent multivitamin transporter (SMVT) or the monocarboxylate transporter type 1 (MCT-1), which is highly expressed throughout the GI tract. Oral bioavailability in monkeys was improved from 25 percent for the parent gabapentin to 85 percent for prodrug XP-13512.

Designing amide prodrugs for *stability enhancement* is based on the greater physicochemical stability of the amide bond in comparison with ester linkages.[33] Recently, amide prodrugs of NSAIDs such as aspirin (Fig. 3.9), ibuprofen, and naproxen were found to be more stable than ester prodrugs, which also improved their absorption and reduced GI irritation.[34–36] In another example, the amide derivatives of cytosine arabinose, a short-half-life pyrimidine nucleoside analogue employed for the treatment of acute and chronic human leukemia, have shown considerably higher stability in plasma than ester prodrugs and consequently showed significant efficacy improvement.[37–40] Various applications of amide prodrugs are summarized in Table 3.3.

3.3.3 Salt-based prodrugs

The design of *salt forms* of prodrugs is associated most commonly with *solubility and stability enhancement* in various dosage forms. Ester prodrugs often are converted to salts to provide the desired balance of hydrophilic/lipophilic properties.[45]

For example, fosphenytoin is designed as a disodium salt ester prodrug of phenytoin (Fig. 3.9) to overcome parenteral delivery problems

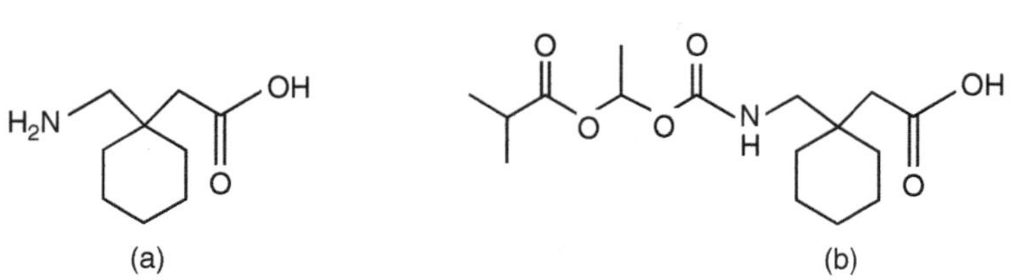

Figure 3.8 Gabapentin (*a*) and XP-3512 (*b*).

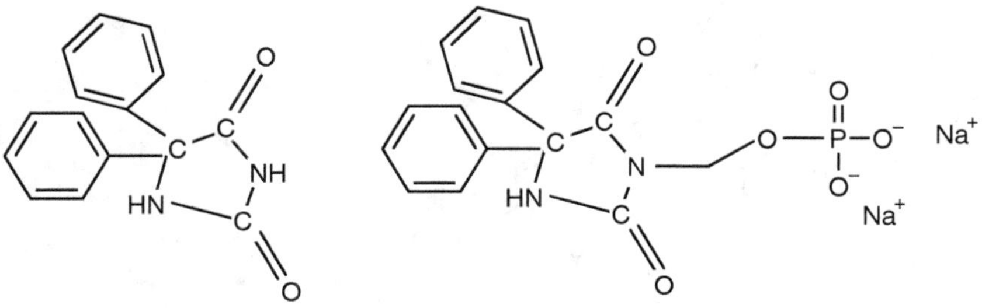

(a) (b)

Figure 3.9 Phenytoin (*a*) and fosphenytoin (*b*).

related to the low aqueous solubility (20 to 25 μg/mL) of the parent compound.[46] The sodium salt of phenytoin provides good solubility enhancement (50 mg/mL) but lacks stability at pH below 12, resulting in rapid precipitation of phenytoin acid from sodium phenytoin solutions. Fosphenytoin provides further improvement of solubility to the level of 142 mg/mL and is stable at pH 7.5 to 8, which essentially results in greater safety, lower irritation, and ease of administration.

Another example of a salt prodrug designed for solubility improvement is fosamprenavir. Its parent compound, amprenavir, is a potent HIV protease inhibitor with limited aqueous solubility, which requires administration of multiple softgel pills to meet the dosing requirements.[20,23,47] Fosamprenavir, the phosphate prodrug of amprenavir (Fig. 3.10) exhibits high water solubility, solution stability, and rapid conversion to the parent amprenavir on the apical side of the intestinal epithelium prior to absorption. Currently, fosamprenavir is administered as a solid dosage form and has lower pill burden (two tablets versus eight capsules of amprenavir).

Yet another successful example of a salt prodrug that improves the solubility and stability of the parent compound is the modification of

TABLE 3.3 Examples of Amide Prodrugs

General design objective	Goal	Drug	Prodrug
Improvement of physicochemical properties	Decrease GI irritation	Nicotonic acid	Nicotinamide[30]
	Decrease GI irritation	Valdecoxib	Parecoxib[41]
Improvement of pharmacokinetic properties	Prolong action	Tolmetin sodium	Tolmetin glycine amide[42]
	Prolong action	Doxorubicin	Doxsaliform[43]
	Increase ocular permeability	Amfenac	Nepafenac[44]

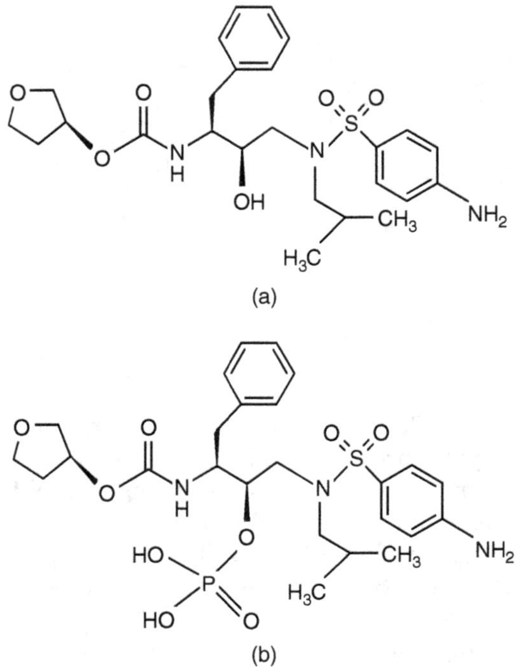

Figure 3.10 Amprenavir (*a*) and fosamprenavir (*b*).

tenofovir, a nucleotide reverse-transcriptase inhibitor (NRTI) particularly useful for treating HIV patients with early therapeutic failure.[48,49] Tenofovir undergoes phosphorylation and forms tenofovir diphosphate, which is a potent and selective inhibitor of viral reverse-transcriptor enzyme. Tenofovir is a dianion at physiological pH with a low partition coefficient, which results in low and erratic oral bioavailability. However, tenofovir disoproxil, a bis-isopropoxyl carbonate derivative, exhibits excellent chemical stability and a bioavailability of approximately 30 percent in dogs. A fumarate salt of tenofovir disoproxil is used to minimize dimerization of the tenofovir disoproxil with degradation product formaldehyde. Tenofovir disoproxil fumarate (Fig. 3.11) is nonhygroscopic and stable and exhibits rapid dissolution and an oral bioavailability of 43 percent in humans. Typical salt prodrugs and their applications are summarized in Table 3.4.

3.3.4 Additional prodrug types

Other prodrugs that are designed to achieve different pharmaceutical goals are summarized in Table 3.5.

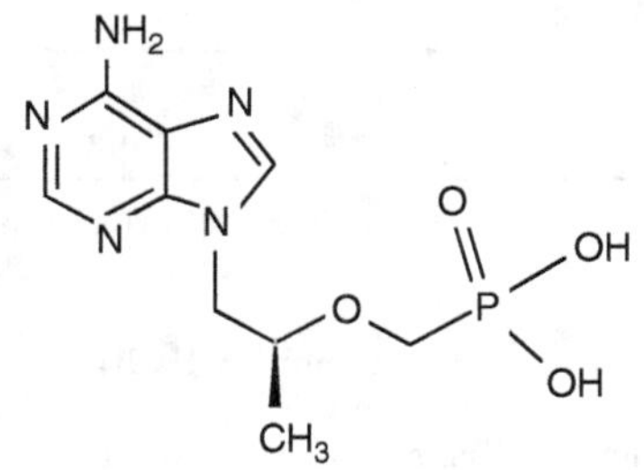

(a)

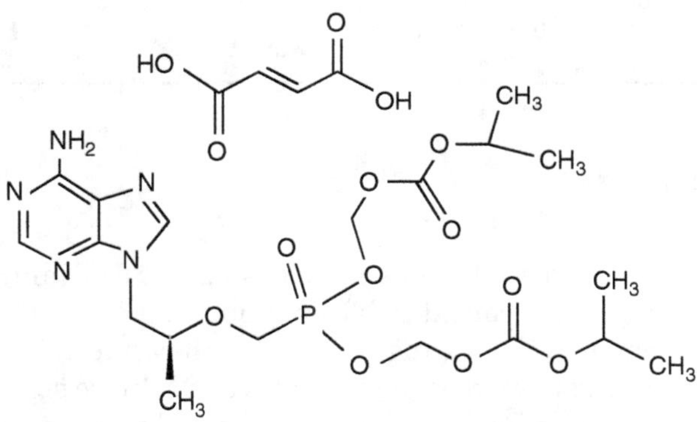

(b)

Figure 3.11 Tenofovir (*a*) and tenofovir disoproxil fumarate (*b*).

TABLE 3.4 Examples of Salt Prodrugs

General design objective	Goal	Drug	Prodrug
Improvement of physicochemical properties	Increase solubility	Phenytoin	Fosphenytoin[47]
	Increase solubility	Amprenavir	Fosamprenavir calcium[50]
	Improve stability	Ganciclovir	Valganciclovir hydrochloride[51]
	Increase solubility	Mesalamine	Balsalazide disodium[52]
	Increase solubility	Trovafloxacin	Alatrofloxacin[53]

TABLE 3.5 Additional Prodrug Types and Applications

General design objective	Goal	Drug	Prodrug	Linkage
Improvement of physicochemical properties	Increase duration of action	Terbutaline	Bambuterol	Carbamate[54]
Improvement of pharmacokinetic properties	Improve targeting	Gemtuzumab	Gemtuzumab ozogamycin	Hydrazone[31]
	Increase topical penetration	Triamicinolone	Triamicinolone acetonide	Ketal[30]
	Decrease side effects	Chloral hydrate	Dichloral-phenazone	Complex[30]
	Decrease side effects	Lovastatin (mevinolinic) acid	Lovastatin	Lactone[55]

3.4 Prodrugs for Prolonged Therapeutic Action

For some therapeutic agents, decreased frequency of dosing and constant plasma concentrations can result in considerable enhancement of safety and efficacy of dosage forms by eliminating the peak-valley effect, as described in Chap. 1. In prodrugs, two design principles can achieve sustained release: (1) The prodrug is incorporated in a controlled release formulation that governs the rate of delivery (input-controlled systems), and (2) the design of the prodrug-drug complex provides a rate-limiting factor of the drug release.[56]

Employing a controlled release formulation is particularly useful for drugs with poor stability, low aqueous solubility, high polarity, or low melting point. In general, these properties are related to the chemical structure of the drug and specifically to the functional groups with high hydrogen bonding potential such as carboxylic acids and alcohols.[57] In these cases, prodrugs can be designed to introduce lipophilicity and mask the hydrogen bonding groups of an active compound by adding another moiety. The classic example is in steroid therapy: lipophilic esters of testosterone. In the case of commonly used testosterone cypionate (Fig. 3.12), the rate of release is determined by the erosion kinetics of the depo formulation, whereas in vivo chemical stability is enhanced by the lipophilic properties of the prodrug.

The second approach to prolonged therapeutic action is based on the controlled rate of conversion of the promoiety into the active compound in vivo. This approach requires particularly detailed study of the kinetics of prodrug-drug conversion. A classic example is bioconversion of azathioprine to 6-mercaptopurine. Azathioprine is used commonly in kidney transplantation, rheumatoid arthritis, and the treatment of various skin disorders. After administration, azathioprine undergoes slow

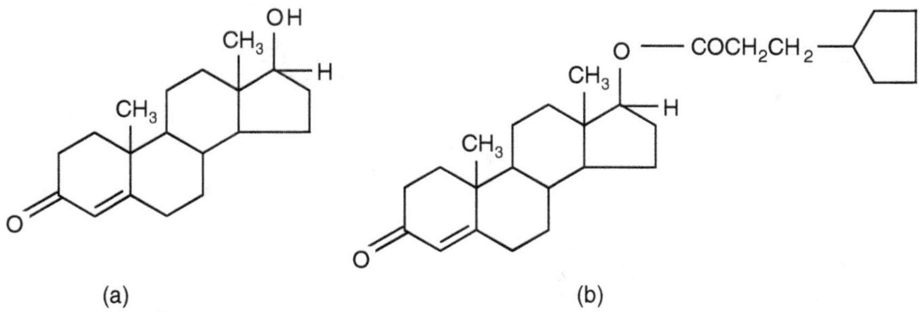

Figure 3.12 Testosterone (*a*) and testosterone cypionate (*b*).

nonenzymatic cleavage governed by glutathione, a tripeptide thiol antioxidant most concentrated in the liver.[58] This cleavage leads to bioconversion to 6-mercaptopurine, a purine analogue that acts as a chemotherapeutic agent interfering with the synthesis of nucleotides, thereby inhibiting T-cell proliferation. The rate of the bioconversion shown in Fig. 3.13 is a major contributor to the release rate of this prodrug-based dosage form.

New water-soluble prodrugs of an HIV protease inhibitor were tested recently; these prodrugs contain two linked units, a solubilizing moiety, and a self-cleavable spacer[59] (Fig. 3.14). These prodrugs convert to the parent drug not enzymatically but chemically via intramolecular cyclization through imide formation in physiological conditions. The release rate of the parent drug is controlled by the chemical structure of both the solubilizing and the spacer moieties.

3.5 Prodrug Design for Various Routes of Administration

Prodrug designs can be used to improve the properties of therapeutic agents, as discussed in the previous sections. Prodrugs also can be designed to be delivered by a particular route of administration.

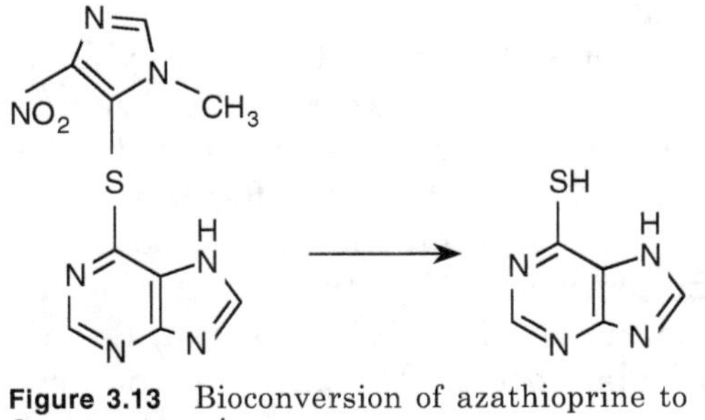

Figure 3.13 Bioconversion of azathioprine to 6-mercaptopurine.

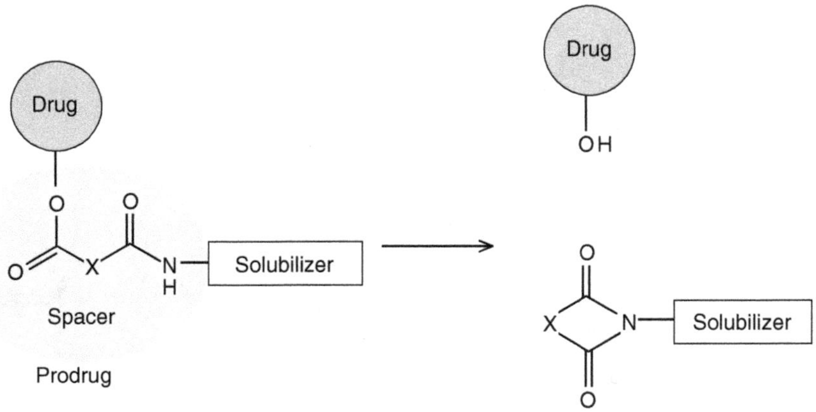

Figure 3.14 Intramolecular cyclization of novel water-soluble prodrugs of HIV protease inhibitor in physiologic fluid.[59]

3.5.1 Prodrugs for nasal delivery

The nasal route offers several advantages, such as high systemic availability and rapid onset of action, as discussed in other chapters.[60] The nasal epithelium allows the transport of both charged and uncharged forms of the drug, and it is rich in several metabolizing enzymes such as aldehyde dehydrogenase, glutathione transferases, epoxide hydrolases, and cytochrome P450–dependent monooxygenases. These enzymes offer another dimension of flexibility in the design of prodrugs for nasal delivery. L-Dopa has been systemically delivered using water-soluble prodrugs through the nasal route.[61] In rats, nasal administration of butyl ester prodrug afforded higher olfactory bulb and cerebrospinal fluid (CSF) L-dopa concentrations (relative to an equivalent intravenous dose) without significantly affecting plasma dopamine levels. These results indicate preferential delivery to the CNS, which suggests a potential to reduce side effects. Another prodrug, Suc-D4T,[62] was shown to be transported directly from the nasal cavity to the CSF, as evidenced by higher CSF concentrations in comparison with those following administration of D4T alone.

A series of 2′-(*O*-acyl) derivatives of 9-(2-hydroxyethoxymethyl)guanine (acyclovir) was synthesized by Shao and coworkers.[63,64] The bioconversion kinetics of the prodrugs appeared to depend on both the polar and the steric properties of the acyl substituents. Rat nasal perfusion studies using the in situ perfusion technique showed no measurable loss of acyclovir from the perfusate. Also, the extent of nasal absorption appeared to depend on the lipophilicity of the prodrugs. All the prodrugs showed enhanced absorption. Branching of the acyl side-chain significantly retarded acyclovir prodrug breakdown, suggesting that a branched-chain prodrug with enhanced lipophilicity may exhibit better absorption and lower presystemic degradation than other designs. The L-aspartate beta-ester prodrug of acyclovir was synthesized to target active transport mechanisms.[65]

Testosterone (TS) used for treating TS deficiency is limited by its low aqueous solubility. With a water-soluble prodrug, TS 17β-*N*,*N*-dimethylglycinate hydrochloride, the aqueous solubility was increased to more than 100 mg/mL, compared with 0.01 mg/mL for TS. The bioavailabilities of both the prodrug and TS after nasal administration of the prodrug were similar to that after equivalent intravenous (IV) doses.[66]

3.5.2 Prodrugs for ocular delivery

For most ocularly applied drugs, passive diffusion is thought to be the main transport process across the cornea.[67] Major challenges in ocular drug delivery include the tightness of the corneal epithelium barrier, rapid precorneal drug elimination, and systemic absorption from the conjunctiva.[68,69] As a result, less then 10 percent and typically less than 1 percent of the instilled dose reaches the intraocular tissues. Many drugs developed for systemic use lack the physicochemical properties required to overcome the previously mentioned barriers.[70–72] Attempts to improve the ocular bioavailability have concentrated on (1) extending the drug residence time in the conjunctival sac and (2) improving penetration of the drug across the corneal barrier. Prodrugs for ocular delivery have been geared mostly to address the latter issue.[73]

Ocular absorption of a drug can be enhanced substantially by increasing its lipophilicity, which can be achieved with prodrug applications. Key requirements for ocular prodrugs involve good stability and solubility in aqueous solutions to enable formulation, sufficient lipophilic properties to penetrate through the cornea, low irritation profile, and the ability to release the parent drug within the eye at a rate that meets the therapeutic need. Prodrugs were introduced into ophthalmology about 15 years ago when ocular absorption of epinephrine was improved substantially by its prodrug, dipivefrine.[74] Currently, it has replaced epinephrine in the treatment of elevated intraocular pressure (IOP) associated with glaucoma. Since the introduction of dipivefrine, numerous prodrugs have been designed to improve the efficacy of ophthalmic drugs, to prolong their duration of action, and to reduce their systemic side effects.

An optimal ocular prodrug should be hydrolyzed by several esterases in order to minimize the effect of individual esterase levels on enzymatic transformation of ocular prodrugs.[75] In rabbits (the typical model used during initial development of ophthalmic drugs), acetylcholinesterase (AChE) and butyrylcholinesterase (BuChE) are considered critical in the enzymatic hydrolysis of ester prodrugs.[76–78] Model ocular tissues also may contain carbonic anhydrase,[79] peptidases,[80] and phosphatases.[81]

In recent years, the following prodrugs have been tested experimentally and clinically for ocular delivery: prodrugs of adrenergic agonists (epinephrine and phenylephrine prodrugs),[82] prodrugs of β-adrenergic antagonists (β-blockers) (timolol, nadolol, and tilisolol prodrugs),[83] pilocaprine prodrugs (mono- and diesters of pilocapric and bispilocapric acids),[84] antiviral prodrugs (acyclovir and idoxuridine esters),[85,86] carbonic anhydrase inhibitor prodrugs, and steroids.[87] Several examples of ocular prodrugs designed to improve ocular bioavailability are summarized in Table 3.6.

3.5.3 Prodrugs for parenteral delivery

Prodrugs often have been employed in parenteral delivery to improve drug solubility, disposition, and patient acceptability (e.g., decreased pain on injection). Ideally, the parenteral prodrug needs to be converted rapidly to the parent drug in the plasma to obtain a rapid response. A classic example of using ester prodrugs to improve the parenteral delivery of sparingly water-soluble drugs is fosphenytoin, a prodrug of phenytoin described in Sec. 3.3.3. Fosphenytoin is water soluble and intrinsically safe, and it readily bioreverts to phenytoin on parenteral administration through the action of phosphatases. Pharmacokinetic and pharmacodynamic studies in animals and humans have shown that fosphenytoin quantitatively releases phenytoin on parenteral administration and provides better absorption, as well as far greater safety, than phenytoin.[92]

Similarly, the sodium salt of the poorly soluble COX-2 inhibitor parecoxib, parecoxib sodium (Dynastat®), was designed as a parenteral analgesic for the management of acute pain to improve solubility of the parent compound. On administration, parecoxib sodium is converted rapidly and essentially completely to the pharmacologically active

TABLE 3.6 Prodrugs for Ocular Drug Delivery

Drug	Prodrug	Advantages	Indication
Epinephrine	Dipivalyl epinephrine	Reduced side effects	Glaucoma[88]
Phenylephrine	Phenylephrine oxazolidine	Improved therapeutic action	Mydriatic[89]
Prostaglandins	Latanoprost, travoprost, bimatoprost, unoprostone	Improved permeability	Glaucoma[90]
Acyclovir	Valacyclovir	Improved bioavailability	HSV keratitis[91]

moiety valdecoxib and propionic acid through enzymatic conversion that occurs primarily in the liver.[93]

An exciting new field of parenteral drug delivery involves oil-based depot formulations for protein delivery. The feasibility of administering such polar drug substances in the form of oil solutions is governed by the attainment of sufficient oil solubility, which can be achieved with prodrugs. Interesting examples of the experimental peptide delivery formulations are 4-imidazolidinone prodrugs of the polar local anesthetic agent prilocaine.[94]

3.5.4 Prodrugs for transdermal delivery

The skin is the major site for noninvasive drug delivery; however, transdermal drug penetration is relatively challenging because of the inherent variability in permeability of the skin. Percutaneous drug absorption is described by Fick's first law of diffusion. Therefore, the transdermal controlled delivery system must alter some of the key mass-transfer parameters, such as partition coefficient, diffusion coefficient, and drug concentration gradient, to increase drug absorption. This can be achieved with the prodrug approach, in which highly absorbable prodrug molecules are activated within the skin.

The successful delivery of prodrug through the skin requires the following sequential steps[95]: (1) dissolution and diffusion of drug molecules in the vehicle into the skin surface, (2) partitioning of the drug into the stratum corneum (SC), (3) diffusion of the drug into the SC, and (4) partitioning of the drug into the epidermis and dermis and uptake into the blood circulation. Based on these requirements, the desired parameters for transdermal prodrugs include low molecular mass (preferably less than 600 Da), adequate solubility in oil and water to maximize the membrane concentration gradient (the driving force for diffusion), optimal partition coefficient, and low melting point.[96]

A good example of a transdermal prodrug is an alkyl ester prodrug of naltrexone designed to improve lipophilicity of the parent compound and increase its delivery rate across the skin. The mean naltrexone flux from the prodrug-saturated solutions exceeded the flux of naltrexone base by approximately two- to seven-fold.

Other examples of prodrug applications for transdermal delivery include using α-(acyloxy) alkyl derivatives to change the physicochemical properties of the parent drug (e.g., nitrofurantoin, benzylpenicillin, and theophylline derivatives),[97] derivatization of the drug with a promoiety that can enhance permeation by covalently conjugating the drug with fatty acids (e.g., propranolol hydrochloride),[98] and novel duplex prodrugs that increase the transdermal drug delivery rate via a permeability increase as a result of the bioconversion-induced steepening of the concentration gradient in the skin (e.g., prodrug of naltrexone).[99] Prodrugs of 5-FU (fluorouracil) and keterolac with a higher flux across the skin also have been reported.[100,101]

3.5.5 Prodrugs for oral delivery

Oral delivery is the most preferred route of drug administration; however, it often entails major challenges, namely, limited solubility of the drug and poor permeation across the GI tract. The major goal of oral drug delivery is to increase the oral bioavailability, which generally is affected by presystemic metabolism (sum of first-pass and intestinal or intestinal membrane metabolisms) and inadequate drug absorption in the GI tract. In both cases, the design of a prodrug must balance the level of stability; premature conversion of prodrug and excessive prodrug linkage both decrease oral bioavailability.

Phosphates or other salts are used often as oral prodrugs to increase the solubility of the parent drugs. Successful examples of this approach include fosphenytoin and hydrocortisone phosphate. In both cases, bioconversion to the parent compound involves rapid prodrug dephosphorylation by intestinal membrane-bound alkaline phosphatase, yielding high concentrations of the poorly soluble parent drug at the apical membrane. The regenerated lipophilic parent drugs are well absorbed compared with their polar, ionized prodrugs.[102]

Ester prodrugs are employed to enhance membrane permeation and transepithelial transport of hydrophilic drugs by increasing the lipophilicity of the parent compound, resulting in enhanced transmembrane transport by passive diffusion. For example, pivampicillin, a pivaloyloxymethyl ester of ampicillin, is more lipophilic than its parent ampicillin and has demonstrated increased membrane permeation and transepithelial transport in in vivo studies.[103]

3.5.6 Prodrugs for buccal delivery

The buccal delivery route has generated interest lately because it offers a noninvasive route of delivery for proteins and peptides that cannot tolerate the harsh acidic environment of the GI tract. Drug delivery by the buccal mucosa prevents the drug loss of first-pass hepatic metabolism. Prodrugs in buccal delivery generally improve drug solubility and stability using polymers. Use of buccal devices incorporating prodrugs provides a constant drug release rate, resulting in a reduced total amount of drug and increased patient comfort.

Buccal delivery of opioid analgesics and antagonists can improve bioavailability relative to the oral route. Esterification of the 3-phenolic hydroxyl group in opioid analgesics such as nalbuphine, naloxone, naltrexone, oxymorphone, butorphanol, and levallorphan improved bioavailability and eliminated the bitter taste. The prodrug of morphine,

morphine-3-propionate, helps to reduce enzymatic degradation in the oral cavity and enhance permeation across biological barriers.[104]

3.6 Recent Advances in Prodrugs as Drug Delivery Systems

The advances in understanding of cellular mechanisms and in combinatorial chemistry have revealed many opportunities for the development of prodrugs. Some of the recent advances made in the development of prodrugs are presented below.

Enzyme-activated prodrug therapy has been used to design specific drug delivery systems for the treatment of cancer. In the initial step, a drug-activating enzyme is targeted and expressed in the tumors. Subsequently, a nontoxic prodrug, which acts as the substrate to the enzyme, is administered systemically, enabling selective activation of the prodrug in the tumor.[105–107] Several strategies have been identified for targeting the tumor.

3.6.1 Antibody-directed enzyme prodrug therapy (ADEPT)

In *antibody-directed enzyme prodrug therapy* (ADEPT), a monoclonal antibody to a cancer-specific antigen is conjugated to an enzyme that is normally absent in body fluid or cell membranes (the antibody enables localization of the conjugate in the tumor cells). First, the antibody-enzyme conjugate is delivered by infusion. After the excess conjugate is cleared from the circulation, a nontoxic prodrug is administered, enabling site-specific activation. For example, Her-2/neu antibody (trastuzumab, Herceptin) recently has received approval for clinical use.[108–110] ADEPT has been used to target a variety of enzyme systems, such as alkaline phosphatases, aminopeptidases, and carboxypeptidases. A slight variation of ADEPT called *antibody-generated enzyme-nitrile therapy* (AGENT) relies on enzymatic liberation of cyanide from cyanogenous glucosides.[111]

The first clinically tested ADEPT prodrug, 4-[(2-chloroethyl)(2-mesyloxyethyl) amino]benzoyl-L-glutamic acid (CMDA), is converted in vivo to the cytotoxic parent drug 4-[(2-chloroethyl)[2-(mesyloxy)ethyl]amino]benzoic acid, as shown in Fig. 3.15.

3.6.2 Gene-directed enzyme prodrug therapy (GDEPT) and viral-directed enzyme prodrug therapy (VDEPT)

Both *gene-directed enzyme prodrug therapy* (GDEPT) and *viral-directed enzyme prodrug therapy* (VDEPT) involve physical delivery of genes encoding prodrug-activating enzymes to the tumor cells for site-specific

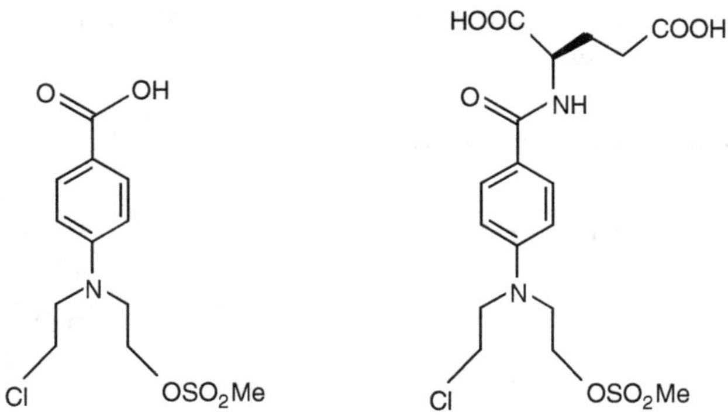

(a) (b)

Figure 3.15 4-[(2-Chloroethyl)[2-(mesyloxy)ethyl]amino]benzoic acid (*a*) and CMDA (*b*).

activation. The only notable difference between the two strategies is that GDEPT uses nonviral vectors for intracellular delivery of genes, whereas VDEPT uses viral vectors for achieving the same purpose. The transfected tumor cells express the enzyme protein, which is further converted to active enzyme and selectively catalyzes intracellular activation of inactive prodrug to the active drug (toxic), resulting in cell death. Another variation of GDEPT is *genetic prodrug activation therapy* (GPAT), which involves the use of transcriptional differences between normal and tumor cells to induce the selective expression of drug-metabolizing enzymes to convert nontoxic prodrug into the active toxic moiety.

Examples of GDEPT include irinotecan (CPT-11), a prodrug of 7-ethyl-10-hydroxy-camptothecin activated by carboxylesterase; 5-fluorocytosine, a prodrug of 5-FU activated by cytosine deaminase; and cyclophosphamide, a prodrug of 4-hydroxycyclophosphamide activated by cytochrome P450, which degrades into acrolein and phosphoramide mustard.[112–115]

3.6.3 Macromolecule-directed enzyme prodrug therapy (MDEPT)

Macromolecule-directed enzyme prodrug therapy (MDEPT) is also referred to as *polymer-directed prodrug therapy* (PDEPT). It is similar to GDEPT and VDEPT, except that it applies a macromolecule conjugate of the drug to enable delivery to the tumor. This method also takes advantage of the enhanced permeation and retention (EPR) of tumors. One of the earliest examples of MDEPT involved *N*-(2-hydroxypropyl) methacrylamide.[116] The modes of delivery and activation of the prodrug by the preceding methods are illustrated briefly in Fig. 3.16.

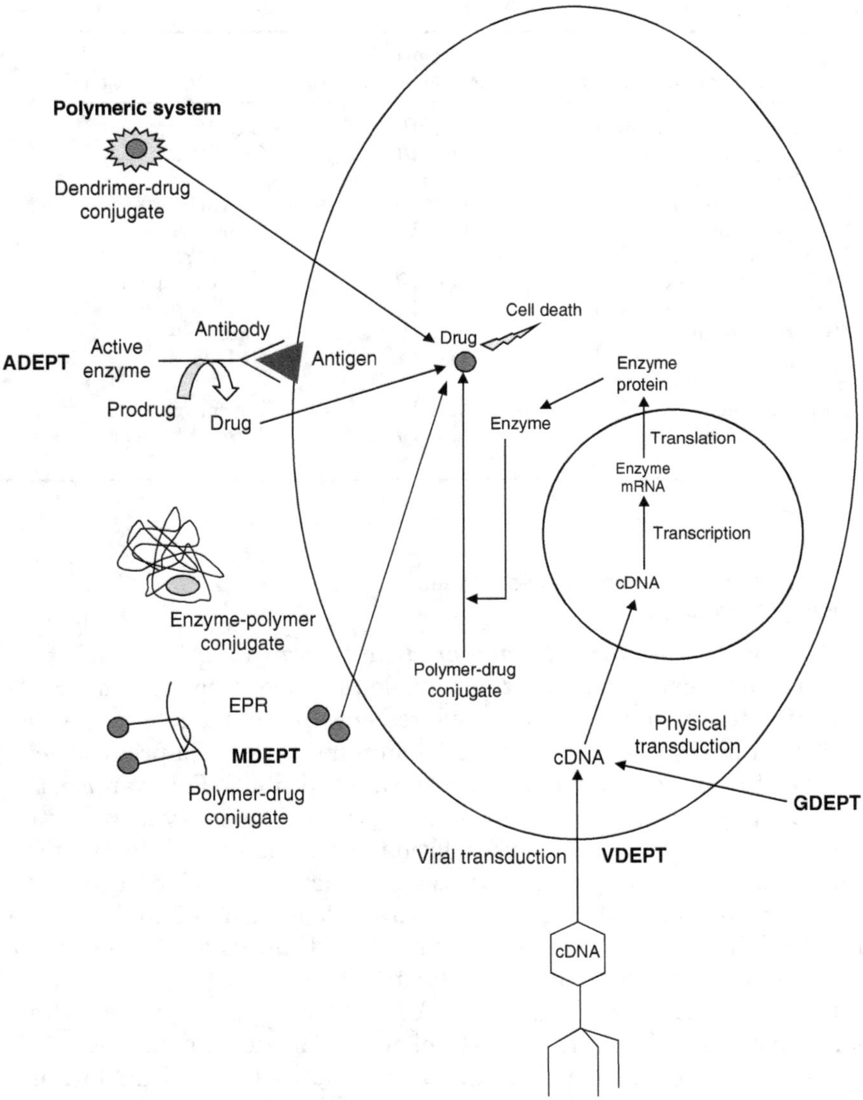

Figure 3.16 Modes of delivery and activation of prodrugs.

Polymeric prodrugs are currently one of the most investigated topics. This research has resulted in breakthrough therapeutics, and many compounds are under clinical development (Table 3.7). Other examples of polymeric prodrug applications include the use of polysaccharides such as dextran, mannan, and pullulan to enable active targeting to tumor cells.[117]

TABLE 3.7 Examples of Polymeric Prodrugs

Drug-polymer conjugate	Stage of development	Organization
HPMA copolymer-doxorbicin	Phase II	CRC/Pharmacia[118]
HPMA copolymer-doxorbicin-galactosamine	Phase I/II	CRC/Pharmacia[119]
HPMA copolymer-paclitaxel	Phase I	Pharmacia[123]
HPMA copolymer-camptothecin	Phase I	Pharmacia[123]
HPMA copolymer-platinate	Phase I	Access Pharmaceuticals[123]
Polyglutamate-paclitaxel	Phase II/III	Cell Therapeutics[120]
Polyglutamate-camptothecin	Phase I	Cell Therapeutics[123]
PEG-camptothecin	Phase II	Enzon[124]
PEG-aspartic acid–doxorubicin micelle	Phase I	NCI, Japan[123]
PEG-paclitaxel	Phase I	Pharmacia[124]
Doxorubicin micelle HPMA platinate	Phase II/III	Access Pharmaceuticals[124]

3.6.4 Lectin-directed enzyme-activated prodrug therapy (LEAPT)

Lectin-directed enzyme-activated prodrug therapy (LEAPT) is a bipartite drug delivery system that first exploits endogenous carbohydrate-to-lectin binding to localize glycosylated enzyme conjugates to specific, predetermined cell types, followed by administration of a prodrug activated by the predelivered enzyme at the desired site.[121] For example, the carbohydrate structure of an α-L-rhamnopyranosidase enzyme was modified through enzymatic deglycosylation and chemical reglycosylation. Ligand competition experiments revealed enhanced, specific localization by endocytosis and a strongly carbohydrate-dependent, 60-fold increase in selectivity toward target cell hepatocytes that generated a greater than 30-fold increase in protein delivery.

Tissue-activated drug delivery (TADD)[122] involves the use of alternating polymers of polyethylene glycol and trifunctional monomers such as lysine. The resulting polymer has a pendant with the PEG forming the chain and lysine providing the reactive carboxylic acid groups at periodic intervals, which can be linked to the drug (Fig. 3.17). The linking group chemistry can be altered to induce activation in specific tissues.

3.6.5 Dendrimers

Dendrimers are highly branched globular macromolecules. Several researchers have exploited the multivalency of dendrimers at the periphery for attachment of drug molecules. A significant advantage of this system is that the drug loading can be tuned by varying the generation

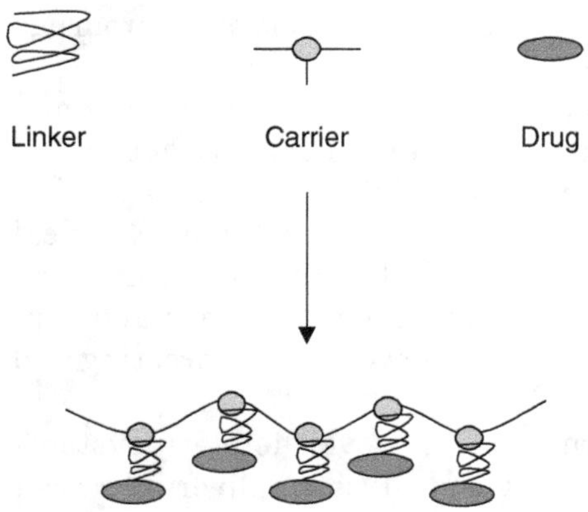

Figure 3.17 Polymeric prodrug with PEG carrier and lysine linker.

of the dendrimer, and the release of drug can be tailored by incorporating degradable linkages between the drug and the dendrimer.[123] Conjugates of poly(amidoamine) (PAMAM) dendrimers with cisplatin have been shown to improve aqueous solubility and to reduce systemic toxicity while simultaneously exhibiting selective accumulation in tumors.[124] Propranolol is a poorly soluble drug and is a known substrate of the P-glycoprotein (P-gp) efflux transporter. A prodrug of propranolol was synthesized by conjugating propranolol to generation 3 (G3) and lauroyl-G3 PAMAM dendrimers. Both derivatives demonstrated improved aqueous solubility and bypassed the efflux transporter with improved bioavailability.[125]

Hep-Direct liver-targeted therapy[126] represents a novel class of phosphate and phosphonate prodrugs that are cyclic 1,3-propanyl esters containing a ring substituent that renders them sensitive to an oxidative cleavage reaction catalyzed by a cytochrome P450 within hepatocytes. This platform may be useful for treating chronic liver diseases such as hepatitis and hepatocellular carcinoma with reduced systemic exposure and side effects. The authors have demonstrated liver targeting of nucleotide analogues adefovir and cytarabine (Ara-C).

3.7 Conclusions and Future Prospects

Prodrugs have been applied successfully to modulate the physicochemical and pharmacokinetic properties of a drug to optimize clinical therapy. Despite the obvious advantages, prodrugs have been used as a last option mainly because of costly preclinical development compared with other nonchemical options. This trend will be reversed in the coming years, fueled largely by discovery pipelines of pharmaceutical companies full of hydrophobic class II molecules in the preclinical stage. The recent approval of a number of blockbusters formulated as prodrugs (e.g., omperazole and simvastatin) seems to confirm the arrival of prodrugs on the drug discovery scene.

The unraveling of the human genome and advances in molecular and cellular biology have improved our understanding of diseases at a molecular level, ushering in the golden era of targeted cellular therapeutics. In this light, prodrugs can no longer be regarded merely as chemical modifications. This direction is aptly evidenced by the increasing complexity of prodrug design aimed at providing gene delivery and controlling cellular expression of enzymes to achieve precise targeting and delivery.

Prodrugs definitely will lead the way in novel delivery systems addressing the need for precise targeting and efficient delivery of cancer chemotherapeutics. The understanding of cellular mechanisms and continued advances in nanotechnology, material science, and engineering will continue to drive innovation in this field.

Adaptation of cutting-edge technology often has been slowed by a lack of cooperation among teams working on different aspects of drug discovery and development. The application of a prodrug strategy will require a change in this mind-set: a multidisciplinary team will combine the cellular understanding of the biologist with the innovative thinking of the medicinal chemist and the systems understanding of the delivery scientist to conjure up imaginative solutions to therapeutic problems.

The ultimate utility of prodrugs may well lie in delivering siRNA (short interfering RNA) and intracellular gene therapeutics to cells to address disease states at the cellular level, resulting in complete cure of disease. The future potential of prodrugs is immense, enabling multimechanistic drug targeting approaches, and it may prove to be the legendary "magic bullet" for delivery of therapeutics.

References

1. Testa, B., and Soine, W. Principles of drug metabolism. In Abraham, D. J. (ed.), *Burgers Medicinal Chemistry and Drug Discovery*, Vol 2, 6th ed. New York: Wiley, 2003, pp. 431–498.
2. Albert, A. Chemical aspects of selective toxicity, *Nature* **182**:421–423, 1958.
3. Harper, N. J. Drug latentiation. *Prog. Drug Res.* **4**:221–294, 1962.
4. Sinkula, A., and Yalkowsky, S. Rationale for design of biologically reversible drug derivatives: Prodrugs. *J. Pharm. Sci.* **64**:181–210, 1975.
5. Higuchi, T., and Stella, V. *Prodrugs as Novel Drug Delivery Systems*. Washington: American Chemical Society, 1975.
6. Roche, E. B. *Design of Biopharmaceutical Properties Through Prodrugs and Analogs*. Washington: American Pharmaceutical Association, 1977.
7. Trefouel, J., Nitti, F., and Bovet, D. Activite du *p*-aminophenyl sulfamide sur les infections streptococciques. *C. R. Seances Soc. Biol. Fil.* **120**:756–762, 1935.
8. Testa, B., and Caldwell, J. Prodrugs revisited: The "ad hoc" approach as a complement to ligand design. *Med. Res. Rev.* **16**:233–241, 1996.
9. Lexxel package insert (Astra Merck, US), new 12/96, rec. 4/97.

10. Bodor, N., and Buchwald, P. Soft drug design: General principles and recent applications. *Med. Res. Rev.* **20**:58–101, 2000.
11. Bodor, N., and Buchwald, P. Designing safer (soft) drugs by avoiding the formation of toxic and oxidative metabolites. *Mol. Biotechnol.* **26**(2):123–132, 2004.
12. Hirayama, F., and Uekama, K. Cyclodextrin-based controlled drug release system. *Adv. Drug Deliv. Rev.* **36**(1):125–141, 1999.
13. Belcheva, N., Smid, J., Lambov, N., et al. Polymeric sustained release formulations of the bronchial dilator vephylline. *J. Control. Release* **37**(1–2):43–48, 1995.
14. Zenebergh, A., Baurain, R., and Trouet, A. Cellular pharmacology of detorubicin and doxorubicin in L1210 cells. *Eur. J. Cancer Clin. Oncol.* **20**(1):115–121, 1984.
15. Robinson, J. R., and Lee, V. H. L., eds. *Controlled Drug Delivery. Fundamentals and Applications*, 2d ed. New York: Marcel Dekker, 1987.
16. Stella, V. J., Michelson, T. J., and Pipkin, J. D. Prodrugs: The control of drug delivery via bioreversible chemical modification. In Juliano, R. L. (ed.), *Drug Delivery Systems: Characteristics and Biomedical Applications*. New York: Oxford University Press, 1980, pp. 110–170.
17. Ranade, V. V., and Hollinger, M. A. *Drug Delivery Systems*, 2d ed. Boca Raton, FL: CRC Press, 2004, pp. 290–299.
18. Hu, L. Prodrugs: Effective solutions for solubility, permeability and targeting challenges. *Idrugs* **7**(8):736–742, 2004.
19. Hu, L. The prodrug approach to better targeting. *Curr. Drug Disc.* **4**(18): 28–32, 2004.
20. Cundy, K. C., Annamalai, T., Bu, L., et al. XP13512 [(±)-1-([(α-isobutanoyloxyethoxy) carbonyl] aminomethyl)-1-cyclohexane acetic acid], a novel gabapentin prodrug: II. Improved oral bioavailability, dose proportionality, and colonic absorption compared with gabapentin in rats and monkeys. *J. Pharmacol. Exp. Ther.* **311**:324–333, 2004.
21. van Gelder, J., Deferme, S., Naesens, L., et al. Intestinal absorption enhancement of the ester prodrug tenofovir disoproxil fumarate through modulation of the biochemical barrier by defined ester mixtures. *Drug Metab. Dispos.* **30**(8):924–930, 2002.
22. Notari, R. E. Prodrug design. *Pharmacol. Ther.* **14**:25–53, 1981.
23. Ettmayer, P., Amidon, G. L., Clement, B., and Testa, B. Lessons learned from marketed and investigational prodrugs. *J. Med. Chem.* **47**(10):2393–2403, 2004.
24. Guo, A., Hu, P., Balimane, P. V., et al. Interactions of a nonpeptidic drug, valacyclovir, with the human intestinal peptide transporter (hPEPT1) expressed in a mammalian cell line. *J. Pharmacol. Exp. Ther.* **289**(1):448–454, 1999.
25. Sinko, P. J., and Balimane, P. V. Carrier-mediated intestinal absorption of valacyclovir, the L-valyl ester prodrug of acyclovir: I. Interactions with peptides, organic anions and organic cations in rats. *Biopharm. Drug Dispos.* **19**(4):209–217, 1998.
26. Kim, I., Chu, X., Kim, S., et al. Identification of a human valacyclovirase: Biphenyl hydrolase-like (BPHL) protein as valacyclovir hydrolase. *J. Biol. Chem.* **278**(28): 25348–25356, 2003.
27. Portolés, A., Terleira, A., Almeida, S., et al. Bioequivalence study of two formulations of enalapril, at a single oral dose of 20 mg (tablets): A randomized, two-way, open-label, crossover study in healthy volunteers. *Curr. Ther. Res.* **65**(1):34–46, 2004.
28. Olkkola, K. T., Brunetto, A. V., and Mattila, M. J. Pharmacokinetics of oxicam nonsteroidal anti-inflammatory agents. *Clin. Pharmacokinet.* **26**(2):107–120, 1994.
29. Bai, A., Meier, G. P., Wang, Y., et al. Prodrug modification increases potassium tricyclo[5.2.1.0(2,6)]-decan-8-yl dithiocarbonate (D609) chemical stability and cytotoxicity against U937 leukemia cells. *J. Pharmacol. Exp. Ther.* **309**:1051–1059, 2004.
30. Brighton, T. A. The direct thrombin inhibitor melagatran/ximelagatran. *Med. J. Aust.* **181**(18):432–437, 2004.
31. Dressler, D., and Potter, H. *Discovering Enzymes*. New York: Scientific American Library, W. H. Freeman, 1991.
32. Bender, M., Kédzy, F. J., and Wedler, F. C. *a*-Chymotrypsin: Enzyme concentration and kinetics. *J. Chem. Educ.* **44**:84–88, 1967.
33. Kim, B.-Y., Doh, H.-J., Le, T. N., et al. Ketorolac amide prodrugs for transdermal delivery: Stability and in vitro rat skin permeation studies. *Int. J. Pharm.* **293**(1–2): 193–202, 2005.

34. Mahfouz, N. M., Omar, F. A., and Aboul-Fadl, T. Cyclic amide derivatives as potential prodrugs II. *N*-Hydroxymethylsuccinimide-/isatin esters of some NSAIDs as prodrugs with an improved therapeutic index. *Eur. J. Med. Chem.* **34**(7–8):551–562, 1999.
35. Bartzatt, R., Suat, L., Cirillo, G., et al. Bifunctional constructs of aspirin and ibuprofen (nonsteroidal anti-inflammatory drugs; NSAIDs) that express antibacterial and alkylation activities. *Biotechnol. Appl. Biochem.* **37**:273–282, 2001.
36. Shanbhag, V. R., Crider, A. M., Gokhale, R., et al. Ester and amide prodrugs of ibuprofen and naproxen: Synthesis, anti-inflammatory activity, and gastrointestinal toxicity. *J. Pharm. Sci.* **81**:149–154, 1992.
37. Fadl, T. A., Hasegawa, T., Youssef, A. F., et al. Synthesis and investigation of *N*4-substituted cytarabine derivatives as prodrugs. *Pharmazie* **50**:382–387, 1995.
38. Choe, Y. H., Conover, C. D., Wu, D., et al. Anticancer drug delivery systems: *N*4-Acyl poly(ethyleneglycol) prodrugs of ara-C. *J. Control. Release* **79**(1–3)41–53, 2002.
39. Hadfield, A. F., and Sartorelli, A. C. The pharmacology of prodrugs of 5-flourouracil and 1-β-D-arabinofuranosylcytosine. *Adv. Pharmacol. Chemother.* **20**:21–67, 1984.
40. Wipf, P., and Li, W. Prodrugs of ara-C. *Drugs Future* **19**:49–54, 1994.
41. Noveck, R. J., and Hubbard, R. C. Parecoxib sodium, an injectable COX-2-specific inhibitor, does not affect unfractionated heparin-regulated blood coagulation parameters. *J. Clin. Pharmacol.* **44**:474–480, 2004.
42. Persico, F. J., Pritchard, J. F., Fisher, M. C., et al. Effect of tolmetin glycine amide (McN-4366), a prodrug of tolmetin sodium, on adjuvant arthritis in the rat. *Pharmacol. Exp. Ther.* **247**(3):889–896, 1988.
43. Cogan, P. S., Fowler, C. R., Post, G. C., and Koch, T. H. Doxsaliform: A novel *N*-Mannich base prodrug of a doxorubicin formaldehyde conjugate (letter). *Drug Des. Discov.* **1**:247–255, 2004.
44. Ke, T. L., Graff, G., Spellman, J. M., and Yanni, J. M., Nepafenac, a unique nonsteroidal prodrug with potential utility in the treatment of trauma-induced ocular inflammation: II. In vitro bioactivation and permeation of external ocular barriers. *Inflammation* **24**(4):371–384, 2000.
45. Bundgaard, H., Larsen, C., and Thorbek, P. Prodrugs as drug delivery systems: XXVI. Preparation and enzymatic hydrolysis of various water-soluble amino acid esters of metronidazole. *Int. J. Pharm.* **18**(1–2):67–77, 1984.
46. Stella, V. J. A case for prodrugs: Fosphenytoin. *Adv. Drug Deliv. Rev.* **19**(2):311–330, 1996.
47. Kazmierski, W. M., Bevans, P., Furfine, E., et al. Novel prodrug approach to amprenavir-based HIV-1 protease inhibitors via O → N acyloxy migration of P1 moiety. *Bioorg. Med. Chem. Lett.* **13**(15):2523–2526, 2003.
48. Goldschmidt, V., and Marquet, R. Primer unblocking by HIV-1 reverse transcriptase and resistance to nucleoside RT inhibitors (NRTIs). *Int. J. Biochem. Cell. Biol.* **36**(9):1687–1705, 2004.
49. De Clercq, E. HIV-chemotherapy and -prophylaxis: New drugs, leads and approaches. *Int. J. Biochem. Cell. Biol.* **36**(9):1800–1822, 2004.
50. Arvieux, C., and Tribut, O. Amprenavir or fosamprenavir plus ritonavir in HIV infection: Pharmacology, efficacy and tolerability profile. *Drugs* **65**(5):633–659, 2005.
51. Czock, D., Scholle, C., Rasche, F. M., et al. Pharmacokinetics of valganciclovir and ganciclovir in renal impairment. *Clin. Pharmacol. Ther.* **72**(2):142–150, 2002.
52. Levine, D. S., Riff, D. S., Pruitt, R., et al. A randomized, double-blind, dose-response comparison of balsalazide (6.75 g), balsalazide (2.25 g), and mesalamine (2.4 g) in the treatment of active, mild-to-moderate ulcerative colitis. *Am. J. Gastroenterol.* **97**(6):1398–1407, 2002.
53. Hall, I. H., Schwab, U. E., Ward, E. S., and Ives, T. J. Effects of alatrofloxacin, the parental prodrug of trovafloxacin, on phagocytic, anti-inflammatory and immunomodulation events of human THP-1 monocytes. *Biomed. Pharmacother.* **57**(8):359–365, 2003.
54. Tunek, A., and Svensson, L. A. Bambuterol, a carbamate ester prodrug of terbutaline, as inhibitor of cholinesterases in human blood. *Drug Metab. Dispos.* **16**(5):759–764, 1988.
55. Halpin, R. A., Ulm, E. H., Till, A. E., et al. Biotransformation of lovastatin: V. Species differences in in vivo metabolite profiles of mouse, rat, dog, and human. *Drug Metab. Dispos.* **21**(6):1003–1011, 1993.

56. Amidon, G. L., Pearlman, R. S., and Lessman, G. D. Design of prodrugs through consideration of enzyme-substrate specificities. In Roche, E. B. (ed.), *Design of Biopharmaceutical Properties through Prodrugs and Analogs*. Washington: American Pharmaceutical Association, Academy of Pharmaceutical Sciences, 1977, pp. 68–97.
57. Beaumont, K., Webster, R., Gardner, I., and Dack, K. Design of ester prodrugs to enhance oral absorption of poorly permeable compounds: Challenges to the discovery scientist. *Curr. Drug Metab.* **4**(6):461–485, 2003.
58. Kaplowitz, N. Interaction of azathioprine and glutathione in the liver of the rat. *J. Pharmacol. Exp. Ther.* **200**(3):479–486, 1977.
59. Matsumoto, H., Sohma, Y., Kimura, T., et al. Controlled drug release new water-soluble prodrugs of an HIV protease inhibitor. *Bioorg. Med. Chem. Lett.* **11**(4): 605–609, 2001.
60. Krishnamoorthy, R., and Mitra, A. K. Prodrugs for nasal drug delivery. *Adv. Drug Deliv. Rev.* **29**:135–146, 1998.
61. Kao, H. D., Traboulsi, A., Itoh, S., et al. Enhancement of the systemic and CNS specific delivery of L-dopa by the nasal administration of its water-soluble prodrugs. *Pharm. Res.* **17**:978–984, 2000.
62. Yajima, T., Juni, K., Saneyoshi, M., et al. Direct transport of 2′,3′-didehydro-3′-deoxythymidine (D4T) and its ester derivatives to the cerebrospinal fluid via the nasal mucous membrane in rats. *Biol. Pharm. Bull.* **21**:272–277, 1998.
63. Shao, Z., Park, G. B., Krishnamoorthy, R., and Mitra, A. K. The physicochemical properties, plasma enzymatic hydrolysis, and nasal absorption of acyclovir and its 2′-ester prodrugs. *Pharm. Res.* **11**:237–242, 1994.
64. Shao, Z., and Mitra, A. K. Bile salt-fatty acid mixed micelles as nasal absorption promoters: III. Effects on nasal transport and enzymatic degradation of acyclovir prodrugs. *Pharm. Res.* **11**:243–250, 1994.
65. Yang, C., Gao, H., and Mitra, A. K. Chemical stability, enzymatic hydrolysis, and nasal uptake of amino acid ester prodrugs of acyclovir. *J. Pharm. Sci.* **90**:617–624, 2001.
66. Hussain, A. A., Al-Bayatti, A. A., Dakkuri, A., et al. Testosterone 17β-*N,N*-dimethylglycinate hydrochloride: A prodrug with a potential for nasal delivery of testosterone. *J. Pharm. Sci.* **91**(3):785–789, 2002.
67. Jvinena, T., and Jrvinenb, K. Prodrugs for improved ocular drug delivery. *Adv. Drug Deliv. Rev.* **19**:203–224, 1996.
68. Lee, V. H. L., and Robinson, J. R. Mechanistic and quantitative evaluation of precorneal pilocarpine disposition in albino rabbits. *J. Pharm. Sci.* **68**:673–684, 1979.
69. Maurice, D. M., and Mishima, S. Ocular pharmacokinetics. In Sears, M. C. (ed.), *Handbook of Experimental Pharmacology,* Vol. 69. *Pharmacology of the Eye*. Berlin: Springer-Verlag, 1984, pp. 19–116.
70. Burstein, N. L., and Anderson, J. A. Corneal penetration and ocular bioavailability of drugs. *J. Ocul. Pharmacol.* **1**:309, 1985.
71. Sloan, K. B., ed. *Prodrugs: Topical and Ocular Delivery (Drugs and the Pharmaceutical Sciences: A Series of Textbooks and Monographs)*. New York: Marcel Dekker, 1992.
72. Jarvinen, T., and Jarvinen, K. Prodrugs for improved ocular delivery. *Adv. Drug Deliv. Rev.* **19**:203–224, 1996.
73. Shoenwald, R. D., and Ward, R. L. Relationship between steroid permeability across excised rabbit cornea and octanol-water partition coefficients. *J. Pharm. Sci.* **67**:786, 1978.
74. Anderson, J. A. Systemic absorption of topical ocularly applied epinephrine and dipivefrin. *Arch. Ophthalmol.* **98**:350–353, 1980.
75. Lee, V. H. L., and Li, V. H. K. Prodrugs for improved ocular drug delivery, *Adv. Drug Deliv. Rev.* **3**:1–38, 1989.
76. Redell, M. A., Yang, D. C., and Lee, V. H. L. The role of esterase activity in the ocular disposition of dipivalyl epinephine in rabbits. *Int. J. Pharm.* **17**:299–312, 1983.
77. Petersen, R. A., Lee, K. J., and Donn, A. Acetylcolinesterase in the rabbit cornea. *Arch. Ophthalmol.* **73**:370–377, 1965.
78. Lee, V. H. L., Chang, S.-C., Oshiro, C. M., and Smith, R. E. Ocular esterase composition in albino and pigmented rabbits: Possible implications in ocular prodrug design and evaluation. *Curr. Eye Res.* **4**:1117–1125, 1985.

79. Stratford, R. E., and Lee, V. H. Aminopeptidase activity in homogenates of various absorptive mucosae in albino rabbit: Implications in peptide delivery. *Int. J. Pharm.* **30**:73–82, 1986.
80. Lee, V. H. Esterase activities in adult rabbit eyes. *J. Pharm. Sci.* **72**:239–244, 1983.
81. Redell, M. A., Yang, D. C., and Lee, V. H. The role of esterase activity in ocular disposition of dipivalyl epinephrine in rabbits. *Int. J. Pharm.* **17**:299–312, 1983.
82. Salminen, L., Krootila, K., Helin, H., and Hakala, T. Corneal vascularization and opacification during long-term use of dipivefrin. *J. Ocul. Pharmacol. Ther.* **11**:37–40, 1995.
83. Sugrue, M. F. The pharmacology of antiglaucoma drugs. *Pharmacol. Ther.* **43**:91–138, 1989.
84. Bundgaard, H., Falch, E., Larsen, C., and Mikkelson, T. J. Pilocarpine prodrugs: I. Synthesis, physicochemical properties and kinetics of lactonization of pilocarpic acid esters. *J. Pharm. Sci.* **75**:36–43, 1986.
85. Colla, L., De Clercq, E., Busson, R., and Vanderhaeghe, H. Synthesis and antiviral activity of water-soluble esters of acyclovir [9-[(2-hydroxy-ethoxy)methyl]guanine]. *J. Med. Chem.* **26**:602–604, 1983.
86. Narurkar, M. M., and Mitra, A. K. Prodrugs of *S*-iodo-2′-deoxyuridine for enhanced ocular transport. *Pharm. Res.* **6**:887–891, 1989.
87. Musson, D., Bidgood, A., and Olejnik, O. In vitro penetration and metabolism studies of prednisolone phosphate, disodium and prednisolone acetate across the cornea of rabbits. *Pharm. Res.* **6**(suppl):S175, 1989.
88. Kohn, A. N., Moss, A. P., Hargett, N. A., et al. Clinical comparison of dipivalyl epinephrine and epinephrine in the treatment of glaucoma. *Am. J. Ophthalmol.* **87**:196–201, 1979.
89. Miller-Meeks, M. J., Farrel, T. A., Munden, P. M., et al. Phenylephrine prodrug report of clinical trials. *Ophthalmology.* **98**(2):222–226, 1991.
90. Cherukury, M., Rix, P., Nguyen, T., et al. Penetration of natural prostaglandins and their ester prodrugs and analogs across human ocular tissues in vitro. *J. Ocul. Pharmacol. Ther.* **14**:389–399, 1998.
91. Satoh, T., Teh, B. S., Timme, T. L., et al. Enhanced systemic T-cell activation after in situ gene therapy with radiotherapy in prostate cancer patients. *Int. J. Radiat. Oncol. Biol. Phys.* **59**(2):562–571, 2004.
92. Gerber, N., Mays, D. C., Donn, K. H., et al. Safety, tolerance and pharmacokinetics of intravenous doses of phosphate-3-ester of hydroxymethyl-5,5-diphenylhydantoin, a new prodrug of phenytoin. *J. Clin. Pharmacol.* **28**:1023–1032, 1988.
93. Sorbera, L., Leeson, P. A., Castaner, J., and Castaner, R. M. Valdecoxib and parecoxib sodium. *Drugs Future* **26**:133–140, 2001.
94. Larsen, S. W., Sidenius, M., Ankersen, M., and Larsen, C. Kinetics of degradation of 4-imidazolidinone prodrug types obtained from reacting prilocaine with formaldehyde and acetaldehyde. *Eur. J. Pharm. Sci.* **20**:233–240, 2003.
95. Hadgraft, J., Guy, R. H., eds. *Transdermal Drug Delivery: Developmental Issues and Research Initiatives.* New York: Marcel Dekker, 1989, pp. 1–22.
96. Barry, B. W. Novel mechanisms and devices to enable successful transdermal drug delivery. *Eur. J. Pharm. Sci.* **14**(2):101–114, 2001.
97. Sloan, K. B., ed. *Prodrugs: Topical and Ocular Drug Delivery, Drugs and the Pharmaceutical Sciences: A Series of Textbooks and Monographs*, Vol. 53, New York: Marcel Dekker, 1992.
98. Namdeo, A., and Jain, N. K. Liquid crystalline pharmacogel–based enhanced transdermal delivery of propranolol hydrochloride. *J. Control. Release* **82**(2–3):223–236, 2002.
99. Hammell, D. C., Hamad, M., Vaddi, H. K., et al. A duplex "gemini" prodrug of naltrexone for transdermal delivery. *J. Control. Release* **97**(2):283–290, 2004.
100. Doh, H. J., Cho, W. J., Yong, C. S., et al. Synthesis and evaluation of ketorolac ester prodrugs for transdermal delivery. *J. Pharm. Sci.* **92**(5):1008–1017, 2003.
101. Roberts, W. J., and Sloan, K. B. Topical delivery of 5-fluorouracil (5-FU) by 3-alkylcarbonyloxymethyl-5-FU prodrugs. *J. Pharm. Sci.* **92**(5):1028–1036, 2003.
102. Heimbach, T., Oh, D.-M., Li, L. Y., et al. Enzyme-mediated precipitation of parent drugs from their phosphate prodrugs. *Int. J. Pharm.* **261**(1–2):81–92, 2003.

103. He, X., Sugawara, M., Kobayash, M., et al. An in vitro system for prediction of oral absorption of relatively water-soluble drugs and ester prodrugs. *Int. J. Pharm.* **263**(2):35–44, 2003.
104. Christrup, L. L., Christensen, B. C., Friis, J. G., and Jorgensen, A. Improvement of buccal delivery of morphine using the prodrug approach. *Int. J. Pharm.* **77**(2): 157–165, 1997.
105. Weyel, D., Sedlacek, H.-H., Muller, R., and Brusselbach, S. Secreted human β-glucuronidase: A novel tool for gene directed prodrug therapy. *Gene Ther.* **7**:224–231, 2000.
106. Springer, C. J., and Niculescu-Duvaz, I. Prodrug-activating systems in suicide gene therapy. *J. Clin. Invest.* **105**:1161–1167, 2000.
107. Hamstra, D. A., and Rehemtulla, A. Toward an enzyme/prodrug strategy for cancer gene therapy: Endogenous activation of carboxypeptidase A mutants by PACE/furin family of propeptidases. *Hum. Gene Ther.* **10**:235–248, 1999.
108. Bagshawe, K. D. ADEPT and related concepts. *Cell. Biophys.* **24–25**:83–91, 1994.
109. Bagshawe, K. D. Antibody directed enzymes revive anticancer prodrugs concept. *Br. J. Cancer* **56**:531–532, 1987.
110. Senter, P. D., Saulnier, M. G., Screiber, G. J., et al. Anti-tumor effects of antibody-alkaline phosphatase conjugates in combination with etoposide phosphate. *Proc. Natl. Acad. Sci. USA* **85**:4842–4846, 1988.
111. Bakina, E., and Farquhar, D. Intensely cytotoxic anthracycline prodrugs: Galactosides. *Anticancer Drug Des.* **14**:507–515, 1999.
112. Danks, M. K., and Potter, P. M. Enzyme-prodrug systems: Carboxylesterase/CPT-11. *Methods Mol. Med.* **90**:247–262, 2004.
113. Kan, O., Kingsman, S., and Naylor, S. Cytochrome P450–based cancer gene therapy: Current status. *Exp. Opin. Biol. Ther.* **2**:857–868, 2002.
114. Brown, N. L., and Lemoine, N. R. Clinical trials with GDEPT: Cytosine deaminase and 5-fluorocytosine. *Methods Mol. Med.* **90**:451–458, 2004.
115. Fillat, C., Carrio, M., Cascante, A., and Sangro, B. Suicide gene therapy mediated by the herpes simplex virus thymidine kinase gene/ganciclovir system: Fifteen years of application. *Curr. Gene. Ther.* **3**:13–26, 2003.
116. Bagshawe, K. D. Tumor therapy, patent no. WO9824478 (1998).
117. Mehvar, R. Recent trends in the use of polysaccharides for improved delivery of therapeutic agents: Pharmacokinetic and pharmacodynamic perspectives. *Curr. Pharm. Biotechnol.* **4**:283–302, 2003.
118. Alexander, C. Synthetic polymers in drug delivery. *Exp. Opin. Emerg. Drugs* **6**(2):345–363, 2001.
119. Duncan, R. The dawning era of polymer therapeutics. *Nature Rev. Drug Discov.* **2**:347–360, 2003.
120. Modi, S., Jain, P. J., and Kumar, N. Polymer-drug conjugates: Recent development for anticancer drugs. *CRIPS* **5**:2–8, 2004.
121. Robinson, M. A., Charlton, T. S., Garnier, P., et al. LEAPT: Lectin-directed enzyme-activated prodrug therapy. *Proc. Natl. Acad. Sci. USA* **101**(40):14527–14532, 2004.
122. Pachene, M. J., Belinka, B., and Simon, P. Novel methods for site-directed drug delivery. *Drug Deliv. Technol.* **3**(1):40–45, 2003.
123. Gillies, R. E., and Frechet, M. J. J. Dendrimers and dendritic polymers in drug delivery. *Drug Deliv. Technol.* **10**(1):35–43, 2005.
124. Malik, N., Evagorou, E. G., and Duncan, R. Dendrimer-platinate: A novel approach to cancer chemotherapy. *Anticancer Drugs* **10**:767–776, 1999.
125. D'Emanuele, A., Jevprasesphant, R., Penny, J., and Attwood, D. The use of a dendrimer-propranolol prodrug to bypass efflux transporters and enhance oral bioavailability. *J. Control. Release* **95**(3):447–453, 2004.
126. Erion, M. D., van Poelje, P. D., MacKenna, D. A., et al. Liver-targeted drug delivery using HepDirect prodrugs. *J. Pharmacol. Exp. Ther.* **312**:554–560, 2005.

Chapter

4

Diffusion-Controlled Drug Delivery Systems

Puchun Liu

Emisphere Technologies, Inc.
Tarrytown, New York

Tzuchi "Rob" Ju

Yihong Qiu

Abbott Laboratories
North Chicago, Illinois

Two basic types of controlled-delivery dosage forms have been designed in which diffusion is the rate-limiting step to generate temporal input profiles for drug delivery: matrix- and reservoir-type systems. A *matrix-type system* consists of a rate-controlling ingredient such as a polymer with drug uniformly dissolved or dispersed in it, and typically, a half-order drug release corresponds to desorption from the preloaded matrix. A *reservoir-type system* separates a drug compartment from a polymer membrane that presents a diffusional barrier to yield drug flux of either zero order (with infinite dose) or first order (by dose depletion). Osmotically controlled systems are a subset of diffusion-controlled systems and often are classified separately (see Chap. 6).

Design of diffusion-controlled systems centers around understanding of the relationship between tailoring physical or polymer chemistry, release of drug from the dosage form, and membrane permeation. Contributions to the drug delivery profile include not only the dosage form itself but also absorption across the relevant biological membrane into the systemic circulation. Transport of a great majority of drugs across the biomembrane for the selected route of access into the body is also governed by diffusion. In particular, when the biomembrane barrier (e.g., skin) is rate limiting, the membrane permeation is often the central interest.

4.1 Diffusion Theory

Diffusion can be defined as a process by which molecules transfer spontaneously from one region to another in such a way as to equalize chemical potential or thermodynamic activity. Although diffusion is a result of random molecular motion, with a wide spectrum of physicochemical properties occurring in various conditions and situations, the diffusion process can be abstracted to a simple system involving molecules of interest, a diffusional barrier, and a concentration gradient. The migrating molecules are termed *diffusants* (also called *permeants* or *penetrants*). The membrane or matrix in which the diffusant migrates is called the *diffusional barrier*. The external phase is called the *medium*. The concentration gradient or profile of the diffusant within the diffusional barrier is the driving force for diffusion. The mathematics of diffusion are discussed briefly in this section, with emphasis on both diffusion across a barrier membrane and diffusional release from a preloaded matrix.[1–4]

4.1.1 Basic equations of diffusion

The mathematical form of the equations describing mass diffusion can be rationalized on the basis of thermodynamic arguments and statistical arguments or by crude analogy to other physical systems such as simple electric circuits. In truth, the basic equations were put forth by Fick in 1855 as an analogy to the heat-conduction equation developed by Fourier in 1822. The theory of diffusion in isotropic substances therefore is based on the hypothesis that the flux J or rate of diffusion (amount Q_t in time t) through a unit area of a barrier section is proportional to the concentration gradient within and normal to the section; that is,

$$J = \frac{dQ_t}{dt} = -D\frac{dC}{dx} \tag{4.1}$$

This is Fick's first law, with the proportionality constant D termed *diffusivity* or *diffusion coefficient.* The negative sign arises because the direction of molecular movement is opposite to the increase in the concentration.

By consideration of the mass balance in a small element of space with a constant diffusivity D, we can obtain Fick's second law in general:

$$\frac{\delta C}{\delta t} = \frac{\delta}{\delta x}\left(D\frac{\delta C}{\delta x}\right) \tag{4.2}$$

This time-dependent partial differential equation relates the change in concentration at any point as a function of time to the change in the concentration gradient with respect to position. This important equation is the starting point for a large amount of literature in mass transfer that deals primarily with solving this equation subject to various initial and boundary conditions.

Although the simple one-dimensional problem is considered here, a more general three-dimensional statement of Fick's second law (Eq. 4.2) is

$$\frac{\delta C}{\delta t} = \frac{\delta}{\delta x}\left(D\frac{\delta C}{\delta x}\right) + \frac{\delta}{\delta y}\left(D\frac{\delta C}{\delta y}\right) + \frac{\delta}{\delta z}\left(D\frac{\delta C}{\delta z}\right) \tag{4.3}$$

For a barrier membrane (thickness h) that is sandwiched between external media (both donor and receiver sides), the surface concentrations within the barrier ordinarily are not known but can be replaced by the membrane-to-medium partition coefficient K multiplied by the concentrations in the medium of donor side C_d or receiver side C_r. In some cases it is not possible to determine D, K, or h independently and thereby to calculate a hybrid constant, permeability or permeability coefficient P. With these considerations, Fick's first law (Eq. 4.1) may be written

$$J = P(C_d - C_r) = \frac{DK}{h}(C_d - C_r) \tag{4.4}$$

When the barrier has two or more independent diffusional pathways present in parallel, the total permeability coefficient P_{total} is the sum of P for each individual pathway as

$$P_{total} = \sum_{i=1}^{i=n} P_i = P_1 + P_2 + \cdots + P_n \tag{4.5}$$

On the other hand, when the barrier consists of two or more physically distinct layers in series, the reciprocal of P_{total} is the sum of reciprocal P for each individual layer as

$$\frac{1}{P_{total}} = \sum_{i=1}^{i=n} \frac{1}{P_i} = \frac{1}{P_1} + \frac{1}{P_2} + \cdots + \frac{1}{P_n} \tag{4.6}$$

With regard to Eq. (4.2), the general diffusion equation describing three standard geometries (slab sheet, cylinder, and sphere) is

$$\frac{\delta C}{\delta t} = x^{-\alpha} \frac{\delta}{\delta x}\left(x^{\alpha} D \frac{\delta C}{\delta x}\right) \tag{4.7}$$

where α is the half-thickness or radius ($x = 0$ at the center and $x = \alpha$ at the surface), and $\alpha = 0$, 1, and 2 for planar, cylindrical, and spherical geometry, respectively. The initial and boundary conditions needed to solve Eq. (4.7) vary depending on the type of system under consideration.

4.1.2 Diffusional release from a preloaded matrix

As diffusional release from the preloaded matrix to external receiver medium, drug desorption out from a matrix involves a diffusional moving boundary.[5] Here we consider the simplest situation of only one moving diffusion front with the time-dependent position *R(t)* in a rigid matrix slab (thickness h) without swelling or erosion of the matrix surface. A perfect skin condition in the receiver medium is also assumed. A dispersed matrix, i.e., drug loading per unit volume A greater than the drug solubility C_s in the matrix, is of particular interest for controlled drug release ($A > C_s$). Under these conditions, the initial and boundary conditions for Fick's second law (Eq. 4.2) are

$$R(0) = h \quad C(h, t) = 0 \quad C(R, t) = C_s \qquad \text{and} \qquad D\frac{\delta C(x, t)}{\delta x}\bigg|_{x=R(t)} = (A - C_s)\frac{dR}{dt} \tag{4.8}$$

The only exact solution is the one available for the semi-infinite slab geometry[6]:

$$C(x,t) = C_s \frac{\operatorname{erf}\left(\dfrac{x}{2\sqrt{Dt}}\right)}{\operatorname{erf}(\xi)}$$

where erf is the error function, and

$$\xi = \frac{R}{2\sqrt{Dt}} \tag{4.9}$$

The expression for Q_t can be shown to be

$$Q_t = \frac{2C_s}{\operatorname{erf}(\xi)}\sqrt{\frac{Dt}{\pi}}$$

where

$$\sqrt{\pi}\xi \exp(\xi^2) \operatorname{erf}(\xi) = \frac{C_s}{(A - C_s)} \tag{4.10}$$

Several useful approximate analytical solutions to Eq. (4.8) were developed. A well-known example is Higuchi's equation, based on a pseudo-steady-state approach[7]:

$$Q_t = \sqrt{C_s(2A - C_s)Dt} \tag{4.11}$$

A more accurate approximation was presented by Lee[8]:

$$Q_t = \left(\frac{1+H}{\sqrt{3H}}\right)\sqrt{Dt}$$

where

$$H = 5\left(\frac{A}{C_s}\right) - 4 + \sqrt{\left(\frac{A}{C_s}\right) - 1} \tag{4.12}$$

Both treatments assume that the dissolution of drug is rapid compared with the diffusion of drug, and both predict release of drug that is linear with $\sqrt{t}$. This is a consequence of the increased diffusional distance and decreased diffusional area at the diffusion front as drug release proceeds. Figure 4.1 illustrates the amount released Q_t versus square-root-time $\sqrt{t}$ plots for two cases: both dispersed (loading greater than saturation,

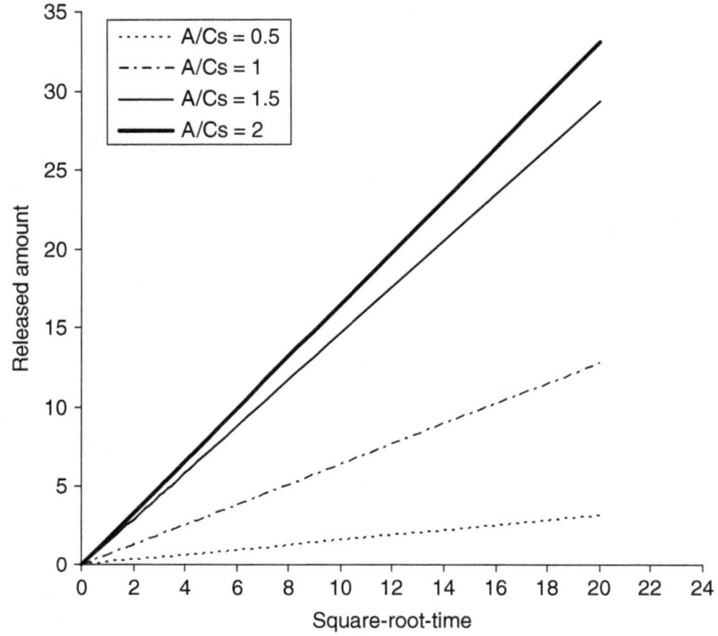

Figure 4.1 Released amount Q_t versus square-root-time $\sqrt{t}$ plots. Illustration of loading less than or equal to saturation (dispersed, $A \leq C_s$) and greater than saturation (dissolved, $A > C_s$) in a matrix-type diffusion-controlled drug delivery system.

$A > C_s$) and dissolved (loading equal to or less than saturation, $A \leq C_s$). For the latter, solutions to Fick's second law (Eq. 4.2) are well known, and the particular expression for semi-infinite geometry is

$$Q_t = 2A\sqrt{\frac{Dt}{\pi}} \tag{4.13}$$

This is exactly the same result from Eq. (4.10) in the limit of $A \to C_s$.

In order to achieve near-zero-order release from the matrix, a unique geometry, a specific nonuniform initial concentration profile, and/or a combined diffusion/erosion/swelling mechanism provide theoretical basis for such an approach.

4.1.3 Diffusion across a barrier membrane

For diffusion through a homogeneous membrane (thickness h) that is sandwiched between external media, an infinite reservoir frequently is assumed in the donor side and a perfect sink in the receiver side. There are two different initial conditions, which give different initial diffusion rates of either lower or higher than steady-state values.

The first "initial condition" is no presence of drug in the membrane, corresponding to a delivery system that is used immediately after manufacture. Mathematically, this problem is stated as Fick's second law (Eq. 4.2) with the following boundary and initial conditions:

$$C(0, t) = C_d \qquad C(h, t) = C_r = 0 \qquad \text{and} \qquad C(x, 0) = C_0 = 0 \tag{4.14}$$

The solution is[1]

$$C(x, t) = C_d\left[1 - \frac{x}{h} - \frac{2}{\pi}\sum_{m=1}^{\infty}\frac{1}{m}\sin\left(\frac{m\pi x}{h}\right)\exp\left(-\frac{Dm^2\pi^2 t}{h^2}\right)\right] \tag{4.15}$$

Then Q_t can be obtained as

$$Q_t = \frac{DC_d}{h}\left[t - \frac{h^2}{6D} - \frac{2h^2}{D\pi^2}\sum_{m=1}^{\infty}\frac{(-1)^m}{m^2}\exp\left(-\frac{Dm^2\pi^2 t}{h^2}\right)\right] \tag{4.16}$$

When $t \to \infty$, Q_t approaches the following asymptotic relation:

$$Q_t = \frac{DC_d}{h}\left(t - \frac{h^2}{6D}\right) \tag{4.17}$$

The intercept on the t axis is the *diffusion lag time* t_L, which is given by

$$t_L = \frac{h^2}{6D} \tag{4.18}$$

The second "initial condition," however, is for a system that has been stored for some time, and the drug saturates the entire membrane. Mathematically, there is a nonzero uniform initial drug concentration C_0 distribution within the membrane. Similarly, Fick's second law (Eq. 4.2) can be solved with the following boundary and initial conditions:

$$C(0, t) = C_d \qquad C(h, t) = C_r = 0 \qquad \text{and} \qquad C(x, 0) = C_0 = C_d \tag{4.19}$$

As the solution, $C(x, t)$ and then Q_t are obtained as

$$C(x, t) = C_d\left\{1 - \frac{x}{h} - \frac{2}{\pi}\sum_{m=1}^{\infty}\frac{1}{m}\sin\left(\frac{m\pi x}{h}\right)\exp\left(-\frac{Dm^2\pi^2 t}{h^2}\right)\right.$$
$$\left. + \frac{4}{\pi}\sum_{n=0}^{\infty}\frac{1}{2n+1}\sin\left[\frac{(2n+1)\pi x}{h}\right]\exp\left[-\frac{D(2n+1)^2\pi^2 t}{h^2}\right]\right\} \tag{4.20}$$

$$Q_t = \frac{DC_d}{h}\left\{t + \frac{h^2}{3D} - \frac{2h^2}{D\pi^2}\sum_{m=1}^{\infty}\frac{(-1)^m}{m^2}\exp\left(-\frac{Dm^2\pi^2 t}{h^2}\right)\right.$$
$$\left. + \frac{4h^2}{D\pi^2}\sum_{n=0}^{\infty}\exp\left[-\frac{D(2n+1)^2\pi^2 t}{h^2}\right]\right\} \qquad (4.21)$$

As $t \to \infty$, Q_t versus t becomes linear as

$$Q_t = \frac{DC_d}{h}\left(t + \frac{h^2}{3D}\right) \qquad (4.22)$$

The *burst time* t_B is defined as the intercept of Q_t on the t axis:

$$t_B = -\frac{h^2}{3D} \qquad (4.23)$$

For an ideal membrane, t_L and t_B emphasize the path length and the diffusion constant. Binding and heterogeneity of the membrane may complicate these simple relationships. In practice, as shown in Fig. 4.2, membrane reservoir systems have a period of constant release, i.e., steady-

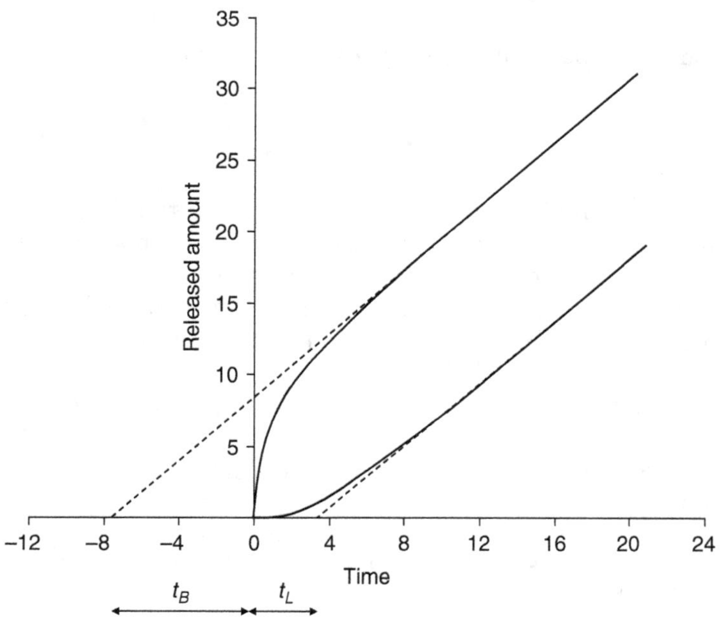

Figure 4.2 Released amount Q_t versus time t plots. Illustration of time lag t_L and burst effect t_B in a reservoir-type diffusion-controlled drug delivery system.

state permeation, preceded by a period of either an increasing (burst) or decreasing (lag time) flux. This initial period may affect the time of appearance of a drug in plasma on the first dose but may become insignificant on multiple dosing. It is noteworthy that the above-mentioned constant or zero-order release is based on the assumption of a constant C_d. If excess solid drug is not present in the reservoir, the thermodynamic activity decreases as the drug diffuses out of the device, the release rate falls exponentially, and the process is referred to as *first-order release*.

4.2 Oral Diffusion-Controlled Systems

Over the past decade we have witnessed the wide spread and availability of a plethora of oral controlled release (CR) products in the marketplace. For example, by 1998, the U.S. Food and Drug Administration (FDA) approved 90 oral CR products for marketing. From 1998 to 2003, in just five years, the FDA approved an additional 29 new drug applications that used CR technologies.[9] Consequently, oral CR technologies are becoming more complex and encompassing multiple presentations. The basic concepts of oral CR systems have been reviewed thoroughly in the literature.[10–14] It is well recognized in the pharmaceutical industry that oral CR dosage forms can be defined based on release-profile characteristic or the underlying release- controlling mechanism.

Two distinct drug release profiles, extended and delayed release, are achievable, and they can be used in various combinations to provide the desired release rate. Three delivery systems dominate today's market of oral CR products: matrix, reservoir, and osmotic systems. Release mechanisms from these dosage forms have been the subjects of extensive studies.[7,13–23] Among them, diffusion plays a key role in both matrix and reservoir systems, whereas osmotic pressure is the predominant mechanism of drug release from osmotic systems and could also play a role in a reservoir system. Owing to technology accessibility, manufacturing, cost, and other considerations, diffusion-based CR products are used more widely than osmotic systems. For example, of the 29 CR products approved by the FDA between 1998 and 2003, 12 were based on matrix systems and 10 based on reservoir technologies compared with 2 osmotic tablets. Oral diffusion-controlled systems are discussed in this section.

4.2.1 Matrix systems

A matrix system consists of active and inactive ingredients that are homogeneously mixed in the dosage form. It is by far the most commonly used oral CR technology, and the popularity of matrix systems can be attributed to several factors. First, unlike reservoir and osmotic systems, products based on matrix design can be manufactured using conventional processing and equipment. Second, development time and cost associated with a matrix system generally are viewed as favorable, and no additional capital investment is required. Lastly, a matrix system is capable of accommodating both low and high drug load and active ingredients with a wide range of physical and chemical properties.

As with any technology, matrix systems come with certain limitations. First, matrix systems lack flexibility in adjusting to constantly changing dosage levels, as required by clinical study outcome. When a new dosage strength is deemed necessary, more often than not a new formulation and thus additional resources are expected. Furthermore, for some products that require unique release profiles (e.g., dual release or delayed plus extended release), more complex matrix-based technologies such as layered tablets (e.g., Allegra D) will be required.[24] Nevertheless, we expect continued popularity of matrix systems because they have demonstrated success across a wide range of product profiles. To further this discussion, we divide matrix systems into two categories, hydrophobic and hydrophilic systems, based on rate-controlling materials.

Hydrophobic matrix systems. This is the only system where use of a polymer is not essential to provide controlled drug release, although insoluble polymers have been used. As the term suggests, the primary rate-controlling components of a hydrophobic matrix are water insoluble in nature. These ingredients include waxes, glycerides, fatty acids, and polymeric materials such as ethylcellulose and methacrylate copolymers. To modulate drug release, it may be necessary to incorporate soluble ingredients such as lactose into the formulation.

The presence of insoluble ingredients in the formulations helps to maintain the physical dimension of a hydrophobic matrix during drug release. As such, diffusion of the active form from the system is the release mechanism, and the corresponding release characteristic can be described by the Higuchi equation (Eq. 4.11). Very often, pores form within a hydrophobic matrix as a result of the release of the active ingredient. For a porous monolithic system, Eq. (4.11) can be further modified as[18]

$$Q_t = A\sqrt{\tau \frac{C_s}{\rho}\left(2 - \frac{C_s}{\rho}\right)Dt} \tag{4.24}$$

where τ and ρ are the tortuosity of the matrix and density of drug particles, respectively. Tortuosity is introduced to account for an increase in diffusion path length owing to branching and bending of the pores. Similar to Eq. (4.11), the square-root-of-time release profile is expected with a porous monolith, whereas the release from such system is proportional to drug loading. In addition, hydrophobic matrix systems generally are not suitable for insoluble drugs because the concentration gradient is too low to render adequate drug release. As such, depending on actual ingredients' properties or formulation design, incomplete drug release within the gastrointestinal (GI) transit time is a potential risk and needs to be delineated during development.

With the growing need for optimization of therapy, matrix systems providing programmable rates of delivery become more important.[5] Constant-rate delivery always has been one of the primary targets of controlled release systems, especially for drugs with a narrow therapeutic index. Over the past 40 years, considerable effort has been and continues to be expended in the development of new delivery concepts in order to achieve zero-order or near-zero-order release. Examples of altering the kinetics of drug release from the inherent nonlinear behavior include the use of geometrical factors (cone shape, biconcave, donut shape, hemisphere with cavity, core in cup, etc.), erosion/dissolution control and swelling control mechanisms, nonuniform drug loading, and matrix-membrane combinations.[25–36] Some of the systems are difficult or impractical to manufacture.

Hydrophilic matrix systems. The primary rate-controlling ingredients of a hydrophilic matrix are polymers that would swell on contact with the aqueous solution and form a gel layer on the surface of the system. Robust swelling/gelling properties and straightforward manufacturing processes are to a large degree responsible for the versatility and performance of the system. As such, hydrophilic matrices dominate today's market of oral CR products.

Hydroxypropyl methylcellulose (HPMC) is the most commonly used hydrophilic polymer. Other polymers include high-molecular-weight polyethylene oxide (Polyox™), hydroxypropyl cellulose (HPC), hydroxyethyl cellulose (HEC), xantham gum, sodium alginate, and polyacrylic acid (Carbopol™). It is well recognized that key formulation variables are matrix dimension and shape, polymer level and molecular weight, and drug load and solubility. Other factors such as tablet hardness, type of inactive ingredients, and processing normally play secondary roles. Choice of manufacturing processes such as direct blending or granulation typically does not affect product performance significantly, although exceptions do exist. Thus, in general, processing and scale-up associated with hydrophilic matrices are more robust than other CR systems.

Extensive studies haven been conducted and significant progress made toward fundamental understanding of drug release from hydrophilic matrices. In summary, polymer dissolution (erosion) and diffusion of drug molecules across the gel layer have been identified as the rate-controlling mechanisms. Unlike a pure diffusion-controlled system, the dual release processes make hydrophilic matrices more suitable for insoluble molecules than other diffusion-controlled systems. That is, through formulation design, polymer erosion can be modulated to further aid release control of insoluble compounds.

Two models that have been established to delineate these dual release mechanisms are highlighted here. Interested readers are encouraged to review the extensive literature available to comprehend the underlying drug release process.[16,17,19–24] Among all the models developed, the semi-empirical "exponent equation" has been used widely to differentiate the contributions of both mechanisms[16,17]:

$$Q_t = kt^n \tag{4.25}$$

where n is a diffusional exponent, and k is a kinetic constant. If diffusion dominates polymer erosion, the value of n would approach 0.5. On the other hand, for erosion-controlled formulations, n would approach the value of unity. Under an "analamous" condition, the value of n falls in between 0.5 and 1 when both diffusion and erosion play roles. Figure 4.3 depicts release profiles under various conditions.

More recently, a "spaghetti" model for a swollen matrix was developed to provide mechanistic understanding of the complex release process (Fig. 4.4). This model treats polymer erosion as diffusion of polymer across a "diffusion layer" adjacent to the gel layer.[19,20] Thus two competitive diffusional processes contribute to overall drug release: diffusion of polymer across the diffusion layer and diffusion of drug across the gel layer. Two parameters have been identified to characterize their relative contributions. Polymer disentanglement concentration $C_{p,\text{dis}}$ gauges the

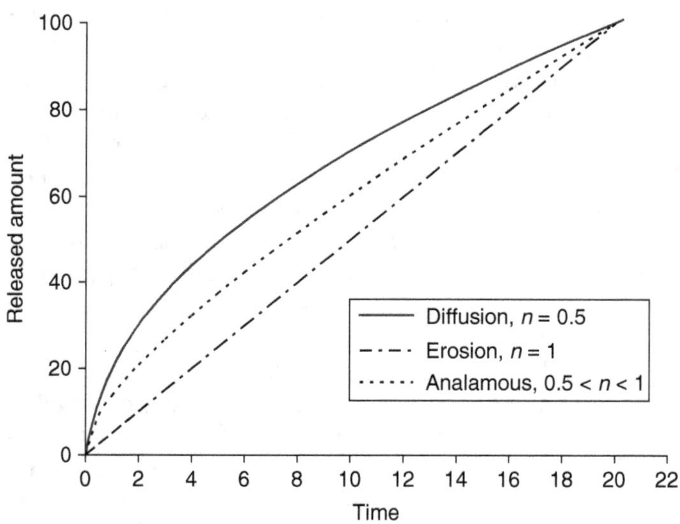

Figure 4.3 Released amount Q_t versus time t plots. Three characteristic drug release profiles from hydrophilic matrices, diffusional, erosional, and analamous.

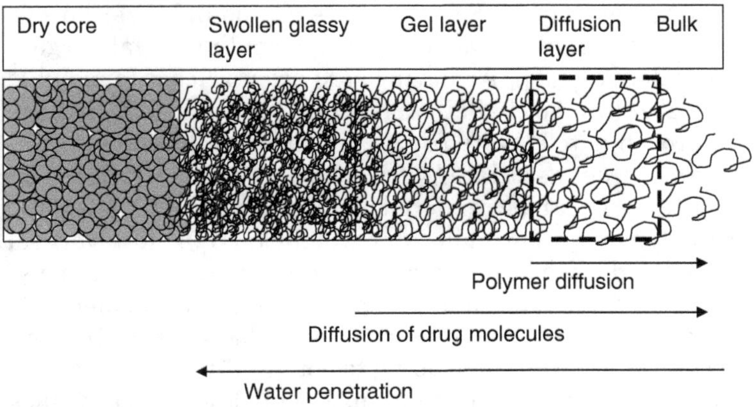

Figure 4.4 "Spaghetti" model for a swollen matrix.

contribution of polymer diffusion/dissolution, whereas solubility of drug molecules defines the diffusion component. The impact of formulation is largely reflected through an equivalent $C_{p,\text{dis}}$. These parameters are defined as

$$(C_{p,\text{dis}})_{\text{eq}} = 0.05\left(\frac{MW_p X_p}{96{,}000}\right)^{-0.8} \quad (\text{g/mL}) \tag{4.26}$$

$$M_p \approx kt^1 \tag{4.27}$$

where MW_p and X_p denote the molecular weight and weight fraction of polymer in the formulation, respectively, and M_p is the polymer release profile. Interestingly, the release profile of polymer (Eq. 4.27) resembles that for the erosion-controlled system of Eq. (4.25). Note that $C_{p,\text{dis}}$ is the intrinsic value of pure polymer, whereas $(C_{p,\text{dis}})_{\text{eq}}$ is an "equivalent" $C_{p,\text{dis}}$ of polymer in a formulation. One can consider $C_{p,\text{dis}}$ as the equivalent "solubility" of polymer because it defines the concentration at which a polymer detaches from the parent matrix. Therefore, conceptually, the relative contribution of both mechanisms can be characterized as the ration of $C_s/(C_{p,\text{dis}})_{\text{eq}}$:

If $C_s/(C_{p,\text{dis}})_{\text{eq}} >> 1$, diffusion-controlled, then

$$Q_t = kt^{0.5} \tag{4.28}$$

If $C_s/(C_{p,\text{dis}})_{\text{eq}} << 1$, erosion controlled, then

$$Q_t = kt^1 \tag{4.29}$$

Equations (4.28) and (4.29) suggest that release profiles from a hydrophilic matrix would vary significantly with formulation and solubility of drug molecules. For insoluble compounds, zero-order release is readily attainable because the corresponding C_s values generally are lower than $(C_{p,\mathrm{dis}})_{\mathrm{eq}}$. Achieving zero-order release for soluble molecules, however, can be challenging owing to their high C_s values. A common approach is to elevate $(C_{p,\mathrm{dis}})_{\mathrm{eq}}$ in the formulation to counteract the high C_s values associated with soluble compounds. Based on Eq. (4.26), high $(C_{p,\mathrm{dis}})_{\mathrm{eq}}$ can be rendered by reducing polymer level (X_p) or employing polymers of low molecular weight (MW_p). However, gel strength within the dosage form can be compromised when fewer polymers or polymers of low molecular weight are used. Weakened gels tend to be more sensitive to environmental variables in the GI tract, such as shear and ionic strength, leading to potentially less robust performance. On the other hand, formulations having low $(C_{p,\mathrm{dis}})_{\mathrm{eq}}$ values are likely to have strong gels and robust in vivo performance. Yet such formulations with low $(C_{p,\mathrm{dis}})_{\mathrm{eq}}$ values would exhibit diffusion-like release. To overcome the dilemma between drug release profile and gel strength, technologies based on the modulation of surface area and shape have found success.[26–28,37]

While remedies do exist, formulating hydrophilic matrices for active ingredients with extreme solubility profiles could be demanding. For very soluble compounds, diffusion of drug molecules is the dominant mechanism of release, and the role of polymer erosion is limited in modulating drug release. Thus, developing a hydrophilic matrix for highly soluble drugs that requires prolonged release (e.g., >12 h) can be challenging. On the other hand, release of less soluble drugs from hydrophilic matrices is expected to be slow because both polymer dissolution and drug diffusion play key roles. This may not be a major problem as long as drug molecules dissolve before polymers erode from the dosage form. However, for highly insoluble compounds, drug particles may not dissolve completely after polymers have eroded. Accordingly, dissolution of drug particles contributes to the overall drug release and needs to be considered during development. Lastly, for active ingredients with solubilities that vary with pH, it is not likely that pH-independent release can be achieved even if the rate-controlling polymer is pH independent. Recent progress to overcome these deficiencies is discussed in Sec. 4.2.3.

4.2.2 Reservoir systems

A typical reservoir system consists of a core (the reservoir) and a coating membrane (the diffusion barrier). The core contains the active ingredients and excipients, whereas the membrane is made primarily of rate-controlling polymer(s). The governing release mechanism is diffusion from the reservoir across the membrane to the bulk solution, and the one-dimensional release rate is described by Eqs. (4.4), (4.17), and (4.22).[10,14] In addition, osmotic pressure could be operative in certain cases, especially for highly soluble drug molecules.

Release profile from a reservoir system depends on both formulation and solubility of drug molecules. In order to maintain zero-order release, all the parameters in Eqs. (4.17) and (4.22) must remain constant. This is possible with soluble compounds when

$$C_d - C_r \approx C_d \qquad \text{and} \qquad Q_t = kt \tag{4.30}$$

For insoluble drugs, the values of C_d can be too low to render adequate and complete drug release. In addition, reduced release is observed often as drug is depleted over time. Similar to matrix systems, developing a reservoir system with pH-independent release is not straightforward unless the solubility of drug molecules is pH independent.

A reservoir system normally contains many coated units (particulates) such as beads, pellets, and minitablets. Unlike single-unit tablets, the number of particulates in a reservoir system often is sufficient to minimize or eliminate the impact of any coating defect associated with a limited number of units. Another attractive feature of reservoir systems is that tailored drug release can be obtained readily by combining particulates of different release rates. An increasing number of products (e.g., Metadate CD and Ritaline LA) have been introduced using such a concept. Lastly, reservoir systems offer the flexibility of adjusting to varying dosage strengths without the need of new formulations. This is highly desirable during clinical development programs, where dose levels frequently are revised based on study outcome.

Key variables for a reservoir system are levels of polymer and pore former in the film coat, drug load, and solubility.[38] The most commonly used materials for constructing the membrane are ethylcellulose (Surelease™ or Aquacoat™) and acrylic copolymers (Eudragit™ RL30D, RS 30D, and NE 30D). Water-soluble polymers such as HPMC and polyethylene glycol (PEG) are employed as pore formers. Typically, special coating equipment such as the Wuster coater is required to apply the coating material uniformly.[39–41] Scale-up of such processing can be challenging and may require changes in formulations between scales in order to maintain similar release characteristics. In addition, it has been recognized that dissolution of certain reservoir system–based products may change on storage. One way to minimize this problem is adding a curing step at the end of the coating process.

4.2.3 Current challenges and future trends

While tremendous progress has been made in both scientific understanding and product development of oral controlled release systems, deficiencies still exist, and challenges lie ahead. In the following section we attempt to highlight some recent development designed to overcome these challenges. Since new findings are evolving constantly, interested readers should search the literature regularly to stay updated.

pH-Independent drug release. It is perceived that pH-independent release profiles could lead to more robust product performance such as decreased food effect. This has been claimed to be one of the reasons behind the lack of food effect observed with certain osmotic systems. While it has some merit, other physiological and biopharmaceutical factors (e.g., GI transit time, ionic strength, secretion of bile salts, regional permeability, GI tract metabolism, transporters, etc.) could further complicate the food effect of a formulation. For example, both matrix and reservoir systems of theophylline were developed that showed pH-independent release, but significant and opposite food effects were observed for the two different formulations.[42]

Use of buffering agents is commonly resorted to render constant local pH within a dosage form and thus pH-independent release profiles. This can be illustrated in both matrix and reservoir systems such as Cardizem CD.[43] However, most buffering agents are soluble small molecules with limited loading and buffering capacity. They are released from the system, losing their effectiveness over time. Recently, the utility of ionic polymers such as Eudragit was demonstrated with a matrix system.[44,45] One also needs to exercise caution when using buffering agents because they may interact with either drug molecules or rate-controlling polymers, resulting in undesired outcomes, such as slow drug release or disruption of gel structure.[46]

Solubility enhancement. Developing CR formulations of poorly soluble drugs could be challenging at time, yet there are some benefits. For moderately insoluble compounds, the corresponding release of drug molecules was found to be similar to that of HPMC.[47] As discussed earlier, it can be straightforward to develop hydrophilic matrices for such molecules to achieve zero-order release because polymer release now can be calculated accurately based on the spaghetti model.

Nevertheless, solubilization still could be required for very insoluble molecules. Several approaches have been reported. Use of cyclodextrin as a complexation agent was shown to be effective and yielded both accelerated and pH-independent release.[48] Further, in situ complexation between drug molecules and cyclodextrin in the matrix was encouraging because it would not require organic processing to render such complexation prior to matrix preparation. Recently, other approaches, such as incorporating a self-emulsifying formulation into a hydrophilic matrix, were reported, whereas the science behind them was not understood yet.[49]

Soluble drugs. While soluble drugs are desirable in the development of conventional immediate-release dosage forms, it can be challenging to retard the release of such molecules. Several approaches that showed promise for very soluble molecules have been reported. First, reservoir systems were able to retard the release of a highly soluble compound more than a hydrophilic matrix.[50] This was attributed to the effectiveness of the membrane in containing drug release. In addition, layered matrix systems consisting of a hydrophobic middle layer and press-coated hydrophilic and/or hydrophobic barrier layer(s) were developed and rendered zero-order release for soluble molecules.[24] The Geometrix™ systems offer similar advantages. More recently, use of starch 1500 in an HPMC-based matrix has resulted in significant retardation of the release of a soluble drug molecule.[51] The mechanism behind the utility of starch 1500 in the HPMC matrix is not fully understood and deserves further investigation. Lastly, hydrophobic matrices can be useful because they retard water penetration into the system, leading to slower drug release.

Robust gels. Maintaining adequate gel strength is a prerequisite to developing a robust hydrophilic matrix. However, as discussed earlier, there is a tradeoff between strong gels and zero-order release. Recently, a self-correcting HPMC-based matrix having strong gels was developed and showed insensitivity to both pH and stirring condition.[52,53] This self-correcting system was based on the use of high levels that achieved the unique product attributes. First, highly concentrated salts help to maintain local pH and thus pH-independent release. Second, electrolytes yield salted-out regions that are resistant to erosion and hydrodynamics. This system could also be designed to render zero-order drug release. As with any new technology, its in vivo performance remains to be tested.

4.3 Transdermal Diffusion-Controlled Systems

Transdermal delivery is a noninvasive "intravenous infusion" of drug to maintain efficacious drug levels in the body for predictable and extended duration of activity. Diffusion-controlled transdermal systems are designed to deliver the therapeutic agent at a controlled rate from the device to and through the skin into the systemic circulation. This route of administration avoids unwanted presystemic metabolism (first-pass effect) in the GI tract and the liver. Patient satisfaction has been realized through decreased side effects, reduced dosing frequency, and improved plasma profiles as compared with conventional oral dosing or painless administration as compared with injection therapy. In the last two decades, among the greatest successes in CR drug delivery is the commercialization of transdermal dosage forms for the systemic treatment of a variety of diseases. To date, nearly 20 drugs alone or in combination have been launched into transdermal products worldwide. Additional drugs are in the late development phases (phase II to registration). Table 4.1 lists components in the market and under development.

As with oral diffusion-controlled systems, there are two basic designs for transdermal diffusion-controlled systems: matrix-type and reservoir-type systems. The matrix-type systems can be further classified as

TABLE 4.1 Compounds in Transdermal Products: Marketed and to Be Marketed

Compounds	Therapeutic indications
Marketed	
Scopolamine	Prevention and treatment of motion sickness
Nitroglycerine Isosorbide dinitrate	Treatment of angina pectoris and second-line therapy in congestive heart failure
Clonidine	Antihypertensive
Estradiol (valerate) Estradiol + norethindronate (acetate) Estradiol (valerate) + levonorgestrel	Hormone-replacement therapy in estrogen-deficiency symptoms and prevention of osteoporosis in postmenopausal women
Nicotine Cytisine	Aid to smoking cessation
Fentanyl Buprenorphine	Chronic pain management in patients requiring opioid analgesia
Testosterone Dihydrotestosterone	Hormone-replacement therapy in hypogonadal males
Ethinyl estradiol + norelgestromin	Female contraceptives
Oxybutynin	Treatment of overactive bladder with symptoms of urge urinary incontinence, urgency, and frequency
Under registration or late clinical development (selective)	
Selegiline	Treatment of depression (Watson/Mylan)
Methylphenidate	Treatment of attention-deficit hyperactivity disorder (Noven/Shire)
Galantamine	Treatment of alcohol and nicotine dependence (HF Arzneimittelforschung)
Testosterone	Treatment of female sexual dysfunction (Watson/Procter & Gamble)

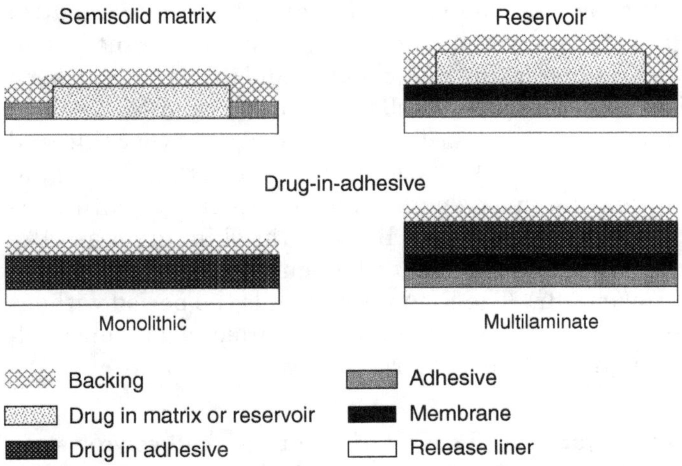

Figure 4.5 Transdermal diffusion-controlled drug delivery systems: four design configurations and their basic elements.

drug-in-adhesive and semisolid matrix systems (Fig. 4.5). Transdermal delivery systems are composed of specialty backing films such as nonwoven and woven materials, porous or nonporous membranes, pressure-sensitive adhesives (PSAs), and corresponding protective silicone- or fluoropolymer-coated release liners. For detailed reviews of transdermal drug delivery systems, readers are referred to several books.[54–58]

4.3.1 Drug-in-adhesive systems

The drug-in-adhesive transdermal product design is characterized by inclusion of the drug directly within a skin-coating adhesive.[59] In this system design, the adhesive fulfills the adhesion-to-skin function and serves as the formulation foundation containing the drug and all excipients. For this type of systems, the cumulative amount of drug released is proportional to the square root of time (Eqs. 4.10 through 4.12), and the activity of the drug in the adhesive may be decreasing constantly as the drug is released gradually. In practice, the burst effect owing to drug partitioned into the adhesive during storage may not be seen in the overall transdermal drug flux, which is mainly limited by skin permeation.

Three classes of PSAs used most widely in transdermal systems are polyisobutylene (PIB), polyacrylate, and polydimethylsiloxane (silicone). More recently, hydrophilic adhesive compositions, "hydrogels" composed of high-molecular-weight polyvinylpyrrolidon (PVP) and oligometric polyethylene oxide (PEO), have been shown to be compatible with a broad range of drugs and are used in several commercial products.[60]

Monolithic adhesive systems. The monolithic drug-in-adhesive matrix system is considered by many to be the state of the art in transdermal system design, with an increasing number of products introduced, such as Minitran, Nitro-Dur, Climara, Vivelle, Testoderm, and Nicotrol. In addition to the efficient use of surface area, this design is very thin and tends to be extremely conformable to the surface of the skin. For example, Minitran, only 200 μm thick, delivers nitroglycerin at a continuous rate of 0.1 mg/h with a 3.3-cm^2 system. Because the skin serves as rate-controlling barrier, Minitran has included an enhancer to increase the skin flux to 0.03 mg/(cm^2·hr). The FDA recommends the period without drug (8 to 12 h) to mitigate the possibility of the patient acquiring a tolerance to the antianginal effects of nitrate therapy.

Multilaminate adhesive systems. A subset of the drug-in-adhesive system design is the multilaminate adhesive system, which encompasses either the addition of a membrane between two distinct adhesive layers or the addition of multiple adhesive layers under a single backing film.[61] Deponit, Catapres-TTS, Transdermal-Scop, and Nicoderm belong to this category. For example, Nicoderm is a multilaminate containing an impermeable backing layer, a nicotine-containing ethyl vinyl acetate (EVA) drug layer, a polyethylene membrane, and an adhesive layer. It is available with delivery rates of 7, 14, and 21 mg/day over 24 hours.

4.3.2 Semisolid matrix systems

The semisolid matrix transdermal system design has the following characteristics: (1) There is no rate-controlling membrane layer, and a drug-containing gel is in direct contact with the skin, and (2) the component of the product responsible for skin adhesion is incorporated in an "overlay" and forms a concentric configuration around the semisolid drug reservoir.[62] The appearance of this system design is similar to that of the reservoir-type system, but drug release from the system is similar to that of the drug-in-adhesive system, where the rate of transdermal delivery is controlled by the skin. This design is the simplest among others and may offer an advantage in terms of excipient choice. However, separation of the adhesive from the semisolid matrix necessitates that the overall patch surface area extend well beyond the semisolid matrix.

Transdermal products with semisolid matrix design include Habitrol, Nitrodisc, Nitroglycerin Transdermal, and Prostep. For example, Habitrol consists of an impermeable backing laminate with a layer of adhesive and a nonwoven pad to which a nicotine solution is applied. Multiple layers of adhesive on a release liner are then laminated on the patch. The systems come in 10-, 20-, and 30-cm^2 sizes corresponding to 7, 14, and 21 mg/day, respectively, delivered over 24 hours.

4.3.3 Reservoir systems

The reservoir-type transdermal system design is characterized by the inclusion of a liquid reservoir compartment containing a drug solution or suspension that is separated from the release liner by a rate-limiting membrane and an adhesive. The vast majority of commercial transdermal products today that use the reservoir-type design have a face adhesive to maximize skin contact, i.e., a continuous adhesive layer between the rate-limiting membrane and the release liner.[63] The primary advantage of this configuration is the zero-order release kinetics of a properly designed system.[64] The membrane plays a critical role in drug release and in overall transdermal drug delivery. In certain cases, the membrane can de designed to also allow passage of one or more of the formulation excipients, which may be incorporated to function as skin penetration enhancers. The rate-controlling membranes used in most products include thin (1 to 3 mil) nonporous EVA or microporous polyethylene films. The diffusion rate is controlled by the polymer property and, as a result, the solubility and diffusivity of the drug in the polymer. Owing to the multiplicity of layers, this design is less conformable. In addition, this configuration is potentially vulnerable to dose dumping owing to rupture of the membrane.

Transdermal products with reservoir-type system design include Duragesic, Transderm-Nitro, Estraderm, and Androderm. For example, Duragesic has a form-fill-and-seal drug reservoir with an EVA membrane designed to deliver the opioid painkiller fentanyl transdermally for up to 72 hours. The reservoir contains an aqueous ethanolic solution of fentanyl to increase its skin permeation yet limit its solubility to minimize the drug contents of this abusable drug. When it is applied for a 24-hour period, the plasma levels increase until the patch is removed and then remain elevated because of a skin depot through 72 hours. Duragesic is available at mean delivery rates of 25, 50, 75, and 100 μg/h, corresponding to surface areas of 10, 20, 30, and 40 cm^2, respectively.

4.3.4 Current challenges and future trends

Despite the advantages of transdermal medication, the merits of each application have to be examined individually in terms of therapeutic rationale, market potential, and technical feasibility. The positive and negative effects need to be weighted carefully before large expenditures for developmental work are committed. Since human skin offers a formidable barrier to the entry of foreign substances, there are limitations, and these have prevented transdermal delivery from achieving its full potential. These limitations include inadequate drug skin permeation to achieve therapeutic effectiveness and skin irritation and sensitization caused by the drug and other excipients when they are placed in contact with the skin. The skin permeation problem can be improved by the use of chemical enhancers as well as physical enhancing methods. Unfortunately, when these enhancements are used at levels of intensity and/or duration adequate to provide a substantial increase in drug permeation, they also potentiate skin reactions.[65] Facing these challenges, future development will continue to address the improvement of both skin permeability and tolerability. Novel system designs are still directed at optimizing the rate and extent of drug delivery, stability, adhesive performance properties, skin tolerability, patentability, and pharmcodynamic profiles. The system design has been applied to topical (excluding dermatological) and transbuccal products.

Enhancement of skin permeation. Although the mechanisms of percutaneous absorption are complex, it is generally accepted that the drug diffusion pathway lies primarily in the lipoidal region within the seams of the cells of the stratum corneum. Thus drugs with low molecular weights, low melting points, and moderate oil and water solubility will permeate skin best. Hydrophilic compounds, however, may diffuse through the follicular appendages and other "pore" pathways with unpredictably large variations.[66] Chemical and physical enhancement of stratum corneum permeability is a major breakthrough for opening opportunities for drug candidates otherwise considered "unsuitable/unfeasible" for transdermal delivery. The common physical methods for facilitated transdermal delivery include electricity (iontophoresis and electroporation) and sound (sonophoresis).[67–69]

A *permeation enhancer* can be defined as a compound that alters the skin barrier function so that a desired drug can permeate at a faster rate. Dozens of enhancers are patented each year, and several books have been written summarizing the work and proposing mechanisms of enhancement.[70–72] The permeation enhancers may be classified simply as polar and nonpolar ones. They can be used individually or in combination, such as binary mixtures. For several drugs, the flux across skin was observed to be linear with that of the most widely used enhancer, ethanol.[73–75] Another polar enhancer, isopropanol, facilitated ion association of charged molecules and enhanced the transport of both neutral and ionic species across the stratum corneum.[76,77] While polar enhancers traverse the skin, nonpolar enhancers are largely retained in the stratum corneum; both aspects make the combination a superior enhancer to the individual enhancers.[78]

Based on Eq. (4.4), the *enhancement factor* E is defined as the enhancement on the maximum flux J_{max} of a drug across skin by increasing the (kinetic) diffusivity and/or the (thermodynamic) solubility in the stratum corneum.[79] Thus

$$E = \frac{\left(\frac{D}{h}\right)^{*}\left(\frac{1}{\gamma_m}\right)^{*}}{\left(\frac{D}{h}\right)^{\circ}\left(\frac{1}{\gamma_m}\right)^{\circ}} = \frac{J^{*}_{\max}}{J^{\circ}_{\max}} \tag{4.31}$$

where D, h, and γ_m are the diffusion coefficient, thickness, and activity coefficient in the stratum corneum, respectively, and the superscripts (* and °) denote with and without the enhancer, respectively. In the case of significant cotransport of an enhancer along with the principal permeant (drug), both the diffusion and activity coefficients of the drug are no longer constant but rather position-dependent within the membrane. A novel theoretical model greatly simplifies solving the problem by mathematically slicing the membrane into n thin elements, each sandwiched with slightly different enhancer concentrations.[80]

Modulation of skin reactions. Skin reactions, including irritation and sensitization, are indeed the Achilles heel of transdermal medication.[81,82] *Irritation* may be defined as a local, reversible inflammatory response of the skin to the application of an agent without the involvement of an immunological mechanism. A large fraction of all drugs, depending on their concentrations, may have potential to cause skin irritation, although the "hidden" redness in the GI tract with oral administration sometimes is more severe than the "visual" one on skin. *Sensitization* results from an immune response to an antigen, which may lead to an exaggerated response on repeated exposure to the antigen. Irritation or sensitization may occur in response to either the drug or a component of the transdermal system. Careful testing of both active and placebo patches is needed.

The concentration-response relationships are important in optimizing drug or enhancer concentration to balance drug permeability versus skin tolerability. More sophisticated strategies involve modulation of the dermal immune system by coapplication of different polymeric formulations, excipients, enhancers, and drugs as "anti-irritants" or "countersensitizers." pH is suggested as the major contributing factor in the irritative response. A pH-controlled aqueous isopropanol counterion composition increases the skin permeation rate and improves the skin irritation profile for a weak-base amine drug.[83] However, much work has been done in trial-and-error approaches and remains to be completed through understanding of the inflammatory processes, the agents responsible for irritation of the arachidonic cascade, and the cellular network and signaling cascade as it pertains to the development of irritation and sensitization. Metabolic modulators, ion channel modulators, and mast cell degranulators are the latest contribution to understanding the mechanisms involved in these processes.[84–86]

Novel system design. One of the many challenges in developing transdermal diffusion-controlled drug delivery products is selection of the most appropriate system design alternative based on a methodical examination of the specific characteristics of the drug candidate. Although simple in concept, reducing this objective to practice involves the simultaneous resolution of many factors. The following questions must be answered to clearly define the desired final product as parameters for system design[87]: (1) What is the range of dosages to be delivered systematically? (2) What is the target population, and what is the maximum patch size accepted by them? (3) What are the preferred site and duration of application? (4) What is the cost/risk associated with residual drug remaining within the patch at the end of the application period? In addition, drug stability, skin adhesion, patentability, and pharmcodynamic profiles are among important aspects for novel system design.

As transdermal product development has matured over the past 25 years, a number of system designs have emerged to deliver drugs across the skin. Many new transdermal systems with physical enhancements (e.g., heat, electricity, and ultrasound) of skin permeation are under development.[67–69] Iontophoresis is merging with biomedical technology in the development of programmable and closed-loop self-regulating transdermal drug delivery devices. The biofeedback loops occur in the same device, not only providing diagnostics but also guiding therapy and delivery of therapeutics.

Will more transdermal drug delivery products become available in the future? The question is no longer should a transdermal delivery system be used but rather how is it best suited for a particular therapeutic agent and a particular disease condition. The technical challenges still remain achieving adequate skin permeability with acceptable tolerability. Identifying compounds is of crucial importance for product development because therapeutic indications and market potential may not be obvious for drugs going the skin route of administration. The success of a potential "niche" transdermal product, whether it is for a new delivery route, a new dosage form, a new indication, or a new combination, relies on its unmet medical needs, technical feasibility and developability, and market viability.

Topical and transbuccal systems. In the current literature, unfortunately, transdermal products usually are labeled "for systemic use" only. It is the time to define "transdermal products for topical (excluding dermatological) use" to those delivered through the intact and healthy skin directly into the local tissues or deeper regions beneath the skin. Dozens of these transdermal systems have been launched for topical use, including analgesics for muscle aches, neuropathic pain, and arthritis and the treatments of breast cancer and erectile dysfunction.[88]

Transbuccal drug delivery systems are a new technology and spinoff of transdermal/topical drug delivery systems. Similar to monolithic adhesive matrix–type skin patches, transbuccal systems are designed to adhere to mucosal tissue in the oral cavity. A number of compounds are being evaluated in clinical studies.[88]

4.4 Other Diffusion-Controlled Systems

The justification for a CR dosage form over a conventional one is either to optimize therapy or to circumvent problems in drug absorption or metabolism. The variety of alternative delivery routes available for drug delivery corresponds to biological membranes in the human body.[12] It is appropriate to distinguish between depot and other formulations that typify sustained release and true CR administration. Briefly introduced in this section are intrauterine devices (IUDs), intravaginal rings (IVRs), intraocular inserts, and parenteral implants as diffusion-controlled systems, which have met with limited success.

4.4.1 Intrauterine devices and intravaginal rings

IUDs and IVRs are two drug delivery systems used primarily for fertility control.[89] Various inert and biocompatible polymers such as polyethylene, EVAc, and Silastic are used widely to construct IUDs and IVRs. Two types of IUDs have been on the market, those releasing copper and those releasing progestines. Progestasert, using silicon in the interior saturated by the reservoir and EVAc as the outer rate-controlling layer, releases progesterone at the rate of 65 μg/day for 1 year, and endometrial proliferation is suppressed.[90] A second hormone-releasing IUD with Silastic design delivers a contraceptive dose of high-potency levonorgestrel for many years.[91] One IUD containing natural estrogen and gestagen is designed for 1 year or longer of hormone-replacement therapy in perimenopausal and postmenopausal women.[92] Intravaginal delivery may have applications beyond birth control. Two existing contraceptive silicone matrix IVRs are for a gestagen (medroxyprogesterone acetate) and for an estrogen-progestin combination.[93] Other IVR products include estradiol for the treatment of postmenopausal symptoms, including atrophic vaginitis (Pfizer/Schering AG), and estradiol acetate as a hormone-replacement therapy (Galen). Under development, an etonogestrel-ethinylestradiol combination contraceptive IVR is based on a coaxial fiber consisting of EVA, and a spermicide IVR uses an inner core of EVA/spermicide surrounded by a thin, permeable layer of Silastic.[94,95]

Given its convenience for self-implantation, the high permeability, and its ease of controlling delivery, the vaginal route may offer opportunities for systemic delivery for the treatment of diseases such as osteoporosis, in which the patient population is predominantly female. Transport may be cyclic because estrogen causes conrnification and thickening of the vaginal mucosa. A number of microorganisms and potential pathogens may proliferate in the vagina, and this may be important for understanding drug metabolism and the constraints for IVR design.

4.4.2 Intraocular inserts

Ocusert, a small intraocular insert, releases pilocarpine to treat glaucoma.[96] This diffusion-controlled reservoir system is effective for 1 week, replacing the need for eyedrops applied 4 times per day. With EVA as the rate-controlling membrane, an initial burst of drug into the eye is seen in the first few hours. The ocular hypotensive effect is fully developed within 2 hours of placement of the insert in the conjunctival sac, and the hypotensive response is maintained throughout the therapy. Vitrasert (Chiron), an intraocular ganciclovir insert for treatment of newly diagnosed cytomegalovirus (CMV) retinitis has been launched recently. The insert contains ganciclovir embedded in a polymer-based system that slowly releases the drug into the eye for up to 8 months. Under research and development, a small intraocular insert delivers 1 μg/day cidofovir to the vitreous humor for treatment of AIDS-induced CMV retinitis for more than 2 years.[97] Tacrolimus may be administered as a surgical insert contained in a diffusible-walled reservoir sutured to the wall of the sclera to provide a slow-release system for the treatment of ocular disease.[98]

Ocular drug delivery, involving localized treatment to the eye, presents several challenges based on anatomy and physiology. The cornea has an outer epithelial layer that is about five cells thick, an aqueous layer, and an inner endothelium. Drugs therefore have to cross two lipid layers and an aqueous layer to enter the eye. The epithelium is rate limiting for most drugs; the aqueous region, i.e., the stroma, is rate limiting for very lipophilic drugs.[89] The eye is highly innervated, and patient comfort is of paramount importance in order to achieve good compliance. Two approaches used to increase the residence time of drugs in the eye, and consequently, the amount of drug absorbed, are increasing the viscosity of the solution and the use of an insert or hydrogel contact lenses loaded with drug. Polymers that undergo a phase change from a liquid to a gel in response to temperature, pH, or ionic strength also show promise in this field.

4.4.3 Subcutaneous implants

Nondegradable subcutaneous implants as diffusion-controlled drug delivery systems, including Norplant, have been reviewed.[89,99] Unlike biodegradable implants with long-term toxicological concern for metabolism of the polymer, nondegradable implants cannot avoid removal of the system after use. Norplant, a female contraceptive implant, contains a set of six flexible closed capsules of levonorgestrel and uses Silastic as the rate-controlling membrane. The capsules are inserted by a physician beneath the skin and removed at the end of 5-year therapy period. Clinical studies have shown plasma concentrations of 0.30 ng/mL over 5 years but are highly variable as a function of individual metabolism and body weight. The typical failure rate with this implant in the first year is only 0.2 percent compared with 3 percent with oral contraceptives. Another synthetic female contraceptive subcutaneous implant is under development (Akzo Nobel) to releases 40 μg/day etonogestrel. A subcutaneous silicone implant of norgestomet has been evaluated in synchronizing estrus and diagnosing pregnancy in ewes, permitting the diagnosis of pregnancy status with 100 percent accuracy with no adverse effects on established pregnancy.[100]

The element of targeting or site-specific delivery is often linked with new CR systems. For example, as fused with a lipoid carrier or encapsulated in microcapsules or in Silastic capsules, breast implants of antiestrogens and prostatic implants of androgen-suppressive drugs render a constant slow release of drugs to the target tumor tissue for extended periods and minimize systemic toxicity.[101,102] There are various approaches that could be used to regulate the growth factor releases from polymer scaffolds in tissue engineering as well as in cell transplantation.[103] Diffusion-controlled DNA delivery from implantable EVAc matrices is useful for DNA vaccination and gene therapy.[104] However, these fields are largely in their infancy and need specific problems identified for them to be developed.

References

1. Crank, J. *The Mathematics of Diffusion*, 2d ed. Oxford, England: Clarendon Press, 1975, p. 44.
2. Carslaw, H., and Jaeger, J. *Conduction of Heat in Solids*, 2d ed. Oxford, England: Clarendon Press, 1959, p. 99.
3. Bird, R., Stewart, W., and Lightfoot, E. *Transport Phenomena*. New York: Wiley, 1964, p. 502.
4. Flynn, G., Yalkowsky, S., and Roseman T. Mass transport phenomena and models: Theoretical concepts. *J. Pharm. Sci.* **63**:479–510, 1974.
5. Lee, P. Diffusion-controlled matrix systems, in Kydonieus, A. (ed.), *Treatise on Controlled Drug Delivery: Fundamentals, Optimization, Applications*. New York: Marcel Dekker, 1992, pp. 155–198.
6. Paul, D., and McSpadden, S. Diffusional release of a solute from a polymer matrix. *J. Membrane Sci.* **1**:33–48, 1976.
7. Higuchi, T. Rate of release of medicaments from ointment based containing drugs in suspension. *J. Pharm. Sci.* **50**:874–875, 1961.
8. Lee, P. Diffusional release of a solute from a polymeric matrix: Approximate analytical solutions. *J. Membrane Sci.* **7**:255–275, 1980.
9. Zhang, G. Oral controlled release dosage forms in new drug applications (NDA). AAPS/CPA Joint Conference, Shanghai, China, November 19–21, 2003.
10. Venkatraman, S., Davar, N., and Chester, A. An overview of controlled release systems, in Wise D. (ed.), *Handbook of Pharmaceutical Controlled Release Technology*. New York: Marcel Dekker, 2000, pp. 431–463.

11. Qiu, Y., and Zhang, G. Research and development aspects of oral controlled-release dosage forms, in Wise, D. (ed.), *Handbook of Pharmaceutical Controlled Release Technology*. New York: Marcel Dekker, 2000, pp. 465–503.
12. Berner, B., and Dinh, S. Fundamental concepts in controlled release, in Kydonieus, A. (ed.), *Treatise on Controlled Drug Delivery: Fundamentals, Optimization, Applications*. New York: Marcel Dekker, 2000, pp. 1–35.
13. Colombo, P., Santi, P., Bettin, R., Brazel, C., and Peppas, N. Drug release from swelling-controlled sytems, in Wise, D. (ed.), *Handbook of Pharmaceutical Controlled Release Technology*. New York: Marcel Dekker, 2000, pp. 183–210.
14. Dressman, J. Mechanism of release from coated pellets, in Ghebre-Sellassie, I. (ed.), *Multiparticulate Oral Drug Delivery*. New York: Marcel Dekker, 1994, pp. 285–306.
15. Theeuwes, F. Elementary osmotic pump. *J. Pharm. Sci.* **64**:1987–1991, 1975.
16. Ritger, P., and Peppas, N. A simple equation for description of solute release: II. Fickian and anomalous release from swellable devices. *J. Contr. Rel.* **5**:37–42, 1985.
17. Korsmeyer, R., Gurny, R., Doelker, E., et al. Mechanism of solute release from porous hydrophilic polymers. *Int. J. Pharm.* **15**:25–35, 1983.
18. Higuchi, T. Mechanism of sustained release medication: Theoretical analysis of rate of release of solid drugs dispersed in solid matrices. *J. Pharm. Sci.* **52**:1145–1149, 1963.
19. Ju, T., Nixon, P., and Patel, M. Drug release from hydrophilic matrices: I. New scaling laws for predicting polymer and drug release based on the polymer disentanglement concentration and the diffusion layer. *J. Pharm. Sci.* **84**:1455–1463, 1995.
20. Ju, T., Nixon, P., Patel, M., and Tong, D. Drug release from hydrophilic matrices: II. A mechanistic model based on the polymer disentanglement and the diffusion layer. *J. Pharm. Sci.* **84**:1464–1477, 1995.
21. Lee, P. Kinetics of drug release from hydrogel matrices. *J. Contr. Rel.* **2**:277–288, 1985.
22. Siepmann, J., and Peppas, N. Mathematical modeling of controlled drug delivery. *Adv. Drug Del. Rev.* **48**:137–138, 2001.
23. Siepmann, J., Streubel, A., and Peppas, N. Understanding and predicting drug delivery from hydrophilic matrix tablets using the 'sequential layer' model. *Pharm. Res.* **19**:306–314, 2002.
24. Qiu, Y., Chidambaram, N., and Lood, K. Design and evaluation of layered diffusional matrices for zero-order sustained-release. *J. Contr. Rel.* **51**:123–130, 1998.
25. Hildgen, P., and McMullen, J. A new gradient matrix: formulation and characterization. *J. Contr. Rel.* **34**:263–271, 1995.
26. Kim, C. Compressed donut-shaped tablets with zero-order release kinetics. *Pharm. Res.* **12**:1045–1048, 1995.
27. Benkorah, A., and McMullen, J. Biconcave coated, centrally perforated tablets for oral controlled drug delivery. *J. Contr. Rel.* **32**:155–160, 1994.
28. Conte, U., Maggi, L., Colombo, P., and Manna, A. Multilayered hydrophilic matrices as constant release devices (Geomatrix systems). *J. Contr. Rel.* **26**:39–47, 1993.
29. Scott, D., and Hollenbeck, R. Design and manufacture of a zero-order sustained-release pellet dosage form through nonuniform drug distribution in a diffusional matrix. *Pharm. Res.* **8**:156–161, 1991.
30. Brooke, D., and Washkuhn, R. Zero-order drug delivery system: Theory and preliminary testing. *J. Pharm. Sci.* **66**:159–162, 1977.
31. Lipper, R., and Higuchi, W. Analysis of theoretical behavior of a proposed zero-order drug delivery system. *J. Pharm. Sci.* **66**:163–164, 1977.
32. Lee, P., and Kim, C. Effect of geometry on solvent front penetration in glassy polymers. *J. Membrane Sci.* **65**:77–92, 1992.
33. Hsieh, D., Rhine, W., and Langer, R. Zero-order controlled release polymer matrices for micro and macro-molecules. *J. Pharm. Sci.* **72**:17–22, 1983.
34. Lee, P. Novel approach to zero-order drug delivery via immobilized nonuniform drug distribution in glassy hydrogels. *J. Pharm. Sci.* **70**:1344–1347, 1984.
35. Lee, P. Effect of nonuniform initial drug concentration distribution on the kinetics of drug release from glassy hydrogel matrices. *Polymer* **25**:973–978, 1984.
36. DiLuccio, R., Duggins, R., and Shefter, E. Eur. Pat. Appl. 200224, 1986.

37. Conte, U., Maggi, L., Colombo, P., and La Manna, A. Multilayer hydrophilic matrices as constant release devices (Geomatrix systems). *J. Contr. Rel.* **26**:39–47, 1993.
38. Porter, S., and Ghebre-Sellassie, I. Key factors in the development of modified-release pellets, in Ghebre-Sellassie, I. (ed.), *Multiparticulate Oral Drug Delivery*. New York: Marcel Dekker, 1994, pp. 217–284.
39. Fukumori, Y. Coating of multiparticulates using polymeric dispersions-formulation and process considerations, in Ghebre-Sellassie, I. (ed.), *Multiparticulate Oral Drug Delivery*. New York: Marcel Dekker, 1994, pp. 79–111.
40. Mehta, A. Valazza, M., and Abele, S. Evaluation of fluid-bed processes for enteric coating systems. *Pharm. Tech.* **10**(4):46–56, 1986.
41. Mehta, A. Processing and equipment considerations for aqueous coating. *Drug Pharm. Sci. (Aq. Polym. Coat. Pharm. Dosage Forms)*, **36**:267–302, 1989.
42. Karim, A., Burns, T., Janky, D., and Hurwitz, A. Food-induced changes in theophylline absorption from controlled-release formulations: II. Importance of meal composition and dosing time relative to meal intake in assessing changes in absorption. *Clin. Pharmacol. Ther.* **38**:642–647, 1985.
43. Ju, T., Cooper, A., Gauri, V., and Patel, M. pH controlled release oral drug delivery: Effects of bulk and iternal pH. Paper no. PDD 7491, AAPS Annual Meeting, San Diego, CA, 1994.
44. Venkatramana, R., Engh, K., and Qiu, Y. Design of pH-independent controlled release matrix tablets for acidic drugs. *Int. J. Pharm.* **252**(1–2):81–86, 2003.
45. Delargy, A., et al. The optimization of a pH-independent matrix for the controlled release of drug materials using in vitro modeling. *Int Symp. Contr. Rel. Bioact. Mater.* **16:**378–379, 1989.
46. Li Wan Po, Wong, L., and Gilligan, C. Characterization of commercially available theophylline sustained- or controlled-release systems: In vitro drug release profiles. *Int. J. Pharm.* **66**:111–130, 1990.
47. Ju, T., Phillip, N., John, S., et al. HPMC-based extended-release matrices containing poorly-soluble compounds: A mechanistic study of the effects of key formulation variables on drug and polymer release. AAPS Annual Meeting, San Francisco, CA, 1998.
48. Rao, V., Haslam, J., and Stella, V. Controlled and complete release of a model poorly water-soluble drug, prednisolone, from hydroxypropyl methylcellulose matrix tablets using (SBE)(7m)-beta-cyclodextrin as a solubilizing agent. *J. Pharm. Sci.* **90**:807–816, 2001.
49. Schwarz, J., Lee, M., Weisspapir, M., Zhang, Q., and Lu, W. Increased bioavailability of coenzyme Q-10 in self-emulsifying controlled release tablet: New type of delivery system for hydrophobic drugs. *Int. Symp. Contr. Rel. Bioact. Mater.* **28**:824–825, 2001.
50. Ju, T. Performance of extended release matrices and multiparticulates: principles and case studies. 38th Annual Eastern Pharmaceutical Technology Meeting, Whippany, NJ, October 16, 1998.
51. Ju, T. U.S. Patent 6197339 B1, 2001.
52. Pillay, V., and Fassihi, R. A novel approach for constant rate delivery of highly soluble bioactives from a simple monolithic system. *J. Contr. Rel.* **67**:67–78, (2000).
53. Hite, M., Federici, C., Turner, S., and Fassihi, R. Novel design of a self-correcting monolithic controlled-release delivery system for tramadol. *Drug Del. Tech.* **3**:48–55, 2003.
54. Kydonieus, A., and Berner, B. *Transdermal Delivery of Drugs*. Boca Raton, FL: CRC Press, 1987.
55. Chien, Y. *Transdermal Controlled Systemic Medications*. New York: Marcel Dekker, 1987.
56. Hadgraft, J., and Guy, R. *Transdermal Drug Delivery: Developmental Issues and Research Initiatives*. New York: Marcel Dekker, 1989.
57. Gurney, R., and Teubner, A. *Dermal and Transdermal Drug Delivery*. Stuttgart: Wissenschaftliche Verlagsgesellschaft, 1993.
58. Ghosh, T., Pfister, W., and Yum, S. *Transdermal and Topical Drug Delivery Systems*. Buffalo Grove, IL: Interpharm Press, 1997.
59. Wick, S. U.S. Patent 4,751,087, 1988.

60. Feldstein, M., Vasiliev, A., and Plate, N. Enhanced drug delivery from transdermal therapeutic systems with hydrophilic polymer matrix. *Int. Symp. Contr. Rel. Bioact. Mater.* **21**:423–424, 1994.
61. Peterson, T., and Dreyer, S. Factors influencing delivery from multilaminate transdermal patch systems. *Int. Symp. Contr. Rel. Bioact. Mater.* **21**:477–478, 1994.
62. Sanvordeker, D., Cooney, J., and Wester, R. U.S. Patent 4,336,243, 1982.
63 Gale, R., and Berggren, R. U.S. Patent 4,615,699, 1980.
64. Good, W. Transdermal drug delivery systems. *Medical Device & Diagnostic Industry.* **2**:35–42, 1986.
65. Kydonieus, A., Wille, J., and Murphy, G. Fundamental concepts in transdermal delivery of drugs, in Kydonieus, A., and Wille, J. (eds.), *Biochemical Modulation of Skin Reactions: Transdermals, Topicals, Cosmetics.* Boca Raton, FL: CRS Press, 2000, pp. 1–14.
66. Liu, P., Nightingale, J., and Bergstrom, T. Variation of human skin permeation in vitro: Ionic vs. neutral compounds. *Int. J. Pharm.* **90**:171–176, 1993.
67. Singh, P., Liu, P., and Dinh, S. Variation of human skin permeation in vitro: Ionic vs. neutral compounds, in Bronaugh, R., and Maibach, H. (eds.), *Percutaneous Absorption: Drugs-Cosmetics-Mechanisms-Methodology*, 3d ed. New York: Marcel Dekker, 1999, pp. 633–657.
68. Prausnitz, M., Bose, V., Langer, R., and Weaver, J. Electroporation, in Smith, E., and Maibach, H. (eds.), *Percutaneous Penetration Enhancers.* Boca Raton, FL: CRC Press, 1995, pp. 393–406.
69. Bommannan, D., Menon, G., Okuyama, H., Elias, P., and Guy, R. Sonophoresis: II. Examination of the mechanism(s) of ultrasound-enhanced transdermal drug delivery. *Pharm. Res.* **9**:1043–1047, 1992.
70. Smith, E., and Maibach, H. *Percutaneous Penetration Enhancers.* Boca Raton, FL: CRC Press, 1995.
71. Hsieh, D. *Drug Permeation Enhancement.* Boca Raton, FL: Marcel Dekker, 1994.
72. Bronaugh, R., and Maibach, H. *Percutaneous Absorption: Drugs-Cosmetics-Mechanisms-Methodology,* 3d ed. New York: Marcel Dekker, 1999.
73. Berner, B., Mazzenga, G., Otte, J., et al. Ethanol-water mutually enhanced transdermal therapeutic systems: II. Skin permeation of ethanl and nitroglycerin. *J. Pharm. Sci.* **78**:402–407, 1989.
74. Liu, P., Bergstrom, T., and Good, W. Co-transport of estradiol and ethanol through human skin in vitro: Understanding permeant/enhancer flux relationship. *Pharm. Res.* **8**:938–944, 1991.
75. Berner, B., and Liu, P. Alcohols, in Smith, E., and Maibach, H. (eds.), *Percutaneous Penetration Enhancers.* Boca Raton, FL: CRC Press, 1995, pp. 44–60.
76. Liu, P., Bergstrom, T., Clarke, F., Gonnella, N., and Good, W. Quantitative evaluation of aqueous isopropanol enhancement on skin flux of terbutaline (sulfate): I. Ion associations and species equilibria in the formulation. *Pharm. Res.* **9**:1036–1042, 1992.
77. Liu, P., and Bergstrom, T. Quantitative evaluation of aqueous isopropanol enhancement on skin flux of terbutaline (sulfate): II. Permeability contribution of equilibrated drug species across human skin in vitro. *J. Pharm. Sci.* **85**:320–325, 1996.
78. Goldberg-Cettina, M., Liu, P., Bergstrom, T., and Nightingale, J. Enhanced transdermal delivery of estradiol in vitro using binary vehicles of isopropyl myristate and short-chain alkanols. *Int. J. Pharm.* **114**:237–245, 1995.
79. Liu, P., Bergstrom, T., and Good, W. Co-transport of estradiol and ethanol through human skin in vitro: Understanding permeant/enhancer flux relationship. *Pharm. Res.* **8**:938–944, 1991.
80. Liu, P., Higuchi, W., Ghanem, A., Bergstrom, T., and Good, W. Assessing the influence of ethanol on simultaneous diffusion and metabolism of β-estradiol in hairless mouse skin for the "asymmetric" situation in vitro. *Int. J. Pharm.* **78**:123–136, 1992.
81. Zurcher, K., and Krebs, A. *Cutaneous Drug Reactions.* Basel: Karger, 1992.
82. Kydonieus, A., and Wille, J. *Biochemical Modulation of Skin Reactions: Transdermals, Topicals, Cosmetics.* Boca Raton, FL: CRC Press, 2000.
83. Bergstrom, T., and Liu, P. U.S. Patent 5,374,645, 1994.

84. Cormiez, M., Ledger, P., Amkraut, A., and Marty, J. U.S. Patent 5,120,545, 1994.
85. Wille, J. U.S. Patent 5,716,987, 1998.
86. Wille, J., and Kydonieus, A. European Patent Application 5,612,525, 1994.
87. Peterson, T., Wick, S., and Ko, C. Design, development, manufacturing, and testing of transdermal dru delivery systems, in Ghosh, T., Pfister, W., and Yum, S. (eds.), *Transdermal and Topical Drug Delivery Systems*. Buffalo Grove, IL: Interpharm Press, 1997, pp. 249–298.
88. Ghosh, T., and Pfister, W. Transdermal and topical delivery systems: An overview and future trends, in Ghosh, T., Pfister, W., and Yum, S. (eds.), *Transdermal and Topical Drug Delivery Systems*. Buffalo Grove, IL: Interpharm Press, 1997, pp. 1–32.
89. Chien, Y. *Novel Drug Delivery*. New York: Marcel Dekker, 1982.
90. Dallenbach-Hellweg, G., and Sievers, S. Die hidtologische reaktion des endometrium auf lokal applizierte gestagene. *Virchows Arch. Pathol. Anat.* **368**:289–298, 1975.
91. Nilsson, C., Lachteenmaki, P., and Luukkainen, T. Patterns of ovulation and bleeding with a low levonorgesterol-releasing device. *Contraception* **21**:225–233, 1980.
92. Stix, J. German Patent Application DE98-19809243, 1999.
93. Duncan, R., and Seymour, L. *Controlled Release Technologies*. Amsterdam: Elsevier Advanced Technology, 1989, p. 11.
94. van Laarhoven, J., Kruft, M., and Vromans, H. In vitro release properties of etonogestrel and ethynylestradiol from a contraceptive vaginal ring. *Int. J. Pharm.* **232**:163–173, 2002.
95. Saltzman, W., and Tena, L. Spermicide permeation through biocompatible polymers. *Contraception* **43**:497–505, 1991.
96. Heilmann, K. *Therapeutic Systems: Rate-Controlled Drug Delivery, Concept and Development*, 2d ed. New York: Thieme-Stratton, 1984, pp. 66–82.
97. Roorda, W., Dionne, K., Brown, J., et al. PCT International Application WO9843611A1, 1998.
98. Peyman, G. U.S. Patent Application 200213340, 2002.
99. Nash, H. Controlled release systems for contraception, in Langer, R., and Wise, D. (eds.), *Medical Applications of Controlled Release*, Vol. 2. Boca Raton, FL: CRC Press, 1984, pp. 35–64.
100. Kesler, D., and Favero, R. The utility of controlled-release norgestomet implants in synchronizing estrus and diagnosing pregnancy in ewes, and factors affecting the diffusion rate of norgestomet from silicone implants. *Drug Dev. Ind. Pharm.* **23**:217–220, 1997.
101. Sahadevan, V. U.S. Patent Application 200314900-8, 2003.
102. Sahadevan, V. U.S. Patent Application 200314793-6, 2003.
103. Lee, K., and Mooney, D. Controlled growth factor delivery for tissue engineering, in Dinh, S. and Liu, P. (eds.), *Advances in Controlled Drug Delivery: Science, Technology, and Products*. ACS Symposium Series 846, Washington, 2003, pp. 73–83.
104. Luo, D., Woodrow-Mumford, K., Belcheva, N., et al. Controlled DNA delivery systems. *Pharm. Res.* **16**:1300–1308, 1999.

Chapter

5

Dissolution Controlled Drug Delivery Systems

Zeren Wang and Rama A. Shmeis

Boehringer-Ingelheim Pharmaceuticals, Inc.
Ridgefield, Connecticut

5.1 Introduction

The dissolution process includes two steps, initial detachment of drug molecules from the surface of their solid structure to the adjacent liquid interface, followed by their diffusion from the interface into the bulk liquid medium. This process could be manipulated to design controlled release delivery systems with desired profiles and a desired rate. In general, either matrix- or barrier/membrane-based controlled release systems are applied to slow down, delay, and control the delivery and release of drugs. In the former, drug is uniformly dispersed in a matrix consisting mainly of polymers or waxes, whereas the latter refers to coated systems. A combination of both (coated matrix) is also possible.[1] The demarcation between a coated and a matrix-type pharmaceutical controlled release product is not always clear. Some of the materials used as coatings to control drug release also may be used for a similar function in matrix-type products.[2]

If the matrix or coated systems are made of water-soluble components, the rate-limiting step governing the release of drug from these systems will be dissolution. For many controlled release drug delivery systems, different mechanisms controlling the release profile and release rate are used in combination. In this chapter, only hydrophilic and water-soluble polymers used for matrix and coated systems are discussed. Release profiles from these systems are usually complicated and controlled by several mechanisms; however, only the effect of the dissolution of drug substances, as well as polymer matrices or polymer coatings, on release will be discussed in detail. Systems employing a mixture of soluble and insoluble coatings (dissolution and diffusion controlled) also will be introduced briefly.[3]

The dissolution controlled release matrix systems provide sustained release profiles; i.e., the active drugs in these systems are released continuously at a slow rate to provide a long-term therapeutic effect. Unlike diffusion controlled release coated systems, release profiles from dissolution controlled release coated systems do not follow zero-order kinetics but fall within the classification of delayed release systems,[4] pulsatile or repeat-action systems,[5] and sustained release systems.[3]

Although examples of delivery systems using the parenteral and oral (solid) routes are presented in this chapter, application of dissolution controlled release matrix and coated systems concepts can extended easily (and has been) used for many other delivery routes.

5.2 Theoretical Considerations for Dissolution Controlled Release Matrix and Coated Systems

5.2.1 Dissolution of solid particles

The dissolution process of solids consists of two steps. First, the molecules at the solid-liquid interface are solvated and detached from the solid surface. Second, the solvated molecules diffuse away from the interface to the bulk solution. It is commonly believed that the first step is much faster than the second step. Therefore, at the steady state of dissolution, the concentration of the dissolving substance at the dissolution interface is equal or close to its solubility (Fig. 5.1). Because the diffusion step is rate limiting, the dissolution flux (the amount of mass dissolved per unit time and per unit dissolving area) can be modeled by Fick's first law of diffusion:

$$\frac{1}{A}\frac{dM}{dt} = \frac{D}{h}(C_s - C_b) \tag{5.1}$$

where M = mass
t = time
A = dissolving surface area
D = diffusion coefficient
h = thickness of diffusion layer
C_s = solubility
C_b = concentration in bulk solution

By simple manipulation, the rate of dissolution, the amount of dissolved solid per unit time, can be calculated by the Noyes and Whitney equation, written as

$$\frac{dM}{dt} = A \times \frac{D}{h}(C_s - C_b) \tag{5.2}$$

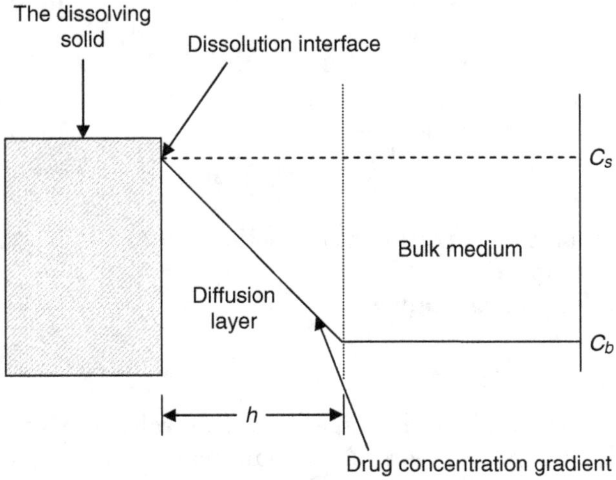

Figure 5.1 Schematic representation of solid dissolution.

Although Eq. (5.2) looks very simple, it actually is complicated owing to the changes of the surface area and the thickness of the hydrodynamic diffusion layer during dissolution. Only several simplified cases can be solved analytically based on Eq. (5.2). For example, for the dissolution of drug powder, only uniformly sized spherical particles can be modeled by solving Eq. (5.2). For such a case, a cube-root relationship was found between the amount of remaining solid mass and time. This is known as the *Hixson-Crowell cube-root law*[6]:

$$M_0^{1/3} - M^{1/3} = \kappa t \tag{5.3}$$

where M_0 = original mass of the drug solid
M = mass of solid remaining at a specific time
κ = cube root dissolution rate constant, which is represented as

$$\kappa = \frac{M_0^{1/3} k C_s}{r\rho} \tag{5.4}$$

where r = radius of the particle
ρ = density
$k = D/h$, which are defined in Eq. (5.1)

Note that in the preceding equations, C_s is the total solubility, which is the same as the intrinsic solubility for a nonionizable compound and is a function of the pH for an ionizable compound, as given by Eqs. (5.5) (for weak acids) and (5.6) (for weak bases):

$$C_s = C_0\left(1 + \frac{K_a}{[\mathrm{H}^+]_s}\right) \tag{5.5}$$

$$C_s = C_0\left(1 + \frac{[\mathrm{H}^+]_s}{K_a}\right) \tag{5.6}$$

where $[\mathrm{H}^+]_s$ = hydrogen ion concentration at the solid surface
C_0 = intrinsic solubility
K_a = acid dissociation constant

5.2.2 Dissolution of coated systems

Modification of the temporal and spatial aspects of drug release using coating involves applying a layer or layers of retardant material between the drug and the elution/dissolution medium. If the coating material is water soluble, drug release will be controlled by dissolution of the coat, which usually consists of a slowly dissolving polymeric material. Once the polymeric membrane has dissolved, the drug inside the membrane is immediately available for dissolution and absorption.[7] At that stage, release will depend on the core properties (drug and excipients) such as porosity, drug solubility, and dissolution rate in the dissolution medium. Cores can be immediate release systems or controlled release matrix systems.

For hydrophilic water-soluble polymers, hydration is the first step of dissolution in aqueous solutions, followed by dissolution of the hydrated phase. The latter step involves disentanglement of polymer molecules. In general, the dissolution kinetics follow Eq. (5.2), suggesting that the solubility of polymers and the viscosity of the hydrated phase are the major variables affecting the dissolution rate. Diffusion of dissolved drug molecules through the hydrated polymer layer also may contribute to the overall release kinetics.

One of the most common designs used for delayed release is the enteric-coated drug delivery system. This is a type of activation-controlled drug delivery system that permits targeting the delivery of a drug only in a selected pH range. Polymers (often esters of phthalic acid) are used as enteric coatings, and they commonly possess carboxylic acid groups that are un-ionized in the relatively low pH of the stomach (normally about 1.5 to 4.5) but ionize and thus repel one another as the pH rises when the delivery system enters the small intestine, thus causing coating disruption. A quantitative model describing the mechanism and kinetics of drug release from enteric-coated tablets was developed by Ozturk et al.[8] Polymers used for enteric coatings are weak acids containing carboxyl groups in a substantial proportion of their monomeric units. Rapid dissolution of these polymers requires pH values of dissolution media much higher than the pK_a values of polymers. However, when hydrated, these polymers are slightly permeable to the confined drug even at pH values lower than the pK_a values of the polymers.

A schematic of an enteric polymer (HP) dissolution and drug release (weak acid, HA) from enteric-coated tablets into a buffered medium (HB) is shown in Fig. 5.2, wherein P^-, A^-, and B^- represent the ionized forms of the polymer, drug, and buffer, respectively. The polymer has an initial thickness of R_2–R_1 surrounding a drug core of radius R_1, and h is the thickness of the stagnant diffusion layer adjacent to the polymer coating. The two interfaces at $r = R_1$ and $r = R$ correspond to the drug-polymer and polymer–stagnant diffusion layer interfaces, respectively; the latter moves with time from the initial position at R_2. The position at time t is represented by R. The drug diffuses first through the polymer and then through the stagnant diffusion layer. During this transfer, the drug simultaneously reacts with the incoming buffer B^- to yield

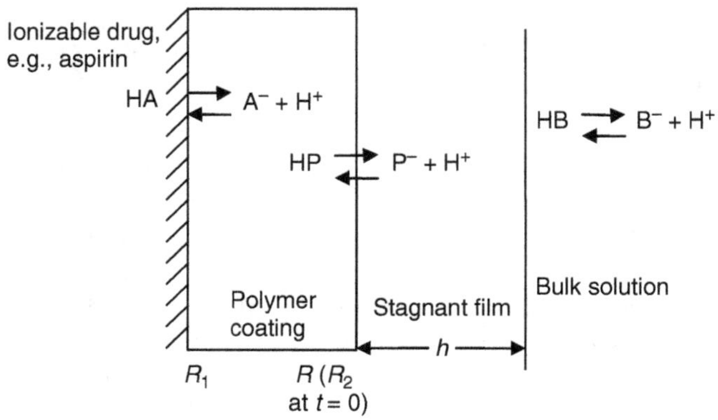

Figure 5.2 Schematic representation of polymer dissolution and drug release from enteric-coated tablets.

the conjugate base A^- and HB. The polymer also diffuses away from the polymer–diffusion layer interface and can react simultaneously with basic species (such as the buffer in the dissolution medium). The bulk is assumed to be well mixed, and it is also assumed that chemical equilibrium is attained instantly throughout. The polymer–stagnant diffusion layer interface moves toward the drug core as the polymer dissolves.

The polymer dissolution flux J_{HP} may be given by Fick's first law of diffusion (similar to Eq. 5.1):

$$J_{HP} = \frac{D_{HP}}{h}([HP]_{T,p} - [HP]_{T,b}) \tag{5.7}$$

where D_{HP} = diffusion coefficient of the polymer in the diffusion layer
h = thickness of the diffusion layer
$[HP]_{T,p}$ = total concentration of polymer (ionized and nonionized) at the interface of the polymer and diffusion layer
$[HP]_{T,b}$ = total concentration of polymer (ionized and nonionized) in bulk dissolution medium

At a pH lower than the pK_a of an enteric coating polymer, $[HP]_{T,p}$ is equal to the intrinsic solubility of the nonionized polymer $[HP]_0$. This solubility is often very low for enteric coating polymers, and thus dissolution of the coating layer at a low pH is very slow. At a pH higher than the pK_a of the polymer, $[HP]_{T,p}$ is given by

$$[HP]_{T,p} = [HP]_0\left(1 + \frac{K_P}{[H^+]_p}\right) \tag{5.8}$$

where K_p is the ionization constant of the polymer, and $[H^+]_p$ is the proton concentration at the interface of polymer and diffusion layer. This proton concentration can be calculated by

$$[H^+]_p^4 + a[H^+]_p^3 + b[H^+]_p^2 + c[H^+]_p + d = 0 \tag{5.9}$$

where a, b, c, and d are constants defined by the ionization constants of the polymer, buffer, and drug; the diffusion constants of the polymer, buffer, and drug; the pH of the dissolution medium; the concentrations of the polymer, buffer, and drug; and the intrinsic solubility values of the polymer, buffer, and drug.

In a similar fashion, the drug (weak acid in this case) release rate J_{HA} can be calculated from

$$J_{HA} = \frac{D_{HA}}{h}([HA]_{T,p} - [HA]_{T,b}) \tag{5.10}$$

where D_{HA} = diffusion coefficient of the drug in the diffusion layer
h = thickness of the diffusion layer
$[HA]_{T,p}$ = total concentration of drug (ionized and nonionized) at the interface of polymer and diffusion layer
$[HA]_{T,b}$ = total concentration of drug (ionized and nonionized) in bulk dissolution medium

A quasi-steady-state approximation may be used to describe the variation of the polymer thickness with time and hence to calculate the time for onset of disintegration:

$$-\rho_M \frac{dR}{dt} = J_{HP} \tag{5.11}$$

where ρ_M is the molal density of the polymer, and R is as defined earlier.

The time required for an enteric coat to be dissolved (or the time for onset of disintegration) may be obtained by substitution of Eq. (5.7) into Eq. (5.11) and integration:

$$t = \frac{h\rho_M}{D_{HP}} \int_R^{R_2} \frac{dR}{[HP]_{T,p}} \tag{5.12}$$

where R_2 is the initial position of the interface of polymer and diffusion layer.

Derivation of the preceding equations involves rigorous mathematical mass balance equations taking into account reactions of all three components (drug, polymer, and buffer), as well as diffusion through the polymer layer and the stagnant diffusion layer that are solved by applying moving boundary conditions.[8]

5.2.3 Dissolution of matrix systems

The delivery from these systems often follows a certain time course determined by the selection of the polymer and the geometry of the matrix. This type of delivery systems is suitable for reducing the frequency of drug administration, reducing toxicity for drugs with a small therapeutic window, and correcting poor pharmacokinetic behavior such as a short half-life. When solid drug particles are embedded in matrix systems, the release mechanism is more complicated than that of solid-powder systems and largely depends on the design of the matrix systems. There are many types of matrix systems where the release can be expressed using different mathematical models. In this section, only three systems in which dissolution plays a significant role will be discussed.

Surface erodible matrix systems. The first system is a solid matrix that does not disintegrate nor swell during dissolution but dissolves from the surface that is exposed to a dissolution medium. In this case, the drug is released from the eroding surface, and the dissolution profile simply follows Eq. (5.2).

Nonerodible systems. In the second matrix system, the matrix does not change during dissolution (insoluble, no disintegration, and no swelling). Polymers that are hydrophobic or cross-linked polymers often are used for the matrix. The drug solid is dissolved inside the matrix and is released by diffusing out of the matrix. Both dissolution and diffusion contribute to the release profile of this type of matrix systems. The mathematical expression for this system can be derived from the following equation:

$$\frac{dC}{dt} = D\frac{d^2C}{dx^x} + K(C_s - C) \tag{5.13}$$

where C = concentration of dissolved drug inside the matrix
C_s = solubility of the drug inside the matrix
D = diffusivity inside the matrix
K = dissolution parameter of the active drug inside the matrix

The first term on the right-hand side of the equation represents diffusion inside a matrix, and the second term corresponds to the particle dissolution. Comparing Eq. (5.13) with Eq. (5.2), one can obtain

$$\frac{dC}{dt} = \frac{A}{V} \times \frac{D}{h}(C_s - C_b) \tag{5.14}$$

$$K = \frac{AD}{Vh} \tag{5.15}$$

where A = total surface area of the active drug
V = volume of the matrix
h = hydrodynamic diffusion layer surrounding solid particles inside the matrix

Soluble matrix systems. The third matrix system is based on hydrophilic polymers that are soluble in water. For these types of matrix systems, water-soluble hydrophilic polymers are mixed with drugs and other excipients and compressed into tablets. On contact with aqueous solutions, water will penetrate toward the inside of the matrix, converting the hydrated polymer from a glassy state (or crystalline phase) to a rubbery state. The hydrated layer will swell and form a gel, and the drug in the gel layer will dissolve and diffuse out of the matrix. At the same time, the polymer matrix also will dissolve by slow disentanglement of the polymer chains. This occurs only for un-cross-linked hydrophilic polymer matrices. In these systems, as shown in Fig. 5.3, three fronts are formed during dissolution[9–11]:

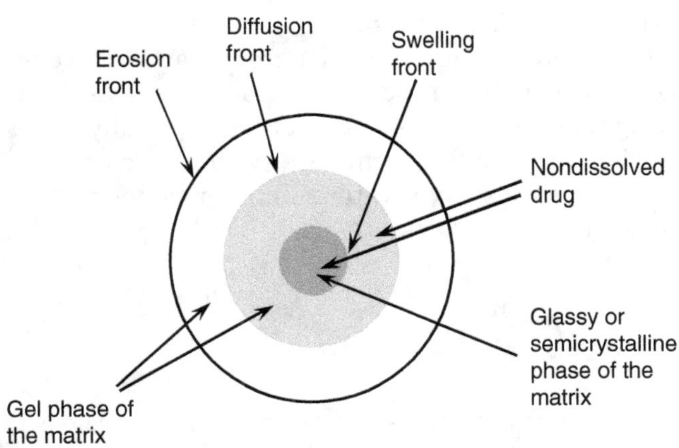

Figure 5.3 Schematic of a swelled hydrophilic polymer matrix.

- The *erosion front* between the dissolution medium and the erosion (or dissolving) surface
- The *diffusion front* between the dissolved and undissolved drug in the gel (or swelled) phase
- The *swelling front* between the gel phase and the glassy (or semicrystalline) phase of the matrix

When such a system is in contact with an aqueous solution, at the early stage of release, the swelling of the matrix causes the erosion front to move outward and the swelling front inward. At the same time, the diffusion front is also receding owing to dissolution of the drug solid in the gel phase and diffusion of the dissolved drug out of the matrix. During the progress of dissolution, the polymer chains at the erosion front begin to disentangle and dissolve away into the dissolution medium. This surface erosion slows down the swelling (outward movement of the erosion front) and causes the erosion front to recede (inward movement of the erosion front) at the later stages of release.

Harland et al.[12] developed a model for drug release based on mass balances of the drug and the solvent at the swelling front and the erosion front. The release profile was found to be a combination of Fickian and zero order, as shown by Eq. (5.16):

$$\frac{M_t}{M_\infty} = A\sqrt{t} + Bt \tag{5.16}$$

where M_t and M_∞ are the amounts of drug released at time t and infinity, and A and B are constants that are functions of the properties of the polymers, drugs, and solvents. The model suggests that at the early stage of dissolution, the diffusion of dissolved drug molecules through the gel layer limits the dissolution, and the release profile is Fickian. At the later stage of dissolution, when the erosion front starts to recede, dissolution (or erosion) of the polymer matrix controls the release profile. Therefore, the drug release approaches zero order, especially when the movements of erosion and swelling fronts are synchronized.

In general, the release profiles from water-soluble polymer matrix systems often are modeled[13,14] simply as

$$\frac{M_t}{M_\infty} = kt^n \tag{5.17}$$

where k is the kinetic constant that measures the rate of drug release, and n is the release exponent indicative of the release mechanism. If

$n = 0.5$, the release is diffusion controlled. This often happens at the early stage of release when the polymer matrix is swelling. As the polymer matrix starts to recede (dissolve), the value of n will increase and approach 1. In this case, polymer dissolution is the release controlling factor.

5.3 Parameters for Design of Dissolution Controlled Release Matrix and Coated Systems

In this section the controlled release delivery systems applying dissolution as the major release mechanism are discussed. Since many of these systems actually apply to both the concepts of dissolution and diffusion in the design, only the dissolution parameters affecting the release profiles and the release rates of these systems are analyzed.

5.3.1 Parameters affecting dissolution of solid particles

From Eq. (5.2), the major parameters affecting the release (dissolution) of a drug are the solubility, the particle size (thus affecting the surface area of the drug solid), and the thickness of the hydrodynamic diffusion layer. These parameters, however, can be influenced by other factors. For example, the diffusion coefficient (diffusivity) is a function of temperature, the viscosity of the solvent, and the molecular size (or weight) of the dissolving material. For drug molecules with molecular weights of several hundreds, the diffusion coefficient in aqueous solutions at 25°C is in the range of 0.5×10^{-5} to 1×10^{-5} cm^2/s. In addition, the thickness of the diffusion layer h is itself affected by temperature, the viscosity of the solvent, the geometry of the dissolving surface, and the hydrodynamics of the stirring solvent.

Although Eq. (5.2) seems to be simple, it is actually very complicated because of the preceding factors. However, it does provide some theoretical basis for rational design of dissolution controlled release systems. Therefore, surface area (related to particle size), solubility, and viscosity may be the parameters that could be regulated or modified to suit certain desired release profiles.

For dissolution of solid particles, the Hixson-Crowell cube-root law (Eq. 5.3) assumes that the thickness of the diffusion layer h is constant during dissolution. However, this is not necessarily true. In addition, most drug particles are nonspherical and nonuniform in size. Therefore, very often the dissolution mechanism of solid drug particles is actually much more complicated. Nevertheless, the Hixson-Crowell cube-root law provides the first approximation to model powder dissolution.

For small spherical particles less than 50 μm in diameter, the thickness of the diffusion layer can be estimated to be the same as the particles' radius if the particles are well agitated in an aqueous solution. By employing this assumption, one can estimate the time for complete dissolution τ for particles with r_0 as the initial radius dissolving under sink conditions ($C_b \approx 0$) using the following equation:

$$\tau = \frac{\rho r_0^2}{2DC_s} \tag{5.18}$$

This equation again demonstrates that particle size and solubility are the main parameters affecting dissolution kinetics of drug powders, which, in turn, could affect the release profile of dosage forms if dissolution is the rate-limiting step of in vivo absorption. Table 5.1 demonstrates several examples of dissolution times of spherical particles (assuming monodispersed systems) as a function of solubility and particle size.

5.3.2 Parameters affecting dissolution of coated systems

For water-soluble coatings that consist mainly of polymers, dissolution or erosion of the coat is the rate-limiting step toward the controlled release. After the coat is dissolved, the drug substance in the core is released, and the release kinetics depend on the core properties. Based on Eq. (5.2), solubility and dissolution/hydration behaviors of the pri-

TABLE 5.1. Dissolution Times for Drug Particles with Different Solubilities and Particle Sizes

(*Note*: Calculated based on Eq. (5.18) using $\rho = 1.5$ g/cm^3 and $D = 0.5 \times 10^{-6}$ cm/s)

Solubility, mg/mL	Diameter of particle size, μm	Time for complete dissolution, min
1	1	6.25×10^{-3}
1	10	0.625
1	50	15.6
0.1	1	0.0625
0.1	10	6.25
0.1	50	156
0.01	1	0.625
0.01	10	62.5
0.01	50	1563

NOTE: The values in this table suggest that dissolution rate (or release rate) can be modified significantly by the solubility and particle size of solid drug particles.

mary coating material (polymer) are the most important formulation variables for modulation of release from these systems. Hydration time is the time required for a polymer to reach maximum viscosity in a solvent. The rate at which this process occurs is the hydration rate. The solubility and hydration/dissolution of a polymer depend on the chemical composition of the polymer and molecular weight, which, in turn, determine the viscosity generated on dissolution.

For low-molecular-weight polymers (e.g., methylcelluloses), a significant hydrated layer is not maintained. As the molecular weight increases, the viscosity of the hydrated layer increases and on contact with water slowly forms a gel (e.g., such as for HMC 90 HG 15,000 cps). Drug release is controlled by penetration of water through a gel layer produced by hydration of polymer and diffusion of drug through the swollen hydrated matrix in addition to dissolution of the gelled layer. The extent to which diffusion or dissolution controls release depends on the polymer selected (molecular weight), as well as on the drug-polymer ratio.[3] It has been proposed that the faster-hydrating polymers are more desirable because rapid gel development limits the amount of drug released initially.[15]

The effect of ions on the degree of hydration of cellulose ethers has been studied. Depending on the polymer, the type and concentration of ions can affect to varying degrees the extent of hydration. Changes in the hydration state result primarily in solution-viscosity and cloud-point changes. These effects were demonstrated with hydroxy propyl cellulose.[16]

The amount of polymer applied as the coat (reflected as coating thickness) and the distribution of the thickness of the polymer coat are also key formulation parameters[3] used to modulate the release profile. Complete dissolution of the coat leads to an abrupt release of contained drug. Based on Eq. (5.12), thickness of a polymer coat affects the onset time for complete dissolution of the coat. For multiparticulate systems, if a dosage form consists of only three or four different-thickness coats, one expects pulsed dosing, i.e., repeat action, to occur. On the other hand, if a spectrum of different particle coats is employed in the dosage form, continuous release of drug is expected.[5] The number of particles included in each group can be manipulated to alter the release pattern for these systems.[3]

Other formulation parameters that may be used to modulate the release include the ratio (relative concentrations) of polymers in the case of incorporation of a mix of two or more polymers as primary coating material, the properties of the core material, and the amount of plasticizers used, which affects the strength of the coat. Plasticizers with low water solubility such as dibutyl sebacate, diethyl phthalate, triacetin, triethyl citrate, and acetylated monoglyceride result in delaying the release as compared with water-soluble plasticizers (polyethylene glycols).[17]

For enteric-coated controlled release systems, the release profiles depend on the pH of the targeted release region. These systems are fabricated by coating the drug-containing core with a pH-sensitive polymer combination.[18] According to Eqs. (5.7) through (5.12), the main parameters affecting the release of the drug, the dissolution of the polymer, and the onset of tablet disintegration are the ionization constants of the drug in the core and the polymer, the intrinsic solubility of the drug and the polymer, the diffusion coefficients of the drug and polymer in the polymer coating layer and in the diffusion layer, and the coating thickness. Ozturk et al.[8] reported that in such a system, the polymer dissolves quickly initially. As the polymer layer thins, the resistance to drug transport decreases, and so the drug is released faster. Meanwhile, the increase in concentration of acidic drug in the coating layer as dissolution progresses results in a reduction in the pH, and this, in turn, leads to a reduction in the polymer dissolution rate. Overall, the dissolution rate of the polymer decreases with time, whereas release rate of the drug increases with time.

Among the parameters discussed earlier, dissolution controlled release dosage forms can be designed to achieve the desired release profiles by manipulating the parameters related to the coating polymer.

5.3.3 Parameters affecting dissolution of matrix systems

Three types of matrix systems were discussed earlier: solid matrix (no disintegration, no swelling), porous matrix (insoluble, no disintegration, no swelling), and water-soluble hydrophilic swellable matrix. For the first matrix system, only the drug at the surface is released. The release profile, often expressed as the amount released versus time, is a function of the change of surface geometry and surface area of the matrix. Therefore, the surface geometry and surface area play a significant role in dissolution. In addition, where water-soluble polymers such as polyethylene glycols are used, the viscosity in the diffusion layer adjacent to the dissolution surface also can contribute to the release profile and release rate. If the dissolution/erosion surface and the viscosity in the diffusion layer can be maintained constant during dissolution, a zero-order release profile is obtained.

Unless the drug loading is very high (>50 percent), the dissolution rate of the matrix often is determined by the properties of excipients, mainly the solubility and viscosity. The more soluble the excipients and the less viscosity generated in the diffusion layer, the faster is the matrix dissolution.

Since only dissolved drug molecules can be absorbed by the body, one should understand the fate of drug solid particles after they are released from the erosion surface of a delivery system. The drug solid particles will start to dissolve in the dissolution medium once they are released. The dissolution rate of these drug particles is a function of the solubility of the drug and their particle size, as shown by Eq. (5.18) and Table 5.1. Based on Table 5.1, for drugs with solubilities of 10 mg/mL or higher, the dissolution rate of drug particles is usually much faster than the dissolution/erosion rate of matrix systems. Therefore, the release rate of dissolved drug molecules is almost the same as the erosion rate of the matrix. On the other hand, for drugs with solubilities of 0.1 mg/mL or less, the dissolution rate of drug solid particles can be very slow (unless the particles are micronized with particle size less than 10 μm) and is the rate-determining step.

For the second type of matrix system, water penetrates the matrix and dissolves the drug particles. In this case, both dissolution and diffusion contribute to the release profile according to Eq. (5.13). The dissolved drug molecules diffuse out of the matrix and are released into the dissolution medium. At pseudo-steady state (i.e., $dC/dt \approx 0$), the drug concentration inside the matrix will be relatively constant or change slowly with time. This pseudo-steady-state concentration inside the matrix will depend on the balance of the dissolution rate of particles and the diffusion rate of dissolved drug substance. If the dissolution rate is much faster than the diffusion rate, the pseudo-steady-state concentration inside the matrix will be close to the solubility of the compound. This situation often happens if the solid particles are small or drug loading is high. On the other hand, if solid particles are large and drug loading is low, the pseudo-steady-state concentration inside the matrix will be lower than the solubility of the drug substance.

Chandrasekaran and Paul[19] have found that for such systems where dissolution is the rate-limiting step [small A/V in Eq. (5.14), where drug loading is low and particle size is large], the release is linear with time, and the release rate is a zero order, as shown by

$$\frac{M_t}{M_\infty} = 2\frac{C_s}{C_0}\sqrt{\frac{DK}{l^2}}\left(\frac{1}{2K}+t\right) \tag{5.19}$$

where M_t and M_∞ = amounts of drug released at time t and infinity
C_0 = drug loading
C_s = solubility
D = diffusion constant of drug molecules in the matrix
K = dissolution constant (a function of A/V)
l = thickness of the slab matrix

On the other hand, for systems where diffusion is the rate-limiting step [large A/V in Eq. (5.14), where drug loading is high and particle size is small], the release is a function of the square root of time, as indicated by the Higuchi equation.[20] At the later stage of drug release, the system is always dissolution controlled because of the low drug loading encountered.

Higuchi[20] developed a model to describe the release profile of drug solids dispersed in a matrix. This model ignores the dissolution of drug particles inside a matrix and assumes that the concentration of a drug inside the matrix is the solubility of the compound. This assumption is only true if a delivery system has high drug loading and is at the early stage of release, where the release is purely diffusion controlled.

Another study[21] also suggested that in dissolution controlled systems (i.e., systems with low drug loading, large drug particle size, and at the later stage of dissolution), the drug release from monodisperse spherical microparticles is a linear function of time, or the release rate is zero order.

The third type of matrix system involves water-soluble and swellable polymers. Their release profile and rate are based on the movements of different fronts (erosion front, diffusion front, and swelling front) during dissolution, as modeled by Eq. (5.16). The solubility, hydration time, and viscosity of the matrix polymers are the parameters that can be manipulated to change the constants A and B in Eq. (5.16) and to modify the release profile and rate of a matrix system. Since the properties of a drug to be delivered (such as solubility and drug loading) are also part of the constants A and B, selection of a polymer with appropriate properties for a certain desired release profile and rate should be considered together with the properties of the embedded drug. Selection of polymers with low water solubility, high viscosity, and slow hydration times results in a slow-moving erosion front. This makes the constant A in Eq. (5.16) much larger than constant B, leading to a slow release rate and Fickian release profile. Similarly, for highly water-soluble drugs and low drug loading, the diffusion front can move as fast as the swelling front.[9] The thickness of the gel layer (distance between the erosion front and the swelling front) controls the release of a drug, and the drug release profile follows Fickian behavior.

For systems with highly water-soluble polymers that hydrate quickly and/or low viscosity, the erosion front moves fast, resulting in a faster drug release rate. In this case, the constant A in Eq. (5.16) is much less than B, leading to a linear time release. This situation also can happen for high drug loading systems or not very soluble compounds, where the movement of the diffusion front of these systems may not be as fast as the swelling front. For these systems, the distance between the diffusion and erosion fronts controls drug release instead of the thickness of the whole gel layer (distance between the erosion and swelling fronts).[10] Synchronization of the movement of the diffusion and erosion fronts leads to a zero-order drug release.

5.4 Applications and Examples of Dissolution Controlled Release Matrix and Coated Systems/Technologies

5.4.1 Delivery systems based on dissolution controlled release solid particles

Modification of physicochemical properties of drug solids is the most traditional method to alter (to enhance or slow down) the dissolution profiles of dosage forms. At the infancy of controlled release techniques, this concept was used in depot-type parenteral controlled release formulations. This type of formulation, which often includes aqueous (or oleaginous) suspensions, acts as a drug reservoir in subcutaneous or muscular tissues that provides constant delivery of dissolved drug molecules, therefore simulating intravenous infusions. In addition to prolonged therapeutic activity with a low frequency of injection, the benefits also include decreasing drug dose, as well as fewer side effects and enhanced patient compliance.

For dissolution controlled release parenteral depot formulations, approaches involving the formation of low-solubility salts or complexes and the control of particle size were used. Typical examples of the first approach are penicillin G benzathine suspension (Bicillin L-A, Wyeth) and penicillin G procaine and penicillin G benzathine combination suspension (Bicillin C-R, Wyeth). These low-aqueous-solubility penicillin G salts can sustain the therapeutic blood level for 24 hours or longer (depending on the concentrations of suspensions), compared with the high-aqueous-solubility salts (such as sodium and potassium salts) of penicillin G that can maintain the therapeutic level for only a few hours. The duration of action of regular insulin is usually only 4 to 8 hours. Therefore, patients may require several injections daily to control their diabetes. Since insulin can react with zinc ion, forming a water-insoluble solid complex, the suspension of this complex, injected subcutaneously, can provide long duration of action. Depending on the pH of the solution, the precipitated solids have different crystallinity and thus different solubility. In acetate buffer at pH 5 to 6, crystalline insulin–zinc complex solid[22] (Humulin U Ultralente, Lilly) can be formed, providing a slower onset and a longer and less intense duration of activity (up to 28 hours) compared with regular insulin. Amorphous insulin–zinc complex solid can be formed at pH 6 to 8 and achieves faster onset and a duration of action that is shorter than crystalline insulin–zinc complex (Untralente). An intermediate-acting insulin (Humulin L Lente, Lilly) with a duration of action of up to 24 hours was made by mixing crystalline and amorphous insulin–zinc complexes. These different release profiles of formulations for single daily injection provide flexible selections for delivery.

An example of the second approach is penicillin G procaine suspensions, where an increase in the particle size resulted in prolongation of the therapeutic level (0.03 unit/mL) from 24 hours for particles at 1 to 2 μm to more than 72 hours for particles at 150 to 250 μm.[23]

Regulation of physicochemical properties of drug powders for controlled release also has been applied in oral drug delivery. Solubility and particle size may be manipulated to a certain extent for this purpose. One example is a controlled release system where the dosage form is required to dissolve or disintegrate within a very short time, often in mouth. Fast-melt dosage forms employ the benefit of high solubility and large surface area of excipients to achieve instantaneous disintegration/dissolution (melt) of dosage forms in the mouth. Zydis[24] (by Cardinal Health) and DuraSolv[25] (by CIMA Laboratories) both apply this concept, with some differences in their manufacturing technologies and selection of excipients.

For poorly water-soluble compounds, release of the active ingredient is often too slow to achieve the desired in vivo exposure. In these cases, selection of highly water-soluble salts and reduction of particle size by micronization or nanosizing have been applied widely to improve dissolution for an immediate release dosage form. As demonstrated in Table 5.1, even for drugs with solubilities of less than 0.1 mg/mL, reducing particle size to the micron or submicron range could achieve dissolution within reasonable times. Patented technologies such as NanoCrystals[26,27] and DissoCubes[28] are just two examples.

For highly water-soluble compounds, dissolution and absorption usually are complete within a few hours or less (given that permeability is adequate). For compounds with short half-lives, repeated dosing may be required to maintain in vivo drug concentrations within therapeutical levels. In these cases, extended release dosage forms are suitable to overcome the frequent dosing problem, leading to better patient compliance.

Unlike extended release parenteral formulations (depot suspensions), application of low-water-solubility salts in oral extended release forms is not used commonly. This probably is due to the fact that for oral delivery, coated and matrix-type systems are easy to design and can generate release profiles that are manipulated more easily than powder systems. Therefore, although it is not impossible, selecting a salt form with low water solubility to achieve a desired extended release profile has not been reported frequently.

5.4.2 Delivery systems based on dissolution controlled release coated technologies

Coating purposes and components. According to *the United States Pharmacopeia* (XXII), three classes of coating are employed commonly in the manufacture of solid dosage forms. The oldest of these, called *plain coatings*, are used to alter the taste and appearance of tablets and to protect them from the detrimental effects of light and moisture, e.g., sugar and hydroxyl propyl methyl cellulose. These are not intended to alter the biopharmaceutical performance of the drug contained within them. The second group of coatings, called *delayed release* or more commonly *enteric coatings*, are insoluble at the low pH of the stomach but dissolve at the higher pH values of the intestine (e.g., cellulose acetate phthalate). Some of the most important reasons for enteric coating are[29]

- To protect acid-labile drugs from the gastric fluids (e.g., enzymes and certain antibiotics)
- To prevent gastric distress or nausea owing to irritation from a drug (e.g., sodium salicylate)
- To deliver drugs intended for local action in the intestine (e.g., intestinal antiseptics could be delivered to their site of action) in a concentrated form
- To deliver drugs to their primary absorption site in their most concentrated form (e.g., drugs that are optimally absorbed in the small intestine or colon)
- To provide a delayed release component for repeat-action tablets

The third group of coatings consists of controlled release coatings.[30] As mentioned earlier, only water-soluble coats will be discussed here. Combinations of those with insoluble coats also will be described briefly. This coating technology has a history of fewer than 55 years. The development of such products depends to a very considerable degree on the availability of chemically modified coatings (especially cellulose derivatives), which are supplied to the pharmaceutical industry as materials of reliable quality. Also, the improvements in coating equipment technology, especially the invention of the Wurster film-coating device, have been essential to improvements in coating technology.[2]

The primary coating materials, usually polymeric (film formers), often require the addition of other excipients such as plasticizers, pore formers, colorants, or antiaggregation agents for the coating to perform in the desired fashion or for the product to be manufactured conveniently.[2] These components have been the subject of numerous studies and reviews.[31,32] The components that affect/modify the release from these systems are film formers, plasticizers, and pore formers and are discussed below.

Single-unit versus multiparticulate coated controlled release systems. The essential elements of coated controlled release pharmaceutical product are a core (consisting of a drug or a drug plus excipients) encased by (a) layer(s) of material(s) that regulate(s) the rate at which drug is released into the surrounding medium.

The core component can be a single-unit dosage form (e.g., tablet or capsule) or a multiple-unit dosage form (e.g., pellets, granules, drug loaded on nonpareil seeds, microparticles, microcapsules, or microspheres) surrounded by one or more layers of regulating coats. The multiparticlulate units then can be placed in capsules or compressed directly into tablets. A multiparticulate unit (tablet or capsule) of this type may contain hundreds of color-coated pellets/granules/beads, etc. divided in three or four groups that differ in the thickness of the time-delay coating. A typical mix consists of pellets/granules//beads providing the release of drug, for example, at 2 or 3 hours, 4 or 6 hours, and 6 or 9 hours to offer pulsed dosing (repeated action) over the desired time. Some units within each group release drug at intervals overlapping with other groups, thus resulting in a smooth rather than discontinuous release profile.

In order to provide loading and maintenance dosing, controlled release dosage forms may consist of two parts: an immediately available dose to establish the required blood levels quickly (loading or initial priming dose) and a sustained part containing several times the therapeutic dose for protracted drug levels (maintenance dose). Several approaches are available to incorporate the immediately available portion with the sustaining part. For single-unit systems, placement of the initial dose in the coat of a tablet or capsule with the sustaining portion in the core has been reported.[5] For multiple units, it is common practice to employ one-fourth or one-third of the particles in nonsustained form, i.e., particles without a barrier membrane to provide for immediate release of drug. Alternatively, a portion of drug can be placed in a faster-dissolving coating membrane (less thickness) to establish therapeutic levels quickly.[1]

Although similar drug release profiles can be obtained with both dosage forms, multiparticulate unit dosage forms offer several advantages.[33] Multiparticulate units spread uniformly throughout the GI tract. High local drug concentrations and the risk of toxicity owing to locally restricted tablets can be avoided. Premature drug release from enterically coated single-unit dosage forms in the stomach, potentially resulting in degradation of the drug or irritation of the gastric mucosa, can be reduced with coated pellets because of a more rapid transit time when compared with enterically coated tablets. The better distribution of multiparticulates along the GI tract could improve the bioavailability, which potentially could result in a reduction in drug dosages and side effects. Inter- and intraindividual variations in bioavailability caused, for example, by food effects are reduced owing to less variation in gastric transit time and gastric emptying. With coated single-dosage forms, the coating must remain intact during the controlled release phase; damage to the coating would result in a loss of the sustained release properties and dose dumping, whereas unwanted dose dumping from pellets is practically nonexistent.[34] There is also statistical assurance of drug release with encapsulated forms because release of drug by a significant fraction of the granules is highly probable. If a core tablet fails to release drug, the entire maintenance dose is lost.[3]

Materials used in controlled release products. Following is a list of the most commonly used water-soluble polymers and plasticizers for controlled release coating, as well as a description of their properties. This list is not meant to be comprehensive of all materials ever proposed for use in such products.

Materials used in enteric coatings

- *Cellulose acetate phthalate (CAP).*[35] This material is used widely in the industry and dissolves only above pH 6, thus delaying absorption of drugs. In comparison with some other enteric polymers, CAP is hygroscopic and relatively permeable to gastric fluids. In addition, it is susceptible to hydrolytic removal of phthalic and acetic acids, resulting in a change of film properties. Because of its brittleness, CAP usually is formulated with hydrophobic film-forming materials or adjuvants to achieve a better enteric film. It is available as an aqueous colloidal dispersion of latex particles.
- *Methacrylic acid polymers (Eudragits).*[35] Polymethacrylates are synthetic cationic and anionic polymers of dimethylaminoethyl methacrylates, methacrylic acid, and methacrylic acid esters in varying ratios. Several different types are available commercially and may be obtained as dry powder, as an aqueous dispersion, or as an organic solution. Eudragit E 12.5 and E 100 are soluble in gastric fluid from pH 5 and are both used in film coatings. For enteric coating, the following polymers can be selected based on the desired release pH range: Eudragit L 12.5 P (soluble in intestinal fluid from pH 6), L 12.5 (soluble in intestinal fluid from pH 6), L 100 (soluble in intestinal fluid from pH 6), L 100-55 (soluble in intestinal fluid from pH 5.5), L 30 D-55 (soluble in intestinal fluid from pH 5.5), S 12.5 P (soluble in intestinal fluid from pH 7), S 12.5 (soluble in intestinal fluid from pH 7), and S 100 (soluble in intestinal fluid from pH 7). Eastacryl, Eastacryl 30 D, Kollicoat, and Kollicoat 30 D and 30 DP are soluble in intestinal fluid from pH 5.5.
- *HPMC phthalate.*[35] This type of polymer has molecular weight ranges of 20,000 to 200,000 Da. Three grades are available, HP-50, -55, and -55S (dissolves in aqueous buffer solutions at pH 5, 5.5, and 5.5, respectively, with S designating a higher-molecular-weight grade that produces films with a greater resistance to cracking).

- *Polyvinyl actetate phthalate (PVAP[35])*. This is used often at concentrations of 9 to 10 percent for tablet enteric film coating. Insoluble in buffer solutions below pH 5 and soluble at pH values above 5, it shows a sharp solubility response with pH at 4.5 to 5. In addition to environmental pH, its solubility also can be influenced by ionic strength.
- *Shellac.*[2] This is a naturally occurring polymer obtained from a gummy exudation produced by female insects. The pH at which drug is released is about 7, which may well be too high for most enteric-coated products. It is not recommended for developing a new product.

Materials used in nonenteric coatings

- *Methylcellulose.*[35] This material swells in cold water and disperses slowly to form a clear to opalescent viscous colloidal dispersion. Various grades with different degrees of polymerization (50 to 1000 with molecular weight averages of 10,000 to 220,000 Da) provide various viscosities at the same concentration (at 2% w/v, aqueous solution viscosity varies from 5 to 75,000 mPa·s). Viscosity of solutions also may be increased by increasing concentration. This material forms a gel at higher temperature (50 to 60°C), which is reversible to a viscous solution on cooling.
- *Hydroxyethylcellulose.*[35] This polymer is soluble in hot or cold water, forming clear, smooth, uniform solutions. It is available in various viscosity grades ranging from 2 to 20,000 mPa·s for a 2% w/v aqueous solution. It can be used as a thickening agent in phthalmic and topical formulations, as a bioadhesive in mucoadhesive patches, as a matrix controlled release polymer in solid dosage forms, and as a binder and film coating agent for tablets.
- *Hydroxyethylmethyl cellulose.*[35] Although this material is practically insoluble in hot water (above 60°C), it can dissolve in cold water to form a colloidal solution and has similar properties to HPMC.
- *Hydroxy propyl cellulose.*[35] This material is soluble in water below 40°C (insoluble above 45°C), GI tract fluids, and many polar organic solvents. It yields flexible films but is not usually used alone (combination with other polymers). Molecular weight is varied by controlling the degree of polymerization (DP) of the cellulose backbone. The DP controls the viscosity such that as the DP increases, the viscosity increases. Low-viscosity grades are used as tablet binders in immediate release dosage forms, and medium- and high-viscosity grades are used in sustained release formulations. Viscosity of a 2% w/v aqueous solution is 150 to 6500 mPa·s. The release rate of a drug increases with decreasing viscosity of HPC.

- *Sodium carboxymethylcellulose (Na CMC).*[35] As an anionic water-soluble polymer, this material's aqueous solubility varies with the degree of substitution (average number of hydroxyl groups substituted per anhydroglucose unit). Various grades differ in viscosity (1% w/v aqueous solution has a viscosity of 5 to 13,000 mPa·s). Increase in viscosity can be achieved by using different grades or by increasing the concentration. This material can be dispersed easily in water at all temperatures, forming clear colloidal solutions.
- *Hydroxy propyl methyl cellulose (HPMC).*[35] HPMC is a partly *O*-methylated and *O*-(2-hydroxypropylated) cellulose available in several grades that vary in viscosity and extent of substitution. It is used widely in pharmaceutical formulations, especially in oral products, as a tablet binder, in film coating, and as controlled release matrix. Soluble in cold water, it forms a viscous colloidal solution. For a 2% aqueous solution (20°C), viscosity can range from 2.4 to 120,000 mPa·s. High-viscosity grades can be used to retard the release of water-soluble drugs from a matrix.
- *Sodium alginate.*[35] This material is insoluble in aqueous solutions in which pH is less than 3. It dissolves slowly in water, forming a viscous colloidal solution. Various grades yield various viscosities (1% w/v aqueous solution has a viscosity of 20 to 400 mPa·s at 20°C). Viscosity may vary depending on concentration, pH, and temperature.

Manufacturing methods for the application of coating materials for controlled release. Pan coating employing solvent evaporation was used for the oldest form of pharmaceutical coating—sugar coating. It is also used extensively for film coating of single-unit core tablets. The coating is sprayed into the tablet bed often with the assistance of an air jet.[2] The relative inefficiency of drying, together with the long period of time for the cores that are required to remain in the conventional pan, may cause discontinuity or irregularity in the film.[36]

Fluidized-bed coating using solvent evaporation[2] is preferred over pan coating for coatings showing minimal defects and tablet-to-tablet variability. It has been used extensively to coat cores to obtain desired properties such as controlled drug release, enteric release, and elegant appearance and taste masking. The fundamental principle behind the fluidized-bed coating in general and the Wurster technique in particular is to suspend tablets in an upward-moving column of warm air during the coating process. This minimizes tablet abrasion and unevenness of film distribution caused by tablet-to-tablet contact in pan coating. The coating is built up in a series of incremental steps; thus, from a processing point of view, although not in terms of composition or function, the coating is multilayered. This cyclic process of spraying, drying, spraying, and drying performed over a period of time can result in optimal conditions for gradual deposition of a coating of uniform thickness and structure. Batches of product from about 0.5 to 500 kg can be coated, and particles as small as 50 μm up to conventional tablets can be coated on this type of equipment. A detailed comprehensive review of Wurster and other fluidized-bed coating technologies has been published by Christensen and Bertelsen.[37]

It is also possible to use compression-coating technique to compress a coating around a preformed (relatively soft) core by using a spherical tablet press. The process basically consists of compression of the core to give a relatively soft compact that is then fed into the die of a tablet press that has already received half the coating material. The core is centered within the die, the remainder of the material is added, and the product is compressed.[38] An example is Smartrix tablets, in which the release profile of a drug is determined by the increase in release surface caused by erosion (dissolution) of the cover layers.[39]

The technique of microencapsulation has been used to encase particles of liquids, solids, or gases. One of the more common approaches is coacervation, which involves the addition of a hydrophilic substance to a solution of a colloid. It starts with a three-phase system consisting of colloidal drug particles, colloidal coating material, and liquid vehicle, followed by deposition of coating material on drug droplets and solidification of the coating material.[5] Microencapsulation has the additional advantage that sustained drug release can be achieved with taste abatement and better GI tolerability. Good examples of microencapsulations are microencapsulated aspirin and potassium chloride. In both cases, drug effects from the microencapsulated dosage forms are more prolonged and less irritating than the same amount taken as ordinary tablets. Both formulations show the same total drug absorbed. One of the disadvantages of this technique is that no single process can be applied to all core material candidates.[40]

Electrostatic coating has been developed recently to allow for the deposition of thin polymeric films without the need for any solvent. Films are formed when a charged particle is attracted to a substrate of opposite charge.[41] An example is the Accudep controlled release system.[42]

The effect of coat mechanical properties and processing parameters on release profiles. Coat mechanical properties and processing parameters can have an indirect effect on the parameters described in the models and thus indirectly affect the release profile and rate.[43] A coat should have good strength to avoid premature breaking. In addition, a coat should be flexible enough to sustain expansion of the core during dissolution. For example, hydroxy propyl methyl cellulose (HPMC) has a very high tensile strength and a very low elongation value. A great deal of force can be applied before an HPMC film breaks, but the film lengthens only a small amount before the break occurs, so to circumvent this problem, plasticizers are added to improve flexibility. Hydroxy propyl cellulose (HPC) has a lower tensile strength and much higher elongation value than HPMC. Not as much force is required to break a film of HPC, but the film itself will stretch a great distance before it breaks. Using a combination of both polymers, bridging of tablet monograms can be eliminated, film adherence problems to tablet substrates is improved dramatically, and the incidence of film cracking on the edge of tablets is reduced greatly.[44]

Processing variables, as well as polymer composition, not only will affect the mechanical properties of a coat but also may alter the hydration time of the coating. Processing variables during the microencapsulation (coacervation) process that may affect releases properties include initial pH, initial temperature, ratio of solid to encapsulating materials, and final pH.[5]

For fluidized-bed coating, processing variables such as temperature, volume, and humidity of fluidizing air; spray rate; and atomization pressure can be adjusted to obtain the required coat characteristics for fluidized-bed coating.[17] The influence of fluidized-bed processing conditions, as well as curing parameters, with and without humidity, on drug release from beads coated with cellulose acetate phthalate aqueous dispersion has been investigated.[45] Theophylline beads prepared by extrusion-spheronization were coated with diethylphthalate-plasitized CAP dispersion. The parameters investigated were plasticizer level, outlet temperature, spray rate during coating application, and fluidizing air velocities using a half-factorial design. The processing temperature during coating applications was identified as a critical factor among the variables investigated. The release rate significantly decreased when the beads were coated at 36°C compared with those coated at 48°C. Higher coating efficiencies and better coalescence of films were obtained at lower coating temperatures. Subsequent removal of moisture absorbed from the beads did not significantly alter the enteric profiles obtained through heat-humidity curing. The extent of coalescence via heat-humidity curing depended on the curing temperature, percent humidity, curing time, and coating temperature. The results demonstrated the importance of the selection of coating temperature for CAP coated beads and the role of moisture on CAP film formation. Curing with humidity was found to be more effective than without.

Storage conditions also could change the release of polymer-coated controlled release systems. The effect of heat and humidity on the coating polymers was studied.[46] Thermal gelation occurred and was viewed as being responsible for changes in the dissolution profiles of some of the tablet products coated with methyl cellulose or cellulose acetate phthalate during aging.

Examples. Two examples are presented in detail to demonstrate how formulation parameters and process variables should be evaluated during the design of a controlled release coated system. In a very recent study, drug release behavior of nifedipine (a calcium channel blocker) from tablets coated with high-viscosity grades of HPMC (100,000 cps)[4] was studied. High-viscosity grades of HPMC were mainly applied to the formulation of sustained release matrix-type dosage forms, such as tablets, pellets, and granules. Although low-viscosity grades of HPMC have been used widely for polymeric film coating as an aqueous basis, high-viscosity grades of HPMC as a coating polymer have not been investigated extensively. The parameters affecting release are determined to be (1) ethanol-water (coating solvent), where a distinct lag time is observed and depends significantly on the ratio, with the higher ratio giving a long lag time of up to 8 and 10 hours (it is assumed that as the amount of water is increased, the gelling and swelling forces of the HPMC-coated film can be decreased, resulting in a short lag time); (2) HPMC concentration in the coating solution, where lower HPMC concentration results in a distinct coat without cracks or pores, which results in increasing lag time (when higher HPMC concentration is used, some pores are observed, leading to fast drug release through these pores); and (3) coating level, where lag time increases as a function of the coating levels. Less than 20 percent coating levels had no significant retarding effect. A 3-hour lag time was obtained at 30 percent coating level and 4 hours at the 40 percent coating level. Based on photoimaging analysis, the coated tablet in the dissolution medium initially swelled and gelled without dissolution and disintegration at least over 5 hours after the release test. The disintegration of the coated tablet occurred approximately 7 hours after dissolution, resulting in a pulsed release of drug.

This time controlled release tablet with a designated lag time followed by a rapid release may provide an alternative to site-specific delivery of drugs with optimal absorption windows or colonic delivery of drugs that are sensitive to low pH or enzyme action for the treatment of localized conditions such as ulcerative colitis, Crohn's disease, and irritable bowel syndrome (IBS). Also, by controlling a predetermined lag time of drug from dosage form, the release behavior can be matched with the body's circadian rhythm pattern in chronotherapy.

The second example involves tablets containing ibuprofen as a model drug and press coated with sodium alginate as the coating polymer.[47] The effect of the following parameters on drug release was evaluated: chemical composition of sodium alginate and the viscosity grade. Bioavailability in humans of several formulations also was evaluated. The conclusion was that the viscosity grade of sodium alginate is not the only parameter that predicts the release rate from this formulation. The chemical structure also has an effect. The in vivo absorption rate was controlled over a range from immediate release, to slow release, to an extended release by using different chemical structures of sodium alginate. Such systems do not provide constant drug levels in the blood, as is the case for the sustained release systems exhibiting zero-order release [relying on diffusion-controlled mechanisms (matrix or barrier) or osmotic systems]. These systems are suitable for diseases that have marked diurnal rhythms, where the therapeutic concentrations should vary during the day. Drug levels should be highest when the symptoms are most severe. For example, in rheumatism, early-morning stiffness is common. In theory, maximum drug levels can be achieved early in the morning if a formulation from which drug release increases with time is administered the previous evening.

5.4.3 Delivery systems based on dissolution controlled release matrix technologies

Materials and processing technologies. The polymers discussed previously for nonenteric coatings such as HPMC (the most widely used), PVP, CMC, and carbomer, xanthin gum, and other naturally occurring polysaccharide polymers may be used for dissolution controlled release matrix systems. Furthermore, conventional processing techniques that were discussed for coating systems also can be used for matrix systems.

Examples. Various designs of matrix systems have been developed for a constant controlled release. By embedding drug powders in a matrix system, the dissolution of drug solids becomes less significant compared with a powder system. The geometric shape of a matrix, the porosity, the dissolution and swelling profile of the matrix components, the solubility of the drug in a matrix, the diffusion of dissolved drug molecules inside the matrix, drug particle size, and drug loading all can contribute to the release profile of a matrix system.

Manipulation of the geometry and surface area of tablet matrices to provide zero-order dissolution has been studied.[48,49] Based on this concept, a patented delivery technology, Procise,[50] was developed. By keeping the dissolving surface area constant during dissolution, a zero-order dissolution is achieved. During dissolution, the diameter D of the tablet core that contains active drug decreases, whereas the thickness H of the core increases. As a result, the surface area that is exposed to a dissolution medium, that is, $H \times D \times \pi$, is constant. Furthermore, by changing the geometry of the core, various drug release profiles can be achieved.

Another example using the geometry concept for a patented controlled release technology is RingCap. During dissolution, the dissolving surface area of RingCap tablets can decrease, remain constant, or even increase with time, achieving desired release profiles. Different placement, number, and width of the bands can give different drug release profiles. This can give formulation scientists the flexibility to meet their different needs.[51]

Polyethylene glycols (PEGs) are water-soluble polymers with low melting points (from 50 to 65°C). After they are molded into a solid matrix, they erode only from the surface and usually do not swell nor disintegrate in aqueous solution. Because of this behavior, the release profiles of the dosage forms can be controlled easily. Specifically, a zero-order release profile may be achieved simply by controlling the erosion of surface geometry. However, it was found[52] that low-molecular-weight PEGs (MW 6000 Da or less) have melting points too low (close to 50°C) to be suitable for use in oral formulations. This is especially true when the active drug or other excipient can further lower the melting point. Therefore, high-molecular-weight PEGs are preferred, with the drawback of the formation of a gel layer at the erosion surface that often impedes the release of active compounds. If this happens, PEG monostearate has been reported to improve the erosion behavior.[52]

Ritger and Peppas[53] studied release from nonswellable systems in the form of slabs, spheres, and disks (or cylinders). Using Eq. (5.17), they found that for diffusion controlled release systems, n = 0.5 only for slabs or for 10 to 15 percent of release from other geometries. Lower values of n (< 0.5) have to be used for release up to 60 percent for cylinders ($n = 0.45$) and spheres ($n = 0.43$).

The effect of matrix geometry on drug release from water-soluble hydrophilic polymer-matrix systems has been reported by several researchers.[54–57] It was demonstrated that the release is faster for systems with higher surface area–volume (SA/vol), ratios, where SA is the surface area of a matrix (or tablet), and vol is the volume. Thus it is important to realize that the release profiles could be different for different size tablets, even though the formulation may be the same. Small tablets will release faster than larger tablets (higher SA/vol ratio for small tablets). Therefore, manipulation of the size and shape of dosage forms could be a way to find a desired release rate. It also was reported that variation of the radius of a matrix tablet has more effect on drug release patterns than variation of the height.[56]

When other water-soluble hydrophilic polymers are applied in matrix systems, the resulting matrices will swell and then erode during dissolution. This type of matrix systems received wide application in controlled release. This is due to the easiness of manufacture, availability of a wide selection of conventional polymer excipients, and flexibility to be fit to different release profiles. Parameters governing the release from such systems (including properties of the polymer, drug, and dissolution medium) were given and discussed earlier in Eq. (5.16). It should be noted that Eq. (5.16) describes the release profiles from one-dimensional systems such as a slab. The effects of polymer properties such as solubility, viscosity of the gel phase, swelling kinetics, and polymer loading (percentage of polymer in a unit dose) were specifically discussed by

Colombo et al.[10] They have shown that drug release from highly water-soluble polymers such as PVP is dissolution controlled and follows a zero-order release, and the rate of release from such polymers is fast. For polymers with lower water solubility, such as HPMC, the release is more diffusion controlled, and the release rate is slow. Polymers with high molecular weights provide high viscosity in the gel layer, leading to small diffusion constants and slow release. These parameters, in combination with polymer loading, provide variables for the rational design of a controlled release drug delivery system. In addition, combinations with other delivery approaches such as coated systems and mixing with less water-soluble polymers can be readily applied.

Drug solubility has a profound effect on the release profile and release mechanism from matrix systems. Highly water-soluble compounds tend to dissolve fast, even inside the gel phase. Thus, for systems with low drug loadings, the diffusion front of the matrix is often close to the swelling front. Therefore, the release profile is diffusion controlled and is a function of a square root of time. Selection of highly water-soluble polymers could help in changing the release mechanism for highly water- soluble compounds from diffusion controlled to dissolution controlled (zero-order release). The high dissolution (disentangling) rate of highly water-soluble polymers leads to early synchronization of the erosion front with other fronts, and thus drug release is a zero-order (or close to zero-order) process.[10] However, the drawback of application of highly water-soluble polymers to highly water-soluble compounds is their fast release rate. This may not be desired. As for poorly water-soluble compounds, the dissolution inside the gel phase will be slow, and the diffusion front often will exist. Therefore, the release is more dissolution controlled, and the release profile is relatively close to zero order. For these compounds, controlled release is only applicable to low doses.

Drug loading (percentage of drug in a unit dosage form) is another factor that can affect the release profile. For example, a dosage form with a higher drug loading releases faster but is closer to zero-order profiles when compared with lower drug loading. In addition, without changing the dose, higher drug loadings mean smaller tablet sizes, which lead to faster dissolution owing to higher SA/vol ratios.

5.5 Future Potential for Dissolution Controlled Release Drug Delivery Systems

5.5.1 Dissolution controlled release coated systems

Classic extrusion, spheronization, and pellitization processes typically result in pellets with irregular surfaces and of varying sizes, which are inherently more difficult to film coat. A recent report described the preparation of almost perfectly spherical particles with a narrow size distribution for improved coating efficiency (Ceform).[58]

A number of patented technologies for multiparticulate dosage forms have been described recently, such as the Micropump system, which is an osmotically driven coated microparticle system designed to increase the absorption time for rapidly absorbed drugs.[59] Combination of water-soluble and water-insoluble polymers could provide enhanced controlled release rates and profiles. A patented technology (COSRx) has been reported to be capable of delivering various sophisticated release profiles. The formulation involves a guar-gum-based tablet and a combination of water-soluble and water-insoluble polymeric tablet coat.[60]

In recent years, interest in multiple-layered tablets as an oral controlled release system has increased. Multiple-layered tablets have some obvious advantages compared with conventional tablets. In addition to avoiding chemical incompatibilities of formulation components by physical separation, release profiles may be modified by combining layers with different release patterns or by combining slow release with immediate release layers. If the core layer of multilayered tablet is completely covered by a surrounding layer, the product is commonly referred to as a *dry-coated tablet.* An example is the Smartrix tablet, in which the release profile of a drug is determined by the increase in release surface caused by erosion (dissolution) of the cover layers.[39]

Examples for spatial control of drug delivery systems coated with water-soluble polymers were reported in the literature recently. If a coating is prepared around a drug delivery system in which the outermost layer contains a bioadhesive material, then control of the location at which drug release will occur becomes possible. A number of materials, including HPC and sodium CMC, have been examined for this application.[61]

5.5.2 Dissolution controlled release matrix systems

Currently, most mature dissolution controlled release systems/technologies are applicable for water-soluble and low-water-solubility compounds (with low doses). For very poorly water-soluble compounds, dissolution controlled release systems/technologies may not be applicable because these compounds have intrinsically slow dissolution/release rates. Recently, several new technologies such as solid dispersions and self-emulsifying drug delivery systems (SEDDS) have been developed to deliver poorly water-soluble compounds at reasonable doses through enhancement of dissolution rate. These technologies have created new potentials for controlled release of poorly water-soluble compounds, often by their combination with the other controlled release technologies described in this book.

Solid dispersion systems often use polymers as stabilizers to prevent the conversion of drug substance from a high-energetic amorphous form to a low-energetic crystalline form during storage. Highly water-soluble polymers such as PVP are used frequently for this purpose for immediate release dosage forms. For controlled release, however, slowly dissolving water-soluble polymers such as HPMC or high-molecular-weight PVP may be used. Manufacturing processes such as melt extrusion can provide polymer matrix systems for controlled release of poorly water-soluble compounds.[62]

Another current trend is to develop controlled release systems with the combination of different release principles (diffusion, dissolution, osmosis, etc.) to meet various needs of different release profiles.

References

1. V. V. Ranade. Drug delivery systems: 5A. Oral drug delivery. *J. Clin. Pharmacol.* **31**:2–16, 1991.
2. C. T. Rhodes and S. C. Porter. Coatings for controlled-release drug delivery systems. *Drug Dev. Ind. Pharm.* **24**(12):1139–1154, 1998.
3. N. J. Lordi. Sustained release dosage forms, in L. Lachman, H. A. Lieberman, and J. L. Kanig, (eds.), *The Theory and Practice of Industrial Pharmacy*, 3d ed. Philadelphia: Lea & Febiger, 1986, pp. 430–478.
4. Q. R Cao, H. G. Choi, D. C. Kim, and B. J. Lee. Release behavior and photo-image of nifedipine tablet coated with high viscosity grade hydroxypropylmethylcellulose: Effect of coating conditions. *Int. J. Pharma.* **274**:107–117, 2004.
5. V. H. L. Lee and J. R. Robinson. Methods to achieve sustained drug delivery: The physical approach: Oral and parenteral dosage forms, in J. R. Robinson (ed.), *Drugs and the Pharmaceutical Sciences*, Vol. 6: *Sustained and Controlled Release Drug Delivery Systems*, 3d ed. New York: Marcel Dekker, 1978, pp. 123–173.
6. A. W. Hixson and J. H. Crowell. Dependence of reaction velocity upon surface and agitation: I. Theoretical considerations. *J. Ind. Eng. Chem.* **23**:923–931, 1931.
7. M. S. Harris. Preparation and release kinetics of potassium chloride microcapsules. *J. Pharm. Sci.* **70**:391, 1981.
8. S. S. Ozturk, B. O. Palsson, B. Donohoe, and J. B. Dressman. Kinetics of release from enteric-coated tablets. *Pharm. Res.* **5**:550–565, 1988.
9. P. I. Lee and N. A. Peppas. Prediction of polymer dissolution in swellable controlled-release systems. *J. Contr. Rel.* **6**:207–215, 1987.
10. P. Colombo, R. Bettini, P. Santi, et al. Analysis of the swelling and release mechanisms from drug delivery systems with emphasis on drug solubility and water transport. *J. Contr. Rel.* **39**:231–237, 1996.
11. P. Colombo, R. Bettini, G. Massimg, et al. Drug diffusion front movements important in drug release control from swellable matrix tablet. *J. Pharm. Sci.* **84**:991–997, 1995.
12. R. S. Harland, A. Gazzaniga, M. E., Sangalli, et al. Drug/polymer matrix swelling and dissolution. *Pharm. Res.* **5**:488–494, 1988.
13. P. L. Ritger and N. A. Peppas. A simple equation for desceiption of solute release: II. Fickian and anomalous release from swellable devices. *J. Contr. Rel.* **5**:37–42, 1985.
14. N. A. Peppas. Analysis of Fickian and non-Fickian drug release polymers. *Pharm. Acta Helv.* **60**:110–111, 1985.
15. S. K Baveja, K. V. Ranga Rao, and K. Padmalatha Devi. Relationship between gum content and half-life of soluble β-blockers from hydrophilic matrix tablets. *Int. J. Pharm.* **47**(1–3):133–139, 1988.

16. E. D. Klug. Properties of water-soluble hydroxyalkyl celluloses and their derivatives. *J. Polymer Sci.* **36**:491–508, 1971.
17. R. K. Chang, C. H. Hsiao, and J. R. Robinson. A review of aqueous coating techniques and preliminary data on release from theophylline product. *Pharm. Technol.* **11**:56–68, 1987.
18. Y. W. Chien. Novel drug delivery systems, in *Drugs and the Pharmaceutical Sciences*, Vol. 50. New York: Marcel Dekker, 1992, pp. 1–49.
19. S. K. Chandrasekaran and D. R. Paul. Dissolution-controlled transport from dispersed matrixes. *J. Pharm. Sci.* **71**:1399–1402, 1982.
20. W. I. Higuchi and T. Higuchi. Theoretical analysis of diffusional movement through heterogeneous barriers. *J. Pharm. Sci.* **49**:598–606, 1960.
21. Ronald S. Harland, Catherine Dubernet, Jean-Pierre Bonoit, and Nikolaos A. Peppas. A model of dissolution-controlled, diffusional drug release from non-swellable polymeric microspheres. *J. Contr. Rel.* **7**:207–215, 1988.
22. K. Hallas-Moller, K. Petersen, and J. Schlichtkrull. Crystalline and amorphous insulin-zinc compounds with prolonged action. *Science* **116**:394–398, 1952.
23. F. H. Buckwalter and H. L. Dickison. The effect of vehicle and particle size on the absorption, by the intramuscular route, of procaine penicillin G suspensions. *J. Am. Pharm. Assoc.* **47**:661–666, 1958.
24. P. Kearney. The Zydis oral fast-dissolving dosage form, in Michael J. Rathbone, Jonathan Hadgraft, and Machael S. Roberts (eds.), *Drugs and the Pharmaceutical Sciences*, Vol. 126: *Modified-Release Drug Delivery Technology*. New York: Marcel Dekker, 2003, pp. 191–201.
25. S. I. Pather, R. K. Khankari, and D. V. Moe. OralSolv and DuraSolv: Efficient technologies for the production of orally disintegrating tablets, in Michael J. Rathbone, Jonathan Hadgraft, and Machael S. Roberts (eds.), *Drugs and the Pharmaceutical Sciences*, Vol. 126: *Modified-Release Drug Delivery Technology*. New York: Marcel Dekker, 2003, pp. 203–216.
26. G. G. Liversidge and K. C. Cundy. Particle size reduction for improvement of oral bioavailability of hydrophobic drugs: I. Absolute oral bioavailability of nonocrystalline danazol in beagle dogs. *Int. J. Pharm.* **125**:91–97, 1995.
27. G. G. Liversidge and P. Conzentino. Drug particle size reduction for decreasing gastric irritancy and enhancing absorption of naproxen in rates. *Int. J. Pharm.* **125**:309–313, 1995.
28. R. H. Muller, C. Jacobs, and O. Kayser. DissoCubes—A novel formulation for poorly soluble and poorly bioavailable drugs, in Michael J. Rathbone, Jonathan Hadgraft, and Machael S. Roberts (eds.), *Drugs and the Pharmaceutical Sciences*, Vol. 126: *Modified-Release Drug Delivery Technology*. New York: Marcel Dekker, 2003, pp. 139–149.
29. J. A. Seitz, S. P. Mehta, and J. L. Yeager. Tablet coating, in L. Lachman, H. A. Lieberman, and J. L. Kanig (eds.), *The Theory and Practice of Industrial Pharmacy*, 3d ed. Philadelphia: Lea & Febiger, 1986, pp. 346–373.
30. *The United States Pharmaceopeia*, XXII ed. Rockville, MD: United States Pharmacopeial Convention, 1990.
31. G. S. Banker. Film coating theory and practice. *J. Pharm. Sci.* **55**(1):81–89, 1966.
32. J. M. Conrad, and J. R. Robinson. Sustained drug release from tablets and particles through coating, in H. A. Lieberman and L. Lachman (eds.), *Pharmaceutical Dosage Forms: Tablets*, Vol. 3. New York: Marcel Dekker, 1982, pp. 149–221.
33. G. A. Digenis. In vivo behavior of multiparticulates versus single-unit dose formulations, in I. Ghebre-Sellassie (ed.), *Multiparticulate Oral Drug Delivery*. New York: Marcel Dekker, 1994, pp. 333–356.
34. R. Bodmeier. Tableting of coated pellets. *Eur. J. Pharm. Biopharm.* **43**:1–8, 1997.
35. R. C. Rowe, P. J. Sheskey, and P. J. Weller. *Handbook of Pharmaceutical Excipients*, 4th ed. Washington: Pharmaceutical Press and the American Pharmaceutical Association, 2003.
36. A. M. Mehta and D. M. Jones. Coated pellets under the microscope. *Pharm. Technol.* **9**(6):52–60, 1985.
37. F. N. Christensen and P. Bertelsen. Qualitative description of the Wurster-based fluid-bed coating process. *Drug Dev. Ind. Pharm.* **23**:451–463, 1997.

38. E. M. Rudnic and M. K. Kottke. Tablet dosage forms, in G. S. Banker, and C.T. Rhodes (eds.), *Modern Pharmaceutics*. New York: Marcel Dekker, 1995, pp. 333–394.
39. H. G. Zerbe and M. Krumme. Smartrix system: Design characteristics and release properties of a novel erosion-controlled oral delivery system, in Michael J. Rathbone, Jonathan Hadgraft, and Machael S. Roberts (eds.), *Drugs and the Pharmaceutical Sciences*, vol. 126: *Modified-Release Drug Delivery Technology*. New York: Marcel Dekker, 2003, pp. 59–76.
40. J. A. Bakan. Microcapsule drug delivery systems, in R. L. Kronenthal, Z. Oser, and E. Martin (eds.), *Polymers in Medicine and Surgery*. New York: Plenum Press, 1975, pp. 213–235.
41. G. V. Savage and C. T. Rhodes. The sustained release coating of solid dosage forms: A historical review. *Drug Dev. Ind. Pharm.* **21**(1):93–118, 1995.
42. S. S. Chrai, D. R. Friend, G. Kupperblatt, et al. Accudep technology for oral modified drug release, in Michael J. Rathbone, Jonathan Hadgraft, and Machael S. Roberts (eds.), *Drugs and the Pharmaceutical Sciences*, Vol. 126: *Modified-Release Drug Delivery Technology*. New York: Marcel Dekker, 2003, pp. 89–99.
43. J. H. Guo, G. W. Skinner, W. W. Harcum, and P. E. Barnum. Pharmaceutical applications of naturally occurring water-soluble polymers. *Pharm. Sci. Technol. Today* **1**(6):254–261, 1998.
44. J. H. Guo, G. W. Skinner, W. W. Harcum, and P. E. Barnum. Pharmaceutical applications of naturally occurring water-soluble polymers. *Pharm. Sci. Technol. Today*. **1**(6):254–261, 1998.
45. R. O. Williams III and J. Liu. Influence of processing and curing conditions on beads coated with an aqueous dispersion of cellulose acetate phthalate. *Eur. J. Pharm. Biopharm.* **49**:243–252, 2000.
46. K. S. Murthy and I. Ghebre-Sellassie. Current perspectives on the dissolution stability of solid oral dosage forms. *J. Pharm. Sci.* **82**(2):113–126, 1993.
47. T. Sirkiä, H. Salonen, P. Veski, et al. Biopharmaceutical evaluation of new prolonged-release press-coated ibuprofen tablets containing sodium alginate to adjust drug release. *Int. J. Pharm.* **107**:179–187, 1994.
48. F. J. Rippie and J. R. Johnson. Regulation of dissolution rate by pellet geometry. *J. Pharm. Sci.* **58**:428, 1969.
49. D. Brooke and R. J. Washkuhn. Zero-order drug delivery systems: Theory and preliminary testing. *J. Pharm. Sci.* **66**:159, 1979.
50. Sham K. Chopra. Procise: Drug delivery systems based on geometric configuration, in Michael J. Rathbone, Jonathan Hadgraft, and Machael S. Roberts (eds.), *Drugs and the Pharmaceutical Sciences*, Vol. 126: *Modified-Release Drug Delivery Technology*. New York: Marcel Dekker, 2003, pp. 35–48.
51. D. A. Dickason and G. P. Grandolfi. RingCap technology, in Michael J. Rathbone, Jonathan Hadgraft, and Machael S. Roberts (eds.), *Drugs and the Pharmaceutical Sciences*, Vol. 126: *Modified-Release Drug Delivery Technology*. New York: Marcel Dekker, 2003, pp. 49–57.
52. D. Bar-Shalom, L. Slot, W. W. Lee, and C. G. Wilson. Development of the Egalet technology, in Michael J. Rathbone, Jonathan Hadgraft, and Machael S. Roberts (eds.), *Drugs and the Pharmaceutical Sciences*, Vol. 126: *Modified-Release Drug Delivery Technology*. New York: Marcel Dekker, 2003, pp. 263–271.
53. P. L. Ritger and N. A. Peppas. A simple equation for description of solute release: I. Fickian and non-Fickian release from non-swellable devices in the form of slabs, spheres, cylinders or discs. *J. Contr. Rel.* **5**:23–36, 1987.
54. J. L. Ford, M. H. Rubinstein, F. McCaul, et al. Importance of drug type, tablet shape and added diluents on drug release kinetics from hydroxypropymethylcellulose matrix tablets. *Int. J. Pharm.* **40**:223–234, 1987.
55. J. Siepmann, K. Podual, M. Sriwongjanya, et al. A new model describing the swelling and drug release kinetics from hydroxypropyl methylcellulose tablets. *J. Pharm. Sci.* **88**:65–72, 1999.
56. J. Siepmann, H. Kranz, N. A. Peppas, and R. Bodmeier. Calculation of the required size and shape of hydroxypropyl methycellulose matrices to achieve desired drug release profiles. *Int. J. Pharm.* **201**:151–164, 2000.

57. T. D. Reynolds, S. A. Mitchell, and K. M. Balwinski. Investigation of the effect of tablet surface area/volume on drug release from hydroxypropylmethylcellulose controlled-release matrix tablets. *Drug Dev. Ind. Pharm.* **28**:457–466, 2002.
58. J. C. Richards, D. V. Prior, S. E. Frisbee, et al. Pharmacoscintigraphic evaluation of novel controlled release microsphere (Ceform) formulations. *Proc. Int. Symp. Contr. Rel. Bioact. Mater.* **25**:920–921, 1998.
59. C. Castan, M. Cicquel, R. Meyrueix, et al. Genvir: The first sustained release dosage form of acyclovir. *Proc. Int. Symp. Contr. Rel. Bioact. Mater.* **27**:1198–1199, 2000.
60. S. A. Altaf and D. R. Friend. MASRx and COSRx sustained-release technology, in Michael J. Rathbone, Jonathan Hadgraft, and Machael S. Roberts (eds.), *Drugs and the Pharmaceutical Sciences*, Vol. 126: *Modified-Release Drug Delivery Technology*. New York: Marcel Dekker, 2003, pp. 21–33.
61. H. R. Chueh, H. Zia, and C. T. Rhodes. Optimization of sotalol floating and bioadhesive extended release tablet formulations. *Drug Dev. Ind. Pharm.* **21**:1725–1748, 1995.
62. J. Breitenbach and J. Lewis. Two concepts, one technology: Controlled-release and solid dispersions with Meltrex, in Michael J. Rathbone, Jonathan Hadgraft, and Machael S. Roberts (eds.), *Drugs and the Pharmaceutical Sciences,* Vol. 126: *Modified-Release Drug Delivery Technology*. New York: Marcel Dekker, 2003, pp. 125–134.

Chapter

6

Gastric Retentive Dosage Forms

Amir H. Shojaei

Shire Pharmaceuticals, Inc.
Wayne, Pennsylvania

Bret Berner

Depomed, Inc., Menlo Park, California

6.1 Introduction

Poor absorption of many drugs in the lower gastrointestinal (GI) tract necessitates controlled release dosage forms to be maintained in the upper GI tract, particularly the stomach and upper small intestine.[1–7]

Other therapies that potentially could benefit from controlled delivery of drug directly to the stomach are treatments for local disorders of the stomach such as *Heliobacter pylori* infections.[8,9]

Over two decades ago, many types of gastric retained drug delivery systems were tested to overcome the limited regions and time for drug absorption in the GI tract.[1–6] Gastric retentive drug delivery systems may be classified as those that use the natural physiology of the GI tract and those that are designed to overcome it. For those that use the inherent physiology, dosage forms that rely on size or flotation for delayed emptying from the stomach depend on the normal duration of the fed state of 4 to 6 hours following a meal[1,2] and a rather reproducible transit time through the small intestine of 2 to 4 hours.

Many gastric retentive dosage forms are designed to overcome the natural physiology and remain in the fasted stomach during the migrating motor complex (MMC). The mechanisms of retention often rely on rapid expansion by either gas generation, mechanical means, or swelling to at least the size of a golf ball, approximately 25 to 30 mm in diameter, followed by a collapse or degradation to a reduced size at some duration after the drug is delivered.[5,6,10–18] These approaches have been somewhat successful. However, additional studies of a larger population are required, especially in light of some variable emptying in smaller studies.

Another design one could take for retention in the upper GI tract has been bioadhesive microparticles that stick to the mucus or the mucosa in the upper GI tract, particularly in the duodenum and jejunum.[19–21] Charged polymers and even antibodies have demonstrated adhesion to the mucosa quite successfully in vitro,[22,23] but these bioadhesives have been less successful in vivo owing to two physiological limitations. The turnover of mucus is rapid and limits the duration of adhesion.[24] Moreover, approximately 2 percent of even the most bioadhesive microparticles with either specific antibody interactions or nonspecific interactions are retained along the stomach or intestinal wall (unpublished data by the authors).

The number of gastric retentive mechanisms attempted over the past 25 years, a lack of reproducibility, and only a few moderately successful products have led to much cynicism regarding gastric retention. Moderate increases in the duration of gastric retention, combined with careful product characterization, are required to advance the technology.

Food is perhaps the most reproducible means of delaying emptying of dosage forms from the stomach. Nevertheless, the range of gastric emptying times after a meal easily can vary 10-fold across individuals and different test diets.[25–36] In the fed state, the closing and contracting of the pylorus with a mean diameter of approximately 1.2 cm[37–39] and regular grinding waves of much smaller amplitude than in the fasted state are a mechanism to digest food by retaining large particles in the stomach until reduced in size. In the fasted state, approximately every 90 minutes a full amplitude series of waves, phase III of the MMC cycle or "housekeeper wave" empties the total contents of the stomach. Food, particularly fatty acids, interrupts the recurrence of this housekeeper wave and prevents emptying of the stomach[37,38] When administered to humans with food, nondisintegrating dosage forms that are at least 1.2 cm in diameter can simulate a large particle of food and may deliver drug to the upper GI tract throughout its residence in the stomach during the fed state and its transit through the small intestine. As with a large food particle, through aligned orientation while in close proximity to the pylorus or through being ground and eroded to a reduced size, the dosage form may slip through the pylorus in advance of complete gastric emptying of the meal, causing variation in dosage form retention times.

One mechanism to avoid this early emptying of dosage forms from the fed stomach is to maintain the dosage form in a position remote from the pyloric opening by creating a floating dosage form, typically by generating gas within the delivery system.[1,3,4] These floating dosage forms stay preferentially on the surface of the fluid gastric contents and avoid early emptying from the stomach. While the subject remains erect, these floating dosage forms remain on the surface of the gastric fluid and away from the pylorus to prevent emptying from the stomach. However, in the supine position, the surface of the gastric contents is located without preference to and can be near the pylorus, and this may result in supine gastric emptying times at least as short as for those dosage forms that just depend on size and food.[1] This may lead to product labeling against a prone position or bed rest, which may be limiting to therapy for certain indications or create poor compliance. Moreover, the amount of fluid in the fed stomach is highly variable, and this may lead to early emptying for floating dosage forms. Both of these approaches employ the inherent physiology of the GI tract, and products have been developed using both mechanisms.

This chapter discusses the GI physiology that defines the limits of gastric retentive technology and a more detailed analysis of the different approaches to design gastric retentive drug delivery systems.

6.2 Physiological Rationale for Gastric Retentive Delivery System Design: GI Motility

To develop gastroretentive devices for drug delivery, an understanding of the motility of the stomach, pylorus, and duodenum under various physiological conditions is essential.

6.2.1 Fasting contractile activity

In the fasted state, the stomach and duodenum exhibit a cyclic pattern of contractile activity known as the *migrating motor complex* (MMC). The MMC cycle consists of four phases. Phase I is quiescent, and phase II activity exhibits small-amplitude intermittent contractions. Phase III activity is a period of maximal contractile activity that lasts from 10 to 15 minutes in the stomach and duodenum.[40] Phase III activity in the antrum is characterized by groups of three to six contractions that gradually increase in force until a couple of contractions of maximal force occur.[41] During these contractions, the pylorus is relaxed, and contractile activity of the duodenum is inhibited.[41,42] This relaxation allows the pylorus to be stretched to its maximal aperture during emptying of indigestible particles. After the end of each set of contractions, contractile activity and tone return to the pylorus and duodenum.[42] This sequence is repeated several times during phase III activity and is responsible for emptying large, indigestible material from the stomach. Phase III activity may be followed by phase IV activity, a brief period of intermittent contractile activity. The MMC recurs every 90 to 120 minutes in the fasting state.

6.2.2 Fed mode

With a meal, the cyclic recurring phase III activity of the MMC cycle is replaced with the fed pattern of contractile activity. In the antrum, the powerful contractions of phase III activity are replaced by small-amplitude propagating contractions. The force of these contractions is only 15 to 25 percent of phase III contractile activity.[43] Moreover, contractions of the pylorus are coordinated with the propagating antral contractions such that the pylorus closes 3 to 4 seconds before the propagating contraction reaches the distal antrum.[44] Additionally, there is an increase in isolated pyloric contraction and tone following a meal.[45] Thus the reduced force of the antral contractions, along with the closure of the pylorus, is likely responsible for retaining nondigestible solids of a critical size in the stomach until the digestive state is complete and fasting contractile activity returns.

Gastric emptying times of plastic spheres ranging in diameter from 0.015 to 5.0 mm when given with a liver meal to dogs were determined by Meyer et al.[46] These investigators found that spheres smaller than a diameter of 1.6 mm emptied either earlier or at the same time as the nutrient part of the meal. However, spheres with a diameter equal to or greater than 2.4 mm emptied slower than the liver meal. Interestingly, at a sphere diameter of 5 mm, very little emptying took place up to about 180 minutes postprandial. This also was the point at which approximately 50 percent of the liver meal had emptied.

In humans, the relationship between gastric emptying and diameter of nondigestible solids has been less clearly defined. However, it is clear that nondisintegrating dosage forms with diameters of up to 10 mm still can be emptied during the postprandial period from the human stomach, whereas larger dosage forms are retained in the stomach until the digestible solids have been emptied. In studies by Mojaverian et al.,[13] radiotelemetric capsules (RTCs) 7 mm in diameter were administered in the fasting state and after a 500-kcal meal. In the fasted state, the RTCs were emptied within 90 minutes, which is consistent with the occurrence of an MMC. In contrast, in the fed state, gastric retention time was increased by about 4 hours, and the RTCs were emptied with the first postprandial phase III activity.

In studies by Davis et al.,[47] a 12-mm-diameter nondisintegrating dosage form was shown to have a longer gastric retention with increasing caloric content of the meal. With a light meal (360 kcal), the tablets were retained in the stomach for about 5 hours, whereas with a heavy meal (720 kcal), gastric retention was increased to nearly 8 hours. In studies using a similar-sized tablet, gastric retention in the fasted state was less than 1 hour and consistent with the MMC occurring during this time.

Once a single-unit dosage form is emptied from the stomach, mean transit through the small intestine is about 3 hours, whether the subject is in the fasted or the fed state. In contrast to gastric emptying of a single-unit dosage form, the size or content of the meal has no effect on small intestinal transit.[47] These studies and many similar studies indicate that to increase gastric retention of reasonably sized dosage forms, it is necessary to administer them in the postprandial state. Furthermore, these studies indicate that delivery systems that extend GI transit time are necessary to exploit the benefits of controlled release technologies for drugs absorbed in the upper GI tract. Therefore, the successful development of an oral controlled release dosage form requires a system that can overcome the limitations resulting from inherent GI physiology.

6.3 Design of Retentive Delivery System Based on Size

6.3.1 Tablet size and the fed mode

Drugs may be administered effectively to the upper GI tract for up to 9 hours by optimization of tablet size and the regimen of drug administration with respect to food. The breakdown of large particles of food in the stomach occurs in the fed stomach through antral grinding motions that reduce the particle size, and this is aided by pyloric closure.[37,38] Sieving of large particles by the stomach mimics sedimentation, with a strong dependence on particle size and a weak dependence on density.[46] Accepted gastric emptying times and small intestinal transit times in the fed mode for single-unit nondisintegrating tablets are, respectively, 2.7 ± 1.5 and 3.1 ± 0.4 hours to provide a total transit time through the upper GI tract of about 6 hours for a drug that is not well absorbed in the colon.[48] While these estimates may be toward the lower range of measurements, this is the mean, and therefore, to treat most of the population, delivery would need to be confined to approximately 4 to 5 hours without optimization to account for the actions of the stomach.[49–51] Pellets, in contrast, with a smaller size were only retained for 1.2 ± 1.3 hours in the fed stomach.[50] Standard teaching would suggest that there is less variation and range of gastric emptying times for pellets, but these data and other literature suggest that pellets are at least as variable as single-unit tablets.[32,33,36,52] Pellets do minimize the variation between fed and fasting, however.

Optimization of drug delivery in the fed mode involves characterization of the dependence of gastric retention on size and duration of the fed state. Meyer et al.,[46] in their studies in beagles, introduced the concept of a gradual cutoff on the dependence on size rather than a sharp decrease above a certain size limit. For dogs, this cutoff is approximately 7 mm,[53] and particles above this size range are retained rather reproducibly in dogs for 4 to 6 hours with about 15 percent of a daily feeding.[54] In humans, the mean pyloric diameter is 12 ± 7 mm,[39] and this is both larger and more variable than in dogs.[39] Timmerman and Moes[39] identified a gradual cutoff of 13 mm for retention in the fed mode, with mean gastric emptying times of approximately 6 hours for particles from 12 to 18 mm and no clear trend in the mean or decreased variation with increasing particle size throughout the size range studied.[39]

The characterization of gastric emptying for smaller particle size ranges clearly trended toward increasing retention with larger particle size. Davis et al.[33] observed a gastric emptying time for 50 percent of the pellets of 1 mm and under ($t_{50\%}$) in the fed state of 2 to 3 hours, whereas O'Reilly[35] found that 7- to 10-mm pellets exited in 3 to 4 hours.

With multiple pellet or bead formulations, gastric emptying times generally increase with particle size. The longest mean gastric emptying times for these pellets were still shorter than the mean gastric emptying times for single-unit nondisintegrating tablets. Some researchers claimed that multiple pellet formulations provide less variation than single-unit dosage forms because emptying from the stomach is a probabilistic event.[46] However, the range observed for multiple pellet formulations is still extensive.[55] The greatest variation in retention of any dosage form results from diet and the interindividual duration of the fed mode in response to a given meal. Multiparticulate formulations would appear to offer comparable variability to sufficiently large single-unit nondisintegrating tablets that are retained during the fed mode, but with a somewhat shorter duration of retention in the fed stomach.

The upper bound to gastric residence for multiparticulate dosage forms is similar to that for large single units. Addition of flotation to multiparticulate systems yields residence times, except in the prone position, that approach the length of residence of nonfloating or floating large single-unit dosage forms. However, addition of flotation to large single-unit tablets does not lengthen the gastric residence time.[1,56]

The potential sources of variability for gastric residence in the fed mode are the size of the pyloric opening versus the tablet size, the rate of reduction in the size of the tablet by dissolution and antral grinding by the stomach, and inter- and intraindividual variations in the duration of the fed mode, particularly as a function of caloric and fat content. The dominant limitation in the use of food for gastric retention is the minimal total fat content.

The typical approach[27,29,57] to studying gastric emptying with food has been to vary the fat and caloric contents simultaneously, in particular, a light breakfast (360 kcal) and a heavy breakfast (720 kcal). With the light breakfast, Davis et al.[27] observed that half the osmotic pumps emptied from the stomach within 3 hours, whereas all the osmotic pumps remained in the stomach for greater than 8 hours with the heavy breakfast. With 12-mm enteric-coated hydroxypropyl methyl cellulose tablets, which are true nondisintegrating tablets, Davis et al.[57] determined that the gastric emptying times after light and heavy breakfasts were 5.1 ± 0.8 and 7.7 ± 0.7 hours, respectively. After the light breakfast, 4 of the 16 tablets emptied from the stomach in less than 3 hours. Similar results have been found by Mojaverian et al.,[25] with a 4.3 ± 1.4 hour gastric emptying time after a 300-kcal light breakfast, and by Coupe et al.,[30,31] who observed gastric emptying times after a 300-kcal light breakfast for a large RTC that measured the onset of the MMC with a range of under 2 to over 7 hours. The type of meal has a dominant influence on the emptying time, and this influence will now be further explored to design the therapeutic regimen to account for this food effect.

The pyloric diameter is 12 ± 7 mm, and in the fed mode, it is closed most of the time.[39] With this large standard deviation, one would expect a tablet size of 13 mm or more to provide reasonable retention. This effect can be best studied by reducing the variation in the duration of the fed mode, i.e., after a high-fat meal. Under these conditions, Timmermans and Moes[39] observed good gastric retention of 6 hours for 12- to 18-mm tablets, with no clear dependence on size. In a pharmacoscintigraphic study, metformin extended release tablets, which swell from 12 mm in the minor dimension to 18 mm at the peak and then erode with a characteristic half-life of 14 to 15 hours, were compared with a nondisintegrating capsule 3.5 cm long and 1.2 cm wide, which was used to measure the onset of the MMC.[58] Under conditions of a high-fat,

1000-cal meal (at least 50 percent of calories from fat), the last time that the tablets were observed in the stomach was at 12.6 ± 6.1 hours. With the exception of one tablet that reached the colon in 20 hours, the tablets disintegrated in the upper GI tract. This is to be compared with a gastric emptying time of 20.9 ± 1.8 hours for nondisintegrating extremely large capsules. Under these conditions, the metformin extended release tablets show excellent retention, with sufficient time to deliver all the drug for all patients to the upper GI tract. However, there is some variability owing to the time of disintegration. The excessively large capsule clearly empties only with the MMC with little variability. With this enormous dosage form and a high-fat repeated-meal condition, the duration of the fed mode and the time for potential drug delivery is 20 hours to the stomach. With three heavy meals, the housekeeper wave does not occur in some subjects for over 24 hours. However, both this high-fat condition and large tablet size are necessary to obtain this extended duration and this reproducibility. Beyond the challenge to the most motivated patient to swallow these large 35-mm capsules, it is impractical and undesirable to use this feeding regimen in therapy. When the fat content is taken into account, a tablet size that swells from 12 mm and delivers to the stomach for 4 to 6 hours and the upper GI tract for 8 to 9 hours is the practical limit to this fed-mode approach to retention.

Under more practical low-fat fed conditions (30 percent fat), these same metformin extended release tablets show similar rates of swelling and disintegration, with mean last times observed in the stomach of 8.5 ± 7 hours.[58] The lower range for this observation was 3 to 4 hours, and the tablets emptied in two populations—under 6 hours and greater than 19 hours. The mean last time observed in the upper GI tract is 11 ± 3 hours, which could cover the entire population with 8 hours of drug delivery. Similarly, the excessively large nondisintegrating capsule emptied in 12.8 ± 8.9 hours, and the emptying times were distributed into the same two populations. Although the drastically increased tablet size may result in a longer mean retention time, the lower range of emptying times remained the same. Thus this large increase in tablet size would provide no real advantage in the duration of drug delivery. Note that the lower fat content meal showed a coefficient of variation of 70 percent owing to the quick emptying group of subjects as compared to a 9 percent coefficient of variation in the high-fat fed state. That is, the variability of the duration of the fed mode decreased with fat content.

Fat content is the limiting factor in gastric retention in the fed mode. In beagles, gastric emptying of a swellable dosage form (initial size 8 × 19 × 6 mm) can be prolonged from the fasting state of 1.4 hours with a coefficient of variation of 23 percent to 2.2 hours with 1.0 g of olive oil, with a coefficient of variation of 56 percent.[54] On the other hand, with 5 g of olive oil, the mean retention time was 3.6 hours, with a coefficient of variation of 18 percent. As in the preceding case shown in humans, increased fat content results in both longer retention and less variable emptying. A combination of low fat (20 g) and low caloric content resulted in insufficient gastric retention to be useful for drug delivery to the upper GI tract. Under these fed conditions in humans, a placebo tablet similar to the metformin extended release tablet described earlier was retained in the stomach for 3.7 ± 0.8 hours, with a range of 2.6 to 4.5 hours,[59] resulting in a couple of tablets exiting the small intestine within 5 hours. The large nondisintegrating capsule also exhibited gastric emptying times of 3.5 ± 0.6 hours. Increased tablet size would not improve the duration because the limitation was due to the duration of the fed mode and the appearance of the MMC. At this quite low fat content, there was no second population with a very long emptying time, and the variability was not large. The duration of the fed mode at a given fat content is critical. The limitation of a lower fat content suggests that these gastric retentive tablets should be administered with the substantial meal of the day, which for most people is dinner.

A complication, and additional source of variability, is the change in tablet size with time in vivo because tablets may swell, crack, disintegrate, or erode during transit through the GI tract. The size of the tablet may be diminished rapidly by erosion, and it may exit the stomach quickly. Erosion of tablets in vivo seldom has been characterized. In vivo erosion can be characterized quantitatively by anterior and posterior imaging during scintigraphy.[60] The grinding action of the stomach in the fed mode that aids digestion may provide rigorous hydrodynamics and erosion of tablets, and the rate of drug delivery in vivo may be more rapid than expected. Hydrodynamics in the human GI tract across the fasted and fed states have exhibited an extreme 20-fold range corresponding to a dissolution stirring speed from less than 10 to over 150 rpm.[61] Since polymeric erosion scales as stirring speed to the 1/3 power, three- to fourfold variability in the rate of disintegration might be expected in vivo. In the characterization of metformin extended release tablets described earlier, the 50 percent in vivo disintegration time was 15 ± 5 hours.[58] That is, in the fed mode, with either low or high fat, the coefficient of variation was 30 percent, with a threefold range from the lowest to the highest, a quite acceptable range for in vivo variability. A similar threefold range was observed in the erosion of Furosemide GR (gastric retentive) tablets in the fed mode.[62] Polymeric erosion in the fed GI tract provides a rather reproducible means of drug delivery for less soluble drugs.[63–65] The rate of erosion persists from the stomach throughout the small intestine and only slows in the colon.[58,62] This disintegration of tablets adds variability to the time when a dosage form becomes a small particle and like a small particle of food exits the stomach with the liquid.

Large, swelling dosage forms administered with food are among the more frequent recent attempts at gastric-retained controlled release pharmaceutical products.[50,51] A swelling metformin extended release dosage form,[66,67] based on a granulation with sodium carboxymethylcellulose blended with HPMC, two swelling and rate-controlling polymers, is marketed for Type 2 diabetes and is administered with dinner.[68] Another metformin extended release tablet using diffusion of the drug from swelling polymers has completed phase III trials.[58,69,70] Characterization and disintegration of this dosage form were discussed in the preceding section. For less soluble drugs, a swelling and eroding ciprofloxacin gastric-retentive tablet that is administered with food has completed phase III trials,[70,71] and a Furosemide GR swelling and eroding tablet is in phase II trials.[62,72]

6.3.2 Retention of expanding systems in the fasted state

Gastric retention in the fasting stomach requires overcoming the existing physiology and has remained an elusive target for practical drug delivery systems. Retention depends on resisting emptying during the MMC, where the pylorus is fully dilated and relaxed with a diameter of 12 ± 7 mm.[39] The MMC contractions in humans can empty particles by pushing them through the relaxed pylorus, which can expand to considerably greater diameters. To resist these contractions and achieve retention, devices have been designed to expand rapidly to the size of a golf ball and then, after the drug has been delivered, based on a different mechanism, to collapse and pass through the digestive tract. The earliest devices tested included a drug delivery system attached to a balloonlike bag that expanded in gastric juice[5] and a highly swelling polymeric coating of Gantrez on a dosage form.[6]

One could design compressed dosage forms that are packed into capsules that expand rapidly after dissolution of the capsule. Curatolo et al.[10,11] described a rolled or coiled-up device that had ribbonlike projections or fingers that were 5 or 10 cm in length. Initial tests were done with 10-cm fibers that showed retention for 24 hours when administered to dogs in the fed state.

Expandable compressed shapes of rings, tetrahedrons, and cloverleaves have been tested in both dogs[12–16] and in humans.[73] These systems were composed of biodegradable polymers, polyorthoesters, blended with polyethylene or polyethylene blends to create a semirigid biodegradable structure that expanded from 1.6 cm in the compressed state to approximately 5 cm in the expanded state. The biodegradable polymers provided the means of erosion to allow the system to collapse and pass through the digestive system after releasing drug. A number of blends were examined in vitro at low pH to study the rate of erosion, and poor in vitro/in vivo correlation probably owing to the wide variation in pH in the human GI tract, poor reproducibility in manufacture, or degradation during storage may have created some developmental problems for this device. In dogs, many of these devices exhibited excellent retention, with all devices remaining 24 hours after administration. In humans, however, the results were less promising, with a mean of 3 hours in the fasted state and 6.5 hours in the fed state.[73]

Expanding hydrogels have been used to achieve sufficient size to resist transit through the pylorus in the fasted state and then gradually degrade to allow passage. To allow for greater and more rapid swelling, Park et al.[74] developed superporous hydrogels, a class of lightly cross-linked hydrogels with large pores greater than 100-μm in diameter. These superporous hydrogels were developed as an open-channel hydrogel foam with a foaming agent, such as a protein or Pluronic, and a gas or a chemical foaming agent. The hydrogels were then dried and formulated as a dosage form. The monomers were selected to allow substantial swelling and may include polyacrylic acid, polyacrylamide, polyhydroxyethyl methacrylate, or hydroxypropyl methacrylate. All these polymers require substantial removal of the monomers and control of the degree of polymerization to address both quality control and toxicology. These hydrogels swell rapidly to a large size and at later times will fall apart owing to weak mechanical strength. Sodium croscarmellose or Ac-Di-Sol has been investigated to improve the mechanical strength.[75] These hydrogels remained in the stomach for 2 to 3 hours in fasted dogs, which is longer than most systems but has limited utility. When given with food to beagles, however, to allow an initial time for swelling in the fed state, these hydrogels remained in the stomach for 24 hours. Through orientation of the dried hydrogels during compression, the interconnected pores were preserved, and swelling was substantial within 10 minutes.[76]

A gas-generating expanding membrane device was investigated by Sinnreich[77] to resist emptying of the dosage form in the fasted state. Several features were incorporated into the device to avoid safety issues. The dosage form consisted of a membrane bag, which was typically polyvinyl alcohol, in to which was placed the medicament, in particular, baclofen, and an agent that generated gas in the presence of gastric acid, such as sodium bicarbonate. Acid also could be incorporated into the dosage form to allow for a time lag in permeation of the acid or variation in gastric pH with food or proton pump inhibitors. This dosage form expanded to approximately 2.5 cm in diameter and remained inflated until the gas source was depleted. When this collapsed, the rolled system was placed in a capsule for administration. The dosage form was intended to be administered with food to allow for swelling sufficiently slow that it did not present a safety issue and to avoid initial emptying before achieving a sufficient size. This dosage form delivering baclofen was then investigated in dogs[78] and in humans.[79] Radiopaque strings were incorporated into the dosage form for visualization by x-ray in dogs, and gas in the dosage form also could be seen with fluoroscopy. All dosage forms inflated in dogs whether in the fed or fasted states within 0.5 hour. Deflation occurred in 3 to 6 hours in the fed state and in 1 to 4 hours in the fasted state. In the fasted state, five of six dosage forms remained in the stomach for at least 7 hours, whereas one emptied at 2 hours. The dosage form remained in the fed stomach for at least 10 hours in five of six dogs and emptied in 6 to 7 hours in the remaining fed beagle. The bioavailability of baclofen from the extended release dosage form was comparable with the immediate release form with a diminished $C_{\max}$ and extended $t_{\max}$.[78]

To study this gas-generating membrane dosage form containing baclofen in humans,[77] samarium was incorporated into the dosage form, and it was then neutron activated[60] to visualize its transit by gamma scintigraphy.[79] The dosage form was administered to 13 healthy volunteers in the fasted low-fat (low calorie) and high-fat states (high calorie) and compared with immediate release baclofen administered fasted and with a high-fat meal. When administered with the high-fat meal, all systems remained in the stomach at 16 hours, and 7 of 13 remained at 24 hours. After a low-fat, low-calorie meal, all except one system remained in the stomach at 4 hours, four had emptied by 6 hours, 60 percent remained at 16 hours, and one remained in the stomach at 24 hours. In the fasted state, three had emptied in 2 hours, and four still remained in the stomach at 16 hours. While the results for the low-fat, low-caloric meal were not entirely consistent, it is the most extended case of gastric retention under these conditions to appear in the literature. The plasma profiles in the high-fat state were 90 percent of that of the immediate release dosage form, and for the low-fat state, the bioavailability was 80 percent. Typical extended release plasma profiles with diminished peak concentrations at longer times were observed from this system.[78]

An unfolding multilayer expandable compressed dosage form that resists gastric emptying in the fasted state has been examined in the laboratories of Hoffman and Friedman.[80–84] The dosage unit consists of an inner polymeric–drug matrix layer with two shielding outer layers and a coating of microcrystalline cellulose to prevent adhesion. The shielding outer layers were composed of hydrolyzed gelatin of 10,000 to 12,000 Da molecular weight cross-linked with glutaraldehyde. Surrounding the inner layer was a rigid biodegradable frame composed of a blend of polylactic acid and ethyl cellulose (9:1).[82] This entire system when unfolded was 2.5 × 5 cm, and it could be folded into a large capsule.

Riboflavin, which is only absorbed in the duodenal cap and exhibits saturable absorption,[83–86] was used as a model drug in this system.[80,83] Radiopaque markers were incorporated, and these riboflavin-containing dosage forms were administered to fasted beagles. The systems remained in the stomach of all six fasted dogs at 13 hours. From this dosage form, riboflavin levels persisted for 48 hours, with a fourfold enhancement in bioavailability, as compared with 10 hours with the immediate release form. Klausner et al.[82] also have administered three prototype dosage forms containing L-dopa to beagles pretreated with carbidopa. The prototype with the longest mean in vitro dissolution time of 4.2 hours resulted in plasma levels exceeding 500 ng/mL for over 9 hours in fasted dogs. More recently, Klausner et al.[81] studied these folding multilayer systems that released 60 mg furosemide over 6 hours in 14 fasted healthy male volunteers. They observed greater efficiency of diuresis and natriuresis than from the immediate release dosage form.

In pilot pharmacokinetic trials in dogs or humans, several prototypes have shown promising results. However, none of these dosage forms have been developed through the clinical phase. The complexity and uniqueness of these fasting dosage forms can result in quite unusual and perhaps more costly manufacturing processes. Variability in larger pharmacokinetic studies and in efficacy studies will need to be characterized. Stability of these complex systems also may require study. Systems that require novel polymers also may require full toxicological packages. The speed of expansion after swallowing could present one set of safety issues for evaluation, and clearance from the GI tract, especially of multiple systems, presents another set of issues. Finally, the safety and performance in special GI populations, e.g., gastroparesis, gastroesophageal reflux, dyspepsia, or diverticulitis, need evaluation. To justify development of any of these dosage forms requires identification of a therapeutic need with a substantial market potential that is uniquely satisfied by these dosage forms. This therapeutic need to date has been achievable at least in part by other controlled release dosage forms, including those using the fed mode for retention.

6.4 Design of Retentive Delivery System Based on Density Difference

Gastric retention of floating and sinking dosage forms is affected by body posture, the fluid content of the gastric lumen, and the rate of emptying of the fluid component of the gastric contents. The fed mode is necessary for the gastric retentive performance of density-based dosage forms.[1,56,87–90] In this class of dosage form, there is a resulting force on the dosage form owing to the difference between the densities of the dosage form and the fluid component of gastric contents, and this force maintains the dosage form away from the pyloric opening. Flotation occurs when the buoyant force exerted on the object by the gastric fluid is larger, and submersion occurs if the force is less than the weight of the dosage form.

6.4.1 Density greater than gastric fluid (submerged)

Dense systems, which sink to the bottom of the gastric luminal contents, have been hypothesized to avoid the peristaltic forces of gastric emptying. Multiple-unit submerged systems have been evaluated by Devereux et al.[91] and Clarke et al.[92] In the former study, gastric emptying of a multiple-unit formulation of density 2.8 g/cm^3 was found to be significantly delayed in both fed and fasted conditions. The latter study evaluated three multiple-unit formulations (1.18 to 1.40 mm in diameter) of density 1.5, 2.0, and 2.4 g/cm^3, respectively, under identical conditions to the former study, except that only the fasted state was evaluated, and revealed no difference in the GI transit of these formulations. These results indicate that the critical density to delay gastric emptying is between 2.4 to 2.8 g/cm^3. Indeed, when Davis et al.[28] studied floating versus nonfloating pellets (density 0.94 versus 1.96 g/cm^3; 0.7 to 1.0 mm) in four normal subjects under fed conditions (light breakfast), mean $t_{50\%}$ gastric emptying times were not different. However, in three subjects, light pellets emptied slower initially (~2 to 4 hours), with two subjects not showing emptying at all, and then emptied quicker thereafter. The investigators indicated that the light pellets ceased floating at later times, but emptying of the gastric fluid may be the actual cause.

6.4.2 Density lower than gastric fluid (floating)

The most in-depth review of this type of oral dosage form to prolong gastric residence is that of Moës.[1] When floating on top of the gastric contents, floating dosage forms are situated high in the stomach, closer to the fundus, and relatively distant from the pyloric opening. Floating units still require the fed state of the stomach to enhance the gastric emptying time significantly. Flotation of the dosage form is determined by the resulting weight,[56] which is the difference between the buoyancy force when entirely submerged and the weight of the dosage form. The object will float if this difference, or resulting weight, is positive. Flotation of the dosage form can be assessed quantitatively in vivo in scintigraphic studies by measuring the ratio of the relative intragastric height—the craniocaudal distance of a floating dosage form measured from the lowest part of the gastric region of interest—and the entire gastric region of interest. However, this value should be determined relative to the fluid level in the stomach. Body posture, whether standing or recumbent, whether supine or reclining, and whether lying on the left side or the right side, and the effect of size in these situations critically influence the gastric retention of these systems. Important factors to predict the in vivo performance should include the in vitro performance with regard to the flotation onset time or lag time for the system to activate or become operable in terms of flotation and the duration of flotation and also the frequency of subsequent food or fluid intake.

Timmermans and Moës[56] evaluated gastric emptying of single-unit floating and nonfloating capsules in fed subjects (breakfast) in upright and supine positions. Several factors, such as size, flotation duration, and posture, were well controlled in this study, in addition to using more frequent imaging than in previous studies. Capsules of both forms (initial densities ~0.5 to 0.7 g/mL for floating versus ~1.1 to 1.2 g/mL for nonfloating) were prepared in three sizes (8 to 9 mm, 11 to 12 mm, and 14 to 15 mm in diameter). On immersion in vitro for 8 hours, the diameters increased during the first hour. Gastric emptying of nonfloating capsules was more variable and directly related to size. In floating capsules, gastric residence times increase from the small to the medium size (mean ~3 to ~4 hours), which showed a retention time similar to the large size. Floating capsules had more prolonged gastric residence times than nonfloating capsules of equal size (mean~3 versus ~1.5 hours, ~4 versus ~2 hours, ~4 versus ~3.5 hours with increasing size, respectively); however, the difference decreases at large sizes. In the supine position, the floating capsules emptied earlier than the nonfloating capsules, which emptied independent of body position.

Monolithic non-gas-generating systems are matrix tablets consisting of hydrocolloids that form an external gel layer when hydrated. The internal tablet core remains dry with an overall density lower than that of the gastric fluid. Hydroxypropylmethycellulose (HPMC) is the most commonly used hydrocolloid. This approach has been developed into marketed drug products as the Hydrodynamically Balanced System (HBS) invented by Sheth and Tossounian.[93] Gastric retention and flotation times up to 6 hours were achieved. Valrelease (diazepam) and Madopar (levodopa and benserazide) were two marketed products developed using this approach.

Gas-generating systems rely on the production of internally trapped carbon dioxide to provide buoyancy. Gas may be generated by incorporation of sodium bicarbonate or calcium carbonate with or without an acidifying agent such as citric acid or tartaric acid. Timmermans and Moës[56] examined monolithic gas-generating systems of different size capsules (nos. 5, 0, and 000) filled with Methocel (HPMC) K4M and sodium bicarbonate (2 percent), with in vitro flotation times of up to 8 hours. In vivo constant relative gastric heights were achieved for up to 180 to 200 minutes in fed upright subjects (650-kcal continental breakfast). Agyilirah et al.[89] prepared tablets consisting of sucrose, lactose, sodium bicarbonate, and magnesium stearate coated with polymethyl vinyl/ethermaleic anhydride that ballooned and floated in gastric media within 15 minutes. The onset time of flotation in vivo in seven healthy male volunteers was approximately 5 to 9 minutes under fasting conditions and 5 to 19 minutes in the fed state, which was induced by a 1000-cal high-fat breakfast followed by a 100-cal lunch 4 hours post-dose. In the fasted state, the tablet emptied within a median time of 100 minutes or less in six subjects and was similar to a nondisintegrating matrix tablet containing calcium phosphate dihydrate, stearic acid, and magnesium stearate. The gastric emptying time of the balloon tablet ranged from 190 to 529 minutes in the fed state. The balloon tablet emptied later than the matrix tablet, from 167.5 to greater than 470 minutes, but with a similar lower range of emptying times.

Layered dosage forms were conceived to overcome constraints imposed on monolithic forms by drug release versus gastric retention. Separate layers of the dosage form are responsible for each individual function. Özdemir et al.[94] developed a bilayer tablet with one layer to provide buoyancy and the other to deliver the active ingredient furosemide. Since furosemide is poorly soluble, a 1:1 inclusion complex with β-cyclodextrin was formed first to enhance dissolution. The drug-release layer consisted of HPMC K100, with or without polyethylene glycol (PEG). The floating layer contained sodium bicarbonate, citric acid, and HPMC 4000. As with all floating tablets based on gas generation, the flotation onset time increased with compression force, i.e., 20 minutes at 16 MPa versus 45 minutes at 32 MPa. An in vivo gastric emptying time of 6 hours was observed in six healthy male volunteers dosed (44 mg furosemide, in vitro release duration of 8 hours) on an empty stomach and then fed a light breakfast 2 hours later. The AUC_{0-24} was 1.8 times that of the immediate release tablet (Furomid) and correlated well with in vitro dissolution.

Multiunit systems, in contrast to single-unit systems, are often stated to avoid all-or-none gastric emptying. This requires that the units remain dispersed and suspended individually in the gastric fluid. Frequently, multiunits tend to aggregate on contact with the gastric fluid and become an agglomerated mass floating on top of the gastric fluid. For example, alginate beads agglomerate by 3 hours after administration.[95,96]

Using gamma scintigraphy, Whitehead et al.[97] tested in vivo hydrophilic polymeric multiple-unit floating-bead systems prepared from freeze-dried calcium alginate. Freeze drying conferred the ability to float to those beads with 2.15-mm mean diameters and 0.33 gc/m^3 density. A positive resulting weight above 0.5 g/per 100 mg beads was maintained for 12 hours in vitro. Floating and nonfloating beads were labeled differently

(110 to 150 beads per dose) and evaluated together in seven healthy males who were standing or sitting. The subjects were maintained in the fed mode by starting with radiolabeled 30 g of cereal and 150 mL of milk for outlining the stomach, followed by a high-fat breakfast and dosing immediately after. All subjects were provided with lunches of varying caloric values, and four subjects had snacks between the main meals. These regimens reflected the subjects' normal eating patterns. Prolonged gastric residence times of over 5.5 hours were achieved in all subjects for the floating beads versus 1 hour for the nonfloating beads. The floating beads maintained their flotation, as shown by relative height measurements. Meal size did not affect gastric retention of the beads.

Ion exchange resin–coated systems containing sodium bicarbonate[98] have been evaluated, and a coating is required for the Dowex 2x10 resin to show significantly prolonged gastric residence time.

6.5 Design of Retentive Delivery System Based on Adhesion: Mucoadhesive Systems

In the context of drug delivery applications, bioadhesives or mucoadhesives are used to enhance the overall efficacy of delivery. In the case of peroral administration, bioadhesives are intended to slow down GI transit of the dosage form, thereby enhancing drug absorption by prolonging contact time at the optimal site of absorption. For transmucosal delivery, in particular, ocular, nasal, or buccal drug delivery, mucoadhesives are used to retain the delivery system at the site of absorption for prolonged periods of time. The bio/mucoadhesive materials employed in these cases are polymeric macromolecules of natural or synthetic origin. In the following discussion, the focus is on bioadhesion of mucoadhesive retentive drug delivery systems, the physiological environment, and the nature of the bioadhesive material used.

6.5.1 Mucus and epithelial layers

Mucus. The cells of internal epithelia throughout the body are surrounded by an intercellular ground substance known as *mucus*. The principal components of mucus are complexes composed of proteins and carbohydrates. These complexes may be free of association or may be attached to certain regions on cell surfaces. This matrix may play a role in cell-cell adhesion, as well as acting as a lubricant, allowing cells to move relative to one another.[99] Moreover, mucus is believed to play a role in bioadhesion of mucoadhesive drug delivery systems.[100]

Mucus exists in the form of a viscous solution or a gel and is a sticky viscoelastic substance with water as its major constituent, accounting for 95 to 99.5 percent.[101] Mucins are large (MW range 1000 to 40,000 kDa) glycoproteins that are responsible for the gel-forming properties of mucus. The mucin glycoproteins are composed of a linear proteinaceous backbone, which is heavily glycosylated by oligosaccharides.[101] The proteinaceous core accounts for nearly 25 percent of the dry weight, and the covalently bound carbohydrate side chains account for the remaining weight.[102] Two different classes of mucins are found—those attached to epithelial cell surfaces are known as *membrane-bound mucins* and those not attached are known as *secretory mucins.*[102] Membrane-bound mucins are likely to have an immunomodulatory role and also may affect cell-cell interactions.[103] Secretory mucins are found in the mucus gels of the gastrointestinal, ocular, respiratory, and urogenital epithelia[104] and are synthesized by epithelial cells or specialized mucus-secreting cells such as goblet cells.

Throughout the body, the thickness as well as the composition of the mucus layer varies depending on the location, surrounding cellular environment, and state of health. In the GI tract, the mucus layer is 50 to 450 μm thick and functions mainly as a protective barrier against damage (enzymatic as well as mechanical) from the luminal contents. At physiological pH, the mucin network behaves like an anionic polyelectrolyte and carries a significant negative charge owing to the presence of sialic acid and sulfate residues. This high charge density may play a role in mucoadhesion.

Recent studies have demonstrated that the mucus is composed of two layers.[105] The inner layer firmly adheres to the gastric mucosa, whereas the outer layer (luminal side) loosely adheres.[105] In the stomach, the loosely adherent layer is between 40 and 60 percent of the total mucus thickness, whereas in the duodenum it makes up about 90 percent of the total mucus thickness.

The outer mucus layer is in a state of continuous flux owing to erosion by luminal proteases and mechanical shear and replacement by secretion of new mucus. Under normal physiological conditions, mucus secretion is equal to its erosion, thus maintaining an adequate protective layer.

Epithelial layers. The epithelial layer lies immediately below the mucus layer and generally provides the major barrier to drug penetration. To reach the systemic circulation, a drug must cross the epithelial barrier to initiate its pharmacological action. The morphology of the epithelial layer differs with location within the body. A commonality among all epithelial layers is the existence of two layers, the apical and the basal regions. The basal end lies on the basal lamina, and the apical end is in contact with the mucus layer.

For regions where absorptive as well as protective functions are crucial, the epithelia are composed of ciliated cells or microvilli to increase the surface area for absorption. The ciliated cells are also protective in that the beating ciliary action works against large-particle deposition.

The stomach has a surface epithelium composed of a single layer of columnar cells with few apical microvilli. The epithelial lining of the small intestine consists of a single layer of columnar cells with densely packed microvilli to promote absorption. The epithelium in the large intestine is similar to that in the small intestine except for the absence of villi in the large intestine.

6.5.2 Polymers as bioadhesives

The notion that bioadhesion enhances the efficiency of drug delivery through an intimate and prolonged contact between the delivery device and the absorption site has resulted in considerable efforts to develop and evaluate bioadhesive polymers. The use of bioadhesive polymers in controlled release drug delivery systems provides potential advantages, including

1. Prolonged residence time at the site of absorption
2. Increased time of contact with the absorbing mucosa
3. Localization in specific regions to enhance drug bioavailability

Diverse classes of polymers have been investigated for their potential use as bioadhesives. These include synthetic polymers such as polyacrylic acid[21,104,106,107] and derivatives,[21,108] hydroxypropylmethylcellulose,[109,110] and polymethacrylate derivatives,[111,112] as well as naturally occurring polymers such as hyaluronic acid[113] and chitosan.[114–117] The mechanisms involved in bioadhesion are not completely understood. However, based on research focused on hydrogel interactions with soft tissue, the process of bioadhesion and the formation of an adhesive bond are believed to occur in three stages.[118] The first is the so-called wetting stage, where the polymer must spread over the biological substrate and create an intimate contact with the surface of the substrate. The surface characteristics and composition of the bioadhesive material and those of the biological substrate play an important role in achieving this intimate contact. The wetting stage is followed by the interpenetration or interdiffusion and mechanical entanglement stages. Physical or mechanical bonds result from entanglement of adhesive material and the extended mucus chains.[119] Secondary chemical bonds are due to electrostatic interactions, hydrophobic interactions and dispersion forces, and hydrogen bonding.[120] Among the secondary chemical bonds, electrostatic interactions and hydrogen bonding appear to be more important owing to the numerous charged and hydrophilic species in the mucus.

Several important physicochemical properties contribute to the adhesive potential of candidate polymers. These properties include the following[99,108,109,111,121–127]:

- High molecular weight (i.e., >100,000 Da) needed to produce interpenetration and chain entanglement
- Hydrophilic molecules containing a large number of functional groups capable of forming hydrogen bonds with mucin
- Anionic polyelectrolytes with a high charge density of hydroxyl and carboxyl groups
- Highly flexible polymers with high chain segment mobility to facilitate polymer chain interpenetration and interdiffusion
- Surface properties similar to those of the biological substrate to provide a low interfacial free energy between the adhesive and the substrate

Although these properties are not all *required* for bioadhesion, they have been found to enhance the bioadhesive characteristics of the polymers.

6.5.3 Factors affecting bioadhesion

Physiological and environmental factors

Effect of pH. Depending on the desired site of application for bioadhesive drug delivery, there are important physiologically relevant factors that affect bioadhesion. Environmental pH is among the most influential issues.[128] Since bioadhesive polymers, natural or synthetic, are polyelectrolytes, slight variations in the pH of the local environment would change the charge density significantly within both the mucus and the polymeric networks. The environmental pH affects the degree of functional group dissociation, which, in turn, affects polymeric hydration. Park and Robinson[129] demonstrated that the pH of the medium is directly related to the force of bioadhesion for Polycarbophil microparticles attached to rabbit stomach tissue. Ch'ng et al.[104] evaluated the bioadhesive performance of some swellable water-insoluble bioadhesive polymers, and it was shown that for lightly cross-linked polymers of acrylic acid, maximum adhesion occurred at pH 5 and minimum at pH 7. Park and Robinson[105] investigated the mechanisms of mucoadhesion of polyacrylic acid, and the pH dependency of bioadhesion in this study was attributed to the change in charge density. The data suggested that mucoadhesion occurred only when carboxyl groups were in acid form. At pH values greater than 6, the electrostatic repulsive forces dominate, and adhesive bonding became negligible.[105]

Mucus turnover rate. The turnover rate of the mucus layer is another physiological factor that affects mucoadhesion. Mucus turnover limits the potential duration of adhesion at the desired site of application. Within the GI tract, mucus is lost continuously secondary to enzymatic degradation (pepsin, lysosomal enzymes, and pancreatic enzymes), acid degradation, bacterial degradation, and mechanical sloughing.[130] Food ingestion also results in gastric mucin loss owing to mechanical friction.[131]

Effect of water/interstitial fluid. A fundamental difference between the adhesion between two solid surfaces and that between a polymer and a soft tissue is the existence of a third interstitial fluid phase in the latter case. This third phase is mostly (>95 percent) composed of water, which has a significant effect on bioadhesion. For the development of a successful semipermanent bioadhesive bond, the polymer needs to withstand the effects of the interstitial fluid. Mortazavi and Smart[132,133] have studied the role of water and hydration on mucoadhesion extensively. Using Carbopol 934P, Carbopol EX55 (polycarbophil), hydroxypropylcellulose, and gelatin as mucoadhesives, they studied water uptake by tablets (dry dosage form) and gels (partially dry dosage form) prepared from each of these materials. The mucoadhesives were placed onto a dialysis membrane in contact with porcine gastric mucus. Despite the presence of the dialysis membrane, significant water uptake was seen for all preparations after 1 minute of contact. For contact times of up to 16 hours, Carbopol C934 tablets caused the greatest dehydration of the mucus gel. The force required for detachment of the adhesive bond increased with increasing mucus dehydration. Rheological studies revealed that the cohesive as well as adhesive properties of the mucus were strengthened by decreasing water content.[132] These findings suggest that for dry and partially dry bioadhesive dosage forms, water movement from the mucus gel into the polymer may be more important than molecular interpenetration.[134] However, it must be noted that excessive hydration of the mucoadhesive material is not optimal for prolonged bioadhesion because the polymer eventually will form a slippery mucilage. Thus, for sustained mucoadhesion the ideal bioadhesive polymer should display a limited hydration profile.

6.5.4 Applications of bioadhesion

There have been numerous studies on gastroretentive bioadhesive formulations; however, most such formulations have proven to be ineffective in prolonging gastric residence.[1] The failures have been associated with two factors: (1) insufficient strength of the bioadhesive bond to overcome the strong propulsive forces of the gastric wall and (2) the continuous mucus production and high mucus turnover within the gastric mucosa.

Some of the most promising data on gastroretentive delivery systems using bioadhesion have resulted from the use of acrylate-based as well as chitosan-based polymers. Poly(acrylates) have been shown to have significant mucoadhesive properties in contact with intestinal mucosal tissues.[104,122,135] Longer et al.[136] demonstrated successful reduction in the gastric emptying rate of chlorothiazide using loosely cross-linked polymers of acrylic acid (polycarbophil) as bioadhesives. The formulations were in the form of microparticles (mean diameter of 505 μm) of polycarbophil[104] mixed with sustained release albumin beads (3:7 w/w ratio of albumin to polycarbophil) loaded into gelatin capsules. Their results showed that 90 percent of the albumin-polycarbophil beads remained in the stomach after 6 hours and that polycarbophil was bound to the gastric mucin–epithelial cell surface.[136]

Among the natural polymers, chitosan and derivatives have shown pronounced mucoadhesion in contact with GI mucosa.[114,115,117,137–141] Intestinal absorption of insulin loaded in chitosan-coated liposomes was demonstrated.[142] Blood glucose levels were reduced significantly after the administration of a single dose of these liposomes in rats. Microspheres of chitosan, prepared by a novel w/o/w emulsion spray drying technique, provided rapid release of model H_2-antagonist drugs (cimetidine, famotidine, and nizatidine).[141] The microspheres displayed significant mucoadhesive properties, as determined by turbidimetric measurements.[140] Bioadhesion studies using rat small intestine have shown prominent retention of chitosan microspheres as compared with ethylcellulose microspheres as controls, where more than 50 percent of the chitosan microspheres were adsorbed on the small intestine.[101] In vivo phase I clinical studies were initiated to evaluate bioadhesive performance of chitosan microspheres in human subjects by gamma scintigraphy.[101]

Microspheres (10 to 200 μm in diameter) of poly(fumaric acid–cosebacic acid) anhydride (20:80) [P(FA:SA)] were shown to exhibit very strong and pronounced mucoadhesive properties both in vitro and in vivo.[143–145] The microspheres were tested for their effect on GI transit of low-molecular-weight drugs salicylic acid and dicumarol. As compared with the control (drug-loaded alginate microspheres of similar size), the P(FA:SA) microspheres significantly delayed the GI transit of these drugs in rats.[145]

6.6 Mechanism or Kinetics of Drug Release

Controlled release dosage forms typically have one of three different dissolution profiles: square root of time or matrix diffusion, zero-order delivery as for erosional dosage forms or osmotic pumps, and zero-order delivery with depletion of the driving force as for a membrane-controlled system. For many controlled release dosage forms, zero-order release may be the "holy grail."

In the case of gastric retentive dosage forms, zero-order delivery may not be as useful. In studying the patient population, certain subjects may show rapid gastric emptying in the fed mode or may be noncompliant and take their medication while fasting, although intended for fed administration. A loading dose of drug in the dosage form or the preponderance delivered up front, as with square-root-of-time release, would at least guarantee that even these exceptional patients receive a substantial portion of their intended medication. Zero-order delivery would not reduce this source of variability.

6.7. Future Potential for Gastric Retentive Delivery Systems

Gastric retentive dosage forms based on flotation have been commercialized in Europe; gastric retentive tablets based on size and food have been and are being developed. Technologies based on retention in the fasted state still are being investigated for various indications, but their complexity, and reproducibility, as well as evaluation of their efficacy and safety in vivo, await identification of a therapy justifying their increased cost and risk of development and manufacture.

References

1. Moes, A. J. Gastroretentive dosage forms. *Crit. Rev. Ther. Drug Carrier Syst.* **10**(2):143–195, 1993.
2. Hwang, S. J. Park, H., and Park, K. Gastric retentive drug-delivery systems. *Crit. Rev. Ther. Drug Carrier Syst.* **15**(3):243–284, 1998.
3. Sheth, P. R., and Tossounian, J. L. Sustained release pharmaceutical capsules, U.S. Patent 4,126,672, 1978.
4. Sheth, P. R., and Tossounian, J. L. Sustained release tablet formulations, U.S. Patent 4,140,755, 1979.
5. Michaels, A. S., Bashaw, J. D., and Zaffaroni, A. Gastro-inflatable drug delivery device, U.S. Patent 3,901,232, 1975.
6. Banker, G. S. Sustained release film-coated preparations, U.S. Patent 3,896,792, 1975.
7. Rouge, N., Buri, P. A., and Doelker, E. Drug absorption sites in the gastrointestinal tract and dosage forms for site-specific delivery. *Int. J. Pharm.* **136**:115–122, 1996.
8. Baumgartner, S., Tivadar, A., Vrecer, V., and Kristl, J. Development of floating tablets as a new approach to the treatment of *Heliobacter pylori* infections. *Acta Pharm.* **51**:21–33, 2001.
9. Hejazi, R., and Amiji, M. Stomach-specific anti-*H. pylori* therapy: II. Gastric residence studies of tetracycline-loaded chitosan microspheres in gerbils. *Pharm. Dev. Technol.* **8**:252–262, 2003.
10. Curatolo, W. J., and Lo, J. Gastric retention system for controlled drug release, U.S Patent 5,002,772, 1991.
11. Curatolo, W. J., and Lo, J. Gastric retention system for controlled drug release, U.S. Patent 5,443,843, 1995.
12. Caldwell, L. J., and Gardner, C. R. Drug delivery device which can be retained in the stomach for a controlled period of time, U.S. Patent 4,767,627, 1988.
13. Cargill, R. C., Caldwell, L. J., Engle, K., et al. Controlled gastric emptying. I. Effects of physical properties on the residence times of nondisintegrating geometric shapes in beagle dogs. *Pharm. Res.* **5**:533–536, 1989.
14. Cargill, R., Engle, K., Gardner, C. R., et al. Controlled gastric emptying. II. In vitro erosion and gastric residence times of an erodible device in beagle dogs. *Pharm. Res.* **6**(6):506–509, 1989.
15. Caldwell, L. J., Gardner, C. R., and Cargill, R. C. Drug delivery device which can be retained in the stomach for a controlled period of time, U.S. Patent 4,735,804, 1988.

16. Caldwell, L. J., Gardner, C. R., Cargill, R. C., and Higuchi, T. Drug delivery device which can be retained in the stomach for a controlled period of time, U.S. Patent 4,758,436, 1988.
17. Fleshner-Barak, M., Lerne, I. E., Rosenberger, V., et al. Rapidly expanding composition for gastric retention and controlled release of therapeutic agents, and dosage forms including the composition, World Intellectual Property Organization, PCT WO, 2002.
18. Asmussen, B., Cremer, K., Hoffman, H., et al. Expandable gastroretentive therapeutic system with controlled active substance release in the gastrointestinal tract, U.S. Patent 6,290,989, 2001.
19. Jimenez-Castellanos, M. R., Zia, H., and Rhodes, C. T. Mucoadhesive delivery systems. *Drug Development and Industrial Pharmacy.* **19**:143–194, 1993.
20. Park, H. On the mechanism of bioadhesion. University of Wisconsin-Madison, 1986.
21. Park, H., and Robinson, J. R. Mechanisms of mucoadhesion of poly(acrylic acid) hydrogels. *Pharm. Res.* **4**:457–464, 1987.
22. Riley, R. G., Smart, J. D., Tsibouklis, J., et al. An in vitro model for investigating the gastric mucosal retention of 14C-labeled poly(acrylic acid) dispersions. *Int. J. Pharm.* **236**:87–96, 2002.
23. Cuna, M., Alonso, M. J., and Torres, D. Preparation and in vivo evaluation of mucoadhesive microparticles containing amoxycillin-resin complexes for drug delivery to the gastric mucosa. *Eur. J. Pharm. Biopharm.* **51**(3):199–205, 2001.
24. Phillipson, M., Atuma, C., Henriksen, J., and Holm, L. The importance of mucus layers and bicarbonate transport in preservation of gastric juxtamucosal pH. *Am. J. Physiol. Gastrointest. Liver Physiol.* **282**:G211–G219, 2002.
25. Mojaverian, P., Ferguson, R. K., Vlasses, P. H., et al. Estimation of gastric residence time of the Heidelberg capsule in humans: effect of varying food composition. *Gastroenterology.* **89**:392–397, 1985.
26. Mojaverian, P., Rocci, M. L., Conner, D. P., et al. Effect of food on the absorption of enteric-coated aspirin: Correlation with gastric residence time. *Clin. Pharmacol. Ther.* **41**:11–17, 1987.
27. Davis, S. S., Hardy, J. G., Taylor, M. J., et al. The in vivo evaluation of an osmotic device (Osmet) using gamma scintigraphy. *J. Pharm. Pharmacol.* **36**:740–742, 1984.
28. Davis, S. S., Stockwell, A. F., Taylor, M. J., et al. The effect of density on the gastric emptying of single- and multiple-unit dosage forms. *Pharm. Res.* **3**:208–213, 1986.
29. Feely, L. C., and Davis, S. S. Correlation of phenylpropanolamine bioavailability with gastrointestinal transit by scintigraphic monitoring of ^{111}In-labeled hydroxypropylcellulose matrices. *Pharm. Res.* **6**:274–278, 1989.
30. Coupe, A. J., Davis, S. S., and Wilding, I. R. Variation in gastrointestinal transit of pharmaceutical dosage forms in healthy subjects. *Pharm. Res.* **8**:360–364, 1991.
31. Coupe, A. J., Davis, S. S., Evans, D. F., and Wilding, I. R. Correlation of the gastric emptying of nondisintegrating tablets with gastrointestinal motility. *Pharm. Res.* **8**:1281–1285, 1991.
32. Itoh, T., Higuchi, T., Gardner, C. R. and Caldwell, L. Effect of particle size and food on the gastric residence time of non-disintegrating solids in beagle dogs. *J. Pharm. Pharmacol.* **38**:801–806, 1986.
33. Davis, S. S., Khosla, R., Wilson, C. G., and Washington, N. Gastrointestinal transit of a controlled release pellet formulation of tiaprofenic acid and the effect of food. *Int. J. Pharm.* **35**:253, 1987.
34. Khosla, R., and Davis, S. S. Effect of tablet size on the gastric emptying of nondisintegrating tablets. *Int. J. Pharm.* **62**:R9–R11 1990.
35. O'Reilly, S., Wilson, C. G., and Hardy, J. G. The influence of food on the gastric emptying of multiparticulate dosage forms. *Int. J. Pharm.* **34**:213, 1987.
36. Blok, D., Arndt, J. W., DeHaan, F. H., et al. Scinitgraphic investigation of the gastric emptying of 3-mm pellets in human volunteers. *Int. J. Pharm.* **123**:145, 1995.
37. Johnson, L. R. *Gastrointestinal Physiology*. St. Louis: Mosby, 1997.
38. Davenport, H. W. A. *Digest on Digestion.* Chicago: Year Book Medical Publishers, 1975.
39. Timmermans, J., and Moes, A. J. The cutoff size for gastric emptying of dosage forms. *J. Pharm. Sci.* **82**(8):854, 1993.
40. Mason, G. R., Kahrilas, P. J., Otterson, M. F., et al. *The Physiologic Basis of Surgery*, 2d ed. Baltimore: Williams and Wilkens, 1996.

41. Sarna, S. K., Otterson, M. R., Cowles, V. E., and Telfor, G. L. *Essentials of Experimental Surgery: Gastroenterology*. Newark, NJ: Harwood Academic Publishers, 1996.
42. Mockiki, E., Kuwano, H., Nakabayashi, T., et al. Pyloric relaxation regulated via intramural neural pathway of the antrum. *Dig. Dis. Sci.* **46**:2307–2313, 2001.
43. Cowles, V. E., Nellans, H. N., Seifert, T. R., et al. Effect of novel motilide ABT-229 versus erythromycin and cisapride on gastric emptying in dogs *J. Pharmacol. Exp. Ther.* **293**(3):1106–11, 2000.
44. Meyer, J. H. *Physiology of the Gastrointestinal Tract*, 2d ed. New York: Raven Press, 1987.
45. Houghton, L. A., Read, N. W., Heddle, R., et al. Relationship of the motor activity of the antrum, pylorus, and duodenum to gastric emptying of a solid-liquid mixed meal. *Gastroenterology* **94**(6):1285–1291, 1988.
46. Meyer, J. H., Dressman, J., Fink, A., and Amidon, G. *Effect of size and density on canine gastric emptying of nondigestible solids. Gastroenterology* **89**(4):805–813, 1985.
47. Davis, S. S., Hardy, J. G., and Fara, J. W. Transit of pharmaceutical dosage forms through the small intestine. *Gut* **27**(8):886–892, 1986.
48. Chawla, G., Gupta, P., Koradia, V., and Bansal, A. K. Gastroretention a means to address regional variability in intestinal drug absorption. *Pharm. Tech.* **27**:50–68, 2003.
49. Shell, J. W., and Louie-Helm, J. Extending the duration of drug release within the stomach during the fed mode. U.S. Patent 6,340,475, 2002.
50. Shell, J. W., Louie-Helm, J., and Markey, M. Extending the duration of drug release within the stomach during the fed mode, U.S. Patent 6,635,280, 2003.
51. Shell, J. W., and Louie-Helm, J. Gastric-retentive oral drug dosage forms for the controlled-release of sparingly soluble drugs and insoluble matter. U.S. Patent 5,972,389, 1999.
52. Sugito, K., Ogata, H., Goto, H. et al. Gastric emptying: Rate of drug preparations: III. Effects of size of enteric microcapsules with mean diameters ranging from 0.1 to 1.1 mm in man. *Pharm. Soc. Japan.* **40**:3343–3345, 1992.
53. Hinder, R. A., and Kelly, K. A. Canine gastric emptying of solids and liquids. *Am. J. Physiol.* **233**(4):E335–340, 1977.
54. Berner, B., Hou, S. Y. E., and Cowles, V. E. Gastric retentive dosage forms: A review with a focus on fed mode. CRS 31st Annual Meeting, Honolulu, Hawaii, 2004.
55. Beten, D. B., VanGesbeke, B. Schoutens, A., and Moes AJ, Evaluation of the gastric behavior of coevaporate particles under fasting and nonfasting conditions. *Int. J. Pharm.* **123**:145, 1995.
56. Timmermans, J., and Moes, A. J. Factors controlling the buoyancy and gastric retention capabilities of floating matrix capsules: New data for reconsidering the controversy. *J. Pharm. Sci.* **83**(1):18–24, 1994.
57. Davis, S. S. Norring-Christensen, F., Khosla, R. and Feely, L. C. Gastric emptying of large single-unit dosage forms. *J. Pharm. Pharmacol.* **40**(3):205–207, 1988.
58. Gusler, G., Hou, S., Connor, A. L., et al. Pharmacoscintigraphic evaluation of metformin ER in healthy volunteers. *Pharm. Sci.* **5**:S1, 2003.
59. Cowles, V. E., Gusler, G., Chen, T. F., Hou, S. Y. E., Wilding, I., Berner, B. Gastric emptying and intestinal transit of gastroretentive tablets in the fed state: Effect of meal content and tablet size. *Gastroenterology* **126**(Suppl 2):A489–A490, 2004.
60. Wilding, I. R., Davis, S. S., Sparrow, R. A., et al. Pharmacoscintigraphic evaluation of a modified release (Geomatrix) diltiazem dosage. *J. Contr. Rel.* **33**:89–97, 1995.
61. Shameem, M., Katori, N., Aoyagi, N., and Kojima, S. Oral solid controlled release dosage forms: Role of GI-mechanical destructive forces and colonic release in drug absorption under fasted and fed conditions in humans. *Pharm. Res.* **12**(7):1049–1054, 1995.
62. Louie-Helm, J., Fell, R., Gusler, G., et al. Dual-labelled pharmacoscintigraphic study of furosemide gastric retentive (GRTM) tablets in healthy volunteers. CRS 30th Annual Meetings Proceedings, Glasgow, Scotland, 2003.
63. Louie-Helm, J., Berner, B., and Urquhart, J. Formulation of an erodible, gastric retentive oral diuretic, U.S. Patent 20030152622A1, 2003.
64. Berner, B., and Louie-Helm, J. Gastric retentive oral dosage forms with restricted drug release in the lower gastrointestinal tract, U.S. Patent 2003014052, 2003.

65. Louie-Helm, J., and Berner, B. Formulation of an erodible, gastric retentive oral dosage form using in vitro disintegration test data, U.S. Patent 20030091630, 2003.
66. Timmins, P., Dennis, A. B., and Vyas, K. A. Biphasic controlled release delivery system for high solubility pharmaceuticals and method, U.S. Patent 6,475,521, 2002.
67. Timmins, P., Dennis, A. B., and Vyas, K. A. Method of use of a biphasic controlled release delivery system for high solubility pharmaceuticals and method, U.S. Patent 6,660,300, 2003.
68. *Physicians' Desk Reference*, 57th ed. Montvale, NJ: Thomson Healthcare, 2003.
69. Gusler, G., Gorsline, J., Levy, G., et al. Pharmacokinetics of metformin gastric-retentive tablets in healthy volunteers. *J. Clin. Pharmacol.* **41**(6):655–661, 2001.
70. Depomed completes dosing in phase III trials of metformin GRTM and Ciprofloxacin GR. Business Wire Press Release, 2003.
71. Dilger, C. Pharmacokinetics of ciprofloxacin gastric retentive tablets in healthy volunteers. CRS 28th Annual Meetings, San Diego, CA, 2001.
72. Depomed initiates phase II trial with once daily diuretic Furosemide GR. Business Wire Press Release, 2004.
73. Fix, J. A., Cargill, R., and Engle, K. Controlled gastric emptying: III. Gastric residence time of a nondisintegrating geometric shape in human volunteers. *Pharm. Res.* **10**(7):1087–1089, 1993.
74. Park, K., and Oark, H. Superabsorbent hydrogel foams, U.S. Patent. 5,750,585, 1998.
75. Chen, J., Blevins, W. E., Park, H., and Park, K. Gastric retention properties of superporous hydrogel composites. *J. Contr. Rel.* **64**(1–3):39–51, 2000.
76. Gemeinhart, R. A., Park, H., and Park K. Effect of compression on fast swelling of poly(acrylamide-co-acrylic acid) superporous hydrogels. *J. Biomed. Mater. Res.* **55**(1):54–62, 2001.
77. Sinnreich, J. Covered retard forms, U.S. Patent 4,996,058, 1991.
78. Wilson, A. M., McKenna, B., Gudipati, M., et al. Evaluation of endogastric therapeutic system (EGTS) containing a GABA receptor agonist in dogs. *AAPS Pharm. Sci.* **3**: W4247, 2001.
79. Cumming, K. I., Wilson, A. M., McKenna, B., et al. The pharmacokinetic and pharmacoscintigraphic evaluation of endogastric therapeutic systems in the fasted and fed state in healthy volunteers. *AAPS Pharm. Sci.* **3**:W4253, 2001.
80. Lavy, E., Klausner, E., Hoffman, A., and Friedman, M. Evaluation of a novel sustained-release gastroretentive dosage form of riboflavin in dogs. *J. Vet. Intern. Med.* **15**:304, 2001.
81. Klausner, E. A., Lavy, E., Stepensky, D. et al. Furosemide pharmacokinetics and pharmacodynamics following gastroretentive dosage form administration to healthy volunteers. *J. Clin. Pharmacol.* **43**(7):711–720, 2003.
82. Klausner, E. A., Eyal, S., Lavy, E., et al. Novel levodopa gastroretentive dosage form: In-vivo evaluation in dogs. *J. Contr. Rel.* **88**(1):117–126, 2003.
83. Klausner, E. A., Lavy, E., Stepensky, D., et al. Novel gastroretentive dosage forms: Evaluation of gastroretentivity and its effect on riboflavin absorption in dogs. *Pharm. Res.* **19**(10):1516–1523, 2002.
84. Klausner, E. A., Lavy, E., Friedman, M., and Hoffman, A. Expandable gastroretentive dosage forms. *J Contr. Rel.* **90**(2):143–162, 2003.
85. Jusko, W. J., Rennick, B. R., and Levy, G. Renal excretion of riboflavin in the dog. *Am. J. Physiol.* **218**(4):1046–1053, 1970.
86. Juskom, W. J., and Levy, G. Pharmacokinetic evidence for saturable renal tubular reabsorption of riboflavin. *J. Pharm. Sci.* **59**:765–772, 1970.
87. Desai, S., and Bolton, S. A floating controlled-release drug delivery system: In vitro–in vivo evaluation. *Pharm. Res.* **10**(9):1321–1325, 1993.
88. Oth, M., Franz, M., Timmermans, J., and Moes, A. The bilayer floating capsule: A stomach-directed drug delivery system for misoprostol. *Pharm. Res.* **9**(3):298–302, 1992.
89. Agyilirah, G. A., Green, M., DuCret, R., and Banker, G. S. Evaluation of the gastric retention properties of a cross-linked polymer coated tablet versus those of a nondisintegrating tablet. *Int. J. Pharm.* **75**:241–247, 1991.

90. Hilton, A. K., and Deasy, P. B. In vitro and in vivo evaluation of an oral sustained-release floating dosage form of amoxycillin trihydrate. *Int. J. Pharm.* **86**:79–88, 1992.
91. Devereux, J. E., Newton, J. M., and Short, M. B. The influence of density on the gastrointestinal transit of pellets. *J. Pharm. Pharmacol.* **42**(7):500–501, 1990.
92. Clarke, G. M., Newton, J. M. and Short, M. B. Comparative gastrointestinal transit of pellet systems of varying density. *Int. J. Pharm.* **114**:1–11, 1995.
93. Sheth, P. R., and Tossounian, J. The hydrodynamically balanced system (HBSTM): A novel drug delivery system for oral use. *Drug Dev. Ind. Pharm.* **10**:319–339, 1984.
94. Özdemir, N., Sefika, O., and Özkan, Y. Studies of floating dosage forms of furosemide: In vitro and in vivo evaluations of bilayer tablet formulations. *Drug Dev. Ind. Pharm.* **26**:857–866, 2000.
95. Wei, Z., Yu, Z., and Bi, D. Design and evaluation of a two-layer floating tablet for gastric retention using cisapride as a model drug. *Drug Dev. Ind. Pharm.* **27**(5):469–474, 2001.
96. Fell, J. T., Whitehead, L., and Collett, J. H. Prolonged gastric retention: Using floating dosage forms. *Pharm. Tech.* **24**:82–89, 2000.
97. Whitehead, L., Collett, J. H., and Fell, J. T. Amoxycillin release from a floating dosage form based on alginates. *Int. J. Pharm.* **210**(1–2):45–49, 2000.
98. Atyabi, F., Sharma, H. L., Mohammad, H. A. H., and Fell, J. T. In vivo evaluation of a novel gastric retentive formulation based on ion-exchange resins. *J. Contr. Rel.* **42**:105–113, 1996.
99. Tabak, L. A., Levine, M. J., Mandel, I. D., et al. Role of salivary mucins in the protection of the oral cavity. *J. Oral Pathol.* **11**:1–17, 1982.
100. Peppas, N. A., and Buri, P. A. Surface, interfacial and molecular aspects of polymer bioadhesion on soft tissues. *J. Contr. Rel.* **2**:257–275, 1985.
101. Harding, S. E. Biopolymer mucoadhesives. *Technol. Genet. Eng. Rev.* **16**:41–86, 1999.
102. Campbell, B. J. Biochemical and functional aspects of mucus and mucin-type glycoproteins, in *Bioadhesive Drug Delivery Systems: Fundamentals, Novel Approaches, and Development.* New York: Marcel Dekker, 1999.
103. Varki, A. Biological roles of oligosaccharides. *Glycobiology* **3**:97–130, 1993.
104. Ch'ng, H. S. Bioadhesive polymers as platforms for oral controlled drug delivery: II. Synthesis and evaluation of some swelling, water-insoluble bioadhesive polymers. *J. Pharm. Sci.* **74**(4):399–405, 1985.
105. Atuma, C., Strugala, V., Allen, A., and Holm, L. The adherent gastrointestinal mucus gel layer: thickness and physical state in vivo. *Am. J. Physiol. Gastrointest. Liver Physiol.* **280**(5):G922–929, 2001.
106. Anlar, S., Capan, Y., and Hincal, Y. A. Physico-chemical and bioadhesive properties of polyacrylic acid polymers. *Pharmazie* **48**:285–287, 1993.
107. Guo, J. H. Investigating the surface properties and bioadhesion of buccal patches. *J. Pharm. Pharmacol.* **46**(8):647–650, 1994.
108. Shojaei, A. H., and Li, X. Mechanisms of buccal mucoadhesion of novel copolymers of acrylic acid and polyethylene glycol monomethylether monomethacrylate. *J. Contr. Rel.* **47**(2):151–161, 1997.
109. Gandhi, R. E., and Robinson, J. R. Bioadhesion in drug delivery. *Ind. J. Pharm. Sc.* **50**:145–152, 1988.
110. Anlar, S., Capan, Y., Guven, O., et al. Formulation and in vitro–in vivo evaluation of buccoadhesive morphine sulfate tablets. *Pharm. Res.* **11**(2):231–236, 1994.
111. Leung, S. S., and Robinson, J. R. Polymer structure features contributing to mucoadhesion. *J. Contr. Rel.* **12**:187–194, 1990.
112. Nakamura, K. Uptake and release of budesonide from mucoadhesive, pH-sensitive copolymers and their application to nasal delivery. *J. Contr. Rel.* **61**:329–335, 1999.
113. Sanzgiri, Y. D., Topp, E. M., Benedetti, L., et al. Evaluation of mucoadhesive properties of hyaluronic acid benzyl esters. *Int. J. Pharm.* **107**:91–97, 1994.
114. Lehr, C. M., Bouwstra, J. A., Schact, E. H., et al. In vitro evaluation of mucoadhesive properties of chitosan and some other natural polymers. *Int. J. Pharma.* **78**:43–48, 1992.

115. Fiebrig, I., Harding, S. E., Stokke, B. T., et al. The potential of chitosan as a mucoadhseive drug carrier: studies on its interaction with pig gastric mucin on a molecular level. *Eur. J. Pharm. Sci.* **2**:185, 1994.
116. Remunan-Lopez, C., Portero, A., Vila-Jato, J. L., and Alonso, M. J. Design and evaluation of chitosan/ethylcellulose mucoadhesive bilayered devices for buccal drug delivery. *J. Contr. Rel.* **55**(2–3):143–152, 1998.
117. Bernkop-Schnurch, A., and Krajicek, M. E. Mucoadhesive polymers as platforms for peroral peptide delivery and absorption: Synthesis and evaluation of different chitosan-EDTA conjugates. *J. Contr. Rel.* **50**(1–3):215–223, 1998.
118. Allen, K. W. A review of contemporary views of theories of adhesion. *J. Adhesion.* **21**:261–277, 1987.
119. Jabbari, E., Wisniewski, N., and Peppas, N. Evidence of mucoadhesion by chain interpenetration at a polyacrylic acid/mucin interface using ATR-FTIR spectroscopy. *J. Contr. Rel.* **26**:99–108, 1993.
120. Chickering, D. E., and Mathiowitz, E. Definitions, mechanisms, and theories of bioadhesion, in *Bioadhesive Drug Delivery Systems: Fundamentals, Novel Approaches, and Development*, New York: Marcel Dekker, 1999.
121. Ponchel, G., Touchard, F., Duchene, D., et al. Bioadhesive analysis of controlled-release systems: I. Fracture and interpenetration analysis in poly(acrylic acid)-containing systems. *J. Contr. Rel.* **5**:129–141, 1987.
122. Leung, S., and Robinson, J. R. The contribution of anionic polymer structural features to mucoadhesion. *J. Contr. Rel.* **5**:223–231, 1988.
123. Bodde, H. E., De Vries, M. E., and Junginger, H. E. Mucoadhesive polymers for the buccal delivery of peptides, structure-adhesiveness relationships. *J. Contr. Rel.* **13**:225–231, 1990.
124. Bodde, H. E., and Lehr, C. M. Bioadhesive polymers: Surface energy and molecular mobility considerations. *Biofouling* **4**:163–169, 1991.
125. Lehr, C. M., Bouwstra, J. A., Bodde, H. E., and Junginger, H. E. A surface energy analysis of mucoadhesion: Contact angle measurements on polycarbophil and pig intestinal mucosa in physiologically relevant fluids. *Pharm. Res.* **9**(1):70–75, 1992.
126. Lehr, C. M., Bodde, H. E., Bouwstra, J. A., et al. A surface energy analysis of mucoadhesion: II. Predication of mucoadhesive performance by spreading coefficient. *Eur. J. Pharm. Sci.* **1**:19–30, 1993.
127. Li, X., and Shojaei, A. H. Novel copolymers of acrylic acid and polyethylene glycol monomethylether monomethacrylate for buccoadhesion: In vitro assessment of buccoadhesion and analysis of surface free energy. *Proceed. Int. Symp. Control. Rel. Bioact. Mater.* **23**:507–508, 1996
128. Knuth, K., Amiji, M., and Robinson, J. R. Hydrogel delivery systems for vaginal and oral applications: Formulation and biological considerations. *Adv. Drug Del. Rev.* **11**:137–167, 1993.
129. Park, H., and Robinson, J. R. Physico-chemical properties of water insoluble polymers important to mucin/epithelial adhesion. *J. Con. Rel.* **2**:47–57, 1985.
130. Forstner, J. F. Intestinal mucins in health and disease. *Digestion* **17**:234–263, 1978.
131. Gupta, P. K., Leung, S. S., and Robinson, J. R. Bioadhesives/mucoadhesives in drug delivery to the gastrointestinal tract, in *Bioadhesive Drug Delivery Systems.* Boca Raton, FL: CRC Press, 1990.
132. Mortazavi, S. A., and Smart, D. An investigation into the role of water movement and mucus gel hydration in mucoadhesion. *J. Contr. Rel.* **25**:197–203, 1993.
133. Mortazavi, S. A., and Smart, D. Factors influencing gel-strengthening at the mucoadhesive-mucus interface. *J. Pharma. Pharmacol.* **46**:86–90, 1993.
134. Smart, J. D. The role of water movement and polymer hydration in mucoadhesion, in *Bioadhesive Drug Delivery Systems: Fundamentals, Novel Approaches, and Development*, New York: Marcel Dekker, 1999.
135. Gu, J., Robinson, J. R., and Leung, S. S. Binding of acrylic polymers to mucin/epithelial surfaces: Structure-property relationships. *Crit. Rev. Ther. Drug Carrier Syst.* **8**(1):21–67, 1988.
136. Longer, M. A., Ch'ng, H. S., and Robinson, J. R. Bioadhesive polymers as platforms for oral controlled drug delivery: III. Oral delivery of Chlorothiazide using a bioadhesive polymer. *J. Pharm. Sci.* **74**(4):406–411, 1985.

137. Takayama, K. Effect of interpolymer complex formation on bioadhesive property and drug release phenomenon of compressed tablet consisting of chitosan and sodium hyaluronate. *Chem. Pharm. Bull.* **38**:1993–1997, 1990.
138. Borchard, G., Lueben, H., de Boer, A. G., et al. The potential of mucoadhesive polymers in enhancing intestinal peptide drug absorption: III. Effects of chitosan-glutamate and carbomer on epithelial tight junctions in vitro. *J. Contr. Rel.* **39**:131–138, 1996.
139. Bogataj, M., and Mrhar, A. Mucoadhesion of polycarbophil and chitosan microspheres on isolated guinea pig vesical and intestinal mucosa. *J. Contr. Rel.* **48**:340–341, 1997.
140. He, P., Davis, S. S., and Illum, L. In vitro evaluation of the mucoadhesive properties of chitosan microsphere. *Int. J. Pharm.* **166**:75–88, 1998.
141. He, P., Davis, S. S., and Illum, L. Chitosan microspheres prepared by spray drying. *Int. J. Pharm.* **187**(1):53–65, 1999.
142. Takeuchi, H., Yamamoto, H., Niwa, T., et al. Enteral absorption of insulin in rats from mucoadhesive chitosan-coated liposomes. *Pharm. Res.* **13**(6):896–901, 1996.
143. Jacob, J., Santos, C., Carino, G., et al. An in-vitro bioassay for quantification of bioadhesion of polymer microspheres to mucosal epithelium. *Proceed. Int. Symp. Control. Rel. Bioact. Mater.* **22**:312–313, 1995.
144. Mathiowitz, E., Carinho, G., Jacob, J., et al. Bioadhesive microspheres for increased bioavailability. *Proceed. Int. Symp. Control. Rel. Bioact. Mater.* **24**:202–203, 1997.
145. Mathiowitz, E., Chickering, D., Jacob, J. S., et al. Bioadhesive drug delivery systems, in *Encyclopedia of Controlled Drug Delivery*. New York: Wiley, 1999, pp. 9–45.

Chapter

7

Osmotic Controlled Drug Delivery Systems

Sastry Srikonda

Xenoport, Inc.
Santa Clara, California

Phanidhar Kotamraj

Thomas J. Long School of Pharmacy and Health Sciences
University of the Pacific
Stockton, California

Brian Barclay

ALZA Corporation, a Johnson & Johnson Company
Mountain View, California

7.1 Introduction

Osmosis can be defined as the spontaneous movement of a solvent from a solution of lower solute concentration to a solution of higher solute concentration through an ideal semipermeable membrane, which is permeable only to the solvent but impermeable to the solute. The pressure applied to the higher-concentration side to inhibit solvent flow is called the *osmotic pressure*. In 1748, Abbe Nollet first reported the osmotic process. In 1877, Pfeffer separated a sugar solution from water using a sugar-impermeable membrane and quantified the water transport. In 1884, Hugo de Vries invoked osmotic concepts to understand the contraction of the contents of plant cells placed in solutions of high osmotic pressure, where the cell membrane acts as a semipermeable membrane. The osmotic pressure difference between inside and outside environments causes osmotic water loss and results in plasmolysis. In 1886, Van't Hoff identified an underlying proportionality between osmotic pressure, concentration, and temperature in Pfeffer's experiment. Later, he revealed a relationship between osmotic pressure and solute concentration and temperature that was similar to the ideal gas equation, where pressure is proportional to concentration and temperature. According to Van't Hoff's equation, the osmotic pressure in a dilute solution is equal to the pressure that the solute would exert if it were a gas occupying the same volume.

Osmotic pressure, a colligative property, depends on the concentration of solute (neutral molecule or ionic species) that contributes to the osmotic pressure. Solutions of different concentrations having the same solute and solvent system exhibit an osmotic pressure proportional to their concentrations. Thus a constant osmotic pressure, and thereby a constant influx of water, can be achieved by an osmotic delivery system that results in a constant release rate of drug. Therefore, zero-order release, which is important for a controlled release delivery system when indicated, is possible to achieve using these platforms. In 1974,

Theeuwes and Higuchi[1] applied the principle of osmotic pressure to a new generation of controlled drug delivery devices with many advantages over other existing controlled drug delivery systems. The first of these devices, the elementary osmotic pump, is considered a typical delivery system that operates on osmotic principles.

7.2 Rationale for Design of Osmotic® Controlled Drug™ Delivery Systems

The ideal situation in most drug therapies is the maintenance of a uniform concentration of drug at the site of action, since the pharmacological action elicited by a drug generally is proportional to its concentration at the site of action. Drugs having narrow therapeutic windows especially require a close monitoring of blood levels to avoid untoward effects. Constant intravenous infusion is a popular method of drug administration in clinical settings to maintain a constant blood concentration. However, with this method, patient compliance is compromised, and infusion is not economically feasible for long-term therapies. Noninvasive routes such as oral administration of multiple doses at a particular frequency can be used to achieve this goal. However, patient compliance is again compromised with the increased frequency of administration. Therefore, the demand has increased for controlled delivery systems that can achieve relatively constant blood concentrations of drug over a prolonged period of time.

The objective of controlled drug release is to deliver a pharmacologically active agent in a predetermined, predictable, and reproducible manner. Since the formulation is metered as a portion of the entire dose at any given time and provides a reduced or once-a-day dosage regimen, controlled drug delivery offers improved patient compliance with reduced side effects. However, providing a constant drug concentration is not necessarily the best treatment regimen. For example, widely fluctuating levels of a drug such as insulin sometimes are required to mimic natural biofeedback mechanisms. Therefore, the term *controlled release* includes modulated release systems as well as zero-order release systems. These systems provide actual therapeutic control despite not providing constant drug concentrations.

Many forms of controlled drug delivery systems are reported in the literature, including delayed release dosage forms, targeted release dosage forms, stimulating circadian rhythm systems, external stimuli controlled dosage forms (including pH changes, magnetic or electrical fields, ultrasound, and chemical responsive systems), and matrix and reservoir systems, including solvent controlled and diffusion controlled systems.[2] All these controlled drug delivery systems suffer from one or more disadvantages, as outlined in their respective chapters. Among all the controlled drug delivery systems, oral controlled drug delivery has received major attention because of its greater popularity and utility.[3–5] Ideal oral drug controlled delivery systems are those that steadily meter out a measurable, reproducible amount of the drug over a prolonged period.[6] Despite this advantage, drug release from oral controlled release dosage forms may be affected by pH, gastric motility, and the presence of food.[7]

One improvement that addresses these disadvantages is the osmotic drug delivery system. In 1974, Theeuwes and Higuchi[1] invoked the principles of osmotic pressure to develop a new generation of controlled drug delivery devices with many advantages over existing drug delivery systems. These delivery systems provide a sustained delivery rate, preventing the sudden increases in plasma concentration that may produce side effects, as well as sharp decreases in plasma concentrations that may lower the effectiveness of the drug. The following sections discuss the different principles necessary for the design of osmotically controlled delivery systems.

7.3 Mechanism of Osmotic Controlled Release

7.3.1 Quantitative aspects of osmosis

Van't Hoff described the relationship between the osmotic pressure of a dilute solution and its concentration as follows:

$$V\pi = nRT \tag{7.1}$$

where π = osmotic pressure in atmospheres
V = volume of the solution in liters
n = number of moles of solute
R = gas constant, equal to 0.082 L·atm/mol·K
T = absolute temperature in K

The Van't Hoff equation also can be written as follows:

$$\pi = \left(\frac{n}{V}\right)RT \tag{7.2}$$

or

$$\pi = cRT \tag{7.3}$$

where c is the concentration of the solute in moles per liter.[8] The preceding equation can be applied satisfactorily to describe the osmotic pressure of dilute solutions of nonelectrolytes such as sucrose and urea. Van't Hoff later observed that the osmotic pressure of electrolyte solutions were two, three, or more times greater than predicted by the general equation. Therefore, a factor i was introduced to account for the behavior of ionic solutions. The corrected equation for electrolyte solutions is written as follows:

$$\pi = icRT \tag{7.4}$$

By application of this equation, it is possible to calculate osmotic pressures for ionic solutions. Van't Hoff also observed that i approaches the number of ions as the molecule dissociates in an increasingly dilute solution. Moreover, the deviations of concentrated electrolyte solutions from ideal behavior can be obtained from Raoult's law.[8]

7.3.2 Release kinetics in elementary osmotic pumps

The elementary osmotic delivery system consists of an osmotic core containing drug and, as necessary, an osmogen surrounded by a semipermeable membrane with an aperture (Fig. 7.1). A system with constant internal volume delivers a volume of saturated solution equal to the volume of solvent uptake in any given time interval. Excess solids present inside a system ensure a constant delivery rate of solute. The rate of delivery generally follows zero-order kinetics and declines after the solute concentration falls below saturation. The solute delivery rate from the system is controlled by solvent influx through the semipermeable membrane.

The osmotic flow of the liquid depends on the osmotic and hydrostatic pressure differences across the semipermeable membrane of the system. This phenomenon is the basic feature of nonequilibrium thermodynamics,

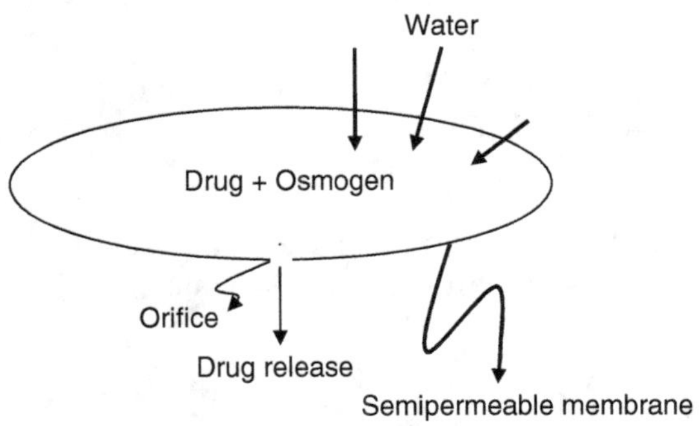

Figure 7.1 Schematic diagram of elementary osmotic delivery system.

which describes the volume flux dV/dt, across the semipermeable membrane in the form of the following equation[9]:

$$\frac{dV}{dt} = \left(\frac{A}{h}\right) L_P(\sigma\Delta\pi - \Delta P) \qquad (7.5)$$

where $\Delta\pi$ and ΔP = osmotic and hydrostatic pressure differences, respectively, across the membrane
L_P = mechanical permeability
σ = reflection coefficient, which accounts for leakage of solute through the membrane
A = surface area of the membrane
h = membrane thickness

The corresponding solute delivery rate dm/dt can be expressed as follows:

$$\frac{dm}{dt} = \left(\frac{A}{h}\right) L_P(\sigma\Delta\pi - \Delta P)C \qquad (7.6)$$

where C is the solute concentration in the delivered fluid.

As size (diameter) of the delivery orifice increases, the hydrostatic pressure of the system decreases, and $(\Delta\pi - \Delta P)$ approximates $\Delta\pi$. The osmotic pressure of the formulation π can be substituted for $\Delta\pi$ when the environmental osmotic pressure is small. Thus the equation can be simplified as follows:

$$\frac{dm}{dt} = \left(\frac{A}{h}\right)(L_P\sigma\pi)C \qquad (7.7)$$

A constant k may replace the product $L_P\sigma$, so the preceding equation further reduces to

$$\frac{dm}{dt} = \left(\frac{A}{h}\right) k\pi C \qquad (7.8)$$

The zero-order release rate of the elementary osmotic pump from $t = 0$ to time t_x, when all the solids dissolve and the solute concentration begins to fall below saturation, can be defined as follows

$$\frac{dm}{dt} = \left(\frac{A}{h}\right) k\pi_S S \qquad (7.9)$$

where S is the solubility at saturation, and π_s is the osmotic pressure at saturation. When the rate of dissolution is not limiting relative to the delivery rate through the aperture, the concentration C can be replaced with solubility S.

7.3.3 Release kinetics in OROS® Push-Pull™ systems

The OROS Push-Pull osmotic delivery system consists of an osmotic core containing two or more compartments. One compartment (the drug compartment) contains drug and, if necessary, osmogens, and another (the "push" compartment) serves as a displacing volume to expel drug through an aperture that penetrates through a semipermeable membrane (SPM) surrounding the osmotic core (Fig. 7.2). A system with constant interval volume delivers a volume of solution (or hydrated suspension in the case of a water-insoluble drug) equal to the volume of solvent uptake in any given time interval. The expansion of the push compartment in conjunction with the dissolution of the drug compartment ensures a constant delivery rate of solute. The rate generally follows zero-order kinetics and then trails off as the solute concentration begins to diminish. Similar to the elementary osmotic pump, the solute delivery rate from the system is controlled by solvent influx through the semipermeable membrane.

In OROS Push-Pull systems, the drug release can be quantified from a modified form of Eq. (7.9). The mass delivery rate dm/dt can be written as follows:

$$\frac{dm}{dt} = \frac{dV}{dt} C_s \tag{7.10}$$

where dV/dt is the total volumetric delivery rate from the dosage form, and C_S is the concentration of drug in the dispensed liquid or suspension.

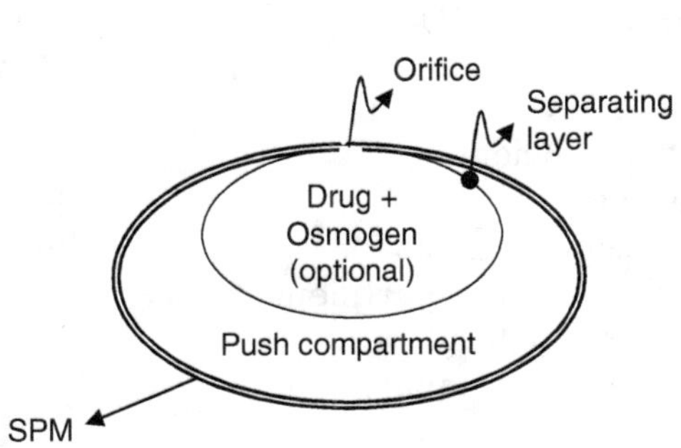

Figure 7.2 Schematic of the OROS Push-Pull system.

The osmotic volumetric flow into the osmotic compartment Q is defined as follows:

$$Q = \left(\frac{dV}{dt}\right)_O \tag{7.11}$$

and the volumetric flow F into drug compartment is defined as

$$F = \left(\frac{dV}{dt}\right)_D \tag{7.12}$$

The concentration of drug released from the formulation can be written as

$$C_S = F_D C_O \tag{7.13}$$

where C_O is the concentration of solids dispensed from the system, and F_D is the fraction of drug formulated in the drug compartment. Therefore, the modified expression for the mass delivery rate from the Push-Pull system is as follows:

$$\frac{dm}{dt} = (Q + F) F_D C_O \tag{7.14}$$

where

$$Q = \frac{k}{h} A_p(H) \pi_P(H) \tag{7.15}$$

and

$$F = \frac{k}{h}[A - A_p(H)] \pi_D(H) \tag{7.16}$$

where k = membrane permeability coefficient
h = thickness of the membrane
A_P = surface area of the push compartment
π_P = imbibition pressure of the push compartment
A = total surface area of the dosage form
π_D = imbibition pressure in the drug compartment

The preceding equations apply to Push-Pull osmotic pumps that deliver water-soluble compounds. In that case, both drug and osmotic agent can exert constant osmotic pressure. However, in case of water-insoluble drugs, the systems are formulated with polymers that exert pressure as a function of the degree of hydration H, which is not a constant. H is expressed as follows:

$$H = \frac{W_H}{W_P} \tag{7.17}$$

where W_H is the weight of the water imbibed per weight of dry polymer W_P. Therefore, the H term in Eqs. (7.15) and (7.16) can be expressed in terms of water uptake and initial polymer weight.[10]

7.3.4 Key parameters that influence the design of osmotic controlled drug delivery systems

Orifice size. To achieve an optimal zero-order delivery profile, the cross-sectional area of the orifice must be smaller than a maximum size S_{max} to minimize drug delivery by diffusion through the orifice. Furthermore, the area must be sufficiently large, above a minimum size S_{min}, to minimize hydrostatic pressure buildup in the system. Otherwise, the hydrostatic pressure can deform the membrane and affect the zero-order delivery rate. Therefore, the cross-sectional area of the orifice S_o should be maintained between the minimum and maximum values. Typically, a diameter of about 0.2 mm through a membrane of 0.2-mm thickness is needed to maintain a delivery rate on the order of 10 mg/h for water-soluble compounds.[11] The minimum cross-sectional area can be estimated from the following equation:

$$S_{min} = 5\left[\left(\frac{L}{P_{max}}\right)\mu\left(\frac{dV}{dt}\right)\right]^{1/2} \tag{7.18}$$

where dV/dt = volume flux through the orifice

L = length of the orifice (usually the same as the thickness of the membrane)

μ = viscosity of the drug solution flowing through the orifice

p_{max} = maximum tolerated hydrostatic pressure difference across the membrane before the occurrence of deformation of the housing

The maximum cross-sectional area of the orifice is obtained by specifying that the diffusional contribution to the release rate must be smaller than a fraction f of the zero-order pumping rate and is defined by the following equation:

$$S_{\max} = \frac{M_{tz} f L}{D_S C_S} \tag{7.19}$$

where M_{tz} is the amount of the drug delivered in zero-order fashion, and D_s is the drug diffusion coefficient in the permeating solvent. In practice, a fraction smaller than 0.025 generally is necessary to minimize diffusional contributions.[12]

Solubility. From Eq (7.9), the release rate depends on the solubility of the solute inside the drug delivery system. Therefore, drugs should have sufficient solubility to be delivered by osmotic delivery. In the case of low-solubility compounds, several alternate strategies may be employed. Broadly, the approaches can be divided into two categories. First, swellable polymers can be added that result in the delivery of poorly soluble drugs in the form of a suspension.[13] Second, the drug solubility can be modified employing different methods such as cocompression of the drug with other excipients, which improve the solubility.[14] For example, cyclodextrin can be included in the formulation to enhance drug solubility.[15] Additionally, alternative salt forms of the drug can be employed to modulate solubility to a reasonable level. In one case, the solubility of oxprenolol is decreased by preparing its succinate salt so that a reduced saturation concentration is maintained.[16]

Osmotic pressure. The osmotic pressure π expressed in Eq. (7.9) directly affects the release rate. To achieve a zero-order release rate, it is essential to keep π constant by maintaining a saturated solute solution. Many times, the osmotic pressure generated by the saturated drug solution may not be sufficient to achieve the required driving force. In this case, other osmotic agents are added that enhance osmotic pressure. For example, addition of bicarbonate salt not only provides the necessary osmotic gradient but also prevents clogging of the orifice by precipitated drug by producing an effervescent action in acidic media.[17]

Semipermeable membrane. Since the semipermeable membrane is permeable to water and not to ions, the release rate is essentially independent of the pH of the environment. Additionally, the drug dissolution process takes place inside the delivery system, completely separated from the environment.[16] The materials used for the preparation of the membrane are described in Sec. 7.4.2.

7.4 Components of Osmotic Systems

The major formulation components of a typical osmotic delivery system include drug, osmotic agents, and a semipermeable membrane.

7.4.1 Osmotic components

Osmotic components usually are ionic compounds consisting of either inorganic salts or hydrophilic polymers. Osmotic agents can be any salt such as sodium chloride, potassium chloride, or sulfates of sodium or potassium and lithium. Additionally, sugars such as glucose, sorbitol, or sucrose or inorganic salts of carbohydrates can act as osmotic agents.[18]

Hydrophilic polymers encompass osmopolymers, osmogels, or hydrogels. These materials maintain a concentration gradient across the membrane. They also generate a driving force for the uptake of water and assist in maintaining drug uniformity in the hydrated formulation. The polymers may be formulated along with poly(cellulose), osmotic solutes, or colorants such as ferric oxide.[19] Swellable polymers such as poly(alkylene oxide), poly(ethylene oxide), and poly (alkalicarboxymethylcellulose) are also included in the push layer of certain osmotic systems. Further, hydrogels such as Carbopol (acidic carboxypolymer), Cyanamer (polyacrylamides), and Aqua-Keeps (acrylate polymer polysaccharides composed of condensed glucose units such as diester cross-linked polygluran) may be used.[18] Finally, tableting aids such as binders, lubricants, and antioxidants may be added to aid in the manufacture of the osmotic systems.[20]

7.4.2 Semipermeable membrane–forming polymers for osmotic pumps

An important part of the osmotic drug delivery system is the semipermeable membrane housing. Therefore, the polymeric membrane selection is key to the osmotic delivery formulation. The membrane should possess certain characteristics, such as impermeability to the passage of drug and other ingredients present in the compartments. The membrane should be inert and maintain its dimensional integrity to provide a constant osmotic driving force during drug delivery.[21] Numerous polymers are currently available to form semipermeable membranes. One class includes cellulosic polymers such as cellulose ethers, cellulose esters, and cellulose ester-ethers. The cellulosic polymers have a degree of substitution (DS) of 0 to 3 on the anhydroglucose unit. The DS is the number of hydroxyl groups present on the anhydroglucose unit being replaced by a substituting group. Examples of this group include cellulose acylate, cellulose diacylate, cellulose triacylate, cellulose acetate, cellulose diacetate, and mono-, di-, and tricellulose alkanylates. Cellulose acetate is available in different grades, such as cellulose acetate having a DS of 1 to 2 and an acetyl content of 21 to 35 percent or cellulose acetate having an acetyl content of 32 to 39.8 percent. Other forms of cellulose polymers with a more specific substitution are cellulose propionate with a DS of 1.8, a propyl content of 39.2 to 45 percent, and a hydroxyl content of 2.8 to 5.4 percent or cellulose acetate butyrate with a DS of 1.8, an acetyl content of 13 to 15 percent, and a butyrate content of 34 to 39 percent.[22] Moreover, the semipermeable membrane may consist of a mixture of cellulose acetates, alkanylates, or acrylates with different degrees of substitution.

Additional semipermeable membrane–forming polymers are selected from the group consisting of acetaldehyde dimethyl cellulose acetate, cellulose acetate ethyl carbamate, cellulose dimethylamino acetate, semipermeable polyamides, semipermeable polyurethanes, or semipermeable sulfonated polystyrenes. Semipermeable cross-linked selectively permeable polymers formed by coprecipitation of a polyanion and a polycation also can be used for this purpose.[22,23] Other polymer materials such as lightly cross-linked polystyrene derivatives, semipermeable cross-linked poly(sodium styrene sulfonate), and semipermeable poly (vinylbenzyltrimethyl ammonium chloride) may be considered.[24,25]

7.4.3 Emulsifying agents

Some patented technologies invoke self-emulsifying agents to deliver liquids from osmotic delivery systems. In one example, an emulsion consisting of up to 65 percent drug, usually hydrophobic, and a surfactant from 0.5 to 99 percent is cited. The surfactant selected for this purpose is a polyoxyethylenated castor oil, polyoxyethylenated sorbitan tristearate, or polyoxyethylenated sorbitan monopalmitate containing different proportions of ethylene oxide. The emulsion initially consists of an oil phase, obtained from vegetable, mineral, or animal origin, in which the hydrophobic drug is dissolved.[19]

7.4.4 Flux-regulating agents

Delivery systems can be designed to regulate the permeability of the fluid by incorporating flux-regulating agents in the layer. Hydrophilic substances such as polyethethylene glycols (300 to 6000 Da), polyhydric alcohols, polyalkylene glycols, and the like improve the flux, whereas hydrophobic materials such as phthalates substituted with an alkyl or alkoxy (e.g., diethyl phthalate or dimethoxy ethylphthalate) tend to decrease the flux. Insoluble salts or insoluble oxides, which are substantially water-impermeable materials, also can be used for this purpose.[18]

7.4.5 Plasticizers

To give the semipermeable membrane flexibility, plasticizers such as phthalates (dibenzyl, dihexyl, or butyl octyl), triacetin, epoxidized tallate, or tri-isoctyl trimellitate are added.[18] In the design of osmotic controlled release systems, these plasticizers help to modulate and achieve the required release rate.

7.4.6 Barrier layer formers

To restrict water entry into certain parts of the delivery system and to separate the drug layer from the osmotic layer, different materials are used as barrier layers. In a multilayered reservoir, the water-permeable coat consists of hydrophilic polymers. In contrast, water-impermeable layers are formed from latex materials such polymethacrylates (Table 7.1). Further, a barrier layer can be provided between the osmotic composition and the drug layer that consists of substantially fluid-impermeable materials such as high-density polyethylene, a wax, a rubber, and the like.[20]

7.5 Osmotic Delivery Systems

7.5.1 Evolution of osmotic delivery systems

About 75 years after discovery of the osmosis principle, it was first used in the design of drug delivery systems. In 1955, Rose and Nelson developed the first osmotic device that represented the forerunner of the modern osmotically controlled drug delivery systems.[26] The unit consisted of three chambers, one each for drug, salt, and water (Fig. 7.3). One of the disadvantages of the Rose-Nelson pump involved the water chamber, which must be charged before use of the pump. A Pharmetrix device overcame this difficulty by separating the water chamber with an impermeable seal, which was broken before administration of the pump.[27] The Higuchi-Leeper pump represented a simplified version of

TABLE 7.1 Materials Used in Different Layer Formulations

Component	Example
Hydrophilic layer (water permeable)	Polysaccharides, hydroxypropymethyllcellulose, hydroxyethylcellulose, poly(vinylalcohol-co-ethyleneglycol)
Water-impermeable layer	Kollicoat, SR latex, Eudragit SR
Barrier layer	Styrene butadiene, calcium phosphate, polysilicone, nylon, Teflon, polytetrafluoroethylene, halogenated polymers

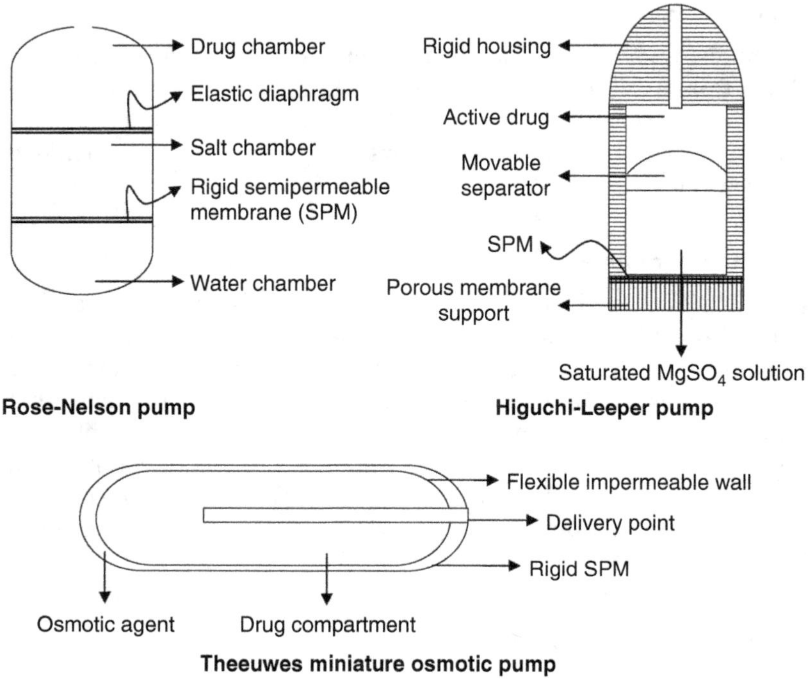

Figure 7.3 Schematic diagrams of different pumps. (*Adapted from Santus and Baker.*[17])

the Rose-Nelson three-chambered pump. The device had no water chamber and was activated by water imbibed from the surrounding environment.[28] In 1970, ALZA Corporation released the first series of Higuchi-Leeper pumps (see Fig. 7.3) that consisted of three compartments: drug, salt, and water. A rigid semipermeable membrane separated the salt and water chambers, and an elastic diaphragm separated the salt and drug chambers. Use of closely fitting half shells to form the pump and a telescopic housing with expandable driving members were two important modifications of the Higuchi-Leeper pump.[29,30] Yet another recent modification in the Higuchi-Leeper pump accommodated pulsatile drug delivery. The pulsatile release was achieved by the production of a critical pressure at which the orifice opens and releases the drug.[28,30]

The simplest version of the Rose-Nelson pump was developed by Higuchi and Theeuwes in 1976. The pump was composed of a rigid, rate-controlling outer semipermeable membrane surrounding a solid sodium chloride layer, below which was placed an elastic diaphragm housing for the drug.[31] In one of the modifications of this pump, a mixture of citric acid and sodium bicarbonate was used in the salt chamber to produce carbon dioxide gas pressure.[32]

A major leap in osmotic delivery occurred as the elementary osmotic pump for oral delivery of medicaments was introduced. The system represented a further simplification of the Higuchi-Leeper pump. In its simplest form, the device consisted of a compressed tablet surrounded by a semipermeable coating, and it delivered the drug solution or suspension through an orifice in the tablet.[33] The concept proved popular, and 135 patents have been issued for various aspects of the system. Since only water-soluble drugs can be delivered via an elementary osmotic pump, another modification was developed to deliver essentially insoluble to highly soluble drugs, thus expanding the concept to a range of drugs. The system consisted of a multichamber osmotic tablet with two compartments supported by either a thin film or movable barrier or fixed as well-defined volume chambers.[34] (Fig. 7.3). However, in these devices, the semipermeable membrane formed the entire shell, and water was drawn into both chambers simultaneously. The concept of osmosis has been used in a number of other systems. Tablets coated with semipermeable membranes containing micropores, polymer drug matrices, or self-formulating in-line systems for parenteral drug delivery are some examples. Other systems featured a tablet coating that was modified to contain defects through which the drug may diffuse.[35]

Recently, osmotic concepts were extended to address delivery of drugs of different physical states. These osmotic systems were uniquely designed to deliver liquid active agents.[18–20]

In one design, the active agent is enclosed in a capsule consisting of a cap, and the body is placed inside the compartment of the main body. Within the inner capsule, distant from the orifice, an osmotic composition resides that acts as a push layer. Alternatively, the drug can be enclosed in a one-piece capsule coated with a semipermeable membrane or placed in a thermoplastic polymer compartment[19] (Fig. 7.4). In this type of system, the drug and osmotic layers are not separated.

In another design, the drug layer is protected from water influx by covering a portion of the delivery system with a water-impermeable coat. The entire dosage form appears as an oval-shaped tablet with an exit orifice formed at one end (Fig. 7.5). The wall or a portion of the wall is semipermeable to facilitate water influx. When the push layer is indirectly in contact with the drug layer, an inert disk is placed between the two layers. The number and arrangement of the inert layers in the drug layer can be designed to accomplish pulsatile delivery. The drug layer consists of a liquid active agent absorbed onto porous particles dispersed in a carrier. Additionally, a flow-promoting layer or a placebo layer can be situated at different locations to control the onset of action.[18]

Usually, the flow-promoting layer is placed between the outer wall of the dosage form and the core chambers, which facilitates the sliding of the drug layer from the cavity during release. The film is applied as a

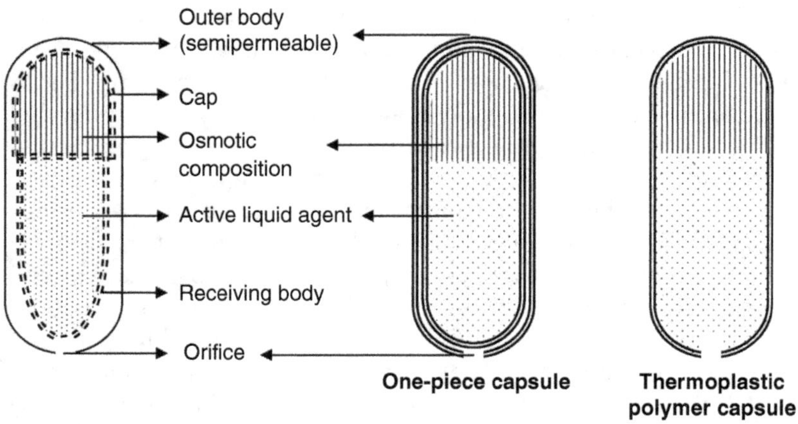

Figure 7.4 Different types of constructions of osmotic devices. (*Adapted from Dong.*[19])

coating over the compacted drug layer and push layer. A semipermeable coating is applied on the composite formed by the drug layer, push layer, and flow-promoting layer.

The construction is slightly different when a placebo layer is present between the drug layer and the orifice to delay onset of release. The extent of the delay owing to the placebo layer depends on the volume such a layer occupies, which must be displaced by the push layer.[18]

The drug layer can consist of a liquid active agent absorbed onto porous particles and a carrier. The preferred characteristics of the particles are presented in Table 7.2. The carrier plays an important role in controlling

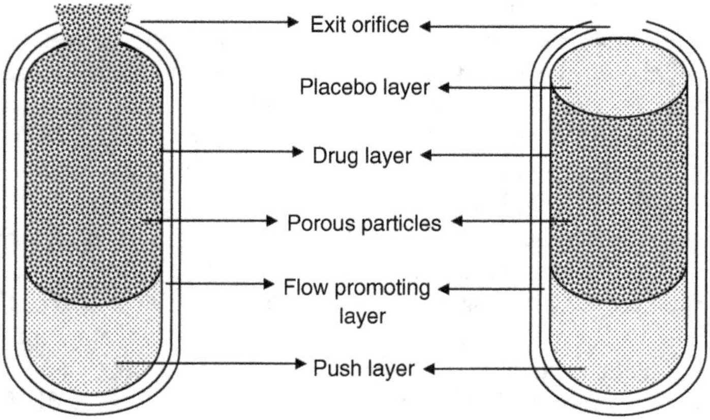

Figure 7.5 Liquid active agent absorbed into the porous particles. (*Adapted from Wong.*[18])

TABLE 7.2 Desirable Properties of Porous Particles

Property	Calcium hydrogenphosphate	Magnesium aluminometasilicate
Mean particle size (μm)	50–150	1–2
Specific surface area (m^2/g)	20–60	100–300
Specific volume (mL/g)	>1.5	2.1–12
Oil absorbing capacity (mL/g)	0.7	1.3–3.4
Hydration state	0–2	0–10
Specific gravity (mL/g)	0.4–0.6 (bulk density)	2
Mean pore size (Å)	>50	Not applicable
Angle of repose (0)	Not applicable	25–45

the release rate of the porous particles. Depending on the desired release characteristics, the carrier can be designed to include ingredients, such as a hydrophilic polymer like hydroxypropylethylcellulose (HPEC), hydroxypropylmethylcellulose (HPMC), or poly(vinylpyrrolidone) (PVP), that are used to enhance the flow properties of the dosage form. Moreover, appropriate bioerodible polymers may be preferred.

The other components of the carrier include surfactants and disintegrants. Surfactants with a hydrophile-lipophile balance (HLB) value between about 10 and 25 are preferred. In some cases, the carrier is completely eliminated or added in small quantities to facilitate rapid release of the drug (Table 7.3).

Suitable materials for a flow-promoting layer include hydroxypropyl, hydroxyethyl, or hydroxypropylmethyl celluloses; povidone; polyethylene glycols (PEGs); or their mixtures [18].

A multireservoir osmotic system is described in which the drug is more protected from the influx of water than in the previously described delivery system. The design consists of a central reservoir, formed from a water-impermeable layer containing a liquid active agent and osmotic agent separated by a barrier layer (Fig. 7.6). The barrier layer prevents mixing of the layer contents and minimizes the residual amount of the active agent after the expandable osmotic composition has ceased its expansion. The layer also provides uniform pressure transfer from the

TABLE 7.3 Ingredients Used to Prepared Carrier Particles

Ingredient	Examples
Core ingredient	Poly(alkylene oxide), polyethyleneoxide, polymethyleneoxide, polycarboxymethylcellulose and its sodium or potassium salts
Surfactants	PEG 400 monostearate, polyoxyehelene-4-sorbitanmonlaurate, polyoxyethelene-20-sorbitan monooleate, sodium oleate
Disintegrants	Starches, cross-linked starches, clays, celluloses, alginates, and gums

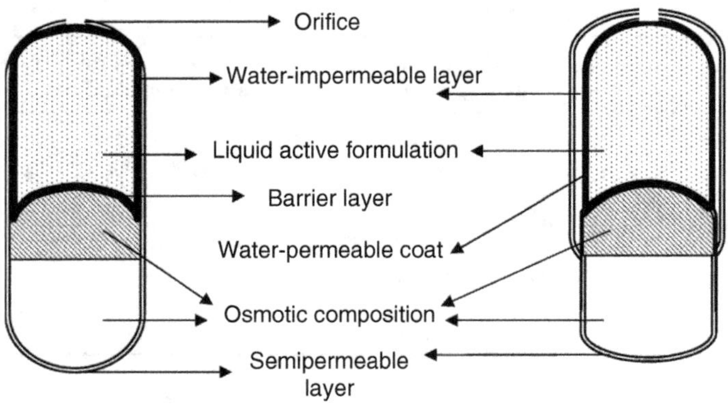

Figure 7.6 Design of an osmotic device for a liquid active agent with a water-impermeable coat. (*Adapted from Dong.*[20])

osmotic composition to the liquid active agent, ensuring a uniform rate of release. Part of the osmotic agent is exposed from the reservoir, which is covered with a semipermeable membrane. The device is provided with an orifice at the opposite end to the osmotic layer to facilitate expulsion of the liquid during delivery.

In a single-layered reservoir, an orifice is formed when the pressure mounts inside the reservoir owing to the expansion of the osmotic composition. In the multilayered reservoir, the water-soluble layer dissolves when the dosage form comes in contact with the release medium. The number of exit orifices depends on the required rate of release.[20]

7.5.2 Classification of osmotic pumps

Based on their design and the state of active ingredient, osmotic delivery systems can be classified as follows:

1. *Osmotic delivery systems for solids*
   a. Type I: Single compartment. In this design, the drug and the osmotic agent are located in the same compartment and are surrounded by the semipermeable membrane (SPM). Both the core components are dissolved by water, which enters the core via osmosis. A limitation is the dilution of drug solution with the osmotic solution, which affects the release rate of the drug from the system. Additionally, water-incompatible or water-insoluble drugs cannot be delivered effectively from a single-compartment configuration (Fig. 7.7).
   b. Type II: Multiple compartments. In this design, drug is separated from the osmotic compartment by an optional flexible film, which is displaced by the increased pressure in the surrounding osmotic compartment, which, in turn, displaces the drug solution or

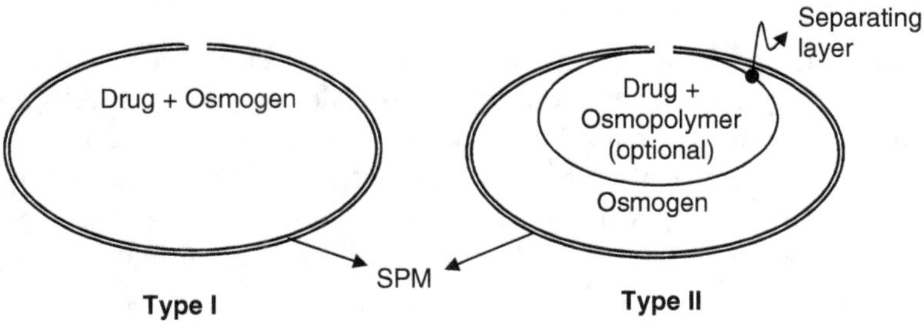

Figure 7.7 Classification of osmotic delivery systems: types I and II.

suspension. The type II system inherently has greater utility than type I systems and can deliver drugs at a desired rate independent of their solubilities in water. One main advantage of these systems is their ability to deliver drugs that are incompatible with commonly used electrolytes or osmotic agents.

2. *Osmotic delivery systems for liquids.* Active ingredients in liquid form are difficult to deliver from controlled release platforms because they tend to leak in their native form. Therefore, liquid active agents typically are enclosed in a soft gelatin capsule, which is surrounded by an osmotic layer that, in turn, is coated with a semipermeable membrane. When the system takes up water from its surroundings, the osmotic layer squeezes the innermost drug reservoir. The increasing internal pressure displaces the liquid from the system via a rupturing soft gelatin capsule (Fig. 7.8). Some patented technologies, discussed in the last section of this chapter, are also being applied to achieve more precise delivery rates.

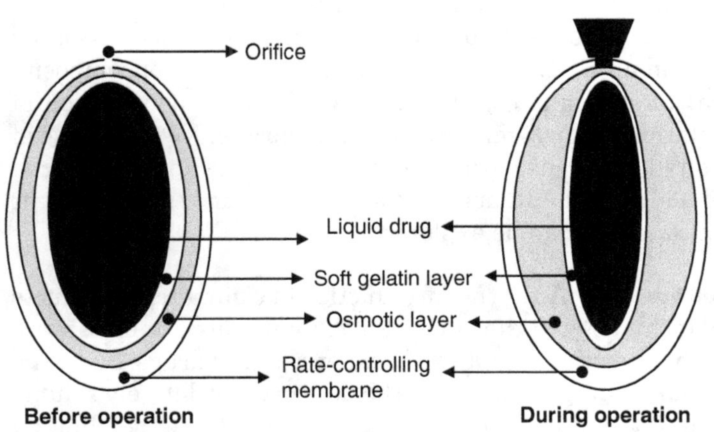

Figure 7.8 Osmotic delivery system for delivery of a liquid active agent.

7.5.3 Oral osmotic delivery systems

The invention that positioned osmosis as a major driving force for controlled drug delivery was the elementary osmotic delivery system. ALZA has developed elementary osmotic delivery systems under the name OROS. A successful modification that overcame the disadvantages of the elementary osmotic pump was the Push-Pull osmotic drug delivery system. The following sections are devoted to the principal features of these systems.

Elementary osmotic pump (EOP). OROS represents the oral osmotically driven dosage forms developed by ALZA. As represented in Fig. 7.3, the basic system is a simplified version of the Higuchi-Theeuwes pump. In the OROS elementary osmotic pump, a tablet core of drug is surrounded by a semipermeable membrane that has one or more openings. After ingestion, the core draws water through the semipermeable membrane from the gastrointestinal (GI) surroundings. The imbibed water dissolves the drug, which is expelled through the orifice in a zero-order fashion. The semipermeable membrane for OROS typically is composed of cellulose acetate. The membrane is nonextendable and preserves the physical dimensions of the dosage form. Drug delivery is zero order until the solid portion of the core is exhausted. Release will then occur in non-zero-order fashion, declining parabolically. The driving force that draws water into the system is the osmotic pressure difference between the outside environment and the saturated drug solution. Therefore, the osmotic pressure of the drug solution must be greater than the GI osmotic pressure. Hence the elementary osmotic pump is suitable for drugs with solubilities greater than about 2 to 10 wt%. Because a thick membrane is required to preserve the shape of the core, the water permeation rate can be unacceptably low, particularly if the drug is moderately soluble and possesses low osmotic pressure.[33]

Different modifications are used to alleviate the limitations associated with delivery from an EOP. One method involves the use of composite structures that form a microporous layer for the easy penetration of water and a relatively thin semipermeable membrane.[34,36,37] The use of bicarbonate salts to prevent blocking of orifice, buffers to modify drug solubility, and addition of adjunct osmotic agents to the core represent other modifications that can be explored.[17]

OROS Push-Pull system. After the introduction of commercial products based on EOP technology in the early 1980s, numerous attempts were made to apply the osmotic concept to a broader range of drugs. Since the elementary osmotic pump is limited to the delivery of relatively soluble drugs with solubilities greater than about 2 to 10 wt%, depending on dose, other modifications were necessary to expand the utility of the EOP design. If an osmogent is incorporated with a poorly soluble drug (<1 percent), it may not be possible to achieve the desired rate of release because the ratio of drug-to-osmogent concentration is not equal to the ratio of drug solubility to osmogent solubility.[38] One approach to this limitation entails the introduction of a second chamber into an elementary osmotic pump. For example, one variant of the Push-Pull platform has two compartments, drug and osmotic, covered with a semipermeable membrane. In some instances, a hydrophilic polymer is incorporated that swells with water uptake. When the tablet is ingested, water is absorbed into the system that expands the osmotic layer, thus pushing the dissolved or suspended drug layer through the orifice in a controlled fashion. The hydrophilic polymers that suspend the drug can be ionic materials such as sodium carboxymethyl cellulose, which can be compressed into a tablet by commonly used machines. Optionally, a barrier layer can be introduced between the drug and polymer layers to ensure a uniform pushing force. Moreover, any potential mixing of the drug layer and salt layer can be circumvented by the use of hydrophilic polymer gels. Thus a push layer composed of hydrophilic polymers and optional adjunct osmogens effectively replaces the barrier and salt layers.

Highly soluble to insoluble drug molecules can be dispensed at a zero-order rate from the Push-Pull platform. For example, a nifedipine device (Procardia XL®) consists of a semipermeable membrane enclosing a drug-polymer bilayer core. The drug layer consists of nifedipine, hydroxypropyl methylcellulose, and poly(ethylene oxide). The polymer (or push) layer consists of hydroxypropyl methylcellulose and poly(ethylene oxide) together with sodium chloride to improve the osmotic pressure profile.[17] Various modifications are possible for improving the capability of the system, including programmable delivery, patterned or pulsatile delivery, and targeted delivery to the colon.[22,34,39]

7.5.4 Recent advances in osmotic drug delivery of liquid active components

Administration of liquid active agents may be preferred over the solid agents because the use of the former may expedite the onset of action. Soft gelatin capsules containing liquid agents have been evaluated for this purpose. However, these delivery platforms lack the ability to deliver the active compound at a zero-order rate. Among the available types of controlled release formulations, osmotic delivery systems (ODS) have been promising in achieving a nearly zero-order release of drug irrespective of potential variables such as gastric pH or motility or nature of the drug. However, ODS were limited by delay in onset of delivery of the liquid active agents. Many designs have been studied to address this deficiency, including varying the number and size of the orifices;

delivering the drug as a slurry, suspension, or solution (where the drug is introduced initially as a solid); or incorporating bioerodable or biodegradable active agent carriers. Among them, some patented technologies have emerged wherein the active drug can be delivered effectively in the liquid form through an osmotic delivery system.

Liquid agent in a capsule with a semipermeable coat. This design consists of a liquid formulation of drug that can self-emulsify in hard gelatin capsules coated with a semipermeable membrane.[19] The main disadvantage of these formulations is that the influx of water or the surrounding medium tends to dilute the active liquid agent because there is no separation of the osmotic composition and the active agent.[19] For details of construction, refer to Sec. 7.5.1 (see Fig.7.3).

Manufacturing of osmotically active capsules begins with preparation of the capsule body and cap by conventional means for gelatin capsules. The capsules then are coated with a semipermeable composition, either before or after the cap and body have been joined. In the case of one-piece capsules, bodies can be produced in various ways, such as the plate process, rotary-die process, reciprocating-die process, or continuous process. In the plate process, warm sheets of prepared capsule lamina-forming materials are placed over the lower mold, and the formulation then is transferred into the mold. A top mold, prepared similarly, is placed over the lower mold to join them together. The entire process is similar to the manufacture of soft gelatin capsules using the rotary-die process. The reciprocating-die process produces a row of pockets across the film as the lamina-forming films traverse between the vertical dies, which open and close alternatively. The pockets are filled with active agent, sealed, and cut to obtain the final capsules. The continuous process is capable of accurately filling both liquids and dry powders into the soft capsules. In the last step, capsules are coated with a semipermeable membrane.[19]

Liquid active agent absorbed into the porous particles. The main principle involved in release of the active agent in this design is the osmotically driven release of particles loaded with liquid active agent, followed by the release of these particles from the carrier. Immediately after placing the dosage form in the release medium, water is taken up through the semipermeable portion of the wall owing to the osmotic pressure generated by the osmotic agent present in the dosage form. The pressure created subsequently leads to expansion of the push layer. As a result, the carrier containing the drug layer is expelled from the dosage form through an orifice. A flow-promoting layer forms a viscous gel by absorbing water and facilitates sliding of the drug layer over the outer wall. The carrier dissolves in the medium to release the particles containing the liquid active agent. The active agent is available immediately as solution, which eventually is absorbed. The porous particles erode with time to release the remaining drug.[18] The construction of the delivery system is shown in Sec. 7.5.1 (see Fig. 7.4).

These dosage forms are prepared using methods very similar to the wet granulation or direct compression used for tablet preparation. Briefly, the liquid active agent is absorbed onto the porous particles, which are later blended with other components of the drug layer. The mixture is either directly compressed or made into granules using anhydrous ethanol, followed by compression. Depending on the requirement, either a trilayer tablet containing a drug layer, a push layer, and a barrier layer (inert layer) or a bilayered tablet with only the push layer and drug layer is compressed using the same technology to produce multilayered tablets. The resulting core is coated with the flow-promoting layer followed by the membrane coating. The exit means can be formed during the manufacture or while delivering the drug by the erosion of an erodible polymer in the membrane coating. The flow-promoting layer is prepared from a solution of ethanol, which is applied onto the tablet bed in a rotating perforated pan-coating unit. The rate-controlling membrane is prepared as a solution of acetone with the necessary ingredients. The end point is detected by determining the weight gain of the tablet at regular intervals, which is then correlated with the time for 90 percent drug release ($T_{90\%}$).

Color and clear overcoats are applied, if required, in a similar way in a temperature range of 35 to 45°C. Properly prepared dosage forms are capable of providing a zero-order release profile, typically from 4 to 24 hours. The platform can be constructed according to the desired application: sustained, delayed, or pulsatile delivery.[18]

Liquid active agents in a reservoir with a water-impermeable coat. These dosage forms consist of a reservoir containing the active agent in liquid form covered by a relatively water-impermeable membrane so as to decrease the contact of water with the liquid active agent during delivery. An osmotic layer at least partially in communication with the drug reservoir pushes the drug out through the orifice while imbibing the water through the semipermeable membrane, as shown in Fig. 7.6.[20]

The hydrophilic layer is prepared by dipping molds in the liquid containing a certain concentration of hydrophilic polymer in a fashion similar to that of gelatin capsule preparation. In the case of a multilayered reservoir, a water-impermeable subcoat is formed either by dip coating or by spray coating. Before spray coating, the opening of the hydrophilic layer is covered with a removable cap to prevent coverage of the internal portion of the hydrophilic layer. Subsequently, the osmotic composition is positioned at the opening of the reservoir using an appropriate assembling apparatus. The tablet usually consists of two layers, including a barrier layer and an osmotic composition. A semipermeable membrane is provided by dip coating or spray coating that covers at least the exposed surface of the osmotic composition. The exit port can be a simple laser-drilled orifice or any porous element.[20]

7.5.5 Marketed products

Many drug products using various osmotic principles have been introduced into the market. OROS designs have been used in more than 10 marketed products. The first OROS product introduced to the U.S. market in 1983 was Acutrim®, a 16-hour appetite suppressant marketed by Ciba-Geigy. Volmax®, a twice-a-day controlled release dosage form for Glaxo's antiasthma drug albuterol, was marketed in 30 countries, including the United States, starting in 1987. Minipress XL®, a once-a-day system for Pfizer's antihypertensive drug prazosin, was launched in 1989. Efidac/24®, the first over-the-counter osmotically driven controlled release cold medication marketed by Ciba Consumer Pharmaceuticals, was introduced in 1992. Glucotrol XL®, a once-a-day orally active hypoglycemic drug delivery system for glipizide, was launched in 1994.

Procardia XL, which is marketed by Pfizer, is a billion-dollar product employing Push-Pull technology. The preparation contains the calcium channel blocker nifedipine.[40] A full list of marketed products developed by ALZA Corporation is shown in Table 7.4.

TABLE 7.4 Marketed Products Based on Osmotic Principles

Elementary osmotic pump (ALZA Corp.)	Once-daily osmotic tablet with solid active agent
Efidac 24®	Chlorpheniraamine
Acutrim®	Phenylpropanolamine
Sudafed 24®	Pseudoephedrine
Volmax®	Albuterol
Push-Pull osmotic delivery system (ALZA Corp.)	**Multilayered tablet for drugs with low to high solubility**
Ditropan XL®	Oxybutynin chloride
Procardia XL®	Nifedipine
Glucotrol XL®	Glipizide
Alpress XL® (France)	Prazosin
Cardura XL® (Germany)	Doxazosin mesylate
Concerta®	Methylphenidate HCl
Covera HS®	Verapamil HCl
DynaCirc CR®	Isradipine

7.6 Conclusions and Future Potential

In osmotic delivery systems, osmotic pressure provides the driving force for drug release. Increasing pressure inside the dosage form from water incursion causes the drug to release from the system. The major advantages include precise control of zero-order or other patterned release over an extended time period—consistent release rates can be achieved irrespective of the environmental factors at the delivery site. Controlled delivery via osmotic systems also may reduce the side-effect profile by moderating the blood plasma peaks typical of conventional (e.g., instant release) dosage forms. Moreover, since efficacious plasma levels are maintained longer in osmotic systems, avoidance of trough plasma levels over the dosing interval is possible. However, a complex manufacturing process and higher cost compared with conventional dosage forms limit their use. Although not all drugs available for treating different diseases require such precise release rates, once-daily formulations based on osmotic principles are playing an increasingly important role in improving patient compliance. Therefore, most of the currently marketed products are based on drugs used in long-term therapies for diabetes, hypertension, attention-deficit disorder, and other chronic disease states. Besides oral osmotic delivery systems, implants that work on osmotic principles are promising for delivery of a wide variety of molecules with a precise rate over a long period of time. Further, with the discovery of newer and potent drugs by the biotechnology industry, the need to deliver such compounds at a precise rate certainly will pave the way for osmotic delivery systems to play an increasingly important role in drug delivery.

References

1. Theeuwes, F., and Higuchi, T. Osmotic dispensing device for releasing beneficial agent, ALZA Corporation, Palo Alto, CA, U.S. Patent 3,845,770, 1974.
2. Chiao, C. S. L., and Robinson, J. R. *Sustained and Controlled-Release Drug Delivery Systems*, 19th ed. Philadelphia: Mark Publishing Company, 1995.
3. Banakar, U. V. Drug delivery systems of the 90s: Innovations in controlled release. *Am. Pharm.* **NS27**(2):39–42, 47–48, 1987.
4. Khan, M. A., Bolton, S., and Kislalioglu, M. S. Optimization of process variables for the preparation of ibuprofen coprecipitates with eudragit S100. *Int. J. Pharm.* **102**(1–3):185–192, 1994.
5. Khan, M. A., Karnachi, A. A., Singh, S. K., et al. Controlled release coprecipitates: formulation considerations. *J. Contr. Rel.* **37**(1–2):131–141, 1995.
6. Sastry, S. V., Reddy, I. K., and Khan, M. A. Atenolol gastrointestinal therapeutic system: optimization of formulation variables using response surface methodology. *J. Contr. Rel.* **45**(2):121–130, 1997.
7. Jonkman, J. H. Food interactions with sustained-release theophylline preparations. A review. *Clin. Pharmacokinet.* **16**(3):162–179, 1989.
8. Marin, A., Bustamante, P., and Chun, A. H. C. *Physical Pharmacy: Physical Chemical Principles in the Pharmaceutical Sciences*, 4th ed. Philadelphia: Lea and Febiger, 1993.

9. Laxminarayanaiah, N. *Transport Phenomena in Membranes.* New York: Academic Press, 1969, pp. 247–248.
10. Swanson, D. R., Barclay, B. R., Wong, P. S. L., et al. Nifedipine gastrointestinal therapeutic system. *Am. J. Med.* **83**(6B): 3–9, 1987.
11. Good, W. R., and Lee, P. I. Membrane-controlled reservoir drug delivery systems, in *Medical Applications of Controlled Release*, ed. R. S. Langer and D. L. Wise, Vol. I. Boca Raton, CRC Press, 1984, pp. 1–39.
12. Theeuwes, F., and Higuchi, T. Osmotic dispensing device with maximum and minimum sizes for the passageway, U.S. Patent 3,916,899, 1975.
13. Khanna, S. C., Therapeutic system for sparingly soluble active ingredients, Ciba-Geigy Corporation, Ardsley, NY, U.S. Patent 4,992,278, 1997.
14. Zentner, G. M., McClelland, G. A., and Sutton, S. C. Controlled porosity solubility- and resin-modulated osmotic drug delivery systems for release of diltiazem hydrochloride. *J. Contr. Rel.* **16**(1–2):237–243, 1991.
15. Okimoto, K., Ohike, A., Ibuki, R., et al. Design and evaluation of an osmotic pump tablet (OPT) for prednisolone, a poorly water soluble drug, using (SBE)7m-beta-CD. *Pharm. Res.* **15**(10):1562–1568, 1998.
16. Theeuwes, F., Swanson, D. R., Guittard, G., et al. Osmotic delivery systems for the beta-adrenoceptor antagonists metoprolol and oxprenolol: Design and evaluation of systems for once-daily administration. *Br. J. Clin. Pharmacol.* **19 (suppl 2)**:69S–76S, 1985.
17. Santus, G., and Baker, R. W. Osmotic drug delivery: A review of the patent literature. *J. Contr. Rel.* **35**(1):1–21, 1995.
18. Wong, P. S. Controlled release liquid active agent formulation dosage forms, Alza Corporation, Palo Alto, CA, U.S. Patent 6,596,314, 2003.
19. Dong, L. C. Dosage form comprising liquid formulation, ALZA Corporation, Mountain View, CA, U.S. Patent 6,174,547, 2001.
20. Dong, L. C. Controlled release capsule for delivery of liquid formulation, U.S. Patent application publication 2004/0058000, 2004.
21. Eckenhoff, B., Theeuwes, F., and Urquhart, J. Osmotically actuated dosage forms for rate-controlled drug delivery. *Pharm. Technol.* **11**:96–105, 1987.
22. Dong, L. C., Dealey, M. H., Burkoth, T. L., et al. Process for lessening irritation caused by drug, ALZA Corporation, Palo Alto, CA, U.S. Patent 5,254,349, 1993.
23. Zobrist, J.C. Cleaning methods and compositions, U.S. Patent 3,173,876, 1965.
24. Loeb, S., and Sourirajan, S. High flow porous membranes for separating water from saline solution, U.S. Patent 3,133,132, 1964.
25. Scott, J. R., and Roff, W. J. *Handbook of Common Polymers.* Boca Raton, FL: CRC Press, 1971.
26. Rose, S., and Nelson, J. F. A continuous long-term injector. *Aust. J. Exp. Biol.* **33**:415, 1955.
27. Baker, R. W., and Castro, A. J. Current U.S. patents. *J. Contr. Rel.* **9**(3):289–292, 1989.
28. Higuchi, T., and Leeper, H. M. Osmotic dispenser, U.S. Patent 3732865, 1973.
29. Higuchi, T., and Leeper, H. M. Improved osmotic dispenser employing magnesium sulphate and magnesium chloride, U.S. Patent 3,760,804, 1973.
30. Wong, P. S. L., Theeuwes, F., Eckenhoff, J. B., et al. Multi-unit delivery system, ALZA Corporation, Palo Alto, CA, U.S. Patent 5,110,597, 1992.
31. Higuchi, T., and Theeuwes, F. Osmotic dispenser with means for dispensing active agent responsive to osmotic gradient, U.S. Patent 3,995,631, 1976.
32. Theeuwes, F. Osmotically triggered device with gas generating means, ALZA Corporation, Palo Alto, CA, U.S. Patent 4203441, 1980.
33. Theeuwes, F. Elementary osmotic pump. *J. Pharm. Sci.* **64**:1987–1991, 1975.
34. Theeuwes, F., and Ayer, A. D. Osmotic devices having composite walls, U.S. Patent 4,077,407, 1978.
35. Theeuwes, F., and Damani, N. C. Osmotically driven active agent dispenser, U.S. Patent 4,016,880, 1977.
36. Theeuwes, F., and Ayer, A. D. Osmotic system having laminar arrangement for programming delivery of active agent, U.S. Patent 4,008,719, 1977.

37. Theeuwes, F. Microporous-semipermeable laminated osmotic system, U.S. Patent 4,256,108, 1981.
38. Kim, C.-J., Osmotically controlled systems, in *Controlled Release Dosage Form Design*, C.-J. Kim, ed. Lancaster, PA: 2000, Technomic Publishing Company, 229–246.
39. Cortese, R., and Theeuwes, F. Osmotic device with hydrogel driving member, U.S. Patent 4,327,725, 1982.
40. ALZA Product Information Brochure, Palo Alto, CA, 1995.

Chapter

8

Device Controlled Delivery of Powders

Rudi Mueller-Walz

SkyePharma AG, Muttenz, Switzerland

8.1 Introduction

There are two areas in drug delivery where the device is of utmost importance for the delivery of a pharmaceutically active solid material. The most prominent one is the area of pulmonary delivery of dry powders for topical or systemic pharmacological action; the other is the emerging technology of intradermal or transmucosal powder injection. Both technologies target an interface between the body and the outside world—the lungs in one case and the skin in the other. Both interfaces are in continuous contact with the environment and provide a huge contact area, which potentially could be applied for drug adsorption.

Targeting the airways themselves as a means to treat local diseases of the respiratory tract has a long history. Historically, the evolution of inhalation therapy can be traced back to India about to the year 2000 B.C., where leaves of *Atropa belladonna* containing the pharmacologically active agent atropine were smoked as a cough suppressant (Grossman 1994). It took a long time from this early beginning of inhalation therapy before the first patent issuance for a dry-powder inhaler (DPI) device in the United States in 1939. The device, however, was never marketed (Clark 1995). A patent in 1949 describes the first DPI for delivery of a pharmaceutical drug that was commercialized by Abbott Laboratories for the delivery of isoprenaline sulfate under the trade name Aerohaler (Clark 1995). In 1967, the Spinhaler was introduced by Fisons as the first unit-dose DPI in Great Britain and in 1971 in the United States (Bell et al. 1971). Since then, several DPIs have been introduced, most, if not all, used for the local therapy of lung diseases (Wasserman and Renzetti 1994). Development of DPI devices gained a big push when it was realized that the chlorofluorocarbons used as propellants in the other portable inhaler system available, the pressurized metered-dose inhaler, were responsible for destruction of the stratospheric ozone layer (Molina and Rowlands 1974) and became restricted and finally were phased out. The number of device-related patent applications filed in this area increased in the following years enormously. However, despite the hundreds of patents, only about 20 devices are currently marketed successfully in major pharmaceutical markets or are in the late phase of development (Greystone Associates 2003). This is the result of the complexity of device and formulation development for pulmonary delivery, which acts as a high barrier to market entry. The complexity of this dosage form also can be illustrated when visualizing the complex network of interdependences in which the patient and the patient's requirements are at the center (Fig. 8.1).

It was only recently that the potential of using the lungs as an entry port to the blood circulation has been recognized: owing to increasing knowledge of pulmonary drug delivery and the demonstration of reasonable bioavailability via the lungs of molecules that otherwise are difficult to deliver (Byron and Patton 1994). Systemic delivery was once the "holy grail" in the area of inhalation delivery that most likely would result in new therapeutic approaches and new medications. If local therapy of the diseased lungs is already a difficult goal, then this is even more the case for systemic applications because systemic bioavailability of a drug is achieved at the end of a sequence of steps, each one having to be controlled carefully to achieve a reliable pharmacokinetic profile (Fig. 8.2).

The challenges of optimizing and controlling the different steps to achieve an effective and reliable therapy result in an unavoidably complex development path. A good visualization of the system-level and

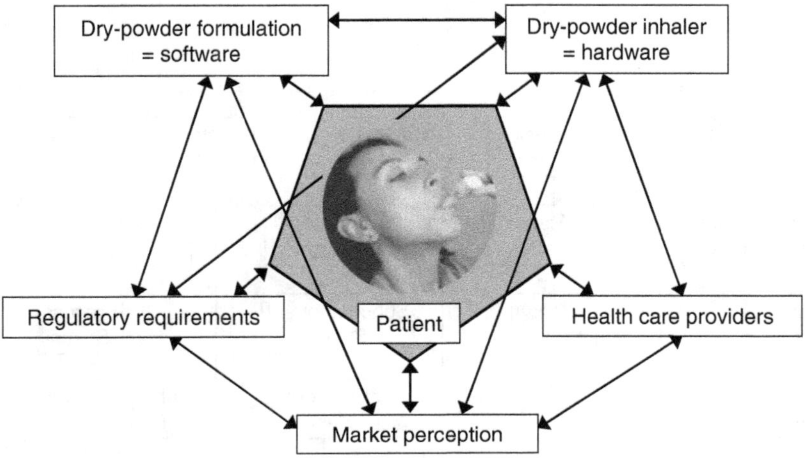

Figure 8.1 Mutually interdependent determining factors in development of a dry-powder inhalation product (*Courtesy of SkyePharma.*)

component-level requirements of such projects is given in Fig. 8.3, which explains why only specialized pharmaceutical and drug delivery companies are successful in the field. The hurdles are significant and evidenced by the fact that it required multi-million-dollar efforts by the pharmaceutical industry for more than 10 years of development to bring the first DPI product for systemic delivery of a peptide (insulin) close to market. In this chapter, the design of devices for controlled delivery of solid dosage forms will be reviewed.

Needle-free injection is an intradermal drug delivery technology that can be considered as a hybrid of transdermal and parenteral technologies. The technology was first proposed in the early twentieth century

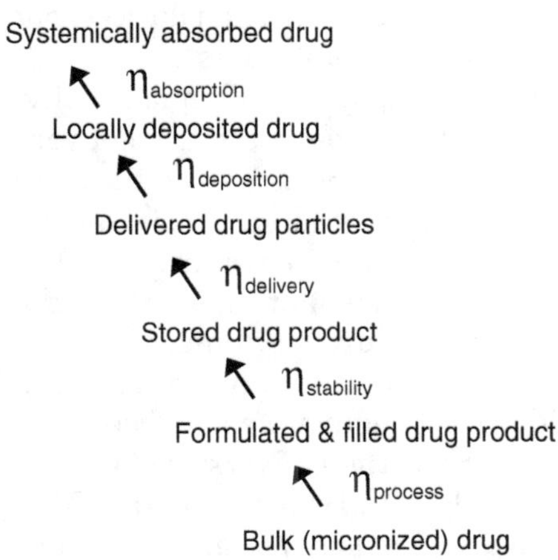

Figure 8.2 Efficiency in topical and systemic pulmonary delivery. [*With modifications from Schultz (2002). Reproduced with permission from RDDOnline.*]

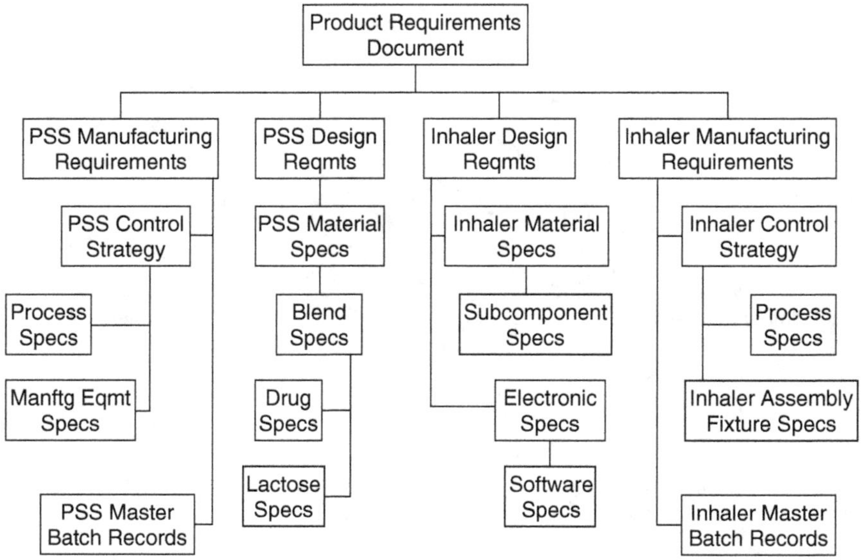

Figure 8.3 Example of the link from system-level requirements to component-level specifications for a DPI product. [*From Phillips and Hill (2002). Reproduced with permission from RDDOnline.*]

for mass vaccination (Hingson et al. 1963). Having been used for vaccination for a long time, the technology is now being promoted for self-administration of parenteral drugs. This should allow for painless injection of solid and liquid drugs without the risk of noncompliance owing to needle phobia and without the risk of needle-stick injuries to health care staff. However, products have not gained wide acceptance so far owing to their high cost, large size, difficulties in manipulation during administration, and the discomfort and fear of patients who believe that something is being forced through their skin. Several companies are at present developing or marketing smaller needleless injection devices, but most of them are aiming for delivery of liquid formulations.

Powder injection applies many of the principles of pulmonary delivery of dry powders to the lungs: The drug has to be in the form of very small particles, is dispensed from a reservoir, and is delivered as an aerosol; i.e., particles are dispersed in a gas. Liquid or dissolved drug can be delivered by precipitation or adsorption onto carrier particles. The big difference with pulmonary delivery is the momentum at which the particles are delivered. Driven by a high-pressure helium gas stream, the particles travel fast enough to penetrate the outer layer of the skin, the stratum corneum. The design of devices to deliver needle-free injection of solids was pioneered by researchers at the University of Oxford who founded PowderJect Pharmaceuticals PLC in 1993 (now PowderMed Ltd.) to develop the only powder-based technology so far. Since that time, the company has shifted its primary focus from drug delivery to injection of vaccines in powder form. Peer-reviewed publications covered in this chapter are rather scarce on this technology owing to the exclusive company-sponsored nature of the research to date.

8.2 Design Rationale for Dry-Powder Delivery

Controlled delivery of a pharmacologically active substance requires a defined temporal and spatial selectivity. In the context of inhalation, the required knowledge and the technologies to achieve this goal are still incomplete but evolving steadily. It is well known in inhalation therapy that the amount of drug reaching the lower airways is much less than the dispensed dose and also a lot less than the part of the dose present in the aerosol cloud as particles small enough to be inhaled. The lung is constructed in such a way that it is an effective filter of airborne particles. The percentage of the metered unit dose of currently marketed products that typically reaches the peripheral region of the respiratory tract is usually between 5 and 20 percent and not only depends on many formulation parameters but also is strongly influenced by the anatomy of the patient's lungs and their physiological and pathological state (Lippmann and Schlesinger 1984). In addition, inhalation maneuver or breathing pattern and patient training and compliance are important factors that influence the dose reaching the targeted site of action (Gonda 1992). Thus site-directed delivery is difficult to achieve, and the effort that has to be made to control this will depend on the therapeutic window, i.e., the difference between the doses required for achieving the desired therapeutic effect and the amount that would cause undesired toxicological side effects. With drugs that are highly potent and/or that have a narrow therapeutic window, the pharmaceutical scientist thus requires more sophisticated means to control and maximize the deposition efficacy.

8.2.1 Deposition mechanisms

Dry-powder inhalers (DPIs) deliver the drug to the respiratory tract in aerosol form. An aerosol is by definition a suspension of free liquid or solid fine particles in a gas phase, which is air in the case of DPIs (and a compressed gas in the case of needle-free injection). The most prominent characteristic that determines the delivery of drug particles to the lungs is the particle size, although particle shape and density are also of considerable importance for the behavior of an aerosol in the respiratory tract (Brain and Blanchard 1993; Gonda 1992; Heyder et al. 1986; Agnew 1984; Heyder et al. 1980).

There are five different physical mechanisms that govern drug deposition in the airways: inertial impaction, sedimentation (gravitational deposition), diffusion, interception, and electrostatic precipitation (Gonda 1992). The relative extent to which each mechanism contributes to the actual deposition of an inhaled aerosol in the airways depends on the particle characteristics, the anatomy of the patient's respiratory tract, and the breathing pattern.

Impaction is caused by the inertial mass of the traveling aerosol particles that forces them to move in a straight-line direction even when the flow of the inhaled air transporting them is bent around a curvature. Hence the particles tend to deposit on obstacles placed in the path of their travel. The inertial mass depends on particle size, density, and velocity. The stopping distance S of a particle having mass m_P and initial velocity $v_{0,P}$ is defined according to

$$S = b \times m_P \times v_{0,P} \tag{8.1}$$

with the correlation factor b being the mechanical mobility of the particle (Hinds 1998). In most dry-powder applications, aerosol particles will travel with the speed of the inhaled air. According to Eq. (8.1), the longer the particles travel in their original trajectories, the greater are their mobility, mass, and velocity. Depending on their inertial mass, they will pass several bifurcations of the conducting airways on their way down into the deep lungs until eventually they will be deposited.

The probability of an inhaled particle being deposited by impaction is a function of the dimensionless Stokes number Stk, which relates particle properties (mass m_P, diameter d_P, and density, ρ_P) to parameters of the airflow (air velocity v_A, viscosity η_A, and airways radius r_A):

$$\text{Stk} = \frac{\rho_P \times d_p^2 \times v_A}{18 \times \eta_A \times r_A} \tag{8.2}$$

From Eq. (8.2) it is obvious that the Stokes number Stk and thus the deposition efficiency by impaction increase with increasing particle size and airflow velocity. Impaction occurs most frequently in the upper respiratory tract (pharynx, larynx, and main trachea), where particles larger than 5 μm are trapped because of their size and the fast and turbulent airflow exerted. Also in the upper tracheobronchial region, impaction is the most prominent mechanism (Hinds 1998).

Sedimentation of airborne particles is caused by gravitational force (Hinds 1998). In a laminar flow, the terminal settling velocity $v_{t,P}$ of a spherical particle of diameter d_P and density ρ_P is

$$v_{t,P} = \frac{{d_P}^2(\rho_P - \eta_A)g}{18\eta_A} \tag{8.3}$$

with g being the gravitational constant. The simple Stokes law is no longer applicable for particles sufficiently small that they can "slip" through the medium with the consequence that they are moving faster than predicted by Stokes law. This is adjusted by incorporating the so-called slip correction factor into the equation. In addition, correction to Stokes law is needed if the flow starts to become turbulent. The Reynolds number Re gives the numerical expression of the laminar or turbulent flow regime. Stokes law is only valid when the Reynolds number Re is sufficiently small (Hidy 1984) according to

$$\mathrm{Re} = \frac{\rho_A L v_A}{\eta_{d,A}} \tag{8.4}$$

where $\eta_{d,A}$ is the dynamic viscosity of the air (which is different from η_A), and L is a characteristic length of the flow path. Airflow in a circular duct like the conducting airways is sufficiently laminar for Re < 2000. Turbulent flow exists with Re > 4000, with an interim regime in between these numbers (Baron and Willeke 2001).

From the preceding it is clear that deposition efficiency by these two mechanisms, impaction and sedimentation, is increasing with an increase in particle mobility (Gonda 1992). This independent parameter governs particle deposition by both mechanisms and depends on particle size, density, and velocity, as explained earlier, whereas it is assumed for most pharmaceutical applications that drug particles have similar density and velocity. With this assumption, particle size remains the most prominent factor to be considered in development. Since the particle is moving in a gas or airflow, the aerodynamic particle diameter is the important parameter. It is defined as the diameter of a sphere of unit density having the same aerodynamic properties as the actual particle, which can be expressed according to Eq. (8.5) (Hinds 1998):

$$d_{\mathrm{Aer}} = \sqrt{\rho_p} \times d_P \tag{8.5}$$

The inhalation airflow comes to a rest in the alveolar region. In still air, the collision of gas molecules with each other results in Brownian motion. The same happens with sufficiently small particles (which can be seen when the dust particles in a nonventilated room are hit by a sunbeam). For very small or ultrafine particles (when the particle size is similar to the mean free path length of the air molecules), the motion is not determined by the flow alone but also by the "random walk" called *diffusion.* The diffusion process is always associated with a net mass transport of particles from a region of high particle concentration to regions of lower concentration in accordance with the laws of statistical thermodynamics. The spatial and temporal net mass transport is calculated according to the first and second of Fick's laws:

$$J_P = -D\frac{\partial n_P}{\partial x} \tag{8.6a}$$

$$\frac{\partial n_P}{\partial t} = D\frac{\partial^2 n_P}{\partial x^2} \tag{8.6b}$$

where D = the diffusion coefficient
n_P = particle number concentration
x = displacement in one dimension (Hinds 1998)

The diffusion coefficient can be calculated according to the Stokes-Einstein equation

$$D = \frac{kT}{3\pi d_P \eta_A} \tag{8.7}$$

In this equation, the diffusion coefficient D is related to air viscosity η_A and particle diameter d_P, with k being the Boltzmann constant and T the absolute temperature. It is clear from this description that diffusion is a rather slow deposition mechanism compared with impaction and sedimentation processes because it depends on the thermal velocity of the particles and not on airflow. It is the primary transport mechanism for small particles and is important when the transport distance becomes small, as in the deep lung. Efficiency of this deposition mechanism can be increased significantly by breath-holding because a portion of the ultrafine particles that are not deposited will be exhaled by the patient.

Other deposition mechanisms are usually of less importance. Although present, their contribution to overall deposition of airborne particles is rather small. Interception frequently occurs when the gravitational center of the traveling particle is aligned within the airflow but one end of the particle touches the airways surface and is caught (Gonda 1992). This situation may be of some importance for the deposition of elongated needle-shaped particles or fibers. These particles exhibit a small aerodynamic diameter because they will be aligned in the direction of airflow (this is the reason why particle shape also may be an important feature to be considered in regional deposition). According to their aerodynamic diameter, fibers would penetrate deep into the small airways. Because the alignment is not ideal and the airflow usually is not laminar, they tend to "wobble" along their long axis, which brings them frequently into contact with the surface and thus increases their deposition efficiency (Gerrity 1990).

Electrostatic precipitation is pertinent when the aerosol particles are charged significantly. Charges may be induced on the particles when the aerosol is being generated. In principle, charged particles induce a charge of opposite sign on the surface of the airways. Although triboelectric generation of charges is seen frequently in DPI powder formulations (Byron et al. 1997), there is little evidence that this mechanism makes an effective contribution to particle deposition in the lungs because charges are likely to be dissipated rather quickly in the high moisture content of typically 99.7 to 99.9 percent relative humidity in the lower airways (Gonda 1992).

8.2.2 Deposition efficiency

The probability of reaching the lower airways increases with decreasing particle diameter. The mathematical description for prediction of lung deposition is the deposition efficiency, which is usual given as a function of aerodynamic particle size. Particles larger than approximately 10 μm are nearly completely removed in the upper conducting airways by impaction, as mentioned earlier. For particles of about 3 μm and smaller, sedimentation becomes more prominent, which results in a drop in the deposition efficiency to about 20 percent. These particles are deposited in the lower airways. When the airborne particles are of submicron size (smaller than 1 μm), they are trapped mainly by diffusion, thus causing the deposition efficiency to increase again (Martonen and Katz 1993). Low to very low velocities, together with the longer residence time of particles in this part of the lung, the respiratory zone, facilitate deposition by this process (Gupta and Hickey 1991). This requires patients to hold their breath in order to allow particles to deposit by diffusion because otherwise a larger percentage of the submicron particles would be exhaled.

It is clear from this that the optimal particle size for inhalation therapy depends on the area that is targeted. In asthma therapy, for example, the optimal particle size for a bronchodilator drug has been found with 2.8 μm to target the β_2-adrenoreceptors, which are located mainly in the peripheral airways and are responsible for dilation of the bronchial smooth muscle cells lining the conducting airways (Zanen et al. 1994). The optimal particle size for the delivery of an anticholinergic drug in asthma therapy was found to be smaller than or equal to 2.8 μm because the muscarinic receptors are located more centrally (Zanen et al. 1995). For steroids, the eosinophils in the proximal airways are the target, which would require a smaller particle size.

For systemic delivery, particles of 1 to 3 μm aerodynamic diameter are assumed optimal because smaller particles usually do not provide enough mass to deliver the required dose within a reasonable time and number of inhalation maneuvers, and particles of much bigger size do not reach the lower regions of the lungs (Hickey and Thompson 1992). It has to be kept in mind, however, that most of these results are provided by studies applying carefully controlled monodisperse or narrow-sized aerosols and controlled inhalation maneuvers, whereas practical inhaled therapies deliver a polydisperse aerosol. The polydispersity of an aerosol is described by the geometric standard deviation derived from aerodynamic particle size distribution data. It has been demonstrated that the degree of polydispersity can influence the deposition in the respiratory tract significantly (Gonda 1992; Hickey and Thompson 1992).

While the physicochemical properties of the delivered drug can be controlled to some extent, there is rather little control of the patient's influence on the deposition of current aerosolized therapeutics. The major contribution of the patient's inhalation maneuver and compliance has been recognized for a long time, but it is only now that engineers and scientists are trying actively to implement sophisticated means of control. The well-known large intersubject variability of deposition within the respiratory tract is problematic if precise delivery is required. Controlling the inspiratory flow rate can reduce the variability significantly. For example, it has been shown that intersubject variability of particle deposition for a controlled slow inspiratory flow is approximately three times smaller than for a spontaneous breathing pattern (Brand et al. 2000). A similar improvement may be achievable by decoupling the inhalation flow rate and the aerosol deposition efficiency to a certain extent.

8.2.3 Physiological and pathological aspects of inhalation drug delivery

The respiratory tract can be divided into distinctive zones (Weibel 1963):

- The *tracheobronchial or conducting airways*, which provide the channels for gas transport but have no gas-exchange function. Their surface is ciliated to provide a particle clearance mechanism called *mucociliary clearance*.
- A *transition zone*, which is partially alveolated.
- An *alveolar region or respiratory zone*, which includes the alveolar ducts and the alveoli where the gas exchange takes place.

The respiratory tract can be described by a simple geometrical model of symmetrically branching tubes (Weibel 1963). The tracheobronchial tract (the conducting zone) consists of branching airways channels with 16 consecutive bifurcations and thus looks similar to a tree turned upside down. It starts at the trachea (generation 0), which divides into the two mainstem bronchi, which further bifurcate into bronchi that enter the two left and the three right lung lobes. Each subsequent pair of branches has a smaller diameter than the previous tract. The conducting zone ends with the sixteenth generation, the terminal bronchi.

The transition zone consists of the respiratory bronchioles (generations 17 to 19), which contain alveoli. At the terminal end, the respiratory zone is composed of parenchyma that contains the alveolar ducts and about 300 million alveoli (alveolar sacs) to provide the gas-exchange surface. Since the surface area expands to such a large extent within the very last generations of bifurcations, the inhalation airflow rapidly slows down to zero velocity so that the movement of gas molecules and the exchange occurs entirely by diffusion (Stocks and Hisloop 2002).

A brief description of the lung and the different barriers posed by respiratory epithelia is given in Chap. 2. The efficacy of inhalation therapy using a DPI or any other device depends on the amount of drug deposited at the target site in the airways. Deposition depends not only on formulation parameters but also on physiology and anatomy of the human airways. The dimensions of the conductive airways vary during breathing and thus are dynamic to some extent, generating turbulences and irregular flow regimes within the inhaled air. Factors such as age, gender, race, and disease state are also relevant to the therapeutic use of aerosols (Byron and Patton 1994). In normal subjects breathing in a highly controlled manner, intersubject variability in lung geometry accounts for a coefficient of variation of the total deposition of 27 percent (Blanchard et al. 1991; Heyder et al. 1988). Even within the same individual, differences in lung morphometry can be found depending on age, lung volume, and diseases or disease state. Quite often when therapeutic aerosols are indicated and applied, the lung is compromised and in a more or less serious state of disease. Whenever the airflow rate is affected, the deposition pattern of the inhaled therapeutic particles is also potentially affected. Differences in deposition thus can be explained as differences in airflow owing to morphological, physiological, or pathological variability.

The clinical signs seen most commonly with diseased lungs are cough, chest tightness, breathing difficulties, and abnormal breathing patterns. Whenever any of these symptoms occur, the continuing obstruction is the primary cause of vital lung functions in disorder. Different diseases can cause acute or chronic lung tissue obstruction. Most commonly diagnosed diseases of the conducting airways include asthma, chronic bronchitis, bronchiolitis, cystic fibrosis, and bronchiectasis.

The pathological conditions may be divided into two classes: obstructive defects and defects in lung compliance (Gerrity 1990). Whereas tissue obstructions and lesions characterize the first, the latter conditions usually result in stiff and compromised lung tissue. Lung obstructive defects include diseases such as chronic bronchitis, asthma, and cystic fibrosis. Lung diseases involving compliance defects include emphysema, pulmonary fibrosis, and bronchiectasis. The term *chronic obstructive pulmonary disease* (COPD) is used for chronic bronchitis and emphysema common among heavy smokers that frequently coexist and thus include both obstructive and compliance defects (Gerrity 1990).

The distribution of diseased parts of the lungs is often heterogeneous or diffuse. Deposition usually is markedly altered by most pathological conditions (Brain and Valberg 1979). Most studies report a significantly increased deposition in the tracheobronchial tract at the expense of the deeper lung, i.e., peripheral deposition. The compromised parts may affect aerosol deposition even in the healthy portions of the lungs because airflow restrictions are compensated, and more airflow is conducted by the unobstructed airways. This may lead to increased particle exposure. Higher airflow eventually results in increased impaction rates and deposition in the healthy airways. Contrary to this deposition, "hot spots" also can be found downstream of sites of obstruction. Such very heterogeneous deposition patterns are seen frequently in patients diagnosed with obstructive diseases such as asthma, chronic bronchitis, and cystic fibrosis; enhanced deposition can be found in parts immediately downstream of the obstructed tissue when high local air velocities together with induced flow turbulences promote impaction (Christensen and Swift 1986).

8.3 Design of Dry-Powder Inhalation Devices

DPIs are versatile devices to which scientists and engineers have devoted a lot of thoughts and ideas. They usually deliver a minute dose in the micro- to milligram range, a metered and aerosolized quantity of drug, into the stream of air inhaled by the patient. Since the inhaler device delivers the dry powder for pulmonary application, the physical principles and design ideas by which the device (i.e., the hardware) operates govern the function and use of the product by the patient. There are many different designs on the market or currently under development. The schematic overview shown in Fig. 8.4 differentiates the devices according to the energy source required to deliver the medication and the principles for metering the dose.

Devices requiring the patient's inspiration effort to aerosolize the powder aliquot are called *passive devices* because they do not provide an internal energy source. *Active devices* may provide different kinds

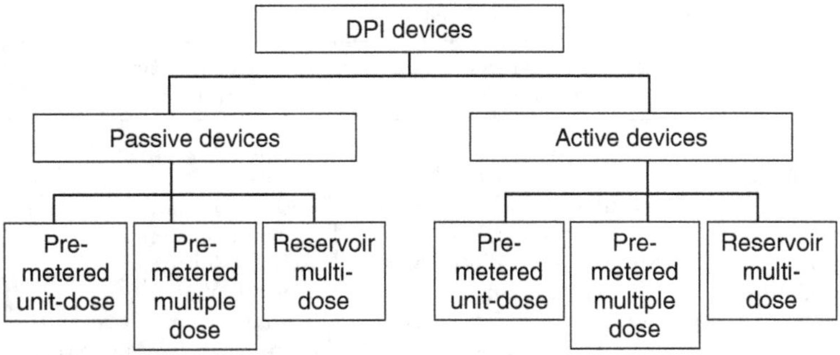

Figure 8.4 Types of DPI devices. [*With modifications from Prime et al. (1997). Reproduced with permission from Elsevier.*]

of energy for aerosolization: kinetic energy by a loaded spring and compressed air or electric energy by a battery. To date, no active device is on the market, although some concepts are in late stage of development.

8.3.1 Passive DPI devices

Common with all passive devices is the requirement of a defined minimal inhalation airflow for delivery and dispersion of the dry-powder formulation. Usually, the efficacy of both processes depends to some extent on the airflow rate in vitro (Srichana et al. 1998) and in vivo (Zanen et al. 1992). The inhalation maneuver and thus the aerosolization energy provided by the patient depend to a large extent on the physiological and pathological state and also on the compliance of the user. This means that passive devices have to be fairly tolerant and have to ensure the delivery of a consistent dose to the lower airways over a wide range of operating airflow rates. Despite tremendous progress in recent years in device development, the lack of control at the patient-device interface constitutes a big disadvantage for controlled delivery. Consequently, passive devices are used so far only for topical delivery of drugs having a fairly broad therapeutic window to the diseased lung. An overview of passive devices currently marketed or in development is given in Table 8.1. The big advantage of these devices is that they are patient triggered; i.e., they operate solely when the patient inhales. To achieve better control on the delivered dose, complex active devices are being developed. Owing to high costs of the complex devices and the availability of novel particle engineering technologies to produce desired small particles, however, passive devices have been getting a second look in recent years.

To maximize drug particle dispersion, mechanical means may be introduced into the flow path to generate a turbulent airflow that exerts

TABLE 8.1 Examples of Passive DPI Devices

Premetered single dose	Premetered multiple dose	Reservoir type multidose
Rotahaler (GlaxoSmithKline)	Diskhaler (GlaxoSmithKline)	Turbuhaler (AstraZeneca)
Spinhaler (Sanofi-aventis)	Accuhaler/Diskus (GlaxoSmithKline)	Clickhaler (Innovata)
Inhalator Ingelheim (Boehringer Ingelheim)	Inhalator Ingelheim M (Boehringer Ingelheim)	Easyhaler (Orion)
Handihaler (Boehringer Ingelheim)	Eclipse (Sanofi-aventis)	Pulvinal (Chiesi)
Cyclohaler (Pharmachemie)	FlowCaps (Hovione)	Novolizer (Viatris)
Aerolizer (Novartis)		Ultrahaler (Sanofi-aventis)
		SkyeHaler (SkyePharma)
		Twisthaler (Schering-Plough)

a greater shear force than a laminar flow. Device engineers propose tortuous flow paths through the device, e.g., spiral pathways in AstraZeneca's Turbuhaler (Wetterlin 1988) and a "swirl nozzle" in Schering-Plough's Twisthaler. A simple grid was incorporated in early devices such as the Spinhaler and Rotahaler that exerts a low internal resistance because of their unrestricted airflow path. A grid is also incorporated in the Diskus/Accuhaler device (GlaxoSmithKline) to generate turbulent flow (Prime et al. 1997). Other manufacturers have built-in baffles as impaction surfaces or exploit a "cyclone principle" to maximize the aerosolization effect, such as, for example, in the Taifun device (Focus Inhalation) and the Novolizer (Viatris).

The more tortuous the pathway and the more elaborated the internal parts in the DPI device, the higher the internal device resistance becomes to inspiration efforts exerted by the patient (Clark and Hollingworth 1993). It has been demonstrated that at "maximum" inhalation effort the flow rate achieved through a device is controlled by the maximum pressure drop exerted by the chest muscles. With "moderate" effort, the relationship is more complex. There has been a strong debate on the appropriate resistance of passive DPI devices. While most patients can generate a constant flow with "moderate" inspiration effort, it is difficult for most to achieve this at "maximum" effort throughout the inhalation cycle. Hence the internal device resistance should be defined in such a range that sufficient airflow could be generated for powder dispersion at moderate inspiratory effort (Dalby et al. 1996). If the internal resistance to airflow is too high because of device design, a device may become unsuitable for certain patient populations such as the elderly and children. On the other hand, the flow dependency of the delivery is limited if the patient needs a large effort because of high device resistance. While this is beneficial in view of better control of the delivered dose, it obviously limits the use of such a device to patient populations that are capable of exerting enough inspiratory effort to actuate the device. Having this in mind, most device engineers nowadays aim to design medium-resistance devices with sophisticated metering and dispersion mechanisms. In addition, minimal flow-rate dependency of particle dispersion is a goal. Nevertheless, patients with severely compromised lungs may not be able to achieve the flow rate and flow volume required for optimal targeting. For example, cystic fibrosis patients may achieve only an inspiratory volume of about 1 L and therefore may be a target population for an active DPI device.

While the devices can be regarded as the hardware by which the drug is applied, the powder formulation is the software. Each formulation is uniquely designed for the device by which it is delivered. Interchange (i.e., the application of a formulation to a device for which it is not designed) is usually neither advisable nor feasible because it may result in an uncontrolled and unreliable dose for the patient. Even when a device is used to deliver more than one drug product, e.g., Turbuhaler (AstraZeneca) and Accuhaler/Diskus (GlaxoSmithKline), the devices and the formulations are both optimized for their intended drug product.

Premetered single-dose devices. Premetered or factory-metered DPIs contain the dose as a dry-powder formulation in a unit-dose containment that is metered and filled in the factory. Since the first generation of these devices, the Spinhaler (Fisons, now Sanofi-aventis) and Rotahaler (Allen and Hanburys, now GlaxoSmithKline), the unit dose typically has been filled into hard gelatin capsules that are opened by an opening mechanism incorporated in the device (Fig. 8.5). This could be either a rotating blade to cut the capsules or needles to pierce them. Unfortunately, hard gelatin capsules change their properties with varying humidity of the surrounding environment. In very dry conditions they tend to become brittle and may shatter, with the risk of the patient inhaling small shreds, or they become more elastic under the influence of high ambient humidity and may fail to be opened properly by the internal mechanism. In addition, these changes may affect the powder characteristics adversely. Given the drawbacks of using gelatin, inhalation-grade capsules using the less-moisture-affected hydroxypropylmethylcellulose (HPMC) are being developed by the main capsule suppliers [Capsugel (Belgium) and Shionogi (Japan)].

Spinhaler uses two perforating pins to puncture the capsule body for opening. To feed the capsule, the patient has to unscrew the device body and push the capsule into a cup that is connected with a vertical spindle holding a propeller. After closing, the capsule is pierced by sliding the outer sleeve down and back again. When the patient tilts his or her

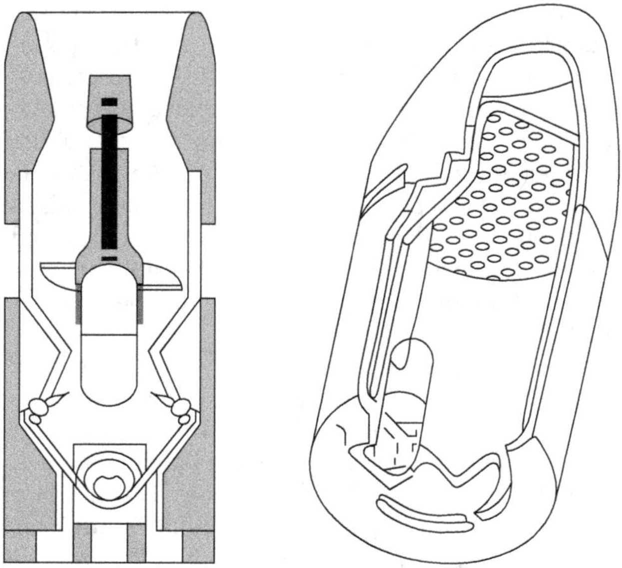

Figure 8.5 Capsule-based single-dose dry-powder inhalers Spinhaler and Rotahaler. [*From Ganderton and Kassem (1992). Reproduced with permission from Elsevier.*]

neck back and inhales, the rotor is propelled by the air inhaled through the rear end and flowing through the device. The rotation causes vibration of the hard gelatin capsule, by which the drug contents are released and aerosolized (Bell et al. 1971). For consistent dosing, the patient is advised to repeat the inhalation once or twice using the same capsule.

The Rotahaler device has a different opening mechanism. The capsule containing the medication is pushed body first into the capsule entry port, a square hole at the rear end of the device. By doing this, the capsule's cap is distorted, and the capsule's shell-locking system is weakened. For opening, the cylindrical sleeve of the device is twisted until it stops, which separates the capsule into two halves with the aid of an internally mounted plastic bar that pushes the capsule body out of the cap (Hallworth 1977). The capsule body falls into the device and releases the drug contents into the airstream, whereas the cap is retained in the capsule entry port and is removed subsequently. Owing to the turbulences generated in the device by the inspiration airstream, the capsule body starts to move randomly, which causes dislodged particles to be dispersed and aerosolized.

The Inhalator Ingelheim (Boehringer Ingelheim, Germany) is a capsule-based single-dose dry-powder device that is fed with one hard gelatin capsule at a time with the mouthpiece opened. After closing the mouthpiece, the capsule is punctured at each end by a needle. The inhaler has to be held vertically with the mouthpiece upward for correct perforation; otherwise, only one side of the capsules will be pierced. When the patient inhales, part of the inhalation airstream is drawn through the capsule, whereas the other part causes the capsule to vibrate and release the dose. Unlike the Spinhaler and the Rotahaler, the internal resistance of the device is relatively high, resulting in a long time requirement to deliver the dose and empty the capsule completely. For this reason, the inhalation has to be repeated twice using the same capsule to complete the dose. Boehringer Ingelheim recently also introduced a modern version of the capsule-based single-dose device operating according to the same working principle with the trade name Handihaler (Fig. 8.6).

In the Cyclohaler device (marketed by Pharmachemie, The Netherlands), the hard gelatin capsule is placed into a cavity located at the bottom of the device that has to be opened. Again, the capsule is opened by piercing the halves with a set of four needles on each side activated by pushing two buttons. After this operation, the patient inhales through the device. The patient is asked to tilt his or her neck back and lift the device during inhalation in order to allow the capsule to fall from the "piercing cavity" into a circular "rotation" chamber." There it starts to rotate under the influence of the inhaled air streaming tangentially into the chamber. The dose is released by the centrifugal force generated by the rotation. A slightly modified device is marketed by Novartis under the trade name Aerolizer.

The common feature of all the devices described is that they require loading each time they are used, which needs some manual dexterity. In

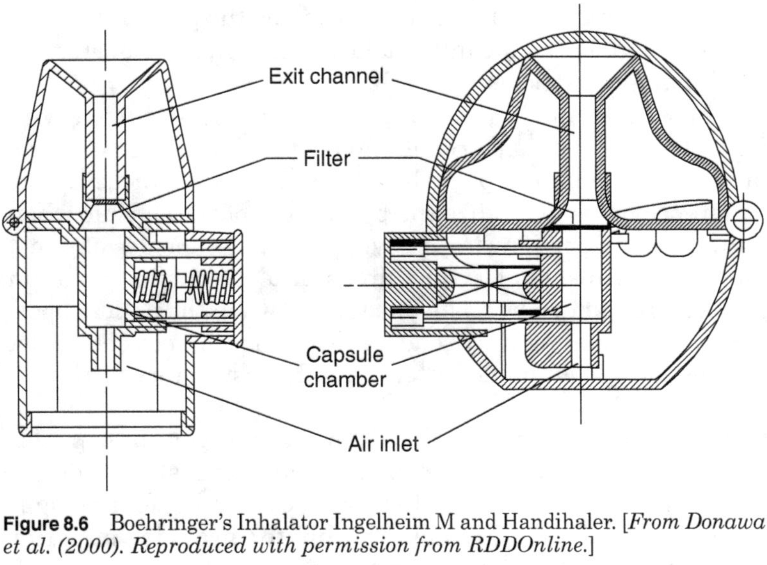

Figure 8.6 Boehringer's Inhalator Ingelheim M and Handihaler. [*From Donawa et al. (2000). Reproduced with permission from RDDOnline.*]

many cases, the capsules are stored in blisters to protect them from humidity. In some cases, such as, for example, with the Spinhaler and the Rotahaler, the capsule has to be loaded in a specific orientation, which may cause additional problems. The complex opening and device loading operations potentially may cause a hazard in emergency situations or may be difficult for handicapped patients. Despite these drawbacks, capsule-based single-dose inhalers are gaining new recognition to date, as demonstrated by the recent approval of tiotropium bromide dry powder in the Handihaler (trade name Spiriva, marketed by Boehringer Ingelheim). A valuable advantage of these devices is the high drug payload that can be dosed, provided that a neat or highly concentrated drug powder is placed in the capsules. A recent example of such an application is the delivery of the antibiotic drug tobramycin by a single-dose capsule device called Turbospin (Newhouse et al. 2003).

Capsule-type multiple-dose devices. Only few examples are known of multiple-dose devices using factory-metered capsules. There are two inhalers marketed to date comprising a small number of capsules per device. The first one is Boehringer's Inhalator Ingelheim M, which contains a revolver-type cartouche with a total of six capsules (see Fig. 8.6). The operating principle is the same as with the single-dose Inhalator Ingelheim.

The other is the Eclipse reusable capsule-based device marketed to date by Sanofi-aventis (France) in Japan and Scandinavia. It contains up to four capsules at a time, loaded body first into the four chambers of the device. Eclipse uses a single blade to cut off the dome of the capsule body when the patient twists the device. Above the capsule, there is a "vortex chamber" in which a ball starts rotating as the patient inhales. The ball spins at high speed during inhalation, helping to disperse the powder into inhalable particles.

Hovione's FlowCaps inhaler is a third example and is currently in development (Hovione, Portugal). Similar to Eclipse, the inhaler uses two blades that make cuts in the cap and body. The cuts are of unequal size, and this difference helps to fluidize the powder out of the capsule when the patient inhales. In the FlowCaps device, up to 14 capsules of size 4 can be loaded. One capsule at a time drops into the cutting area, an accurately sized tube guided by an inclined ramp, when the patient operates the device. Twisting the body relative to the mouthpiece cuts a slot into the capsule for release of its content into the inspiratory airflow.

Blister-type multiple-dose devices. The drawbacks of gelatin capsules were overcome by GlaxoSmithKline when the Rotadisk/Diskhaler device was introduced. Rotadisk is a small blister disk containing four or eight unit doses of medication at a time. The blister provides superior protection of the medication from environmental stress such as high humidity. To load a blister cassette, the body part of the device is pulled out and separated until the mouthpiece tray is open and a disk can be placed. To open a blister, the patient has to lift the lid containing a plastic pin fully upright to puncture both the top and the bottom to allow the air to flow through the open blister. Gently pulling out the tray and pushing it once turns the blister cassette to the next dose. The individual blisters are numbered, with the blister numbers appearing in a small window on top of the tray, but the patient has to be aware of the number of doses he or she already has used in order to replace the empty disk at the appropriate time.

Further improvement of this principle of factory-metered blistered unit doses was achieved when GlaxoSmithKline developed the Accuhaler device (also named Diskus in some countries (Fig. 8.7). This multiple-dose DPI device contains a blister strip of 60 unit doses that is transported to the next filled unit dose by pulling a small ergonomic lever prior to inhalation. The powder is presented for aerosolization by peeling back the foil from the blister, which is superior to simple piercing because it is not associated with variability in foil flap shape and device retention. The dose indicator on top of the device tells the patient how many doses are left and decreases each time the lever is pulled. This means that as with most other devices, operation of the dose indicator is associated with the loading operation and not with the inhalation maneuver.

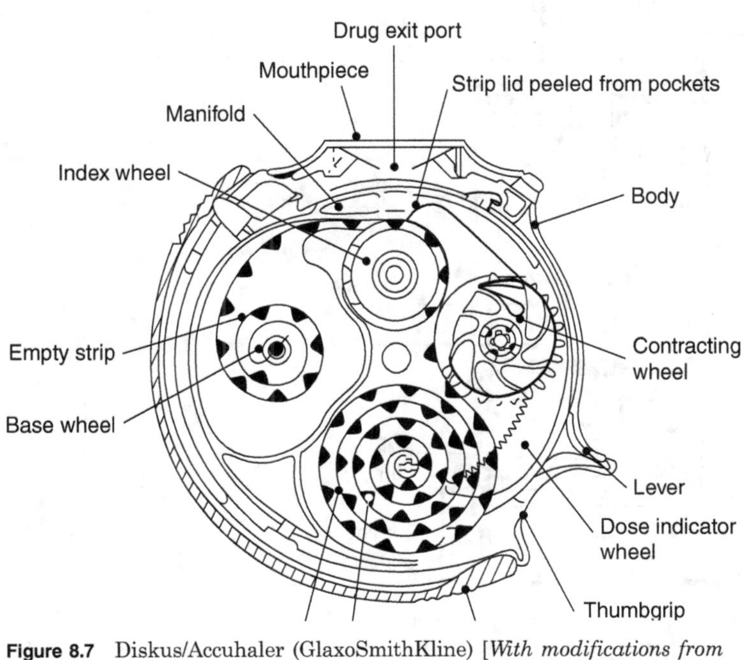

Figure 8.7 Diskus/Accuhaler (GlaxoSmithKline) [*With modifications from Crompton (1997). Reproduced with permission from Blackwell Science.*]

Recent examples of device developments using the principle of multiple blistered doses include the Technohaler by Innovata Biomed, the respiratory division of ML Laboratories (now Innovata plc, United Kingdom), and the C-Haler by Microdrug (Switzerland). Both devices contain a plastic cartridge with a number of factory-metered unit doses sealed with a protective aluminum laminate foil. The single-dose and multiple-dose inhalers developed by Respirics (United States) use foil-foil laminate blisters for maximum moisture protection of the filled medication and a dual piercing mechanism to allow maximum emptying and dispersion of the dose.

Reservoir-type multiple-dose devices. A completely different approach was first introduced by AstraZeneca with the Turbuhaler (or Turbohaler in some countries), a powder reservoir from which the device meters each dose by the patient turning the knob at the bottom side of the device (Fig. 8.8).

Since the device itself meters the dose, it must contain a precisely working measuring mechanism to ensure dose accuracy and consistency. The metering unit of the Turbuhaler consists of groups of truncated

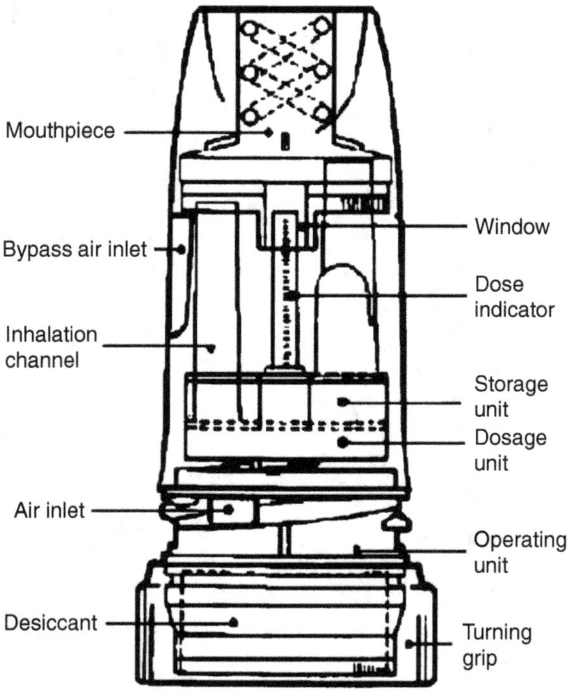

Figure 8.8 Turbuhaler (AstraZeneca). [*From Wetterlin (1988). Reproduced with permission from Springer Science and Business Media.*]

conical holes, with their larger diameters toward the powder reservoir to achieve consistent filling of the cavities (Wetterlin 1988). The powder is filled volumetrically into these holes when the patient twists the knob at the bottom of the device. Excess amounts are scraped off by a specially designed scraper blade. The device also has a colored dose indicator that shows a red mark in a small window when the patient approaches the last nominal doses.

The volumetric metering of a dry-powder formulation poses some unique problems. Of course, the powder has to be freely flowing and optimized for the specific metering mechanism used. The formulation strategy to achieve this with the Turbuhaler is described in the paragraph on dry-powder formulations. Another imminent issue is that dry-powder beds contained in a reservoir may be sensitive to moisture ingress into the device. In order to prevent changes in the physical properties of the powder formulation, resulting in suboptimal dose reproducibility and dispersion, the Turbuhaler contains a desiccant for moisture protection.

Similar reservoir-type multidose devices include the Twisthaler (by Schering-Plough, United States) and the Pulvinal inhaler (by Chiesi Pharmaceutici, Italy). While the Pulvinal is used by following the same sequence of operations as the Turbuhaler (i.e., removing the protective cap and turning a knob or part of the device for metering and inhalation), the Twisthaler is designed to automatically meter a unit dose when the cap is removed. Eventually putting back the protective cap resets the metering plate and indexes the numerical dose counter.

The Easyhaler (by Orion Pharma, Finland) and the Clickhaler (by Innovata plc, United Kingdom) are available at present in some European markets. Unlike the DPIs described earlier, these two reservoir-type inhalers meter the dose when the patient presses the top of the device similar to actuation of a pressurized metered-dose inhaler. Both devices contain a dose indicator, which is standard for reservoir multidose DPIs. Recently, Innovata presented the Twinhaler for asthma combination therapy, a new development based on the Clickhaler. This device does not require the combined drugs to be formulated in one powder blend but delivers two powder formulations from two reservoirs into one airflow path.

The Novolizer (Viatris, Germany) is unique among the devices because it is reloadable. It uses plastic cartridges filled with up to 200 doses of the medication that are inserted into the DPI body. The device is currently marketed in some European countries. As with most of the recent developments, the device has both acoustic and visual control features.

The SkyeHaler by SkyePharma (Switzerland), currently in a late stage of development, is an example of a modern reservoir-type device that offers novel features, including a numerical counter that counts only on successful inhalation and a locking mechanism when the last dose

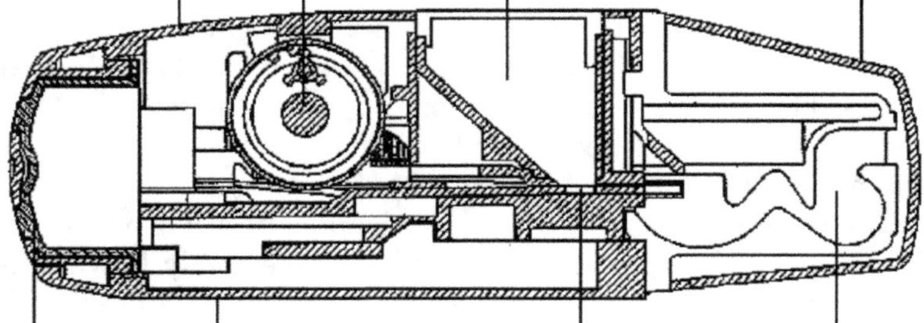

Figure 8.9 Cross-sectional view of the novel SkyeHaler DPI developed by SkyePharma. (*Reproduced with permission from SkyePharma.*)

is delivered (Fig. 8.9). The dose is metered from the powder reservoir by a vibratory-assisted gravity feed with volumetric metering, which takes place on opening of the protective cover. This device is designed for intuitive handling by the patient, resulting in a correct and efficient use owing to its interactive design and sensory feedback mechanisms. The dose is delivered only if sufficient inspiration energy is provided by the airflow exceeding a preset minimum actuation flow threshold.

Compacted powder reservoir devices. There are two examples to date of this group of DPI devices. The Ultrahaler currently developed by Sanofi-aventis exploits a completely different type of reservoir and metering principle. Instead of the freely flowing powder bed, the dry-powder formulation in the Ultrahaler is compacted on the internal walls of a barrellike drug container from which the unit dose is scraped off by means of a sharp blade and aerosolized in the airstream. Obviously, this technology requires a formulation that can be compressed into a consistent plug that can be redispersed into the original primary drug particles by means of the patient's inhalation. The challenge is to make a "soft compact" of consistent physicochemical properties that do not change during shelf life. The device includes a dose counter, a locking mechanism activated on device exhaustion, and like many other novel devices in development, features to aid patient use and improve patient compliance. The other example in this category, the Jethaler by PulmoTec (Germany), is already marketed in Germany with the steroid drug budesonide. This device comprises a spring-driven ceramic millwork for mechanical aerosol generation.

8.3.2 Active DPI devices

In active DPI devices, the energy for delivery and/or aerosolization of the dry-powder medication is provided by a source stored in or derived from the device itself. This may provide better control and improved accuracy and reproducibility of the delivery independent of the patient's capabilities (Crowder et al. 2001). Most often this is at the expense of increased complexity and device cost. Thus these devices are considered mainly for controlled, most often systemic delivery of expensive medication or drugs with a narrow therapeutic window. Active devices can provide deep lung deposition of drugs because the energy input for aerosolization can be controlled to generate an aerosol of optimal particle size distribution.

No active devices are currently marketed, and examples of successful developments are fewer than for passive devices. There are some promising device developments in the late stage but also examples of concepts that failed in development. Also, some less expensive developments are currently being pursued, intended for the local therapy of lung diseases.

Air-pressure-driven active devices. Air-pressure-driven aerosolization is the concept employed in a number of devices currently in different stages of development with drugs for local or systemic action. These devices rely on a small patient-operated air pump. Air is compressed by mechanical means (piston or bellows) and is released on the external trigger given by the patient's inspiratory cycle. Because of the use of this air pump, these devices have an active aerosolization mechanism and are assumed to be less flow-rate-dependent than passive DPI devices.

The active inhaler made by Nektar Therapeutics (formerly Inhale Therapeutic Systems, United States), called Pulmonary Delivery System (PDS), mechanically compresses a fixed volume of air required for delivery and dispersion of a premetered dry-powder unit dose by a spring-loaded pump (Fig. 8.10). Generation of the respirable aerosol cloud thus is independent of the inspiration effort exerted by the patient. The aerosol is generated in a transparent holding chamber that acts as a spacer from which the patient inhales the "standing cloud" of particles (Patton 1997). The PDS device is actually close to market for inhaled delivery of insulin under the trade name Exubera.

Other examples in this group include the Airmax (Ivax Pharmaceuticals, United Kingdom), the Aspirair (Vectura, United Kingdom), and the Prohaler (Valois Pharm, France).

Battery-powdered active devices. An example of a battery-powered device is the Spiros developed by Dura Pharmaceuticals, now Elan Drug Delivery. The Spiros DPI is a premetered multiple-use device that relies on a battery-powered motor-driven high-speed impeller for drug dispersion and aerosolization (Hill 1994). Because of this mechanism for particle dispersion, the device is claimed to achieve efficient deposition in the patient's lungs using a slow deep inspiration maneuver and a low inspiratory effort. However, it is a highly complex device, and sufficient battery power is critical for all functions performed by the device. For this reason, Elan currently has a less complex Spiros S2 in development,

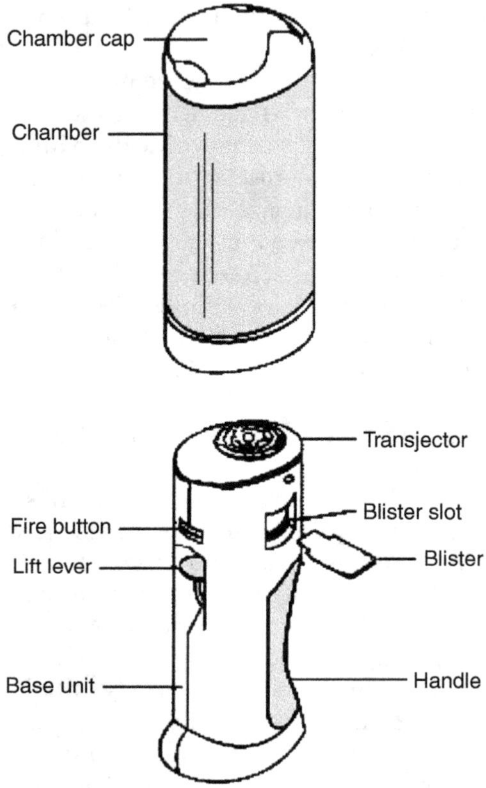

Figure 8.10 PDS device developed by Nektar Therapeutics for pulmonary delivery of insulin.

which is a nonmotorized *passive* device designed for both unit-dose and multidose delivery. Instead of the motor-driven impeller, free-floating beads in a dispersion chamber create the shear force to disperse the powder in much the same way as in the Eclipse device by Sanofi-aventis. This enables the manufacturer to use rather simple drug formulations with lactose as the carrier excipient. Spiros S2 is proposed either as a unit-dose or a multiple-dose blisterpack device.

The MicroDose DPI (MicroDose Technologies, United States) is a breath-activated device that includes a piezoelectric vibrator that converts electrical energy from a battery to mechanical motion that is then transferred into the dry powder. The vibration energy deaggregates and aerosolizes the dose. By controlling the energy input, i.e., the amplitude and frequency of the vibration, the DPI is claimed to be usable for various compounds. As with the devices from Nektar and Dura, the MicroDose DPI uses accurately filled unit-dose blisters.

Spring-loaded active devices. Another active device patented by 3M uses mechanical agitation by a hammer to disperse the drug from an especially microstructured tape. In this device, pure micronized drug powder is filled into very small dimples embossed on the tape. The powder is held in place by Lifshitz–van der Waals forces. A hammer agitates the tape and releases the drug. As with passive devices, the released drug will be dispersed by the patient's inspiratory effort, which makes this proposed DPI a sort of hybrid between an active and a passive device. Obviously, the tape lacks the additional moisture protection given to other formulations by blistering.

8.4 Powder Formation

It is evident that drugs for application via the inhaled route have to be provided in the form of very small particles. As already described, the aerodynamic particle size should be in the range of 1 to 5 μm for topical delivery to treat the respiratory system (Moren 1987) and ideally below 3 μm to reach the alveoli and to enter the circulation for systemic delivery (Byron and Patton 1994; Patton and Platz 1992; Gupta and Hickey 1991).

Such small particles usually are generated by air-jet micronization and less frequently by controlled precipitation or spray drying. As bulk powder, they usually tend to be very cohesive and exhibit poor flow and insufficient dispersion because of large interparticle forces such as van der Waals and electrostatic forces (Zeng et al. 2001; Podczeck 1998; Hickey et al. 1994). The control of sufficient powder flow and deaggregation (dispersion) is thus of utmost importance to ensure efficient therapy with a dry-powder aerosol. Two different formulation approaches are used currently in marketed DPI preparations to fulfill the requirements. Most often, coarse particles of a pharmacologically inactive excipient, usually α-lactose monohydrate, are added that act as a "carrier" and provide sufficient powder flow to the mixture. Other carbohydrates, amino acids, and phospholipids have been suggested frequently (Crowder et al. 2001).

The coarse carrier particles blended with micronized drug form an ordered or interactive mixture (Fig. 8.11) (Hersey 1975) stabilized by adhesive Lifshitz–van der Waals and electrostatic forces (Podczeck 1998; Hickey et al. 1994). The shear forces exerted in the airflow of a DPI device must be greater than the adhesive forces in order to provide sufficient deaggregation and dispersion of the drug particles. Unfortunately, however, this process is more or less incomplete and disperses only a proportion of the agglomerated drug particles depending on the inhalation airflow (Zanen et al. 1992).

The second formulation strategy includes the "pelletization" or "spheronization" of micronized drug particles (Olsson and Trofast 1998). Weak agglomerates (sometimes referred to as soft pellets) are formed under carefully controlled process conditions (Fig. 8.12) and may consist either

Figure 8.11 Scanning electron microscopic view of an ordered powder blend of micronized salbutamol sulfate with lactose carrier. (*Courtesy of SkyePharma.*)

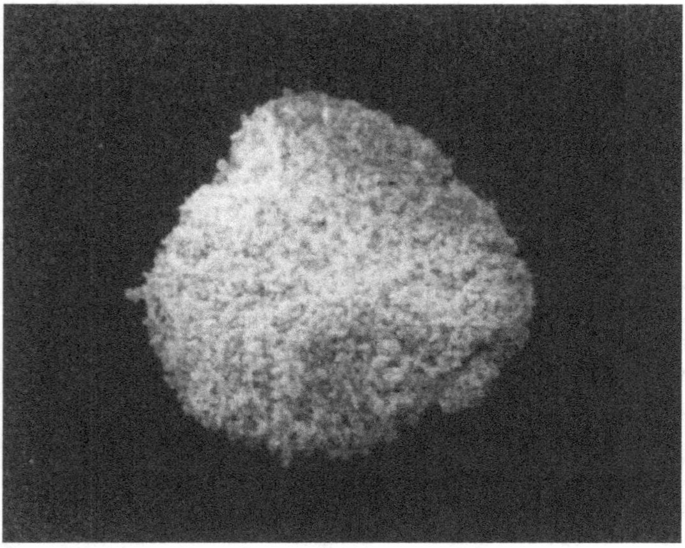

Figure 8.12 Scanning electron micrograph of a spheronized pellet of budesonide. [*From Dunbar (2002). Reproduced with permission from Euromed Communications.*]

of neat drug particles or, alternatively, of a blend with α-lactose monohydrate of *similar* microfine size distribution (Clarke et al. 1998).

The carrier properties have been studied in numerous in vitro studies to understand the influence on powder performance, especially drug detachment. The particle size distribution of the carrier is of paramount importance for the delivery and dispersion of drug particles by a device at given flow-rate conditions (Steckel and Mueller 1997; French et al. 1996; Kassem et al. 1989). An increased proportion of fine particles results in more efficient dispersion (Podczeck 1999), which has led to the proposal to deliberately add microfine lactose as a ternary agent (Lukas et al. 1998).

Surface roughness has been found to influence the strength of the adhesion forces between drug and carrier particles (Podczeck 1999; Kassem and Ganderton 1990). Lactose particles with smooth surfaces were prepared by recrystallization (Kassem and Ganderton 1990) and more recently by diffusion-controlled recrystallization (Zeng et al. 2000), but it was found that a certain microscopic roughness is beneficial in reducing the contact area with microfine particles and results in reduced interaction (Podczeck 1999). Other authors proposed the use of granulated or composite lactose because it is assumed that the surface characteristics can be modified in a number of ways to optimize drug particle adherence and detachment behavior. *Corrasion*, a sort of mild milling process, was proposed to improve carrier detachment because asperities of the lactose crystals are cleared off by this process, assuming that high-energy binding sites on the lactose surface are occupied by fine particles generated in situ (Staniforth 1996).

The use of ternary agents, i.e., additional excipients, was proposed for improving the drug particle detachment of interactive powder blends including L-leucine (Staniforth 1996), magnesium stearate (Ganderton and Kassem 1992; Kassem 1990), and lecithin (Staniforth et al. 2002). Magnesium stearate is notorious for destabilizing the ordered powder mixture by "stripping" off the drug from the carrier particles (Lai and Hersey 1979). It was found that this can be avoided by careful control of the blending conditions while achieving a significant improvement in the physical stability and dispersion properties of the powder blend (Keller et al. 2000).

Spray-drying techniques are proposed frequently to generate ideally spherical microparticles of homogeneous structure and surface. Improved efficiency and improved physicochemical stability have been achieved by co-spray-drying protein or peptide drugs with excipients such as salts, phospholipids, carbohydrates, and/or amino acids forming a high T_g matrix and providing a hydrophilic environment that stabilizes the drug molecule's ternary structure by hydrogen bonding. Such excipients were used previously for stabilization of lyophilized powders for parenteral delivery. Most promising because of its high T_g

is the use of the glass-stabilizing excipient trehalose in the formation of spray-dried microspheres of protein and peptide drugs (Clark et al. 1996; Maa et al. 1999). Protein drugs formulated as dry powder tend to be amorphous particles with higher molecular mobility and reactivity and need to be stabilized. The compound particles are formed in an amorphous glass state with high T_g and minimal moisture content but with reduced molecular mobility and hence increased stability. Stability of this meta-stable glass is guaranteed as long as the T_g of the compound material is significantly higher than the environmental temperature (Patton 1997). Nektar Therapeutics (formerly Inhale Therapeutic Systems) pioneered the use of this technology (named PulmoSol) (Fig. 8.13) for the application of inhaled insulin delivered by the PDS device (Patton et al. 1999). Other excipients proposed include poly(L-lactic acid) (PLLA), poly(D, L-lactic-coglycolic acid) (PLGA), dextran, starch, and human serum albumin (HSA), with the added benefit of an assumed sustained release of the drug from the polymer matrix (Philip et al. 1997; Haghpanah et al. 1994). Most of these excipients have not been used in inhalation dosage forms so far and require a full toxicological characterization to establish their safety profile.

A recent application of particle formation by solvent evaporation and spray-drying techniques is based on the concept of the aerodynamic diameter. According to Eq. (8.5), the aerodynamic diameter d_{Aer} is correlated with the true particle diameter d_P and the particle density $\rho_p^{0.5}$. It is evident that particles formed in a particle-formation process can be much bigger, provided that their density is very small. Increased bioavailability of such large porous insulin particles (Fig. 8.14) has been demonstrated on inhalation by rats and has been correlated with a

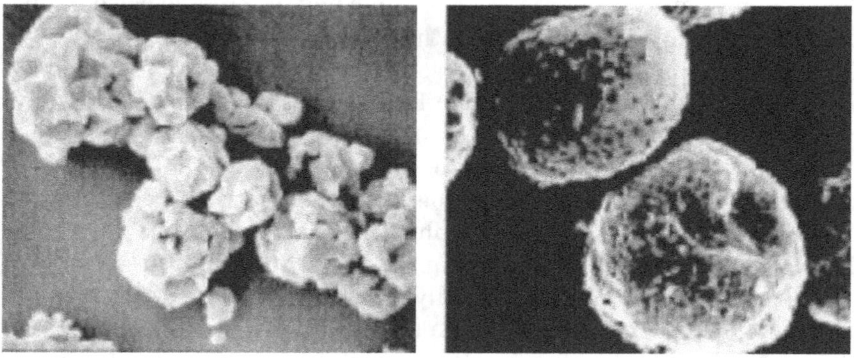

Figure 8.13 Scanning electron micrograph of spray-dried PulmoSol particles (*left*) and PulmoSpheres particles (*right*), both developed by Nektar Therapeutics. [*From Peart and Clarke (2001). Reproduced with permission from Russell Publishing.*]

Figure 8.14 Scanning electron micrograph of spray-dried large porous particles (airborne particles) developed by Alkermes. [*From Peart and Clarke (2001). Reproduced with permission from Russell Publishing.*]

much higher in vitro fine particle fraction than for conventional particles (Edwards et al. 1997). The porous particle structure of these Air particles (developed by Alkermes, United States) was obtained by spray drying the constituents at conditions producing thin-walled hollow particles first that collapse during drying and yield a structure similar to crumpled paper (Fig. 8.14). The large porous particles show reasonably good powder flow owing to their favorable geometric particle diameters in the range of 5 to 30 μm (Vanbever et al. 1999a). It has been demonstrated in animal models for drugs such as insulin, estradiol, and others, that fairly long resident times can be achieved by deposition of large porous particles in the respiratory zone (i.e., the alveolated and distal nonciliated regions of the peripheral lungs) to avoid mucociliary clearance (Vanbever et al. 1999b; Wang et al. 1999). The disadvantage of porous particle formation techniques is their inability to provide formulations of adequate drug loading, which may limit this approach to highly efficient drugs.

PulmoSpheres (Fig. 8.13) (pioneered by Nektar Therapeutics) are small porous particles of less than 5 μm geometrical diameter and low density formed in a proprietary spray-drying process of a submicron oil-in-water emulsion using a perfluorocarbon as a "blowing agent" (Tarara et al. 2000). Improved lung delivery has been demonstrated in proof-of-concept studies against well-characterized comparators (Venthoye et al. 2001). The company is currently developing the antibiotic tobramycin for inhalation delivery via a dry-powder device using this technology.

Spray freeze drying also has been proposed as an alternative technology to produce light and porous particles for peptide and protein delivery. Liquid nitrogen is used as recipient agent, into which the formulation is sprayed. The formed microparticles are harvested and lyophilized eventually. DNase and monoclonal anti-IgE antibodies have been used to demonstrate the feasibility of this concept (Maa et al. 1999). Promaxx microspheres are manufactured in a phase-separation process between water-soluble polymers and therapeutically active protein that results in particles having a high protein payload of up to 90 percent (Brown et al. 1999).

Particle-formation methods using supercritical fluids can be regarded as an evolution of the simple principle of solvent-based crystallization and precipitation or of spray drying a solution containing the drug molecules (and potentially additional formulations aids). The big difference is in the nature of the solvent, which is usually carbon dioxide in the supercritical state. Unfortunately, it is also a rather poor solvent to most pharmacologically active compounds, which limits its use to processes that require the microparticle constituents to be dissolved in the supercritical phase. In a process named *rapid expansion by supercritical solutions* (RESS), microparticles can be formed by solvating drug compound and excipients in a mixture of supercritical carbon dioxide with organic solvent and rapidly depressurizing through an adequate nozzle (Bodmeier et al. 1995). The limitation of solubility in the supercritical fluid is overcome by the *gas antisolvent process*, (GAS) for which the microparticle constituents are dissolved in a polar liquid solvent, e.g., ethanol or isopropanol. Saturating this solution with supercritical carbon dioxide as antisolvent decreases the solubility and forces the substrate to precipitate or crystallize. To produce composite microparticles, a variation of the process named the *aerosol solvent extraction system* (ASES) has been developed in which the drug solution is sprayed into a container consisting of compressed supercritical carbon dioxide, where microparticles are formed by precipitation caused by extraction of the solvent into the CO_2 phase (Bleich et. al. 1993). The SEDS process, which stands for *solution enhanced dispersion by supercritical fluids*, is a specific implementation of ASES, which has the advantage of processing a drug solution into a micron-sized particulate product in a single operation because solution and supercritical fluid are both sprayed together through a coaxial nozzle (York et al. 1998). The SEDS process can be operated under a wide range of working conditions and allows a controlled change of particle size and morphology. Careful optimization and control of process parameters yields uncharged, freely flowing particles of narrow size distribution and high degree of crystallinity (York and Hanna 1996). The PGSS process for *particles from gas-saturated solutions* (and suspensions) is based on the capability of the supercritical fluid to soften and swell certain biocompatible matrix polymers such as polyethylene glycol (Weidner et al. 1996). At this state, the polymer may disperse microfine drug particles and can be depressurized through a nozzle to form monodisperse microparticles without the use of any organic solvent.

Particle-formation technologies such as spray drying are also used for modification of the pharmacokinetic characteristics. Modified or sustained release of drugs in the lungs is a major challenge because the size of inhalable particles provides a large surface area for instantaneous diffusion-controlled dissolution. Slowly degrading particles are subjected to mucociliary clearance if they are deposited in the tracheobronchial airways, which cannot be influenced. Particles deposited in the alveolated airways for sustained release of drugs for systemic action may be subject to alveolar phagocytosis. However, it seems possible to avoid phagocytic engulfment by alveolar macrophages by using large porous particles for systemic delivery (Vanbever et al. 1999b; Wang et al. 1999). It also has been demonstrated that particles smaller than a few hundred nanometers often escape both mucociliary and macrophage clearance after deposition and are present in the lung lining fluid for a prolonged period of time (Renwick et al. 2001). Surface coating of microparticles is currently also being evaluated, aiming to present a defensive cloak to alveolar macrophages. The solvent-less Nanocoat process, now exploited by Nanotherapeutics, Inc., produces a continuous coating around drug particles by pulsed laser deposition (Talton et al. 2000). The coating may provide sustained release profiles with a low excipient load, tailored bulk-powder properties, and improved product stability owing to reduced moisture uptake.

8.5 Devices for Powder Injection

The unique form of needle-free injection of powders into the epidermis or mucosa has been developed by researchers at the University of Oxford and Powderject Pharmaceuticals PLC (now PowderMed Ltd., United Kingdom). Drugs in microparticulate form are accelerated to sufficient velocities to enter the skin or mucosa and achieve a therapeutic effect (Burkoth et al. 1999). Provided the drug particles are sufficiently small to avoid skin lesions and pain, the concept has been shown to be clinically effective, pain-free, and applicable to a range of therapies. Use is pain-free because the penetration depth of the particles is typically less than 100 μm into the epidermis, and thus the sensory nerve endings lying in the papillary dermis usually are not excited (Fig. 8.15).

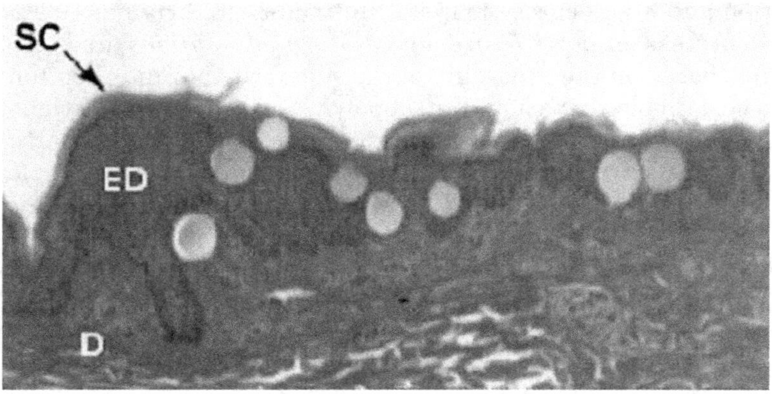

Figure 8.15 Histological demonstration of dermal powder injection in the pig. The stratum corneum (SC), dermis (D), and particles delivered to the epidermis (ED) are clearly shown. The particles consist of swellable, slowly soluble polysaccharide microspheres (50 μm diameter when dry). [*With modifications from Hickey (2001). Reproduced with permission from Euromed Communications.*]

Studies have been performed for microfine particles in the size range of 20 to 100 μm and have demonstrated that the patient has no adverse effects from particle penetration but only feels the sensation of the compressed gas impact (Hickey 2001).

Optimized targeted delivery of microparticles to a defined site within the epidermis or to the mucosa requires a profound understanding of the factors affecting penetration. The penetration depth x is governed by the Petry equation derived through empirical penetration models and application studies:

$$x = \frac{P \times m_P}{A} \times \log\left[\frac{1 + (v_{0,P} - v_{t,P})^2}{\phi}\right] \tag{8.8}$$

where P = the Petry constant, which is related to biomechanical properties of the tissue
m_P = particle mass
A = targeted area
$v_{0,P}$ and $v_{t,P}$ = impact and threshold velocity of the particles
ϕ = penetration coefficient

From Eq. (8.8) it is evident that particle momentum depending on mass and velocity of the particles is important to control the delivery. The particle acceleration and impact velocity are defined by particle properties such as size, density, and morphology and device properties such as pressure of the compressed gas source, nozzle geometry, and others.

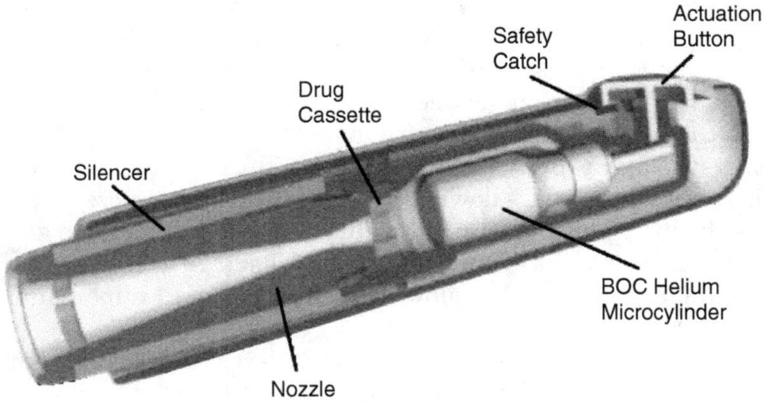

Figure 8.16 Cross-sectional diagram of a single-use disposable powder injection system highlighting the major components. When the actuator button is depressed, the driver gas (He) is released into the surrounding rupture chamber. At a specific pressure, the plastic membranes of the drug cassette burst, and the drug particles are entrained in the gas flow, which is accelerated through the convergent-divergent nozzle. [*From Hickey (2001). Reproduced with permission from Euromed Communications.*]

Compressed helium gas is used to accelerate the particles, which is fast enough to allow them to penetrate the stratum corneum (Fig. 8.16).

The transient gas-particle dynamics of the earlier prototypes were found to deliver microparticles with a range of velocities and a nonuniform spatial distribution. For targeted delivery, however, especially in the area of gene and peptide delivery, the system should deliver particles with a narrow and controllable velocity range and a uniform spatial distribution. This was achieved with a certain embodiment called the *contoured shock tub*e configured to achieve uniform particle impact conditions by entraining particles within a quasi-steady gas flow (Kendall et al. 2002).

To give the particles the required momentum, they should be densely packed and rigid and have a well-defined narrow particle size distribution. Friable and oblique particles are not desirable because the penetration depth will increase if the particle characteristic is more variable (Hickey 2001). Studies have been performed with particles ranging from 20 to 40 μm in size and 1.1 to 7.9 g/cm^3 in density impacting human cadaver skin (Kendall et al. 2000). Velocities of up to 260 m/s were applied to particles of this size range. For many applications, smaller particles of about 1 to 4 μm diameter may be required for an optimized delivery. To deliver particles of this size into the skin, higher densities and impact velocities are required. For this reason, gold particles are used as a carrier material for the delivery of plasmid DNA vaccines (Kendall et al. 2001).

A schematic diagram of a single-use powder injection system is given in Fig. 8.16. When a valve within the device opens the high-pressure ampule, the compressed driver gas hits a cassette that holds a single dose of the medication between two membranes. The gas pressure causes the membranes to rupture instantaneously at a defined rupture pressure and rapidly expands the gas, thus forming a shock wave. This shock wave travels down a convergent-divergent nozzle and accelerates the drug particles until they hit the skin surface. The gas does not penetrate the skin but is reflected back into the device through a "silencer" required to slow down the transient supersonic flow (Hickey 2001).

The total mass that can be delivered by Powderject's powder injection technology is about 1 to maximum 3 mg per application. This limits the application to potent drugs, e.g., certain biotechnology drug molecules and vaccines. It had been found that the delivered drug actually reaches the systemic circulation faster than if the same dose was administered by subcutaneous injection, which is probably caused by the increased permeability and increased flux of water across the skin for up to 24 hours after injection as a reaction of the stratum corneum to the microscopic lesions. With the limits described earlier, the device is especially suitable for the delivery of macromolecular drugs and vaccines. Drug targeting to different layers of the skin may be achieved by controlling particle momentum according to Eq. (8.8). Most macromolecular drugs for systemic adsorption are targeted to the deep dermis because of the high density of blood capillaries in that region. Currently in development is a novel DNA vaccine against hepatitis B that is ideally targeted to the germinal cells of the deep epidermis, where the highest level of gene expression is generated. Conventional vaccines are rather directed to the basement membrane between epidermis and dermis, the stratum basale.

8.6 Future Potential of Device Controlled Delivery of Powders

What can be expected in the near and middle future for inhaled powder delivery? There certainly will be some more new *passive* devices to reach the market and the patient, but it seems that a lot of purely mechanical design ideas have been implemented already in the newest generation of multiple-dose DPIs. Most important are audible and visible feedback features that help to improve delivery efficiency and patient compliance. Future improvements also seem possible in reducing dependency of the particle dispersion on actuation flow rate and internal device resistance. The next or over-next generation of multiple-dose devices also may incorporate lean, inexpensive, but rugged electronics with the potential to reduce the number of mechanical parts and to implement additional features such as patient alarms and logbook functions. There is certainly also a need for small, inexpensive DPIs delivering a single dose or a very limited number of doses. The potential dose range may expand at both ends for accurate delivery of extremely small doses and—even more demanding—of large doses of more than 25 mg.

A major breakthrough is expected with the launch of the first *active* device, Nektar's PDS device, developed for the systemic delivery of insulin (Exubera, jointly marketed by Pfizer and Sanofi-aventis). This product also will set the development and regulatory standards for systemic delivery of peptides and other biomolecular drugs since. It is expected that no future product could be delivered with less efficacy and less control, although the relative bioavailability of inhaled insulin is only about 10 percent of the intravenous forms. Inhaled insulin is to date only for instant release. No depot or sustained release form has progressed much for the reasons described earlier in this chapter. However, novel and emerging particle-formation technologies intended to achieve "better" dry-powder formulations by specific particle composition and tighter control of process conditions also may show the potential of controlled release. It has to be kept in mind that most composite particles consist of materials that never before have been delivered to the lungs and hence require full toxicological clearance to establish their safety as excipients.

Device development for powder injection also may gain much more momentum with the successful market introduction and penetration of a first product, most likely a hepatitis B vaccine. Of course, this product has to compete with other dosage forms and has to prove superiority to gain acceptance and become more than a niche product. Multiple-dose devices for powder injection probably will take much longer to market. Similar to inhalation devices, such systems will require advanced features such as dose counting and logbook functions.

References

Agnew, J. E. Physical properties and mechanisms of deposition of aerosols, in *Aerosols and the Lung: Clinical and Experimental Aspects*. London: Butterworths, 1984.

Baron, P. A., and Willeke, K. Gas and particle motion, in *Aerosol Measurement: Principles, Techniques and Applications*, 2d ed. New York: Wiley, 2001.

Bell, J. H., Hartley, P. S., and Cox, J. S. G. Dry powder inhalers: I. A new powder inhalation device. *J. Pharm. Sci.* **60**:1559–1564, 1971.

Blanchard, J. D., Heyder, J., O'Donnel, C. R., and Brain, J. D. Aerosol-derived lung morphometry: Comparisons with a lung model and lung function indexes. *J. Appl. Physiol.* **71**:1216–1224, 1991.

Bleich, J., Mueller, B. W., and Wassmus, W. Aerosol solvent extraction system: A new microparticle production technique. *Int. J. Pharm.* **97**:111–117, 1993.

Bodmeier, R., Wang, H., Dixon, D. J., et al. Polymeric microspheres prepared by spraying into compressed carbon dioxide. *Pharm. Res.* **12**(8):1211–1217, 1995.

Brain, J. D., and Valberg, P. A. Deposition of aerosol in the respiratory tract. *Am. Rev. Respir. Dis.* **120**:117–156, 1979.

Brain, J. D., and Blanchard, J. D. Mechanisms of particle deposition and clearance, in *Aerosols in Medicine. Principles, Diagnosis and Therapy.* New York: Elsevier, 1993.

Brand, P., Friemel, I., Meyer, T., et al. Total deposition of therapeutic particles during spontaneous and controlled inhalations. *J. Pharm. Sci.* **89**:724–731, 2000.

Brown, L., Blizzard, C., Rashba-Step, J., et al. Versatile bioerodible microsphere technology, in *Proceedings of the 26th International Symposium on Controlled Release of Bioactive Materials.* San Diego, CA: Controlled Release Society, 1999.

Burkoth, T. L., Bellhouse, B. J. Hewson, G., et al. Transdermal and transmucosal powdered drug delivery. *Crit. Rev. Ther. Drug Carrier Syst.* **16**(4):331–384, 1999.

Byron, P. R., and Patton, P. S. Drug delivery to the respiratory tract. *J. Aerosol Med.* **7**:49–75, 1994.

Byron, P. R., Peart, J., and Staniforth, J. N. Aerosol electrostatics: I. Properties of fine powders before and after aerosolization by dry powder inhalers. *Pharm. Res.* **14**(6):698–705, 1997.

Christensen, W. D., and Swift, D. L. Aerosol deposition and flow limitation in a compliant tube. *J. Appl. Physiol.* **60**:630–637, 1986.

Clark, A. R., Dasovich, N., Gonda, I., and Chan, H. K. The balance between biochemical and physical stability for inhalation protein powders: rhDNAse as an example, in *Proceedings of Respiratory Drug Delivery V.* Buffalo Grove, IL: Interpharm Press, 1996.

Clark, A. R. Medicinal aerosol inhalers: Past, present, future. *Aerosol Sci. Technol.* **22**:374–391, 1995.

Clark, A. R., and Hollingworth, A. M. The relationship between powder inhaler resistance and peak inspiratory condition in healthy volunteers: Implications for in vitro testing. *J. Aerosol Med.* **6**:99–110, 1993.

Clarke, M. J., Layzell, G., Tobyn, M. J., and Staniforth, J. N. The role of fine particle lactose in agglomerated dry powder aerosol formulations. *J. Pharm. Pharmacol.* **50**(suppl):186, 1998.

Crompton, G. K. Delivery systems, in *Allergy and Allergic Diseases.* London: Blackwell Science, 1997.

Crowder, T. M., Louey, M. D., Sethuraman, H. D. C., and Hickey, A. J. 2001: An odyssey in inhaler formulation and design. *Pharm. Technol.* **25**(7):99–113, 2001.

Dalby, R. N., Hickey, A. J., and Tiano, S. L. Medical devices for the delivery of therapeutic aerosols to the lungs, in *Inhalation Aerosols: Physical and Biological Basis for Therapy* (*Lung Biology in Health and Disease*, Vol. 94). New York: Marcel Dekker, 1996.

Donawa, M. E., Hochrainer, D., and Horhota, S. T. Controlling inhaler design: An important tool for avoiding regulatory pitfalls, accelerating product development and increasing user acceptance, in *Proceedings of Respiratory Drug Delivery VII.* Raleigh NC: Serentec Press, 2000.

Dunbar, C. Dry powder formulations for inhalation. *Drug Del Syst. Sci.* **2**(3):78–80, 2002.

Edwards, D., Hanes, J., Caponetti, G., et al. Large porous particles for pulmonary drug delivery. *Science* **276**:1868–1871, 1997.

French, D. L., Edwards, D. A., and Niven, R. W. The influence of formulation on emission, deaggregation and deposition of dry powder for inhalation. *J. Aerosol Sci.* **27**:769–783, 1996.

Ganderton, D., and Kassem, N. M. Dry powder inhalers, in *Advances in Pharmaceutical Sciences.* London: Academic Press, 1992.

Gerrity, T. R. Pathophysiological and disease constraints on aerosol delivery, in *Respiratory Drug Delivery.* Boca Raton, FL: CRC Press, 1990.

Gonda, I. Targeting by deposition, in *Pharmaceutical Inhalation Aerosol Technology* (*Drugs and the Pharmaceutical Sciences Series*, Vol. 54). New York: Marcel Dekker, 1992.

Greystone Associates. *Dry Powder Inhalation: Advancing Technology—Emerging Therapies.* Amherst, NH: Greystone Associates, 2003.

Grossman, J. The evolution of inhaler technology. *J. Asthma* **31**(1):55–64, 1994.

Gupta, P. K., and Hickey, A. J. Contemporary approaches in aerosolized drug delivery to the Lung. *J. Contr. Rel.* **17**:129–148, 1991.

Haghpanah, M., Marriott, C., and Martin, G. P. Drug delivery to the lung using albumin microparticles. *J. Pharm. Pharmacol.* **46**(suppl):1075, 1994.
Hallworth, G. W. An improved design of powder inhaler. *Br. J. Clin. Pharmacol.* **4**:689–690, 1977.
Heyder, J., Gebhart, J., and Stahlhofen, W. Inhalation of aerosols: Particle deposition and retention, in *Generation of Aerosols and Facilities for Exposure Experiments*. Ann Arbor, MI: Ann Arbor Science Series, 1980.
Heyder, J., Gebhart, J., Rudolf, G., et al. Deposition of particles in the human respiratory tract in the size range 0.005 to 15 μm. *J. Aerosol Sci.* **17**:811–825, 1986.
Heyder, J., Gebhart, J., and Scheuch, G. Influence of human lung morphology on particle deposition. *J. Aerosol Med.* **1**:81–88, 1988.
Hersey, J. A. Ordered mixing: A new concept in powder mixing practice. *Powder Technol.* **11**:41–44, 1975.
Hickey, A. J. and Thompson, D. C. Physiology of the airways, in *Pharmaceutical Inhalation Aerosol Technology* (*Drugs and the Pharmaceutical Sciences Series*, Vol. 54). New York: Marcel Dekker, 1992.
Hickey, A. J., Concessio, N. M., van Oort, M. M., and Platz, R. M. Factors affecting the dispersion of dry powders as aerosols. *Pharm. Technol.* **Aug**:58–64, 1994.
Hickey, P. L. Powder injection: A novel mode of drug delivery opening the way for new modes of therapy. *Drug Del Syst. Sci.* **1**:25–29, 2001.
Hidy, G. M. *Aerosols: An Industrial and Environmental Science*. Orlando, FL: Academic Press, 1984.
Hill, M. R. Characteristics of an active multiple dose dry powder inhaler, in *Proceedings of Respiratory Drug Delivery IV*. Buffalo Grove, IL: Interpharm Press, 1994.
Hinds, W. C. *Aerosol Technology: Properties, Behavior, and Measurement of Airborne Particles*, 2d ed. New York: Wiley, 1998.
Hingson, R. A., Hamilton, S. D., and Rosen, M. The historical development of jet injection and envisioned uses in the mass immunization and mass therapy based upon two decades' experience. *Milit. Med.* **128**:516–524, 1963.
Kassem, N. M., Ph.D thesis, King's College, Department of Pharmacy, University of London, 1990.
Kassem, N. M., and Ganderton, D. The influence of carrier surface on the characteristics of inspirable powder aerosols. *J. Pharm. Pharmacol.* **42**(suppl):11P, 1990.
Kassem, N. M., Ho, K. K. L., and Ganderton, D. The effect of air flow and carrier size on the characteristics of an inspirable cloud. *J. Pharm. Pharmacol.* **41**(suppl): 14P, 1989.
Keller, M., Mueller-Walz, R., Gilchrist, P., et al. Effects of storage on the in vitro performance of formoterol powder blends in the SkyePharma DPI, in *Proceedings of Respiratory Drug Delivery VII*. Raleigh, NC: Serentec Press, 2000.
Kendall, M. A. F. The delivery of particulate vaccines and drugs to human skin with a practical, hand-held shock tube–based system. *Shock Waves J.* **12**(1):22–30, 2002.
Kendall, M. A. F., Mitchell, T. J., Hardy, M. P., and Bellhouse, B. J. The ballistic delivery of high density, high velocity micro-particles into excised human skin, in *Proceedings of Bioengineering Conference of the American Society of Mechanical Engineers*. Snowbird, UT: 2001.
Kendall, M. A. F., Wrighton-Smith, P. W., and Bellhouse, B. J. Transdermal ballistic delivery of micro-particles: Investigation into skin penetration, in *Proceedings of World Congress on Medical Physics and Biomedical Engineering*. Chicago, IL: 2000.
Lai, F. K., and Hersey, J. A. A cautionary note on the use of ordered powder mixtures in pharmaceutical dosage forms. *J. Pharm. Pharmacol.* **31**:800, 1979.
Lippmann, M., and Schlesinger, R. B. Interspecies comparisons of particle deposition and mucociliary clearance in tracheobronchial airways. *J. Toxicol. Environ. Health* **13**(2–3):441–469, 1984.
Lukas, P., Anderson, K., and Staniforth, J. N. Protein deposition from dry powder inhalers: Fine particle multiplets as performance modifiers. *Pharm. Res.* **15**(4):562–569, 1998.
Maa, Y., Nguyen, P., Sweeney, T., et al. Protein inhalation powders: Spray drying vs. spray freeze drying. *Pharm. Res.* **16**(2):249–254, 1999.
Martonen, T. B., and Katz, I. M. Deposition patterns of poly-disperse aerosols within human lungs. *J. Aerosol Med.*, **6**:251–274, 1993.
Molina, M. J., and Rowlands, F. S. Stratospheric sink for chlorofluoromethane: Chlorine atom catalyzed destruction of ozone. *Nature* **249**:810–812, 1974.

Moren, F. Dosage forms and formulations for drug administration to the respiratory tract. *Drug Dev. Ind. Pharm.* **13**(4–5):695–728, 1987.

Newhouse, M. T., Hirst, P. H., Duddu, S. P., et al. Inhalation of a dry powder tobramycin pulmosphere formulation in healthy volunteers. *Chest* **124**:360–366, 2003.

Olsson, B., and Trofast, J. Dry powder inhalers: A powerful drug delivery system. *Eur. Pharm. Rev.* **3**(1):71–76, 1998.

Patton, J. S. Deep-lung delivery of therapeutic proteins. *Chemtech* **27**(12):34–38, 1997.

Patton, J. S., Bukar, J. and Nagarajan, S. Inhaled insulin. *Adv. Drug Del. Rev.* **35**:235–247, 1999.

Patton, J. S., and Platz, R. M. Routes of delivery: Case studies: 2. pulmonary delivery of peptides and proteins for systemic action. *Adv. Drug Del. Rev.* **8**:179–196, 1992.

Peart, J., and Clarke, M. J. New developments in dry powder inhaler technology. *Am. Pharm. Rev.* **4**:37–45, 2001.

Philip, V. A., Mehta, R. C., Mazumder, M. K., and DeLuca, P. P. Effect of surface treatment on the respirable fractions of PLGA microspheres formulated for dry powder inhalers. *Int. J. Pharm.* **151**:165–174, 1997.

Phillips, E. M., and Hill, M. Developing a powder inhalation system for insulin delivery: Lessons learned, in *Proceedings of Respiratory Drug Delivery VIII.* Raleigh NC: Davis Horwood International Publishing, 2002.

Podczeck, F. *Particle-Particle Adhesion in Pharmaceutical Powder Handling.* London: Imperial College Press, 1998.

Podczeck, F. The influence of particle size distribution and surface roughness of carrier particles on the in vitro properties of dry powder inhalations. *Aerosol Sci. Technol.* **31**:301–321, 1999.

Prime, D., Atkins, P. J., Slater, A., and Sumby, B. Review of dry powder inhalers. *Adv. Drug Del. Rev.* **26**:51–58, 1997.

Renwick, L. C., Donaldson, K., and Clouter, A. Impairment of alveolar macrophage phagocytosis by ultrafine particles. *Toxicol. Appl. Pharmacol.* **172**:119–127, 2001.

Schultz, R. K. Systemic drug delivery: Issues and challenges in a changing environment, in *Proceedings of Respiratory Drug Delivery VIII.* Raleigh, NC: Davis Horwood International Publishing, 2002.

Srichana, T., Martin, G. P., and Marriott, C. Dry powder inhalers: The influence of device resistance and powder formulation on drug and lactose deposition in vitro. *Eur. J. Pharm. Sci.* **1**:73–80, 1998.

Staniforth, J. N. Improvements in dry powder inhaler performance: Surface passivation effects, in *Proceedings of Drug Delivery to the Lungs VII.* London: Aerosol Society, 1996.

Staniforth, J. N., Patel, N., Morton, D. A. V., et al. Biotech DPI strategies: Accomplishing affordable control and effective action, in *Proceedings of Respiratory Drug Delivery VIII.* Raleigh, NC: Davis Horwood International Publishing, 2002.

Steckel, H., and Mueller, B. W. In vitro evaluation of dry powder inhalers: II. Influence of carrier particle size and concentration on in vitro deposition. *Int. J. Pharm.* **154**:31–37, 1997.

Stocks, J., and Hisloop, A. A. Structure and function of the respiratory system, in *Drug Delivery to the Lung* (*Lung Biology in Health and Disease Series*, Vol. 162). New York: Marcel Dekker, 2002.

Talton, J., Fitz-Gerald, J., Singh, R., and Hochhaus, G. Nano-thin coatings for improved lung targeting of glucocorticoid dry powders: In vitro and in vivo characteristics, in *Proceedings of Respiratory Drug Delivery VII.* Raleigh NC: Serentec Press, 2000.

Tarara, T. E., Weers, J. G., and Dellamary, L. A. Engineered powders for inhalation, in *Proceedings of Respiratory Drug Delivery VII.* Raleigh, NC: Serentec Press, 2000.

Vanbever, R., Mintzes, J. D., Wang, J., et al. Formulation and physical characterization of large porous particles for inhalation, *Pharm. Res.* **16**:1735–1742, 1999a.

Vanbever, R., Ben-Jabria, A., Mintzes, J. D., et al. Sustained release of insulin from insoluble inhaled particles. *Drug Dev. Res.* **48**:178–185, 1999b.

Venthoye, G., Weers, J. G., and Tarara, T. E. Pulmosphere particle engineering: Technology development to pilot scale and commercial viability, in *Proceedings of Drug Delivery to the Lungs XII.* London: Aerosol Society, 2001.

Wang, J., Ben-Jabria, A., and Edwards, D. A., Inhalation of estradiol for sustained systemic delivery. *J. Aerosol. Med.* **12**:27–32, 1999.

Wasserman, M. A., and Renzetti, L. M. Therapeutic targets in lung disease: The state of the art, in *Proceedings of Respiratory Drug Delivery IV*. Buffalo Grove, IL: Interpharm Press, 1994.

Weibel, E. R. *Morphometry of the Human Lung*. Berlin: Springer-Verlag, 1963.

Weidner, E., Steiner, R., and Knez, Z. Powder generation from polyethyleneglycols with compressible fluids, in *High Pressure Chemical Engineering* (*Process Technology Proceedings*, Vol. 12). Amsterdam: Elsevier, 1996.

Wetterlin, K. Turbuhaler: A new powder inhaler for administration of drugs to the airways. *Pharm. Res.* **5**(8):506–508, 1988.

York, P., and Hanna, M. Particle engineering by supercritical fluid technologies for powder inhalation drug delivery, in *Proceedings of Respiratory Drug Delivery V*. Buffalo Grove, IL: Interpharm Press, 1996.

York, P., Hanna, M., Shekunow, Y., and Humphreys, G. O. Microfine particle formation by SEDS (solution enhanced dispersion by supercritical fluids): Scale up by design, in *Proceedings of Respiratory Drug Delivery VI*. Buffalo Grove, IL: Interpharm Press, 1998.

Zanen, P., Go, L. T., and Lammers, J.-W. J. The optimal particle size for β-adrenergic aerosols in mild asthmatics. *Int. J. Pharm.* **107**:211–217, 1994.

Zanen, P., Go, L. T., and Lammers, J.-W. J. The optimal particle size for parasympathicolytic aerosols in mild asthmatics. *Int. J. Pharm.* **114**:111–115, 1995.

Zanen, P., van Spiegel, P. I., van der Kolk, H., et al. The effect of the inhalation flow on the performance of a dry powder inhalation system. *Int. J. Pharm.* **81**:199–203, 1992.

Zeng, X. M., Martin, G. P., and Marriott, C. *Particulate Interactions in Dry Powder Formulations for Inhalation*. London: Taylor & Francis, 2001.

Zeng, X. M., Martin, G. P., Marriott, C., and Pritchard, J. Crystallization of lactose from carbopol gels, *Pharm. Res.* **17**(7):879–886, 2000.

Chapter

9

Biodegradable Polymeric Delivery Systems

Harish Ravivarapu

SuperGen, Inc.
Pleasanton, California

Ravichandran Mahalingam and Bhaskara R. Jasti

Thomas J. Long School of Pharmacy and Health Sciences
University of the Pacific
Stockton, California

9.1 Introduction

Development of suitable carrier systems for pharmaceutical products remains a major challenge. Historically, polymeric devices for implant were prepared from silicon, rubber, and polyethylene. A serious drawback of using these inert polymers as parenteral devices is their nonbiodegradability, and this requires their surgical removal after depletion of the drug. To overcome this problem, the concept of biodegradable polymers was first introduced in early 1970s for sustained release parenteral drug delivery. Use of biomaterials has provoked considerable interest, especially after the successful introduction of bioresorbable surgical sutures three decades ago. Since then, biodegradable polymers have become increasingly popular, and several new polymers were synthesized and employed for drug delivery applications. These polymers degrade in vivo either enzymatically or nonenzymatically to produce biocompatible or nontoxic by-products, and therefore, surgical removal of the exhausted delivery device can be avoided. These well-characterized and widely available polymers can be fabricated via well-established processes into various delivery systems such as prefabricated implants, in situ forming implants, and particulate carriers such as microspheres and nanoparticles. In addition to subcutaneous or intramuscular administrations, these particulate carriers also can be injected intravenously as long as their particle size is within physiologically acceptable range. This chapter is a generalized attempt at capturing and summarizing the information on available polymers, devices, and drug release from them with some relevant examples.

9.2 Rationale for the Use of Biodegradable Systems

Conventional drug therapy typically involves periodic dosing of a therapeutic agent that has been formulated in a manner to ensure its stability, activity, and bioavailability. For most drugs, conventional dosage forms are quite effective. However, in some cases continuous administration of the drug is desirable to maintain therapeutic plasma drug levels. The concept of drug delivery therefore was, introduced to overcome this limitation of conventional therapy. Many oral sustained release products have been formulated successfully and are available on the market, but these products usually are unsuitable for delivering a drug for more than 24 hours owing to the physiological limitations of gastrointestinal system. Approaches such as mucoadhesion to ensure long residence of delivery device in the gastrointestinal (GI) tract did not yield satisfactory results. In addition, drugs such as peptide and protein molecules are poorly absorbed and unstable in the GI tract, cannot be delivered by the oral route, and require parenteral administration. Infusion of drugs by the parenteral route for chronic treatment is not desirable because of patient discomfort and noncompliance. Additionally, development of injectable sustained release products is difficult, especially for drugs such as hormones and peptides that have very short half-lives. Benzathine penicillin, medroxyprogesterone acetate, zinc insulin, etc. are typical examples of sustained release parenteral delivery systems, but they are developed by either chemical modification or physical conjugation of the drug. Such design is useful only for limited drugs.

Application of polymer systems as an alternative approach for sustained drug delivery is rapidly gaining acceptance scientifically as well as commercially. This is especially true given to the recent progress in biotechnology, where many peptide and protein drugs are made available, and many are expected to be available as treatment options. It is well known that these biological molecules, with large molecular weights and high water solubility, are not amenable to conventional formulations owing to their GI instability and poor bioavailability. Their known short plasma half-lives require frequent injections, which may be acceptable for short-term use but not chronic treatments. On developing a polymeric device or carrier system that is biodegradable, it is possible that one single injection will replace a regimen of an injection a day for 30, 60, or more days with similar efficacy. This would result in manifold convenience for patients, reduced health care costs owing to fewer hospitalization, and in many instances reduced drug doses for similar or better therapeutic effect.

9.3 Biodegradable Polymers Used in Drug Delivery

Biodegradable polymers may be synthetic or natural in origin. Natural biodegradable polymers include human serum albumin, low-density lipoproteins (LDLs), bovine serum albumin, gelatin, collagen, hemoglobin, polysaccharides, etc.[1] Use of the natural polymers is limited by difficulties in purification and large-scale manufacture. They are also known to cause immunogenic adverse reactions. Of these natural biodegradable polymers, LDLs offer a unique opportunity for targeting drugs to tumors because tumor cells overexpress LDL receptors. With the advances in polymer science, tremendous knowledge has been gained over the past 30 years on the synthesis, handling, and mechanism of degradation of many biodegradable polymers. Today, many synthetic biodegradable polymers are being employed successfully for drug delivery applications. Irrespective of their source and chemistry, all biodegradable polymers possess some common characteristics, such as (1) stability and compatibility with the drug molecule, (2) biocompatible and biodegradable, (3) ease of manufacture on a larger scale, (4) amenability to sterilization, and (5) flexibility to yield multiple release profiles.

Biodegradable polymers can be divided into water-soluble and water-insoluble polymers. The water-soluble biodegradable polymers can be used as drug carriers for targeting drugs, which is covered in other chapters. Because of their versatility, the rest of this chapter will deal only with synthetic biodegradable polymers. Some of the widely used synthetic biodegradable polymers in drug delivery technology are summarized in Table 9.1. Among these polymers, polyesters, polylactic acids, polylactones, poly(amino acids), and polyphosphazenes predominantly undergo bulk erosion, and polyorthoesters and polyanhydrides undergo surface erosion. Polymer biocompatibility and lack of toxicity are important considerations in the design of a drug delivery system, especially those designed for systemic application.

9.3.1 Polyesters and polyester derivatives

Polyesters (polylactic acid and polyglycolic acid) are the first polymeric materials that have been used successfully as sutures over last two decades, and their degradation products are known to be nontoxic and their metabolic pathways are well established. The homopolymers of polylactic acid (PLA) and polyglycolic acid (PGA) are also known as polylactides and polyglycolides. Biodegradable polylactic acid polymers used for controlled release applications are stereoregular and available as D-, L-, and racemic DL-polylactide.[2] Polyesters and their copolymers have been tested extensively as implants, nanoparticles, and microspheres for the delivery of various drugs, such as narcotic antagonists, contraceptives, local anesthetics, cytotoxics, and antimalarial agents.[3,4] Polylactic acid has been reported to exhibit excellent biocompatibility at subcutaneous and other injection sites.[5] However, PLA sutures are reported to produce a mild inflammatory response when compared with a nondegradable material.[6] A study also indicated a mild inflammatory response of polylactic acid microparticles when tested in vivo by the intraarticular route.[7] This could be overcome in part by incorporating antiinflammatory agents in the formulation.[8] Use of polyesters as rate-controlling membranes and erodible polymeric excipients for injectable drug delivery systems, therefore, holds considerable promise in providing efficacious formulations.

Polylactic acid has been studied extensively for controlled release applications ranging from the oral delivery of simple drugs such as indomethacin[9] to the parental administration of complex proteins such as insulin.[10] Polylactic acid of different molecular weights has been studied as matrix material for parenteral administration. Seki et al.[11] used polylactic acid 6000 and Smith and Hunneyball[8] used polylactic acid 100,000 for the controlled delivery of drugs by the parenteral route. Several polylactic acid systems have been studied for the controlled

TABLE 9.1 Biodegradable Polymers and their Degradation Products

Polymers	Structures	Degradation products	References
Polyester	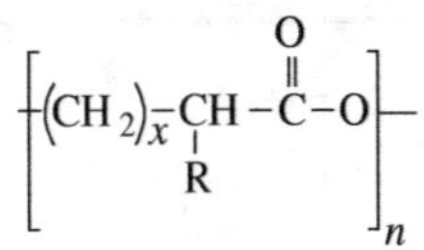	Hydroxyalkyl acids: lactic acid, glycolic acid, ε-hydroxy-caproic acid	28–32
Poly(orthoester)	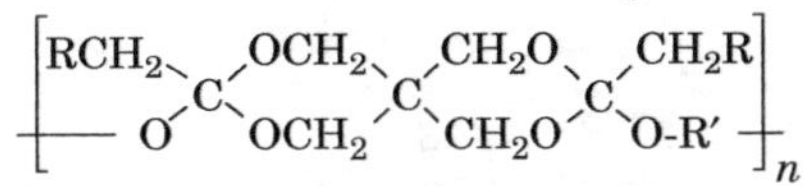 (a)	Pentaerythitol, propionic acid	21–24, 33, 34
	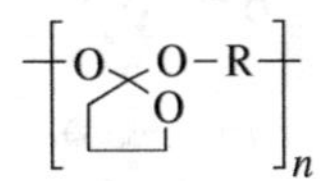 (b)	Diol, γ-hydroxybutyric acid	
Poly(caprolactone)	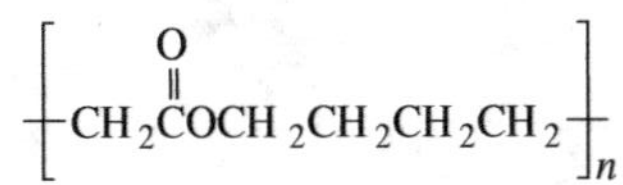	ε-Hydroxycaproic acid	35, 36
Poly(α-amino acids)	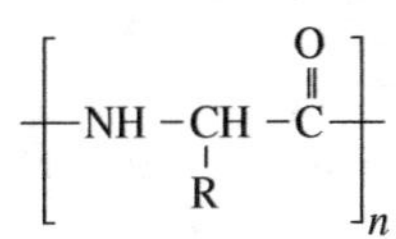	Amino acids	37–46
Pseudo-poly(amino acids)	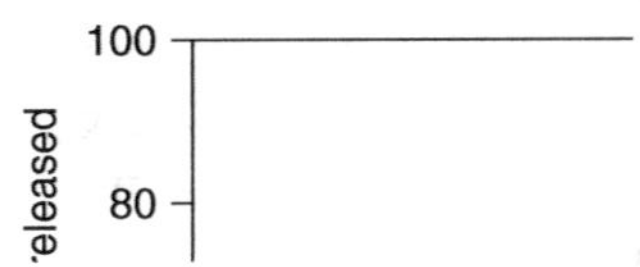	Amino acids (with trifunctional groups)	47–50

TABLE 9.1 Biodegradable Polymers and their Degradation Products (*Continued*)

Polymers	Structures	Degradation products	References
	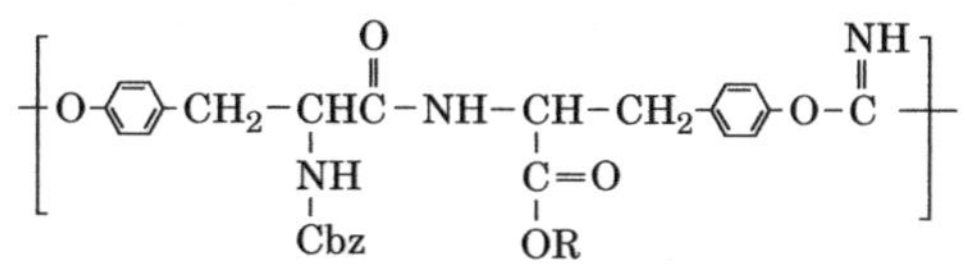	Polyiminocarbonate base on tyrosine	
Polydepsipeptides	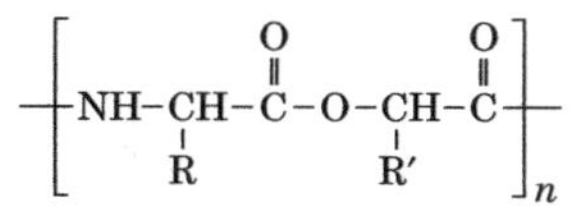	Amino acids and α-hydroxy carboxilic acids	51–54
Polyphosphazene	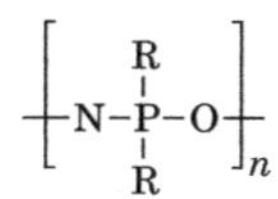	Ammonia, phosphate, water, and R	55–58
Polyphosphoester	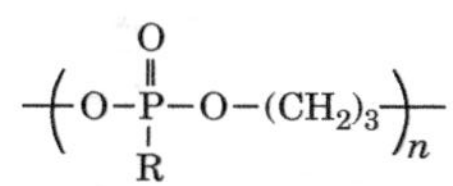	Phosphate, diol and R	59,60
Polyanhydrides	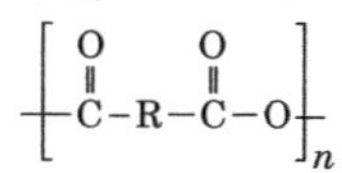	Diacids	50,59, 61–65
Polycyanoacrylates	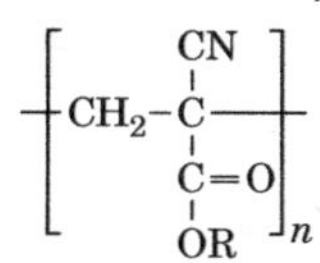	Formaldehyde, alkyl cyanoacetates	66, 67

(Note: $n = 1,2,3,\ldots$; $x = 1,2,3,\ldots$; R = alkyl)

delivery of anticancer drugs such as lomustine,[12] cisplatin,[13] and mitomycin C[14] by parenteral administration. Polylactic acid 60,000 microparticles completely released cisplatin in 144 hours at 27.2 percent drug loading,[12] and there was a near-complete release of mitomycin C over a 24-hour period from the microcapsules prepared from polylactic acid 33,000.[14]

A relatively new series of thermoplastic biodegradable hydrogels (TBHs) based on star-shaped poly(ether-ester) block copolymers that are easy to process with better biocompatibility for injectable drug delivery has been studied.[15] The thermoplastic properties and biodegradability of star-shaped poly(ethylene oxide)-poly(lactic acid) (PEO-PLA) and poly(ethylene oxide)–poly(ϵ-caprolactone) (PEO-PCL) block copolymers are based on their molecular architecture. A broad spectrum of performance characteristics can be obtained easily by manipulation of various monomers, number of arms, polymer composition, and polymer molecular weight. Their unique physical properties are due to the three-dimensional hyperbranched molecular architecture. These properties also may influence microsphere fabrication, drug release, and degradation profile. A biodegradable triblock copolymer consisting poly(ethylene glycol)-poly(DL-lactic acid coglycolic acid)–poly(ethylene glycol) (PEG-PLGA-PEG) forms a solution at room temperature and becomes a gel at body temperature within few seconds. The gelation mechanism appears to be micellar packing driven by hydrophobic interactions.[16]

9.3.2 Polylactones

The successful use of polymers of lactic acid and glycolic acid as biodegradable drug delivery systems and as biodegradable sutures led to an evaluation of other aliphatic polyesters and to the discovery of poly(ϵ-caprolactone). The homopolymer of ϵ-caprolactone degrades slower than polyglycolic acid and polyglycolic acid–co-lactic acid and hence is most suitable for long-term delivery systems. In addition, high permeability to many therapeutic agents and lack of toxicity have made poly(ϵ-caprolactone) and its derivatives well suited for controlled drug delivery. Another property of poly(ϵ-caprolactone) that has stimulated much interest is its exceptional compatibility with a variety of other polymers.

Poly(ϵ-caprolactone) is a semicrystalline polymer, melting in the range of 59 to 64°C, depending on the crystallinity. Polymerization of ϵ-caprolactone can be carried out by four different mechanisms categorized as anionic, cationic, coordination, and free-radical polymerization. Each method has unique attributes, providing different degrees of control of molecular weight and molecular weight distribution, end-group composition, chemical structure, and sequence distribution of copolymers. The biodegradation rate can be reduced by decreasing accessible ester bonds. Crystallinity is also known to play an important role in determining both permeability and biodegradability. That is, an increase in crystallinity reduces the permeability by both reducing the solute solubility and increasing the diffusional pathway. Model compounds such as chlorpromazine and L-methadone have been microencapsulated in poly(ϵ-caprolactone)–cellulose propionate blends by the emulsion solvent evaporation technique. Zero-order kinetics were achieved for 6 days for both these drugs. Polycaprolactone-bopolyethyleneoxide (PCL-b-PEO) block copolymers are used as polymeric micelles to improve the solubility of lipophilic therapeutic agents.[17]

9.3.3 Poly(amino acids)

Poly(amino acids) have been investigated extensively as biomaterials. The degradation products of poly(amino acids) are nontoxic for human beings because they are derived from simple nutrients. One of the major limitations for the medical use of synthetic poly(amino acids) is their pronounced antigenicity when they contain three or more different amino acids. For this reason, the search for biomaterials among the synthetic poly(amino acids) is confined to polymers derived from one or two different amino acids. Another limitation is related to the fact that synthetic poly(amino acids) have rather unfavorable material properties. For example, most synthetic poly(amino acids) derived from a single amino acid are insoluble, high-melting materials that cannot be processed into shaped objects by conventional fabrication techniques. Finally, the costs to make high-molecular-weight poly(amino acids) are high, even if they are derived from inexpensive amino acids.

9.3.4 Polyphosphazenes

Polyphosphazenes are a relatively new class of biodegradable polymers. Their hydrolytic stability or instability is determined not by changes in the backbone structure but by changes in the side groups attached to an unconventional macromolecular backbone. Synthetic flexibility and versatile adaptability of polyphosphazenes make them unique for drug delivery applications. For example, Veronese et al.[18] prepared polyphosphazene microspheres with phenylalanine ethyl ester as a phosphorous substituent and loaded it with succinylsulphathiazole or naproxen. The kinetics of release from these matrices were very convenient in yielding local concentrations of the two drugs that are useful per se or when mixed with hydroxyapatite for better bone formation. Polyphosphazene matrices are also considered as potential vehicles for the delivery of proteins and vaccines.[19]

9.3.5 Poly(orthoesters)

Poly(orthoesters) (POEs) are prepared by transesterification using diethyl orthoester and a diol. Since 1980, both linear and cross-linked POEs have been investigated successfully as biodegradable and biocompatible carriers for the delivery of many drugs, such as 5-fluorouracil,[20] levonorgestrel,[21,22] norethindrone,[23] cyclobenzaprine hydrochloride,[24] and insulin.[25] Among four different POE families (I, II, III, and IV), POE IV, which contains lactic acid units in the polymer backbone, is promising for drug delivery applications because of its mechanical and thermal properties. It is available as solid as well as semisolid forms. The solids are useful for preparing particulate carriers, and the semisolids are useful for preparing injectable formulations. Drug incorporation into semisolid POE is especially attractive because it can be prepared by simple mixing and does not require any organic solvents or heating. This feature makes it more suitable for formulating delivery systems for protein peptides and other thermolabile drugs. Drug release from a POE matrix is predominantly controlled by surface erosion. Since erosion is readily controlled by modulation of its molecular weight and structure, drug release rates from POE matrices can be tailored according to need, from few days to months. They have been employed in ocular drug delivery and treatment of veterinary and periodontal diseases. In recent years, the targeting potential of block copolymers of POE and poly(ethylene glycol) have been investigated. POEs are stable under anhydrous conditions at room temperature and are suitable for radiation sterilization.[26]

9.3.6 Polyanhydrides

Polyanhydrides are polycondesates of diacid monomers, and the resulting anhydride polymers have the ability to degrade into biocompatible by-products. They are surface-eroding polymers with a variety of applications. Hydrolysis of the highly labile anhydride bonds results in the release of drugs from delivery devices prepared from polyanhydrides. The homopolyanhydrides display a zero-order hydrolytic degradation profile and drug release profile. The second type of polymer, unsaturated polyanhydrides with the structure $[—(OOC—CH{=}CH—CO)_x—(OOC—R—CO)_y—]_n$ have the advantage of being amenable to cross-linking. This is important for enhancing the physical strength. The erosion behavior and drug release from polyanhydride matrices can be manipulated by changes in the hydrophobicity of polymer, changes in the physical properties of the final matrix such as method of fabrication, geometry, and addition of hydrophobic and hydrophilic components. The surface-eroding property of polyanhydrides helps in preventing protein and peptides from being exposed to physiological conditions owing to the restricted entry of fluids into the core. In addition, it provides a better microclimate for proteins because the degradation products do not significantly change the pH. Degradable poly(anhydride ester) implants in which the polymer backbone breaks down into salicylic acid also were investigated.[27]

9.4 Design Principles [Diffusion versus Erosion (Surface versus Bulk)]

Temporal drug release, delivering drug over extended time and/or at a specific time, is advantageous for many clinical conditions and for many classes of drugs such as chemotherapeutics, anti-inflammatory agents, antibiotics, opioid antagonists, etc. This would avoid peak and trough plasma drug levels and maintain constant levels of drug in the therapeutically effective range. In order for drugs to exert their therapeutic action, they need to be made available in the bloodstream to reach the target site unless deposited at the site itself. However, when administered as a part of polymeric delivery system, the drug is inhibited from being readily available for dissolution because it is surrounded by protective polymer. Knowledge of the control mechanisms of drug release from polymeric systems therefore is essential to design a successful delivery system.

In general, drug release from biodegradable polymeric devices is controlled by diffusion of drug and/or polymer erosion. In practice, both these release mechanisms play a role in controlling the release rate; one dominates the other depending on the drug, morphology of the carrier, and other physicochemical characteristics. Release of small drug molecules from polymeric systems is mainly attributable to diffusion. Diffusion of drug, in general, closely follows the Fickian diffusion equation with appropriate boundary conditions. In contrast, release of macromolecules such as proteins and peptides from polymer systems is more complex because it depends largely on polymer degradation. In this section the design of biodegradable systems based on the drug release mechanisms diffusion and erosion (bulk and surface) is discussed.

9.4.1 Diffusion Controlled Systems

In diffusion controlled systems, the carrier usually retains its structural integrity even after the drug is depleted. Polymer degradation may take place throughout the drug release process, during only a part of the drug release time, or only after delivery system is exhausted. The rate of diffusion in such systems is controlled by the following factors:

1. Solubility of the drug in surrounding medium, including aqueous and polymer solubility
2. Concentration gradient across the delivery system
3. Drug loading
4. Morphological characteristics such as porosity, tortuosity, surface area, and shape of the system
5. Hydrophilicity/hydrophobicity of the system
6. Chemical interaction between drug and polymer
7. Polymer characteristics such as glass transition temperature and molecular weight
8. External stimulus such as pH, ionic strength, and thermal and enzymatic action

Most of these factors are self-explanatory and are covered extensively in Chap. 4. The effect of particle size, drug loading, porosity, molecular weight, of the polymer, and ionic interactions on the release of drugs from the biodegradable polymers are discussed in the following sections.

Size. Size is one of the important physical parameters that may be altered in order to acquire a desired release rate. Rate of diffusion from a biodegradable system with large physical dimension may be slow; however, it can be very high from colloidal or very small particulate carries that have enormous surface area because they have shorter distances to diffuse. Microspheres of etoposide prepared by oil/oil suspension and solvent evaporation technique using polylactide (PLA) of molecular weight 50,000 Da were divided into size ranges of less than 75 μm, 75 to 180 μm, and 180 to 425 μm by passing through series of standard sieves, and their drug release was evaluated. Particles that are less than 75 μ showed faster release rates compared with larger size fractions, as shown in Fig. 9.1. The difference in the rate of release is attributed to the difference in the surface area. Alterations in drug release rates therefore could be attained by simple mixing of different size fractions of microspheres.

Drug loading. One could design a biodegradable delivery system where drug release rate can be controlled by the initial drug loading. The rate of diffusion will be higher for drugs with higher aqueous and polymer solubility, as well as for those not chemically interacting with the polymer. Higher drug loading will mean higher amounts of drug present on the surface or proximal to the surface that will lead to higher initial release. In addition, the rate of pore formation can be higher on drug depletion because the drug-polymer ratio is higher.

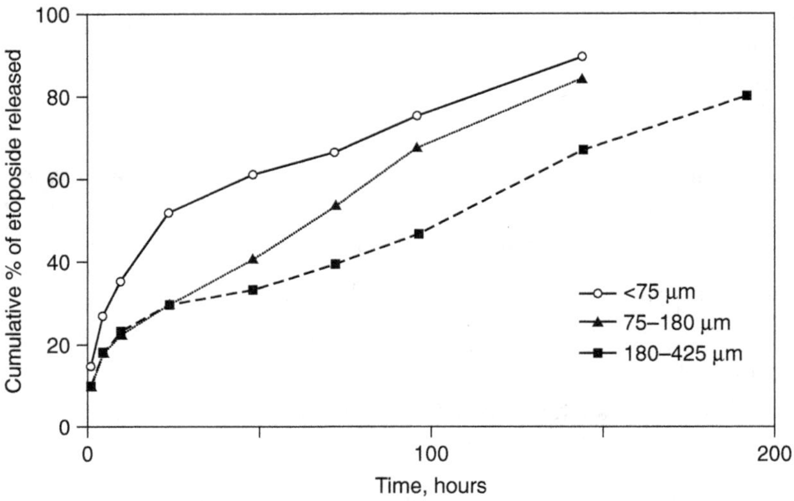

Figure 9.1 Effect of size on the in vitro drug release of etoposide from PLA microparticles.

An example illustrating the effect of leuprolide acetate loading on the physicochemical properties and in vitro drug release of PLGA microspheres is shown in Table 9.2. Formulations *A* and *B* with 11.9 and 16.3 percent of drug loads were prepared by a solvent-extraction-evaporation method. Higher drug incorporation resulted in a substantial increase in specific surface area and a decrease in bulk density. When observed under the scanning electron microscope, higher-drug-loaded microspheres showed a higher surface porosity. This resulted in higher initial release from microspheres with higher drug incorporation. As shown

TABLE 9.2 Physicochemical Characteristics of PLGA Microspheres Containing Leuprolide Acetate

Formulation code	Drug loading, %w/w	Encapsulation efficiency, %	Surface area m^2/g	Size μm	Bulk density, g/cc
A	11.9	95.2	0.387	18.0	0.54
B	16.3	81.5	7.278	27.0	0.29
C	9.79	78.3	1.249	24.7	0.48
D	9.08	72.6	1.480	24.2	0.42
E	10.1	80.8	1.023	20.5	0.57
F	10.9	87.2	0.913	28.5	0.56
G	11.0	88.0	0.778	24.9	0.64

NOTE: Formulations *C* through *G* contain different concentrations of calcium chloride; their target drug load is 12.5 percent.

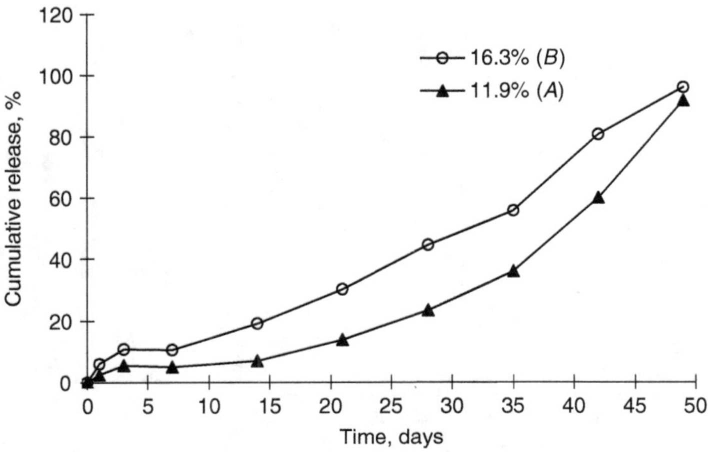

Figure 9.2 Effect of drug load on the in vitro release of leuprolide acetate from PLGA microspheres.

in Fig. 9.2, following the initial phase, similar drug release from both formulations was observed.

Porosity. Maneuvering the porosity is another way of designing biodegradable delivery systems. As shown in Table 9.3, various concentrations of water-soluble calcium chloride were added as a porosigen into the aqueous and organic (methanol) phases during the preparation of microspheres.[68] The aqueous-to-organic phase composition of these formulations (*C* through *G*) is given in Table 9.3, and the release profiles are shown in Fig. 9.3. When the porosigen was added to the organic phase, very porous microspheres with poor drug encapsulation were observed (see Table 9.2). Concentrations of calcium chloride in organic

TABLE 9.3 Manufacturing Parameters of Leuprolide Acetate–Loaded PIGA Microspheres

Formulation code	Molar concentration of calcium chloride		
	Organic phase	Aqueous phase	Ratio (organic:aqueous)
C	0.268	0.100	2.68
D	0.268	0.060	4.46
E	0.155	0.060	2.58
F	0.117	0.060	1.95
G	0.117	0.045	2.60

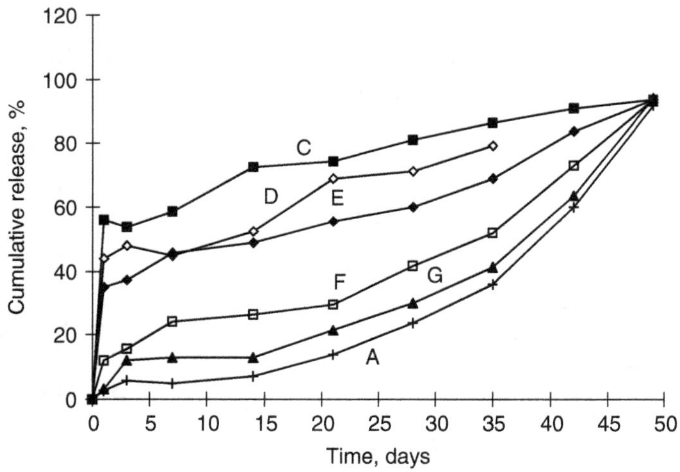

Figure 9.3 In vitro release of leuprolide acetate from PLGA microspheres containing different loading and porosigen concentrations.

and aqueous phases were adjusted to lower the osmotic gradient favoring the organic phase, and it improved the entrapment efficiency. The porosigen containing microspheres showed higher surface area (versus formulation *A*) with higher surface porosity and slightly lower drug entrapment. All formulations appeared to have similar release profiles except for the initial release. This would mean that only diffusional release is affected, whereas the erosional process is not affected. Faster onset of action compared with control microsphere formulation was observed when microsphere formulations containing higher leuprolide loading and calcium chloride were evaluated in vivo. In this case, serum testosterone levels as a biomarker were measured, resulting in lowering of the testosterone levels to chemical castration levels as desired.

Molecular weight of polymer. Molecular weight of polymer offers an attractive opportunity to design biodegradable delivery systems with tailored drug release rates. This is evident from leuprolide acetate–loaded microspheres, shown in Fig. 9.4, where the microspheres were prepared using various combinations of two different molecular weights of PLGA. PLGA polymers (50:50) with molecular weights of 28.3 and 8.6 kDa showed different porosity and associated specific surface areas. Particles prepared from lower-molecular-weight polymer were very porous and of lower bulk density and higher specific surface area, even though their mean particle sizes were comparable. As expected, in vitro release of leuprolide was rapid, with approximately 60 percent release within the first 24 hours. Microspheres from the 28.3-kDa polymer were nonporous, and only about 3 percent of the entrapped drug was released within the same time frame.

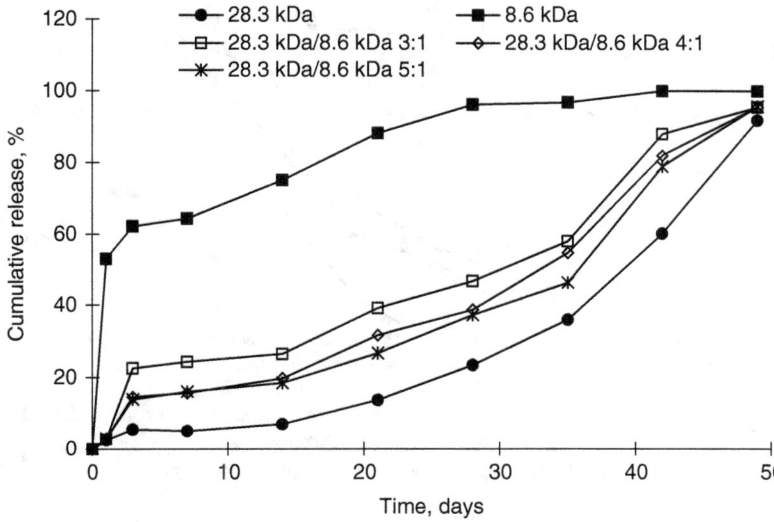

Figure 9.4 In vitro release of leuprolide acetate from microspheres made from various PLGA blends.

Two different approaches were evaluated for obtaining quicker onset of therapeutic action (in this case, lowering testosterone levels), as well as avoiding possible acute toxicity (which is not a concern in the case of leuprolide). In the first approach, polymers were premixed in various ratios (3:1, 4:1, and 5:1) of the 28.3-kDa and 8.6-kDa polymers, and microspheres were made. In vitro drug release from these microspheres is shown in Fig. 9.4. In the second approach, microspheres were prepared from these two molecular weight PLGAs separately and were physically mixed in a 3:1 ratio. Their in vitro drug release was compared with that of microspheres made from a 3:1 polymer mixture and theoretical release (Fig. 9.5). As compared with the 28.3-kDa microspheres, microspheres prepared with the 8.6-kDa polymers, obtained with both the preceding approaches, yielded a faster onset of therapeutic action, validating the feasibility of these approaches in designing delivery systems with required release characteristics.

Ionic interactions. Biodegradable polymers such as PLGA or PLA contain terminal carboxylic groups, which may interact with drugs and alter their degradation rate and hence release kinetics. Neutralization reactions between these terminal carboxyl groups and basic drugs may minimize the autocatalytic effect of the acidic chain and reduce the polymer degradation rate. In contrast, these drugs may act as base catalysts and enhance polymer degradation by cleaving the ester bonds.

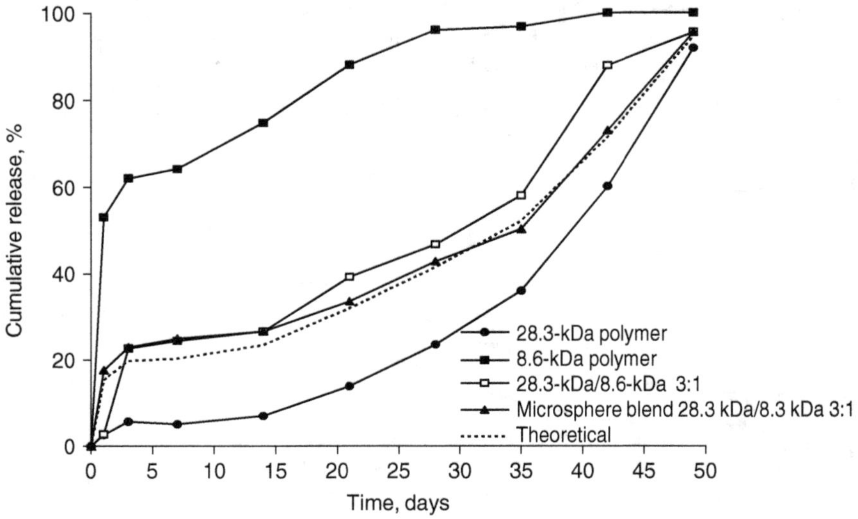

Figure 9.5 In vitro release of leuprolide acetate from blended microspheres and microspheres made from PLGA blends.

Such ionic interactions therefore should be paid careful attention during the design of delivery systems. An example that investigates the relationship of the ionic property of drugs to their release profiles from the PLGA matrix was reported by Makoto et al.[69] Cylindrical PLGA matrices containing basic, acidic, and neutral drugs were prepared by the heat-compression method using a low-molecular-weight PLGA with relatively rapid degradation, and their release was evaluated in phosphate-buffered saline at pH 7.3. Both erosion and diffusion controlled release of basic drugs were suppressed because of their strong interaction with the polymer. They shielded the polymer carboxyl residues and formed a more rigid and less hydrophilic matrix. In the case of acidic and neutral drugs, their weaker ionic interaction with the terminal carboxylic group resulted in precipitation of drugs as crystals in the matrix within a day after immersion in the phosphate-buffered solution (PBS). This has transformed the rods into drug dispersed matrix, and hence matrix erosion was not affected by the drug. It was inferred that the solubility in the hydrated matrix is the primary rate-limiting factor for drugs that show weaker ionic interaction with the polymer matrix.

9.4.2 Erosion and degradation controlled systems

Erosion refers to the dissolution and/or degradation of the polymer to soluble fragments and the progressive weight loss of the matrix. A thorough understanding of the erosion mechanism of particulate carriers is essential for modulating their release characteristics. A number of parameters may be altered in controlling the erosional release of drug. Two terms, *biodegradation* and *erosion*, are commonly associated with this kind of release. Erosion means loss of mass or physical depletion of material and subsequent loss of carrier structure. Biodegradation, however, refers to the bond cleavage and shortening of the polymer chain length that is caused by hydrolytic or enzymatic reaction or both. Thus, for most of biodegradable systems, biodegradation of polymers precedes the erosion of the carrier system.

Erosion. Erosion can be classified further as *bulk* or *surface* erosion based on the characteristics of polymer. For example, polyesters undergo bulk erosion. On water uptake, the random scission of polymer chains yields lower-molecular-weight oligomers and monomers. As they reach critical molecular weights, they become water soluble and erode the matrix. In general, water intake is faster than the polymer chain scission. pH also plays an important role in this bulk erosion process. The degradation products of polyesters are acidic in nature, lowering the microenvironmental pH of the device or carrier. This lowered pH can cause autocatalytic polymer degradation that is faster in bulk than at the surface.

Polyanhydrides (PA) and polyorthoesters (POE) undergo predominantly surface erosion. In general, polymer degradation is much faster than water intake, and it causes the outermost layers of the device or carrier to start eroding first. This could mean that the drug release is more uniform and that the drug is protected from the aqueous environment until it is released. POEs are pH sensitive, and incorporation of salts into the polymer matrix can manipulate the rate of surface erosion with a random hydrolysis mechanism. Drug release is facilitated by the diffusion of water into the polymer. On dissolution of the incorporated salt, the ideal pH is provided for matrix erosion and dissolution of the incorporated drug from the polymer.

Polymer erosion and release of 5-fluorouracil were found to depend on the nature and concentration of the incorporated acidic excipient used, namely, itaconic acid, adipic acid, or suberic acid.[20] Researchers also prepared POEs that do not afford acidic products and thus undergo autocatalytic biodegradation. Inorganic salts, such as sodium sulfate, can be added to these matrices to promote both matrix erosion and drug release. Restriction of water permeation during polymer swelling also facilitates the diffusional release of drug from the matrix at the swelling point.

Degradation. Biodegradable polymers generally undergo three types of degradation. *Type I degradation* refers to polymer where degradation occurs on the main chain of the polymer. The cleavage of the linkages between the monomers results in the disassembled polymer. Most of the linear biodegradable polymers belong to this category. If the water-soluble polymer is made insoluble initially by cross-linking with a hydrolyzable covalent bond, degradation of the polymer can proceed by the cleavage of these bonds. Such degradataion is referred to as *type II degradation*. *Type III degradation* involves a degradation of the polymer side chain. A hydrophobic or water-insoluble polymer can become water soluble by hydrolysis, ionization, or protonation of the side chains. This type of degradation is observed with partially esterified maleic anhydride copolymers.[70,71] The polymer becomes soluble in water as the side chain carboxylic groups are ionized. Besides these three basic types of degradation, combinations of any two types may give a hybrid mode of degradation. In general, degradation of polymers is influenced by various factors, which include

1. Chemical structure and composition
2. Molecular weight
3. Polymer concentration
4. Hydrophilicity/hydrophobicity
5. Carrier morphological properties such as size, shape, and porosity
6. Additives in the system (acidic, basic, monomers, drugs)
7. Microenvironmental climate such as pH
8. Method of preparation
9. Sterilization

Examples of designing controlled release delivery systems by varying some of these factors are illustrated briefly in the following section.

Molecular weight and copolymerization. To date, the largest body of literature exists on polyesters such as poly(DL-lactide) or poly(DL-lactide-coglycolide). These polyesters are available from a variety of vendors on a commercial scale with varied molecular weights and monomer ratios of lactide and glycolide. They are also available with acid end groups to impart higher hydrophilicity. Addition of low-molecular-weight poly(DL-lactide) (MW 2000 Da) increases drug release from a biodegradable poly(DL-lactide) (MW 120,000 Da) drug delivery system. Bodomeier et al.[72] found that the duration of action could be varied over a range of several hours to months by varying the amount of low-molecular-weight poly(DL-lactic acid). Degradation of these polymers occurs by hydrolysis of ester linkages causing random scission and mainly depends on the polymer concentration, ratio of comonomers, and hydrophilicity.

Degradation leads to the formation of lactic acid and glycolic acid, which are normal intermediates in carbohydrate metabolism.

Pitt et al.[73] showed that DL-polylactic acid undergoes first-order degradation with respect to initial molecular weight, according to the equation

$$M_n^t = M_n^0 e^{-kt}$$

where M_n^t = molecular weight at time t
M_n^0 = initial molecular weight of the polymer
k = degradation rate constant

DL-Polylactic acid, for the most part, was found to erode in about 12 months. Slow degradation of DL-polylactic acid often becomes a limitation on its use. This rate can be accelerated appreciably by copolymerizing with up to 50 mol% glycolide to yield complete erosion in as fast as 2 to 3 weeks. Incorporation of glycolide into the polylactide chain alters crystallinity, solubility, biodegradation rate, and water uptake of the polymer.

The effect of glycolic acid units in polylactic–coglycolic acid copolymer microparticles on etoposide release is shown in Fig. 9.6. Drug release from lactide microparticles was slower compared with microparticles that contain a combination of lactide and glycolide. A proportional increase in drug release was observed with systems that contain higher levels of glycolide because they have faster degradation rates.

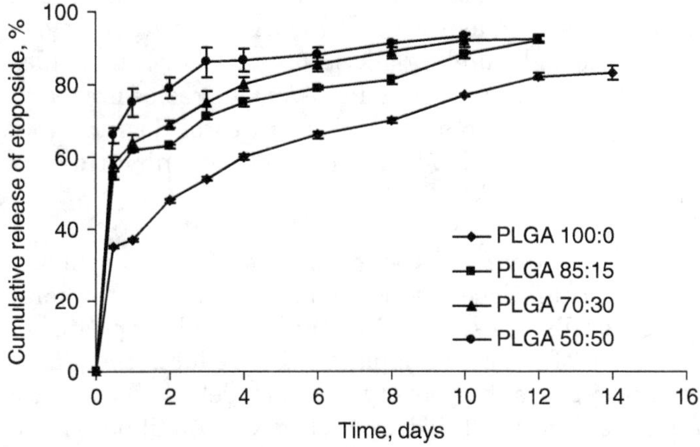

Figure 9.6 Effect of lactide-to-glycolide ratio on the in vitro release of etoposide from PLGA copolymer microparticles.

Molecular weight of drug. In case of macromolecular drugs, a major portion of drug is released by polymer degradation and erosion, and a small portion is released by the diffusion mechanism. Polypeptides usually have limited solubility in the polymer, which greatly prevents their diffusion. In addition, the aqueous channels present in the delivery system could be too narrow or tortuous for these macromolecules. Reports from the literature indicate that the drug release is multi-or triphasic, which is characterized by higher initial release (can be termed *burst* in some cases), a lag phase where minimal amount of drug is released, and finally, release of drug at a higher rate until depletion. Physiologically, this may mean therapeutic activity immediately after dosing (or even acute toxicity depending on the drug), no therapeutic activity corresponding to the varied length of the lag phase, and finally, sustained activity. The initial release is attributed to the release of drug that is adsorbed onto the surface or present close to the surface in the pores by diffusion. If the carrier system contains a nonporous structure, the initial release of drug may be reduced. Then the degradation of polymer starts slowly, but without losing its mass or structure, during which the drug is immobile. As the degradation of polymer reaches a critical level, it triggers erosion of the carrier structure and leads to continuous drug release. It is desirable that the lag phase of drug release be eliminated or minimized.

Crystallinity. Many classes of polymers are typically hydrophobic and crystalline in nature and need certain modifications to have acceptable biodegradation and drug release. As an example, polycaprolactone is highly crystalline and hydrophobic polymer, and it can take years for complete degradation in the biological environment. Similar to polylactides, the biodegradation rates of polycaprolactones can be hastened by blending with polymers such as PLA and PLGA. The rate of degradation also can be increased by adding alkylamines.[74] Polyanhydrides are very hydrophobic in nature. Sebacic acid in various ratios is copolymerized with polyanhydrides to obtain various rates of degradation. Copolymerization can hasten the biodegradation of polyanhydrides as much as from 8 months to 2 weeks.[75]

Additives. The flexibility of polymeric systems can be improved by modifying their physical properties. Copolymerization techniques or incorporation of plasticizers generally is used to alter the physical properties of polymeric films. Addition of appropriate plasticizers reduces the glassy nature of the polymers, reduces the porosity, and may deform the surface owing to dehydration, resulting in altered drug release.[76,77] Addition of more hydrophilic polymers with acid end groups also can accelerate the degradation because these polymers would allow faster wetting and hydrolysis.

Poly(orthoester) films containing 0.5, 1.0, 2.0, and 4.0 percent of oleic acid/palmitic acid were prepared in one of the authors' laboratories to study the effect of acid-catalyzed degradation on drug release and polymer degradation. Films of three different thicknesses were prepared to study the effect of thickness on drug release and polymer degradation. The influence of palmitic acid at 10 percent metronidazole loading and 125-μm film thickness is shown in Fig. 9.7. It was observed that with palmitic acid, the rate of device disappearance lagged behind the drug release considerably at 10 percent drug concentration. The influence of oleic acid at 10% drug loading and 440- to 450-μm film thickness is shown in Fig. 9.8. It was observed that with the increase in oleic acid concentration, the rate of drug release increased, and the time for complete disappearance of the device decreased. In absence of oleic acid, the release was slow and took 24 days to complete. The release profiles in the absence of oleic acid showed distinct phases: the rapid phase with burst release owing to dissolution of drug on the surface or those slightly below the surface (up to 7 days), the zero-order release phase where polymer degradation is established and the drug diffuses out of the polymer (between 8 and 20 days), followed by depletion phase when the drug in the device is depleted (20 to 25 days).

Method of preparation and solvent effects. The method of preparation of the delivery system and the solvents employed also may influence drug release. Degradation of the polymer matrix, as well as stability of the drug, should be considered for selecting an ideal manufacturing process. Vapreotide, a somatostatin analogue, incorporated PLGA implants have been formulated by two different manufacturing techniques, extrusion

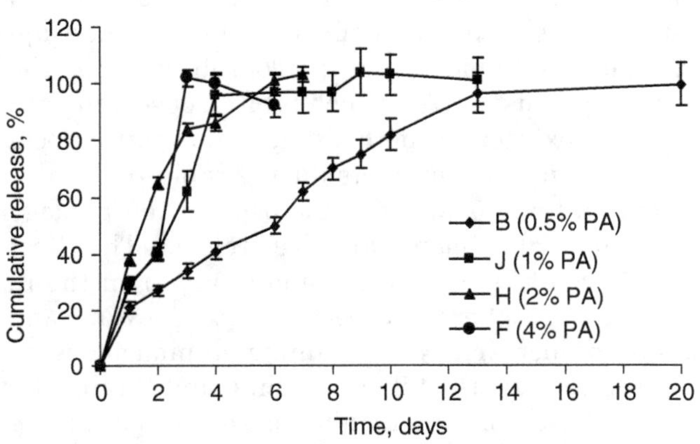

Figure 9.7 Effect of palmitic acid (PA) on metronidazole release at 10 percent loading and 125 μm film thickness.

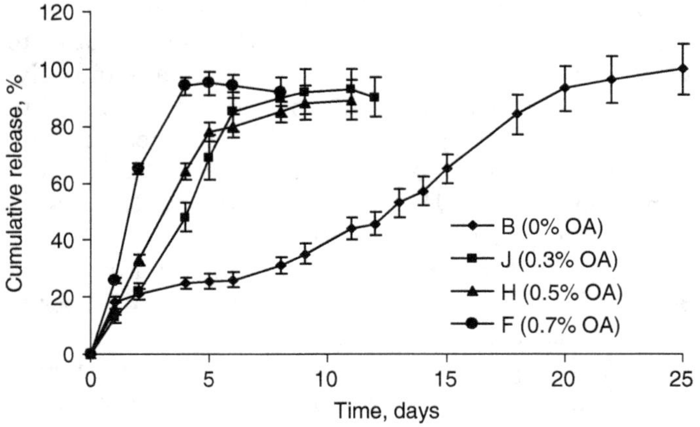

Figure 9.8 Effect of oleic acid (OA) on metronidazole release at 10 percent loading and 400- to 450-μm film thickness.

and injection molding, and the influence of processing methods on the in vitro degradation of the polymeric matrix was studied.[78] Both methods decreased the molecular weight and polydispersity of the polymer; however, there was no change in the crystalline network of polymer. The extruded implants degraded more rapidly in vitro than the injection-molded ones and showed higher release rates from 2 to 6 days. Significant change in matrix porosity and breakdown of the matrix and an increase in the surface area were reasoned for the faster matrix degradation and drug release. Although injection-molded implants showed slow release, the degradation of drug was higher owing to higher temperature and greater shearing forces employed during the manufacturing process.

Physicochemical properties of solvents such as their boiling point, volatility, and miscibility with other solvents also should be considered when micro- or nanospheres are prepared by spray drying, solvent evaporation or emulsion techniques. Rapid evaporation of solvents from the dispersion usually forms a porous matrix, which is often observed with microparticles prepared by extraction methods. It is due to the local explosion that takes place inside the droplets during rapid removal.[79] All these solvent effects that tend to alter the microsphere morphology will change the drug entrapment efficiency and drug release. The effect of two nonhalogenated solvents (ethyl acetate and ethanol) on the in vitro drug release of PLGA microspheres has been reported by Wang and Wang (2002).[80] Spray-dried microspheres containing etanidazole were prepared using ethyl acetate and ethanol as solvents, and the results were compared with those of microspheres prepared using dichloromethane (DCM). Use of a large proportion of ethyl acetate along with DCM decreased the initial burst and sustained the release of drug.

This occurred because of the rapid phase transition and slower evaporation and formation of nonporous microspheres in presence of ethylacetate. DCM produced porous microspheres and hence showed initial burst release. Use of ethanol along with DCM further increased the initial burst because of the structure of microspheres and inhomogeneous drug distribution. This feature may be very useful during the design of sustained release particulate carriers for highly water-soluble drugs.

Sterilization. Radiation sterilization is an ideal technique to sterilize biodegradable particulate carriers or implants. PLGA has been used extensively in the preparation of biodegradable drug delivery systems. However, it undergoes degradation when the system is sterilized by radiation.[81–83] Faisant et al.[84] have investigated the effect of different gamma-radiation doses ranging from 4 to 33 kGy on the in vitro release of 5-fluorouracil from PLGA microspheres. A dose-dependent acceleration in the initial release was observed in microspheres undergoing gamma irradiation.

9.5 Delivery Devices

Delivery devices made of biodegradable polymers may be implanted in a specific biological location or delivered to the systemic circulation for sustained drug release or to reach to a specific target. Particulate carriers such as micro- and nanoparticles are suitable for both implantation and circulation, whereas cylindrical devices are meant only for implantation. Even though intravenous injections of particulate carriers are feasible with good control of the particle size, in many instances the development is intended for subcutaneous (SC) or intramuscular (IM) applications. On SC or IM injection, these formulations become depot systems for the entrapped or associated drug, which is released at the desired rate based on the optimized system parameters. The release can be varied or controlled according to the polymeric material employed or the design of the delivery system. Several polymer-based drug delivery systems are now in commercial use.

9.5.1 Microparticles

Microparticles or microspheres, as they are interchangeably called, are fine spheres usually less than 1000 μm in diameter.[85] Microparticles can be prepared by well-established manufacturing processes. The drug can be distributed homogeneously throughout the polymer matrix (microparticles), or it can be encapsulated into a polymer surrounding to form a drug reservoir (microcapsules). It is also possible to adsorb drug onto the particle surface by ionic or chemical interactions depending on the application. In practical terms, all these morphological attributes are present in all microparticle delivery systems. A brief description of some of the widely used methods follows.

Coacervation phase separation. This technique is used to microencapsulate water-soluble drugs. The core material (drug) is suspended in a nonaqueous polymer solution (coating material), and the polymer is made to form a uniform coat by various approaches, such as temperature change, addition of an incompatible polymer, addition of a nonsolvent, or addition of a salt.

Emulsion phase separation. Water-soluble drugs are fabricated in the form of microcapsules by this method. An aqueous phase containing dissolved drug and an organic phase containing polymer are emulsified. Then polymer is phase separated using the techniques such as temperature change, addition of salts, etc. A nonsolvent then is used to harden the microspheres.

Solvent evaporation. This is the most commonly used method for microencapsulation of the drugs that are soluble or suspended in the organic phase. In this method, a solution or suspension of drug in an organic solvent containing dissolved polymer is emulsified to form o/o or o/w dispersion, possibly with the aid of a surfactant. The organic phase is then evaporated by heating or applying vacuum, leaving microspheres.[86–88]

Spray drying. Microencapsulation by spray drying is an ideal method for poorly water-soluble drugs. The drug is dispersed in polymer (coating) solution, and then this dispersion is atomized into an airstream. The air, usually heated, supplies the latent heat of vaporization required to remove the solvent and forms the microencapsulated product. This technique is employed most commonly when microcapsules are intended for oral use because the resulting microspheres are porous in nature, and large batch sizes are required.[89]

Biodegradable microspheres can be used for targeted delivery of drugs in addition to sustained release applications. When given intravenously, the drug-loaded particles can cause arterial chemoembolization and deliver drugs at the desired site. A number of clinical requirements determine the parameters of the developed microsphere system. The total required dose, practical injection volume, and desired duration of action would determine the extent of drug loading. For administration of suspensions by the parenteral route, the preferred needle size is 18 gauge, which has an internal diameter of 838 μm.[90] One therefore would expect a great deal of latitude in the size range of microspheres for parenteral administration. Certain limitations do exist in practice use. Above 100 μm, interparticulate interactions seem to clog the tip of the needle. Use of more dilute solutions may ameliorate this problem. On the other hand, very small microspheres present their own problems. For example, high-surface-area particles may be difficult to wet with the suspending vehicle and may require high concentrations of surfactant in the vehicle, which is unacceptable for other reasons. Additionally, even if a good suspension is readily formed, non-Newtonian rheologic properties such as dilatancy may ensue. Other factors that restrict the use of extremely small microspheres include poor solid-state flow and hence poor filling into vials and poor handling properties owing to rapid generation of electrostatic charges. Physiological parameters also must be considered for the parenteral administration of microspheres because particle size plays an important role in deciding their biofate. It was shown that microspheres smaller than 7 to 8 μm pass through the lungs and concentrate in the reticuloendothelial system (RES), whereas particles larger than 7 to 8 μm usually become entrapped in the lungs.[91] Microspheres of 15 to 25 μm were used for lung targeting without endangering the function of the lung.[92]

Leupron Depot, an injectable microsphere depot formulation of leuprolide acetate (a leutinizing hormone–releasing hormone analogue), developed by TAP Pharmaceutical Products, Inc., is one of the successful examples. In addition to being a commercial success, this product reduced the dose to one-fourth to one-eighth compared with daily SC injections of the analogue solution and increased patient compliance and convenience owing to reduced frequency of dosing.[93] This PLGA-based microsphere product delivers leuprolide acetate over 1 to 4 months with one single injection. Alkermes, a company formed to commercialize sustained release microsphere systems, has developed PLGA-based Nutropin Depot for recombinant human growth hormone somatotropin (one or two injections a month) and Risperdal Consta for an antipsychotic drug risperidone (one injection every 2 weeks). Alkermes is also developing a long-acting naltrexone formulation (Vivitrex) for treating alcohol dependence. All these products are injected subcutaneously or intramuscularly and form depot systems to yield sustained release of drugs.

There are a number of reports in the literature supporting the use of polyester microspheres for intravenous route. Similarly, combinations of polyamino acid and sodium alginate for the delivery of pancreatic cells, release of FD&C Blue No.1 for 3 to 7 days from polyglycolic acid, and PLGA polymer microspheres for intravenous administration were reported.[94] Several lipophilic drugs such as norethinsterone and thioridazone have been delivered using polylactic and polylactic–coglycolic acid microspheres. A large number of reports on the delivery of antineoplastic drugs in the form of microspheres using biodegradable matrices have been published. Injectable carboplatin or BCNU polymeric microspheres that release drug for 2 to 3 weeks for enhancing the survival in a rodent model of surgically resurrected glioma were documented.[95] Use of biodegradable polymeric carrier systems for the local delivery of antibiotics for prolonged duration in bone infections has been reviewed.[96] Further development of such biodegradable systems will provide a novel approach in future for the eradication of chronic osteomyelitis.

9.5.2 Nanoparticles

Nanoparticles are morphologically similar to microspheres, but the sizes of particles are in the submicron range. These particles can be prepared by different methods and using a variety of starting materials that include biopolymers (e.g., gelatin, albumin, casein, and polysaccharides) and synthetic polymers (e.g., polyesters, polyanhydrides, polycaprolactone, and alkyl-2-cyanoacrylates). The choice of polymer depends on the therapeutic application of the system, biocompatibility, desired drug release profile, etc. Drug release ranging from few hours to several months has been obtained from nanoparticle formulations.[97]

Similar to the earlier discussion on the effect of microparticle size on their biodistribution, nanoparticles, when given intravenously, are taken up by the reticuloendothelial system (RES). Unless the RES is the desired target site, this can mean a higher rate of plasma clearance for the drug. When given intraperitoneally, nanoparticles are preferentially taken up by the lymphatic system and can be useful in certain pathological conditions that affect the immune system such as AIDS. PEGylation of nanoparticles, as shown with "stealth" liposomes, can circumvent the rapid uptake by the RES, providing longer plasma half-lives.

Based on the available literature, nanoparticles appear to be more suitable for targeted release applications than for sustained release. Biodegradable polycaprolactone nanoparticles have been shown to increase the oral bioavailability and control the biodistribution of cyclosporine, thereby potentially reducing the drug toxicity.[98] Similarly, nanoparticle-bound antitumor agents have shown to prolong drug retention in tumors, reduce tumor growth, and prolong survival of tumor-bearing animals compared with free drug treatment. Oligonucleotides adsorbed onto polyalkylcyanoacrylate nanoparticles showed enhanced stability against nucleases and good cellular disposition.[99] The in vitro release studies of ganciclovir-loaded albumin nanoparticles have shown a burst release of the drug in 1 hour followed by sustained release for 5 days and constant release for 30 days.[100] Incorporation of several inhibitors of sterol biosynthesis into long-circulating polyethylene glycol–polylactide (PEG-PLA) nanospheres in order to improve the bioavailability of these poorly soluble compounds also has been reported.[101] Mice infected with CL and Y strains of *Trypanosoma cruzi* and treated for 30 consecutive days with D0870-loaded nanospheres at doses 3 mg/kg/per day by the intravenous route showed higher cure rate (60 to 90 percent) for both strains.

Nanoparticles with drug dispersed into the polymer matrix or nanocapsules with the core filled with the therapeutic agent or simply adsorbed onto the preformed nanoparticle surface can be manufactured. Vauthier et al.[102] have employed alkylcyanoacrylate as a starting material for manufacturing nanoparticles. In addition to its biodegradability, the polymerization process, namely, emulsion polymerization, is simple and does not require any energy input that possibly can destabilize the drugs. Water-insoluble monomer droplets are added into a pH-controlled aqueous polymerization medium that may contain other functional excipients such as dextran (steric stability) and glucose (isotonicity). Polymerization is initiated by hydroxyl (OH^-) ions and controlled by adjusting the pH. Drugs can be combined with the polymerization medium either before or after the polymerization. Owing to their enormous surface area and porous nature, nanoparticles can adsorb a wide variety of drugs with high association efficiency.

As of today, there are no commercially available pharmaceutical products of this technology. The pharmaceutical industry however, is involved in developing nanoparticle-based delivery systems. Use of nanospheres to modify the blood-brain barrier (BBB)–limiting characteristics of the drug enables targeted brain delivery via BBB transporters and provides a sustained release in brain tissue and vaccine delivery systems to deliver therapeutic protein antigens into the potent immune cells are under investigation.[103]

9.5.3 Implants

A biodegradable implant can function by releasing a drug over a period of time with a simultaneous or subsequent degradation of polymer in the tissues to harmless constituents and avoid surgical removal. The approach to develop such a system is to use hydrolytically labile polymers into which a drug could be physically dispersed under mild conditions and in which release of the drug could be controlled by hydrolytic erosion of the polymer matrix over a therapeutically meaningful period of time. Compared with particulate systems such as micro- or nanoparticles, a unique advantage of these systems would be ease of removal for the discontinuation of therapy at any moment.

Implants can be prefabricated and administered with the help of specialized injection devices such as trocars, often under local anesthesia. These implants can be manufactured by standard hot-melt extrusion, compressing polymer and drug mixture, or injection molding to yield different sizes and shapes (flat films, rolled implants, rods, etc.). The discomfort associated with implantation procedure can be alleviated by use of in situ forming implants that are formed at the site of injection on injecting a polymeric solution. This approach was adapted by Atrix Laboratories in developing a proprietary technology called Atrigel.[104] In this system, biodegradable and water-insoluble polymers such as PLGA, PLA, polycaprolatones, etc. are dissolved in biocompatible organic solvents such as *N*-methyl-2-pyrrolidone (NMP), triacetin, etc. along with the drug. The drug-polymer solution or suspension is then injected subcutaneously or intramuscularly using conventional 21- 22-guage needles. On coming in contact with physiological surroundings, NMP slowly dissipates into the surrounding tissues, and water permeates into the polymer solution. This process leads to phase separation and subsequent coagulation of the polymer to form an implant in situ that traps the drug. Release of the drug can be tailored by choosing the optimal system parameters, such as polymer and solvent characteristics. A family of Atrigel products delivering leuprolide acetate over 1 to 4 months has been approved for commercial use recently. In addition to ease of administration, this system is simple to manufacture and cost-effective. The perceived limitations of using organic solvents in these systems are limitation of injection volume, limitation of polymer concentration, and higher initial release (burst effect) of drugs during formation of the implant. Burst effect, with as high as 30 to 40 percent of the loaded drug being released within the first few hours, could lead to acute toxicity and limits the application to certain drugs such as leuprolide acetate that have a wide therapeutic index. On the other hand, it could be beneficial if the treatment regimen calls for acute drug delivery followed by smaller availability of drug for maintenance therapy. The burst effect seen with in situ implant systems is usually higher than that from microspheres. Alzamer Depot, developed by Alza, is closely related to the Atrix system and claims to have reduced burst effect by use appropriate solvent systems.

Zoladex, a PLGA-based biodegradable implant developed by AstraZeneca PLC, delivers goserelin acetate in palliative treatment of prostate carcinoma. Another commercially available implant, Gliadel Wafer, made of polyanhydrate copolymer matrix poly[bis(p-carboxyphenoxy) propane: sebacic acid] delivers the chemotherapeutic agent carmustine for the treatment of brain tumors. This novel approach of placing the drug-loaded wafers at the tumor site after surgical removal of the tumor circumvents the BBB, which does not allow drugs to reach brain on systemic administration. This approach is expected to reduce toxic effects and affords higher local concentrations. Biodegradable implants also have found application in treating macular edema and other degenerative ocular conditions. Dexamethasone delivered through a PLGA-based implant, Posuredex developed by Oculex Pharmaceuticals/ Allergan, had shown significant improvement in curing the signs and symptoms of macular edema.[104] Eligard, which delivers leuprolide acetate over extended time frames, is an example of an in situ forming implant system that was developed by Atrix Laboratories, Inc.

9.6 Future Potential

Numerous synthetic biodegradable polymers are available and still being developed for sustained and targeted drug delivery applications. An enormous amount of literature is available on various means of altering the performance of these polymers or the delivery system. Development of such an optimized drug delivery system using biodegradable polymers can offer significant improvement in patient comfort and compliance. These systems in many cases reduce the dose intake and thus unwanted toxicities, as well as providing better therapeutic efficacy owing to continuous availability of drug in the therapeutic ranges over a long period of time. Microspheres and implant systems have taken the lead in realizing the potential of biodegradable polymeric delivery systems. Development of these systems may prove to be a turning point for a large number of macromolecules such as proteins and peptides, as well as for other molecules that are deemed to be active but not deliverable or too toxic. Issues related to inflammation at the site of injection, reproducibility of drug release, and scale-up of laboratory preparation batches to industrially feasible production batches, however, need to be addressed to extend the benefits of biodegradable drug delivery systems to a large group of drugs and therapeutic conditions.

References

1. Heller, J. Hydrogels in medicine and pharmacy, in *Properties and Applications*, ed. N. Peppas, Vol. 3. Boca Raton, FL: CRC Press, 1987.
2. Beck, L. and Tice, T. Poly(lactic acid) and poly(lactic acid-co-glycolic acid) contraceptive delivery systems, in *Long-Acting Steroid Contraception*, ed. R. Daniel and J. Mishell. New York: Raven Press, 1983, pp. 175–199.
3. Wood, D. Biodegradable drug delivery systems. *Int. J. Pharm.* **7**:1–18, 1980.
4. Deasy, P. Polymerization procedures for biodegradable micro- and nanocapsules and particles, in *Microencapsulation and Related Drug Processes. Drugs and the Pharmaceutical Sciences*, ed. J. Swarbrick, Vol. 20. New York: Marcel Dekker, 1984, pp. 219–240.
5. Kulkarni, R. K., Pani, K. C., et al. Polylactic acid for surgical implants. *Arch Surg* **93**(5):839–843, 1966.
6. Cutright, D. E., and Hunsuck, E. E. Tissue reaction to the biodegradable polylactic acid suture. *Oral Surg Oral Med Oral Pathol* **31**(1):134–139, 1971.
7. Ratcliffe, J. H., Hunneyball, I. M., et al. Preparation and evaluation of biodegradable polymeric systems for the intra-articular delivery of drugs. *J. Pharm. Pharmacol.* **36**(7):431–436, 1984.
8. Smith, A., and Hunneyball, I. Evaluation of poly(lactic acid) as biodegradable drug delivery system for parenteral administration. *Int. J. Pharm.* **30**:215–220, 1986.

9. Ammoury, N., Fessi, H., et al. In vitro release kinetic pattern of indomethacin from poly(D,L-lactide) nanocapsules. *J. Pharm. Sci.* **79**(9):763–767, 1990.
10. Kwong, A., Chow, S., et al. In vitro and in vivo release of insulin from poly(lactic acid) microbeads and pellets. *J. Contr. Rel.* **4**:47–62, 1986.
11. Seki, T., Kawaguchi, T., et al. Controlled release of 3′,5′-diester prodrugs of 5-fluoro-2′-deoxyuridine from poly-L-lactic acid microspheres. *J. Pharm. Sci.* **79**(11):985–987, 1990.
12. Benita, S., Benoit, J. P., et al. Characterization of drug-loaded poly(D,L-lactide) microspheres. *J. Pharm. Sci.* **73**(12):1721–1724, 1984.
13. Spenlehauer, G., Veillard, M., and Benoit, J. P. Formation and characterization of cisplatin loaded poly(D,L-lactide) microspheres for chemoembolization. *J. Pharm. Sci.* **75**(8):750–755, 1986.
14. Tsai, D. C., Howard, S. A., et al. Preparation and in vitro evaluation of polylactic acid–mitomycin C microcapsules. *J. Microencapsul.* **3**(3):181–193, 1986.
15. Jeong, B., Choi, Y. K., et al. New biodegrabable polymers for injectable drug delivery systems. *J. Contr. Rel.* **62**:109–114, 1999.
16. Jeong, B., Bae, Y. H., and Kim, S. W. In situ gelation of PEG-PLGA-PEG triblock copolymer aqueous solutions and degradation thereof. *J. Biomed. Mater. Res.* **50**(2):171–177, 2000.
17. Allen, C., Eisenberg, A., et al. PCL-b-PEO micelles as a delivery vehicle for FK506: Assessment of a functional recovery of crushed peripheral nerve. *Drug Del.* **7**(3):139–145, 2000.
18. Veronese, F. M., Marsillo, F., et al. Polyphosphazene membranes and microspheres in periodontal diseases and implant surgery. *Biomaterials* **20**(1):91–98, 1999.
19. Payne, L. G., and Andrianov, A. K. Protein release from polyphosphazene matrices. *Adv. Drug Del. Rev.* **31**(3):185–196, 1998.
20. Maa, Y., and Heller, J. Controlled release of 5-fluorouracil from linear poly(ortho esters). *J. Contr. Rel.* **13**:11–19, 1990.
21. Heller, J., Fritzinger, B. K., et al. In vitro and in vivo release of levonorgestrel from poly(ortho esters): I. Cross-linked polymers. *J. Contr. Rel.* **1**:233–238, 1985.
22. Heller, J., Fritzinger, B. K., et al. In vitro and in vivo release of levonorgestrel from poly(ortho esters): II. Linear polymers. *J. Contr. Rel.* **1**:225–232, 1985.
23. Heller, J., Sparer, R., and Zentner, C. *Biodegradable Polymers as Drug Delivery Systems*, ed. M. Chasin and R. Langer. New York: Marcel Dekker, 1990, pp. 121–161.
24. Sparer, R., Shih, C., et al. Controlled release from erodible poly(ortho ester) drug delivery systems. *J. Contr. Rel.* **1**:23–32, 1984.
25. Heller, J., Chang, A. C., et al. Release of insulin from pH-sensitive poly(ortho esters). *J. Contr. Rel.* **13**:295–302, 1990.
26. Suzanne, E., Sergio, C., et al. Therapeutic applications of viscous and injectable poly(ortho esters). *Adv. Drug Del. Rev.* **53**(1):45–73, 2001.
27. Erdmann, L., Macedo, B., and Uhrich, K. E. Degradable poly(anhydride ester) implants: Effects of localized salicylic acid release on bone. *Biomaterials* **21**(24):2507–2512, 2000.
28. Dittrich, V., and Schulz, R. Kinetics and mechanism of the ring-opening polymerization of L-lactide. *Angew. Makromol. Chem.* **15**:109, 1971.
29. Kulkarni, R., Moore, E. G., et al. Biodegradable polylactic acid polymers. *J. Biomed. Mater. Res.* **5**:169–181, 1971.
30. Lewis, D. Controlled release of bioactive agents from lactide/glycolide polymers. *Drug Pham. Sci.* **45**:1–42, 1990.
31. Pitt, C. Poly-e-carprolactone and its copolymers. *Drugs Pharm. Sci.* **45**:71–120, 1990.
32. Brode, G. Lactone polymerization and polymer properties. *J. Macromol. Sci. Chern.* **A6**:1109–1147, 1972.
33. Heller, J., Penhale, D., and Helwing, R. Preparation of poly(ortho esters) by the reaction of diketene acetals and polyols. *J. Poly. Sci. Polym. Lett. Ed.* **18**:619–624, 1989.
34. Heller, J. Development of poly(ortho esters): A historical overview. *Biomaterials* **11**:659–665, 1990.
35. Elfick, A. Poly(e-caprolactone) as a potential material for a temporary joint spacer. *Biomaterials* **23**:4463–4467, 2002.

36. Sinha, V., Bansal, K., et al. Poly-caprolactone microspheres and nanospheres: An overview. *Int. J. Pharm.* **278**:1–23, 2004.
37. Anderson, J. M., Gibbons, D. F., et al. The potential for poly-alpha-amino acids as biomaterials. *J. Biomed. Mater. Res. Symp.* **5**(1):197–207, 1974.
38. Anderson, J. M., Hiltner, A., et al. Biopolymers as biomaterials: Mechanical properties of gamma-benzyl-L-glutamate-L-leucine copolymers. *J. Biomed. Mater. Res. Symp.* **4**:25–35, 1972.
39. Simons, E. R., and Blout, E. R. The effect of proteolytic enzymes on synthetic polypeptides. *Biochim. Biophys. Acta.* **92**:197–199, 1964.
40. Miller, W. G., and Nylund, R. E. The stability of the helical conformation of random L-leucine-L-glutamic acid copolymers in aqeous solution. *J. Am. Chem. Soc.* **87**:3542–3547, 1965.
41. Miller, W. G. Degradation of synthetic polypeptides: III. Degradation of poly-alpha, L-lysine by proteolytic enzymes in 0.20 M sodium chloride. *J. Am. Chem. Soc.* **86**:3918–3921, 1964.
42. Miller, W. G. Degradation of synthetic polypeptides: II. Degradation of poly-alpha, L-glutamic acid by proteolytic enzymes in 0.20 M sodium chloride. *J. Am. Chem. Soc.* **86**(19):3913–3922, 1964.
43. Green, M., and Stahmann, M. A. Synthesis and enzymatic hydrolysis of glutamic acid polypeptides. *J. Biol. Chem.* **197**:771–782, 1952.
44. Waley, S. G., and Watson, J. The action of trypsin on polylysine. *Biochem. J.* **55**(2):328–337, 1953.
45. Sarid, S., Berger, A., and Katchalski, E. Proline iminopeptidase. *J. Biol. Chem.* **234**(7):1740–1746, 1959.
46. Sarid, S., Berger, A., and Katchalski, E. Proline iminopeptidase. II. Purification and comparison with iminopeptidase (prolinase). *J. Biol. Chem.* **237**(7):2207–2212, 1962.
47. Kohn, J. Current trends in the development of synthetic materials for medical applications. *Pharm. Technol.* **14**(10):32–37, 1990.
48. Kohn, J. Pseudopoly(amino acids). *Drugs Pharm. Sci.* **45**:195–229, 1990.
49. Pulapura, S., Li, C., and Kohn, J. Structure-property relationships for the design of polyiminocarbonates. *Biomaterials* **11**:666–678, 1990.
50. Domb, A., Ron, E., and Langer, R. Poly(anhydrides): 2. One-step polymerization using phosgen or diphosgen as coupling agents. *Macromolecules* **21**:1925–1929, 1988.
51. Yoshida, M., Asano, M., et al. Sequential polydepsipeptides as biodegradable carriers for drug delivery systems. *J. Biomed. Mater. Res.* **24**:1173–1184, 1990.
52. Helder, J., Feijen, J., et al. Copolymers of D,L-lactic acid and glycine. *Makromol. Chem. Rapid. Commun.* **7**:193–198, 1986.
53. Helder, J., Dijkstra, P., and Feijen, F. J. In vitro degradation of glycine/DL-lactic acid copolymers. *J. Biomed. Mater. Res.* **24**:1005–1020, 1990.
54. Schakenraad, J. M., Nieuwenhuis, P., et al. In vivo and in vitro degradation of glycine/DL-lactic acid copolymers. *J. Biomed. Mater. Res.* **23**:1271–1288, 1989.
55. Allcock, H., Inorganic macromolecules. *Chem. Eng. News.* **63**:22–36, 1985.
56. Allcock, H. Polyphosphazenes as new biomedical and bioactive materials. *Drugs Pharm. Sci.* **45**:163–193, 1990.
57. Allcock, H., and Scopelianos, A. Syntheis of sugar-substituteed cyclic and polymeric phosphazenes and their oxidation, reduction, and acetylation reactions. *Macromolecules.* **16**:715–719, 1983.
58. Laurencin, C., Koh, H. J., et al. Controlled release using a new bioerodible polyphosphazene matrix system. *J. Biomed. Mater. Res.* **21**:1231–1246, 1987.
59. Leong, K. Synthetic biodegradable polymer drug delivery systems, in *Polymer for Controlled Drug Delivery*, ed. P. Tarcha. Boca Raton: FL: CRC Press, 1991, pp. 127–148.
60. Pretula, J., Kaluzynski, K., and Penczek, S. Synthesis of poly(alkylene phosphates) with nitrogen-containing bases in the side chains: 1. N- and C-substituted imidalzoles. *Macromolecules* **19**:1797–1799, 1986.
61. Leong, K., Brott, B., and Langer, R. Bioerodible polyanhydrides as drug-carrier matrices: I. Characterization, degradation, and release characteristics. *J. Biomed. Mater. Res.* **19**:941–955, 1985.

62. Leong, K., Simonte, V., and Langer, R. Synthesis of polyanhydrides: Melt-polycondensation, dehydrochlorination, and dehydrative coupling. *Macromolecules* **20**:705–712, 1987.
63. Domb, A., and Langer, R. Polyanhydrides: I. Preparation of high molecular weight polyanhydrides. *J. Polym. Sci. Polym. Chem.* **25**:3373–3386, 1987.
64. Domb, A., and Langer, R. Poly(anhydrides): III. (Polyanhydrides) based on aliphatic-aromatic diacids. *Macromolecules* **22**:3200–3204, 1989.
65. Maniar, M., Xie, X., and Domb, A. Polyanhydrides: V. Branched polyanhydrides. *Biomaterials* **11**:690–694, 1990.
66. Vezin, W., and Florence, A. In vitro heterogeneous degradation of poly(*n*-alkyl *a*-cyanoacrylates). *J. Biomed. Mater. Res.* **14**:93–106, 1980.
67. Hegyeli, A. Use of organ cultures to evaluate biodegradation of polymer implant materials. *J. Biomed. Mater. Res.* **7**:205–214, 1973.
68. Ravivarapu, H. B., Lee, H., and DeLuca, P. P. Enhancing initial release of peptide from poly(D,L-lactide-co-glycolide) (PLGA) microspheres by addition of a porosigen and increasing drug load. *Pharm. Dev. Technol.* **5**(2):287–296, 2000.
69. Makoto, M. A., Jun'ichi, K., Akira, O., K. Masaru, I. The effects of drug physicochemical properties on release from copoly (lactic:glycolic acid) matrix. *Int. J. Pharm.* **169**:255–263, 1998.
70. Heller, J., Baker, R. W., et al. Controlled drug release by polymer dissolution. I. Partial esters of maleic anhydride copolymers—Properties and theory. *J. Appl. Polym. Sci.* **22**: 1991–2009, 1978.
71. Heller, J., and Trescony, P. Controlled drug release by polymer dissolution: ll. Enzyme-mediated delivery device. *J. Pharm. Sci.* **68**:919–921, 1979.
72. Bodomeier, R., Oh, K., and Chen, H. The effect of the addition of low molecular weight poly(DL-lactide) on drug release from biodegradable poly(DL-lactide) drug delivery systems. *Int. J. Pharm.* **51**:1–8, 1989.
73. Pitt, C., Gratzl, M. M., et al. Aliphatic polyesters: II. The degradation of poly(DL-lactide), poly (e-caprolactone) and their copolymers in vivo. *Biomaterials* **2**:215–220, 1981.
74. Lin, W. J., Flanagan, D. R., and Linhardt, R. J. Accelerated degradation of poly(epsilon-caprolactone) by organic amines. *Pharm. Res.* **11**(7):1030–1034, 1994.
75. Shieh, L., Tamada, J., et al. Erosion of a new family of biodegradable polyanhydrides. *J. Biomed. Mater. Res.* **28**(12):1465–1475, 1994.
76. Crawford, R. R., and Esmerian, O. K. Effect of plasticizers on some physical properties of cellulose acetate phthalate films. *J. Pharm. Sci.* **60**(2):312–314, 1971.
77. Pradhan, R., and Vasavada, R. Formulation and in vitro release study of ply(DL-lactide) microspheres containing hydrophilic compounds, glycine and its homopeptides. *J. Contr. Rel.* **30**:143–154, 1994.
78. Rothen-Weinhold, A., Besseghir, K., et al. Injection-molding versus extrusion as manufacturing technique for the preparation of biodegradable implants. *Eur. J. Pharm. Biopharm.* **48**(2): 113–121, 1999.
79. Arshady, R. Preparation of biodegradable microspheres and microcapsules: 2. Polylactides and related polyesters. *J. Contr. Rel.* **17**:1–22, 1991.
80. Wang, F. J., and Wang, C. H. Sustained release of etanidazole from spray dried microspheres prepared by non-halogenated solvents. *J. Contr. Rel.* **81**(3):263–280, 2002.
81. Hausberger, A. G., Kenley, R. A., and DeLuca, P. P. Gamma irradiation effects on molecular weight and in vitro degradation of poly(D,L-lactide-CO-glycolide) microparticles. *Pharm. Res.* **12**(6):851–856, 1995.
82. Montanari, L., Costantini, M., et al. Gamma irradiation effects on poly(DL-lactictide-coglycolide) microspheres. *J. Contr. Rel.* **56**(1–3):219–229, 1998.
83. Bittner, B., Mader, K., et al. Tetracycline-HCl-loaded poly(DL-lactide-co-glycolide) microspheres prepared by a spray drying technique: influence of gamma-irradiation on radical formation and polymer degradation. *J. Contr. Rel.* **59**(1):23–32, 1999.
84. Faisant, N., Saipmann, J., et al. The effect of gamma-irradiation on drug release from bioerodible microparticles: A quantitative treatment. *Int. J. Pharm.* **242**:281–284, 2002.

85. Lin, Y., and Paschalis, A. Physicochemical aspects of drug delivery and release from polymer-based colloids. *Curr. Opin. Colloid Interfac. Sci.* **5**:132–143, 2000.
86. Wada, R., Hyon, S. H., and Ikada, Y. Lactic acid oligomer microspheres containing hydrophilic drugs. *J. Pharm. Sci.* **79**(10):919–924, 1990.
87. Itoh, M., Nakano, M., and Juni, K. Sustained release of sulfamethizole, 5-flourouracil, and doxorubicin from polylactic acid microcapsules. *Chem. Pharm. Bull.* **28**(4):1051–1055, 1980.
88. Fujimoto, S., Miyazaki, M., et al. Biodegradable mitomycin C microspheres given intraarterially for inoperable hepatic cancer: With particular reference to a comparison with continuous infusion of mitomycin C and 5-fluorouracil. *Cancer.* **56**(10):2404–2410.
89. Watts, P. J., Davies, M. C., and Melia, C. D. Microencapsulation using emulsification/solvent evaporation: An overview of techniques and applications. *Crit. Rev. Ther. Drug Carrier Syst.* **7**(3):235–259, 1990.
90. Sanders, L. Controlled delivery systems for peptides, in *Peptide and Protein Drug Delivery*, ed. V. Lee. New York: Marcell Dekker, 1991, pp. 785–807.
91. Bissery, M., Valeriote, F., and Thies. C. *Microspheres and Drug Therapy: Pharmaceutical, Immunological and Medical Aspects*, ed. S. Davies, et al. New York: Elsevier Science Publishers, 1984, pp. 217–227.
92. Tomlinson, E., Burger, J., and Schoonderwoerd, E. Human serum albumin microspheres for intraarterial drug targeting of cytostatic compounds, pharmaceutical aspects and release characteristics, in *Microspheres and Drug Therapy. Pharmaceutical, Immunological and Medical Aspects*, ed. S. Davies et al. New York: Elsevier Science Publishers, 1984, pp. 217–227.
93. Okada, H., and Toguchi, H. Biodegradable microspheres in drug delivery. *Crit. Rev. Ther. Drug Carrier Syst.* **12**(1):1–99, 1995.
94. DeLuca, P., Kanke, M., and Sayo, T. Biodegradable microspheres for injection and inhalation, in *Microspheres and Drug Therapy: Pharmaceutical, Immunological and Medical Aspects*, ed. S. Davies et al. New York: Elsevier Science Publishers, 1984, pp. 217–227.
95. Emerich, D. F., Winn, S. R., et al. Injectable chemotherapeutic microspheres and glioma: I. Enhanced survival following implantation into the cavity wall of debulked tumors. *Pharm. Res.* **17**(7):767–775, 2000.
96. Kanellakopoulou, K., and Giamarellos-Bourboulis, E. J. Carrier systems for the local delivery of antibiotics in bone infections. *Drugs* **59**(6):1223–1232, 2000.
97. Labshetwar, V., Song, C., and Levy, R. Nanoparticle drug delivery system for restenosis. *Adv. Drug Del. Rev.* **24**:63–85, 1997.
98. Molpeceres, J., Aberturas, M. R., and Guzman, M. Biodegradable nanoparticles as a delivery system for cyclosporine: Preparation and characterization. *J. Microencapsul.***17**(5):599–614, 2000.
99. Akhtar, S., Hughes, M. D., et al. The delivery of antisense therapeutics. *Adv. Drug. Del. Rev.* **44**(1):3–21, 2000.
100. Merodio, M., Arnedo, A., et al. Ganciclovir-loaded albumin nanoparticles: Characterization and in vitro release properties. *Eur. J. Pharm. Sci.* **12**(3):251–259, 2001.
101. Molina, J., Urbina, J., et al. Cure of experimental Chagas' disease by the bis-triazole DO870 incorporated into "stealth" polyethyleneglycol-polylactide nanospheres. *J. Antimicrob. Chemother.* **47**(1):101–104, 2001.
102. Vauthier, C., Dubernet, C., et al. Poly(alkylcyanoacrylates) as biodegradable materials for biomedical applications. *Adv. Drug Del. Rev.* **55**(4):519–548, 2003.
103. Randall, C. Good things in small packages: Nanotech advances are producing mega-results in drug delivery. *Mod. Drug Disco.* **July**:30–36, 2004.
104. Wong, V. G., and Hu, W. L. Methods for treating inflammation mediated conditions of the eye. U.S. Patent 6,726,918, 2004.

Chapter

10

Carrier- and Vector-Mediated Delivery Systems for Biological Macromolecules

Jae Hyung Park

College of Environment and Applied Engineering, Kyung Hee University
Gyeonggi-do, South Korea

Ick Chan Kwon

Biomedical Research Center, Korea Institute of Science and Technology
Seoul, South Korea

Jin-Seok Kim

College of Pharmacy, Sookmyung Women's University
Seoul, South Korea

10.1 Introduction

Over the past few decades, there have been considerable advances in immunology and molecular genetics that enable us to understand the genetic and cellular basis of many diseases. At the same time, remarkable progress in recombinant DNA technology and proteomics has made it possible to develop a number of potential protein and genetic materials-based therapeutic molecules for treating diseases. To realize the clinical potential of such biological macromolecules, effective delivery technologies obviously are required. These delivery technologies may offer the ability to improve the stability of therapeutic agents against enzymatic degradation (e.g., proteases and nucleases), to enhance oral bioavailability, to augment therapeutic effect by targeting the macromolecular drugs to a specific site, and to sustain the therapeutic effect at the target site.

The delivery systems for biological macromolecules must be tailor-made according to the biophysical, biochemical, and physiological characteristics of the therapeutic agents and disease states. In this chapter, the delivery systems are classified into carrier- and vector-mediated systems, where the former refers to delivering therapeutic agents such as peptides and proteins and the latter is for the delivery of genetic materials, including DNA and RNA. The use of appropriate carriers or vectors for effectively delivering a specific macromolecule into its site of action may pave the way for the treatment of human diseases. In this chapter, the design principles and recent progress in carrier- and vector-mediated delivery systems for biological macromolecules such as peptides, proteins, and genes will be discussed.

10.2 Carrier-Mediated Delivery Systems

The great progress in the areas of biotechnology and biochemistry has allowed the discovery of enormous therapeutic macromolecules based on peptides and proteins. Most such macromolecular drugs currently are administrated by the parenteral route to achieve therapeutic effects. However, many diseases that require chronic therapy make the repetitive dosing of macromolecular drugs necessary, which is attributed primarily to their short half-lives in biological fluids. The frequent administration of macromolecular drugs may result in fluctuating blood concentrations and is poorly accepted by patients. Therefore, much effort has been devoted to developing alternatives for administration of macromolecular drugs such as oral, nasal, buccal, rectal, vaginal, and transdermal routes.[1–5]

Considering patient acceptability and ease of administration, there is no doubt that oral administration is the most favored route, even if there have been reports on successful delivery of macromolecular drugs across nonperoral mucosal routes.[6,7] Despite such advantages in oral administration, various barriers are encountered in the gastrointestinal (GI) tract that should be surmounted in order to gain sufficient bioavailability of macromolecular drugs. This has stimulated researchers to develop suitable carrier systems. For example, protein drugs administrated orally are exposed to an acidic environment in the stomach and proteolytic enzymes in the gut lumen and the brush border membrane, which may degrade the proteins into biologically inactive subunits that are sufficiently small to be absorbed. In an attempt to achieve protein delivery via oral administration, several approaches have been investigated, including chemical modification of protein drugs to improve their physicochemical properties,[5,8] coadministration of an absorption enhancer,[9,10] and the use of protease inhibitors to minimize protein degradation by proteolytic enzymes.[11,12] In recent years, carrier-mediated delivery systems have received increasing interest for oral delivery of protein drugs because they can effectively protect proteins from enzymatic degradation and significantly improve bioavailability.[4,5,13–15]

Herein, various barriers to oral delivery of macromolecular drugs will be discussed in brief, after which carrier-mediated delivery systems on the basis of chemical modification of the drugs and particulate drug carriers (e.g., nanoparticles, microcapsules, and liposomes) will be introduced as representative of promising approaches currently being investigated.

10.2.1 Barriers to oral delivery of macromolecular drugs

The limitations in oral delivery of macromolecules primarily are attributed to their large molecular size, susceptibility to enzymatic degradation, and hydrophilic characteristics. In general, oral bioavailability is considered to sharply decrease as the molecular size increases owing to poor permeability through the mucosal surface and cell membrane, especially for drugs with molecular weights higher than 500 to 700 Da.[16] Further, without the minimum degree of lipophilicity, most macromolecules cannot be absorbed transcellularly through passive diffusion because of a lack of interaction with the cell membrane.[17] Unfortunately, most macromolecular drugs currently being evaluated are rather hydrophilic. In subsequent subsections, various barriers to oral delivery of macromolecular drugs are described, accompanied by representative approaches to overcoming those barriers.

Physical barrier. Following oral administration of macromolecular drugs, their potential absorption pathways from the intestinal lumen to the bloodstream can be classified into transcellular transport associated with adsorptive or receptor-mediated endocytosis and paracellular transport (Fig. 10.1). The GI tract surface consists of a tightly bound single layer of epithelial cells covered with thick and viscous mucus, which serves as a defensive deterrent against permeation of xenobiotics and harmful pathogens. The epithelial cells in the GI tract are

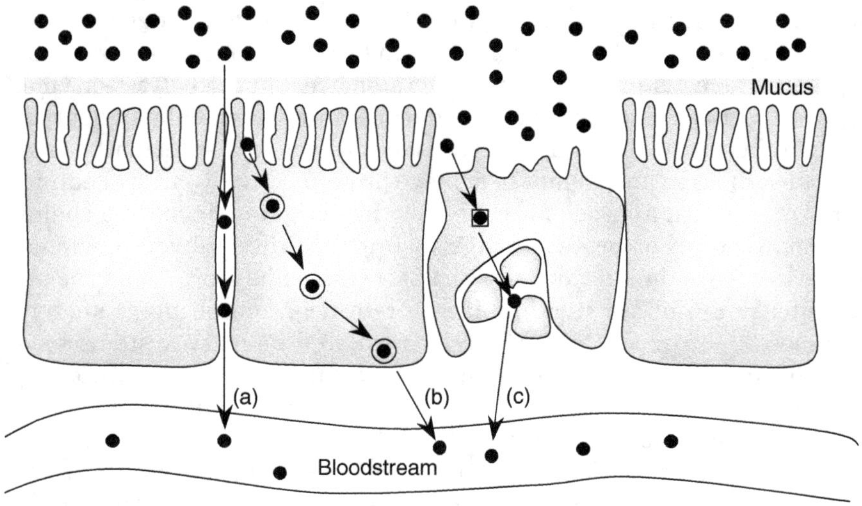

Figure 10.1 Pathways for intestinal absorption of macromolecular drugs. (*a*) Paracellular transport of macromolecules can be achieved by altering or disrupting the tight junctions that exist between cells and are only permeable to small molecules (<100 to 200 Da). (*b*) Adsorptive enterocytes and (*c*) M cells of Peyer's patches allow transcellular transport of macromolecules involving transcytosis and receptor-mediated endocytosis.

bound to one another by tight junctions, contributing less than 1 percent of the mucosal surface area. Since the pore size of tight junctions is reported to be less than 10 Å,[18,19] paracellular transport may not be an option for macromolecules. However, it is now established that tight junctions have dynamic structures, rendering them modifiable by absorption enhancers.[20–22] Therefore, a number of candidates with absorption-enhancing properties have been developed, such as low-molecular-weight compounds and high-molecular-weight polymers, although some of them may cause serious damage to the mucosal epithelium.[23] For example, chitosan, a polysaccharide composed of glucosamine and *N*-acetylglucosamine, has emerged as one of promising enhancers for oral absorption of macromolecular drugs because of its ability to open tight junctions, outstanding biocompatibility, and moderate biodegradability.[24] The electrostatic interactions between the positively charged chitosan and the negatively charged surfaces of epithelial cells may be responsible for a structural reorganization of tight junction–associated proteins, thus enhancing paracellular transport of poorly absorbable drugs.[25]

Cellular internalization of macromolecules by endocytosis is an important biological process for their transcellular transport. Endocytosis can be categorized into adsorptive and receptor-mediated endocytosis (RME). RME involves specific binding of ligand to the receptor on the apical cell membrane, followed by pinching of membrane vesicles and intercellular trafficking. In contrast, adsorptive endocytosis does not require the receptor and is initiated by nonspecific physical interactions between adsorptive materials and cell surfaces, such as electrostatic interaction, hydrogen bonding, and hydrophobic interaction.[14] These endocytotic pathways may provide opportunities to increase bioavailability of macromolecular drugs; e.g., the transport of proteins from the GI tract to the bloodstream can be enhanced by chemical conjugation of specific ligands for cellular absorption by RME.[26]

One of the interesting aspects of successful oral drug delivery is the direct transport of endocytotic vesicles to the basolateral membrane (i.e., bypassing the lysosomes), commonly referred to as *transcytosis*.[26–28] This pathway effectively may deliver macromolecules into the bloodstream without significant degradation. Although the mechanisms of transcytosis are not fully understood yet, M cells of Peyer's patches, existing in intestinal tissues, have been known to possess a significant ability to transcytose macromolecules. M cells exhibit unique morphological features that are different from other enterocytes: (1) immature microvillous structures, (2) the presence of apical microfolds, and (3) the absence of mucus (see Fig. 10.1).

Besides the absorption barrier on the basis of epithelial permeability, a layer of mucus can be another limiting factor that may be responsible for poor bioavailability of macromolecules. Mucus covers the intestinal epithelial cell layer to serve as a lubricant and protective barrier. It is a thick, viscous, constantly changing mixture of glycoproteins (mucins), enzymes, electrolytes, and exfoliated epithelial cells. Although the barrier effect of mucus is considered to be minimal for low-molecular weight drugs, the diffusion rates of macromolecular drugs can decrease significantly owing to the presence of mucin, accounting for gel formation and the high viscosity of mucus,[29] which macromolecules with molecular weights higher than 5 kDa are estimated to be hardly able to penetrate.[21] However, it has been revealed that this mucus barrier can be attenuated by the use of mucolytic agents such as proteases that degrade the protein core of mucin glycoproteins, sulfhydryl compounds breaking disulfide bonds of mucoproteins, and detergents disturbing physical interactions between constituents of the mucus.[21,30] Proteases can reduce mucus viscosity via enzymatic action toward mucin glycoproteins and thus allow easy penetration of macromolecular drugs, whereas extended applications have been limited because they also may act on protein drugs, leading to loss of biological activity.[30] Therefore, it is preferred to employ sulfhydryl compounds and detergents for strong liquefying action but minimizing degradation or denaturation of protein drugs. Since sulfhydryl compounds cleave disulfide bonds, they are only applicable to macromoleules without cysteine moieties in their primary structure or protein drugs bearing disulfide bonds that are not accessible owing to the compact tertiary structure.

Enzymatic barrier. Apart from the physical barrier based on the epithelial cell layer covered with mucus, degradation of macromolecular drugs should be considered owing to the presence of various enzymes in the GI tract. Enzymatic degradation of macromolecular drugs is carried out by luminally secreted, brush-border membrane-bound, or cytosolic proteases and peptidases. On reaching the duodenum, macromolecular drugs can be exposed to and degraded by pancreatic proteases; these luminal enzymes account for approximately 20 percent of enzymatic degradation of ingested proteins.[18] Although proteolytic enzymes in lysosomes and other organelles within the enterocytes are partially responsible for degradation of macromolecular drugs, the biggest deterrent to absorption is known to be brush-border peptidases that are active against tri-, tetra-, and higher peptides with up to 10 amino acid residues.[18,31] As a promising strategy to overcome the enzymatic barrier, the use of inhibitory agents has gained considerable recognition recently because they have been demonstrated to improve the bioavailability of therapeutic peptides and proteins.[12,32,33] Table 10.1 shows representative enzymes in the GI tract and inhibitory agents. However, the effects of inhibitory agents on the regular digestion process of nutritive proteins remains questionable, especially for protein drugs that should be administrated for a long period of time.

A number of enzymes have been investigated for improving the bioavailability of macromolecular drugs, including organophosphorus inhibitors, amino acids, peptides, polypeptides, and mucoadhesive polymers.[12] Of these, polypeptide-based protease inhibitors and mucoadhesive polymers have emerged as promising approaches because of their nontoxicity and potent inhibitory activity. For example, Sjostrom et al.[34] investigated the mechanism behind the increased bioavailability of a thrombin inhibitor inogatran during coadministration of a trypsin inhibitor aprotinin (bovine pancreatic enzyme with a molecular mass of 6.5 kDa). They demonstrated that the presence of the trypsin inhibitor resulted in increased bioavailability owing to the competitive displacement of inogatran from trypsin, although inogatran is highly susceptible to intestinal trypsin and trypsinlike enzymes. Mucoadhesive polymers such as polyacrylate derivatives have been shown to enhance the membrane permeability of macromolecular drugs.[35,36] Several mechanisms have been suggested to describe their permeability-enhancing effect: (1) They can bind to essential enzyme cofactors such as Ca^{2+} and Zn^{2+}, causing a conformational change resulting in loss of enzymatic activity, (2) owing to the Ca^{2+}-binding properties of such polymers, the local concentration of extracellular Ca^{2+} decreased, which makes the

TABLE 10.1 Enzymes in the GI Tract and Their Inhibitory Agents

Enzymes	Inhibitory agents
Luminally secreted proteases	
Carboxypeptidase A	Chitosan-EDTA conjugates, EDTA, polyacrylate derivatives
Carboxypeptidase B	Chitosan-EDTA conjugates, EDTA, polyacrylate derivatives
Chymotrypsin	Aprotinin, benzyloxycarbonyl-Pro-Phe-CHO, Bowman-Birk inhibitor, chicken ovoinhibitor, chymostatin, 4-(4-isopropyl piperadinocarbonyl)phenyl 1,2,3,4-tetrahydro-1-naphthoate methanesulphonate, soybean trypsin inhibitor
Elastase	Bowman-Birk inhibitor, chicken ovoinhibitor, diisopropyl fluorophosphates, elastatinal, phenylmethylsulfonyl fluoride (PMSF), soybean trypsin inhibitor
Pepsin	Dioctylsodium sulfosuccinate, bovine uterine serpin, ovine uterine serpin, pepsinostreptin, pepstatin
Trypsin	*p*-Aminobenzamidine, antipain, aprotinin, Bowman-Birk inhibitor, camostat mesylate, chicken ovoinhibitor, chicken ovomucoid, human pancreatic trypsin inhibitor, soybean trypsin inhibitor, polyacrylate derivatives
Brush-border membrane-bound proteases	
Aminopeptidase A	α-Aminoboronic acid derivatives (ACDs), EDTA, phosphinic acid dipeptide analogues (PADAs), 1,10-phenanthroline, puromycin
Aminopeptidase N	Amastatin, amino acids, ACDs, bacitracin, bestatin, chitosan-EDTA conjugates, di- and tripeptides, EDTA, Na-glycocholate, PADAs, puromycin
Aminopeptidase P	ACDs, bestatin, PADAs, PMSF
Aminopeptidase W	ACDs, PADAs
Carboxypeptidase M	D,L -2-Mercaptomethyl-3-guanidinoethylthiopropanoic acid
Carboxypeptidase P	Enterostatin, EDTA
Dipeptidyl peptidase IV	Diisopropylfluorophosphate, *N*-peptidyl-*O*-acylhydroxyl-amines, boronic acid analogues of proline and alanine,
γ-glutamyl transpeptidase	Acivicin (amino-(3-chloro-4,5-dihydro-isoxazol-5-yl)-acetic acid, L-serine-borate
Leucin aminopeptidase	Amastatin, ACDs, bestatin, flavonoid inhibitors, PADAs
Neutral endopeptidase	1,10-Phenanthroline, phosphoramidon, thiorphan [(2-mercaptomethyl-3-phenyl-propionylamino)-acetic acid]

SOURCES: From refs 12 and 21.

integrity of tight junctions loose, and (3) their mucoadhesive properties provide a prolonged residence time of the dosage form in the intestine. Moreover, such properties decrease the distance between the macromolecular drugs released from the dosage form and the epithelial cell membrane, which attenuates drug degradation by luminally secreted enzymes. Despite these advantages, it is likely that simple formulations based on mucoadhesive polymers do not provide a sufficient protective effect on enzymatic degradation of macromolecular drugs.[37] Therefore, to improve the inhibitory activity of mucoadhesive polymers, covalent attachment of enzyme inhibitors has been attempted in recent years.[32,38]

10.2.2 Design of macromolecular drugs through chemical modification

It is evident that the enhanced oral absorption of macromolecular drugs can be achieved by chemical modification to endow them with appropriate physicochemical properties for permeation throughout the epithelial lining of the intestine. Of particular interest is covalent attachment of small molecules that are recognized by endogenous cellular transport systems. Such systems involve various transporters and receptors, triggering endocytosis. The former has been explored for oral drug absorption, based on the glucose transporter,[39] the di- and tripeptide transporters,[31] and the bile acid transporter.[40,41] In particular, the intestinal transport system for bile acids is gaining recognition for macromolecular drug delivery because it plays an important role in preserving the total amount of bile acids in the body through the *enterohepatic circulation*, mediated by the high efficacy of both intestinal and liver absorptions.[40–43] Kramer et al.[42] reported a series of small linear model peptides that were modified by chemical conjugation to the 3-position of a bile acid. The resulting peptide–bile acid conjugates were found to be significantly less susceptible to enzymatic degradation compared with peptides bearing cephalexin, a metabolically stable substrate of the intestinal H^+-oligopeptide transport system. The conjugates were transported in their intact forms from the intestinal lumen, whereas the parent peptides could not be transported, indicating superior intestinal absorption of the conjugates over the parent peptides. The principal limitation in the use of bile acid is that the transporter for bile acid is located in the terminal ileum region of the GI tract, and thus peptide- and protein drug–bile acid conjugates should explore the harsh environments along the entire region of the small intestine before reaching the terminal ileum. On the other hand, chemical conjugation of various ligands to macromolecular drugs, inducing RME, has received increasing attention because this approach does not seem to be limited with regard to the size of the drugs.[26,44]

Design of ligand-bound macromolecules for receptor-mediated transport. In order to bind to ligands, macromolecules should contain appropriate functional groups for chemical conjugation and have the ability to enhance drug absorption by RME. Of different ligands that can mediate RME, lectins and vitamin B_{12} have been investigated intensively because these ligands are believed to deliver macromolecules via the transcytotic pathway.[44–46]

Lectin. Lectins are plant proteins that bind reversibly to specific sugar residues found on the luminal surfaces of the intestinal epithelial cells. Therefore, it may be possible to exploit sugar residues on the surfaces of enterocytes and M cells as targets for lectin-mediated delivery of macromolecular drugs. Several studies, however, have revealed that the use of certain lectins (e.g., kidney bean lectins) is not suitable for oral drug delivery because they can induce strong systemic immune responses after oral administration.[44,47] A lectin derived from tomato (TL, specific to *N*-acetylglucosamine), unlike many other lectins, offers the following advantages: (1) It is relatively nontoxic, [48,49] (2) its resistance to enzymatic degradation has been proven,[48,50] and (3) it specifically binds to human intestinal Caco-2 cells.[51,52] Recently, Hussain et al.[53] investigated the usefulness of a surface-conjugated TL to enhance intestinal uptake of nanoparticles. After TL-coated nanoparticles were administrated to rats by oral gavage daily for 5 days, analysis of their intestinal uptake was carried out. The results indicated that nanoparticles conjugated with TL exhibit 50-fold higher intestinal uptake than those blocked with specific sugar residues (*N*-acetylchitotetraose). Carreno-Gomez et al.[54] prepared TL-conjugated polystyrene microparticles to evaluate their intestinal uptake. They found that the uptake rate of microparticles coupled with TL was four-fold higher than that with bovine serum albumin, indicating a significant role of TL in the intestinal uptake of microparticles. Since TL specifically interacts with *N*-aceylglucosamine residues presenting both in intestinal mucus and on the intestinal epithelial cell surfaces, the intestinal uptake of TL conjugates can be inhibited by competitive interactions. In fact, the uptake of TL conjugates by intestinal mucus has been observed.[53,55] However, the results showed that large numbers of TL conjugates can penetrate the mucus barrier and be absorbed by enterocytes, indicating the positive effect of mucus entrapment.[53,55]

It is well known that intestinal M cells have the ability to efficiently transcytose macromolecules and inert particles, which has stimulated researchers to target M cells for oral delivery of macromolecular drugs and particles.[26,52] For example, *Ulex europaeus* 1 (UEA1), a lectin specific for α-L-fucose residues, selectively binds to mouse Peyer's patch M cells.[56] It has become apparent that UEA1 can be used to target macromolecules to mouse Peyer's patch M cells and to enhance macromolecular absorption across the intestinal epithelial barrier.[52,57] Also, there is evidence that particulates (e.g., polymeric nanoparticles and liposomes) can be targeted to intestinal M cells by coating UEA1.[58,59] Foster et al.[59] prepared carboxylated polystyrene microspheres (0.5 μm in diameter) covalently coated with various proteins, including UEA1, concanavalin A, *Euonymus euopaeus*, human immunoglobulin A, and bovine serum albumin. Of the proteins studied, only the UEA1 coating exhibited selective binding to intestinal M cells and resulted in rapid endocytosis. Recently, Clark et al.[60] developed UEA1-conjugated liposomes (200 nm in diameter) as potential oral vaccine delivery vehicles, and they suggested that UEA1-mediated M cell targeting may permit the efficacy of mucosal vaccines to be enhanced. It should be emphasized, however, that more profound studies are needed to use lectins as targeting ligands to intestinal M cells because the glycosylation pattern of M cells shows significant variations in affinities among animal species,[61] and the overall contribution of M cells to the absorptive surface area of the GI tract is only 10 percent,[26] which may limit widespread application.

Vitamin B_{12}. Vitamin B_{12} is a larger molecule than the other vitamins, and it can be absorbed via the intestine, which involves binding to specialized transport proteins.[44] After oral administration, vitamin B_{12} binds to intrinsic factor (IF) produced from the parietal cells in the stomach and proximal cells in the duodenum. The vitamin B_{12}–IF complex passes down the small intestine until it reaches the ileum, where the complex binds to a specific IF receptor located on the apical membrane of the villous enterocyte. The complex is then internalized via RME, vitamin B_{12} is released from IF by the action of cathepsin L on IF, and free vitamin B_{12} consequently forms the complex with transcobalamin II to be delivered into the basolateral side of the membrane via the transcytotic pathway.

The conjugation of vitamin B_{12} has been shown to increase oral bioavailability of peptides, proteins, and particles.[44–46,62,63] Russsell-Jones and coworkers have attempted to exploit RME of vitamin B_{12} for the enhanced intestinal uptake of macromolecules such as luteinizing hormone–releasing factor (LHRH), granulocyte colony-stimulating factor, erythropoietin, and α-interferon.[44,46,63] Also, they demonstrated that surface modification of nanoparticles with vitamin B_{12} can increase their intestinal uptake.[44,62] The extended applications of this unique transport system, however, appear partially hampered by its limited uptake capacity. In human being, the uptake of vitamin B_{12} is only 1 nmol per intestinal passage.

Design of lipophilic macromolecules for facilitated transport. Most macromolecular drugs are hydrophilic because they contain a number of polar and ionizable groups, such as primary amino and carboxyl groups, which may decrease membrane permeability of the drugs. Thus chemical modification to improve the lipophilicity of the drugs can be a promising approach to increase their oral bioavailability. The lipophilicity of various drugs, often expressed as a log *P* (logarithm value of octanol-water partition coefficient), has been reported to correlate with cell membrane permeability.[64] In most cases it may be preferred that the lipophilic moieties are attached with a labile bond that will be cleaved after oral absorption, which delivers macromolecular drugs to the systemic circulation in their intact forms.[5] If it is not possible to remove attached molecules before entering the systemic circulation, the modified drugs would be considered as new drugs, requiring safety and toxicological data. Also, before clinical use, the effect of chemical modification should be evaluated with regard to pharmacokinetics and pharmacological activity.

There are a number of reports in which the increase in lipophilicity by chemical modification is shown to improve the intestinal uptake of macromolecular drugs.[65–68] Hashizume et al.[69] examined the intestinal absorption of insulin that was chemically modified with palmitic acid following administration into closed large intestinal loops of rats. The results indicated that chemical modification of insulin with palmitic acid not only increases the lipophilicity of insulin but also attenuates its degradation by proteolytic enzymes, leading to the increased transfer of insulin across the intestinal mucous membrane. Heparin cannot be absorbed in the GI tract owing to its high molecular weight and hydrophilicity. Byun et al.[70] attempted to develop heparin derivatives that could be administrated orally. Heparin derivatives were prepared by coupling reaction between the amino group of heparin and the carboxylic acid of deoxycholic acid (DOCA). The resulting derivatives showed comparable bioactivity to parent heparin and could be absorbed via the intestinal membrane of rats, whereas they did not cause any damage to the microvilli and the cell layer of the GI tract. Subsequent study demonstrated that chemical conjugation of DOCA into heparin enhances the intestinal uptake, by which oral bioavailability of heparin reaches 7.8 percent.[68] These promising results were estimated to be involved in the increased hydrophobicity of heparin by chemical modification with DOCA. In addition, the interaction between DOCA and bile acid transporter in the ileum might strengthen the adhesion of heparin to the intestinal membrane, thereby increasing the possibility of heparin transport into systemic circulation.

10.2.3 Design of colloidal drug carriers

Although chemical modification of macromolecular drugs with ligands or lipophilic moieties offers the opportunity for increasing oral bioavailability, this strategy cannot protect the drugs enough from the hostile environment of the GI tract, such as the low pH of the stomach and various enzymes in the intestine that may degrade the drugs into biologically inactive forms. One promising approach to shield such drugs from the hostile environment of the GI tract is the use of the colloidal carrier system, primarily based on liposomes, microparticles, and nanoparticles.[13,21] In addition to macromolecular drugs, the colloidal carrier system may contain enzyme inhibitor and absorption enhancers to increase the intestinal uptake of drugs. Another advantage of colloidal carriers is that they can be tailored for site-specific delivery of drugs; e.g., polymers dissolving at a defined pH may selectively deliver the drugs into the desired part of the GI tract with the specific pH.

Liposomes. Owing to their biodegradable and nontoxic nature, liposomes have been investigated intensively as potential carriers for macromolecular drugs.[71–74] Although liposomes have the ability to encapsulate both hydrophilic and hydrophobic drugs, their chemical and physical instability in the GI tract has limited extended applications as carriers for oral drug delivery. To surmount this drawback of liposome, Iwanaga et al.[75] prepared insulin-loaded liposomes coated with the sugar-chain portion of mucin or polyethylene glycol (PEG). The coated liposomes completely suppressed enzymatic degradation of insulin in the intestinal fluid of rats, thus causing a gradual decrease in the glucose level after oral administration. Formation of a thick water layer on the coated liposome surfaces may play a vital role in improving their stability, which minimizes the direct interaction between the lipid membrane and bile salts in the GI tract, responsible for disruption of the uncoated liposomes.[75,76] The improved stability of PEG-coated liposomes also was demonstrated by Li et al.,[74] who used them as carriers for oral delivery of recombinant human epidermal growth factor. Recent studies have suggested that in comparison with uncoated liposomes, PEG-coated liposomes are distributed widely in the GI tract with a high transit time, primarily attributed to the interaction of PEG with the intestinal surface.[21,77] Mucoadhesive polymer-coated liposomes also have received considerable attention as carriers for macromolecular drugs such as insulin and calcitonin owing to their enhancing effect on the intestinal uptake of the drugs and outstanding stability in intestinal fluids.[78] Takeuchi et al.[79] monitored the blood glucose level of rats following oral administration of chitosancoated liposomes containing insulin. The coated liposomes significantly decreased the blood glucose level of the rats compared with uncoated liposomes or insulin solution. Moreover, the lowered glucose level was maintained for more than 12 hours, suggesting strong mucoadhesive properties of the coated liposomes in the GI tract. Besides insulin, these coated liposomes have been demonstrated to effectively deliver calcitonin to the bloodstream by oral administration.[80]

Polymeric particulates. A wide range of polymers has been investigated as nano- and microsized reservoirs of various drugs that will be released at a specific site of action. For oral delivery of macromolecular drugs, the important technologies to prepare polymeric particulates are reviewed extensively.[4,13] Particles smaller than 5 μm in diameter were found to permeate the GI tract through M cells of Peyer's patches, normal epithelial cells (enterocytes), and paracellular routes.[53,54,81,82] Mathiowiz et al.[82] prepared polymeric microparticles made of biologically erodable polymers such as poly(fumaric anhydride) and poly(lactide-co-glycolide) (PLGA) and evaluated the ability of microparticles to traverse both the mucosal absorptive epithelium and M cells of Peyer's patches. It was found that microparticles maintain contact with intestinal epithelium for extended periods of time and penetrate it through and between cells, thus increasing the absorption of incorporated substances with different molecular sizes, e.g., dicumarol, insulin, and plasmid DNA. With a similar concept, Jiao et al.[83] exploited polymeric nanoparticles composed of poly(ε-caprolactone) and PLGA as potential carriers of oral heparin, by which significant increases in both anti–factor Xa activity and activated partial thromboplastin time for a long period of time were observed in the rabbits after oral absorption of heparin released from nanoparticles. Although these approaches using polymeric particulates sound promising, it should be noted that the intestinal uptake of most particulates has been considered to be lower than 8 percent of the ingested dose.[13,21,81] One of the recent strategies to improve the intestinal uptake of particulates is chemical attachment of ligands on their surfaces, such as lectins [53,54,58,59] and vitamin B_{12},[44,62] which stimulate RME. This, however, needs complex sample preparation and several obstacles to be surmounted, as described in earlier sections. In other aspects, increased particle uptake may be achieved by optimizing the physicochemical factors of the particulates themselves, such as size, surface charge, surface chemistry, and hydrophobicity.

Many researchers have investigated the size effects of the particles on the intestinal uptake by using different polymers, including polystyrene,[84–87] poly(D,L-lactic acid),[88] and PLGA.[89–91] Overall, there are general agreements[14]: (1) Particles smaller than 100 nm exhibit higher uptake by absorptive enterocytes than those larger than 300 nm, (2) small nanoparticles (<100 nm in size) are efficiently transported through M cells of Peyer's patches, and (3) the size of particles should be lower than 500 nm in diameter to reach the bloodstream.

Apart from particle size, surface properties of particulates also have been considered to affect uptake by the intestinal epithelia. Although the effect of surface hydrophobicity is not fully understood, the uptake of nanoparticles prepared from hydrophobic polymers appears to be higher than that of those with hydrophilic surfaces. For example, it has been demonstrated that surface coating of hydrophobic nanoparticles with hydrophilic poloxamer causes a significant decrease in intestinal uptake in vivo.[92] With regard to surface charge, its effect on intestinal uptake is the subject of much discussion. Jani et al.[93] tested polystyrene nanoparticles with different charges to evaluate their intestinal uptake, indicating that carboxylated particles are taken up to a lesser degree than positively charged and uncharged particles. In contrast, Mathiowitz et al.[82] observed that negatively charged particles, based on poly(anhydride) copolymers, are highly adhesive to the intestinal epithelia, thus increasing the oral bioavailability of the incorporated macromolecular drugs. The negatively charged nanoparticles, prepared from poly(acrylic acid), also have been shown to interact strongly with the intestinal epithelia.[78,94] Taking these results into account, it can be suggested that surface charge and hydrophobicity of particles should be compromised to augment their intestinal uptake, for which more profound studies are needed.

10.3 Design of Vector-Mediated Delivery Systems for Genetic Materials

Gene therapy has drawn increasing attention for a variety of biomedical applications. During the past few decades, numerous vectors have been developed to deliver genes into specific cells or organs. Most of gene delivery approaches have used viral vectors based on adenoviruses,[95,96] retroviruses,[97, 98] and adeno-associated viruses.[98,99] Such viral vectors have shown high transfection efficiency, whereas their extended use has been limited by their toxicity and gene transfer into the cells with their native receptors, resulting in poor targeting efficiency. Recently, nonviral vectors, composed of polymers or lipids, have emerged as promising candidates for gene delivery because of several advantages, such as low risk, ease of chemical modification, and physical stability. In subsequent subsections, various barriers to vector-mediated gene delivery will be discussed in brief, and the attributes of vectors currently attracting significant interest will be reviewed.

10.3.1 Barriers to vector-mediated gene delivery

For effective transfection of the cells, DNA has to be delivered into the nucleus in a transcribable form. In this process, the vectors may improve cellular uptake, facilitate escape from the endosomal compartment, allow migration through the cytoplasm, and enhance uptake into the nucleus. Most of all, the attributes of cells and vectors may play a vital role in delivering genes. Although a number of vectors have shown potential for delivering genes efficiently in vitro, most of them are not able to deliver genes in vivo primarily because of their interactions with biological fluids and the extracellular matrix.[100,101] In particular, when the vectors are administrated systemically, they have to travel from the injection site to the target cells. Before reaching the target cells, they may encounter a number of proteins, cell types, and organs. Also, they can be inactivated or removed by the natural defense mechanisms of the organism (e.g., the complement, reticuloendothelial, and immune systems). Extravasation of the vectors out of the blood vessels at target sites and migration into the cells via the extracellular matrix are additional obstacles that should be surmounted. Recently, the ligand-directed targeting of genes has received attention as an in vivo gene delivery system because this approach can minimize gene transfer into nontarget cells.[101] Local in vivo applications such as direct injection of the vectors into the target site or the surrounding region can be alternative strategies to evade or minimize the effect the preceding barriers arising from systemic applications,[100,102] although there are still obstacles to be faced, such as the extracellular matrix and immune responses.

10.3.2 Viral vector

Retroviral and adenoviral vectors represent the most widely used vector system for gene therapy of various diseases from cancer to infectious diseases. Retrovirus is a single-stranded RNA virus that replicates via DNA intermediates that are synthesized by reverse transcriptase. The advantages of retroviral vectors, mostly derived from the Moloney murine leukemia virus (MoMuLV), lie in the ease of manipulation of the virus in vitro with moderate to high titers and a wide range of target cells to be transfected. However, disadvantages include that cell division is required for integration, and more seriously, safety issues, such as insertional mutagenesis, still remain. Historically, a retroviral vector was used successfully for treatment of adenosine deaminase (ADA) deficiency syndrome using an ex vivo approach, even though the judgment of therapeutic efficacy of this gene therapy was very difficult because PEG-ADA protein was coadministered to the same patient at the time of the gene therapy.[103] Another application of retroviral vector was for Gaucher's disease, which is a lysosomal storage disease, using genetically modified glucosidase expressing CD34+ cells. Suicide genes, such as those encoded by the herpes simplex virus (HSV) thymidine kinase (*tk*) or cytosine deaminase (*cd*) genes, which convert the nontoxic prodrugs ganciclovir and 5-fluorocytosine, respectively, to cytotoxic active drugs, also were delivered to the tumor site via retroviral vectors. The efficacy of this approach, however, was limited by the poor infection of tumor cells in vivo. In recent studies, the use of bispecific bridging molecules has shown substantial progress toward improving targeting efficiency.[104, 105] These molecules, composed of a cell-specific ligand and the native retrovirus receptor, are noncovalently bound to the vector and thus mediate specific targeting to cells. Another interesting approach employing the retroviral vector is to use a coat protein from another enveloped virus, as suggested by Hatziioannou et al.[106] These authors demonstrated that retroviral vectors can efficiently incorporate influenza hemagglutinin glycoproteins that are engineered to display different peptides such as epidermal growth factor, an antihuman major histocompatability complex (MHC) class I molecule scFV, an antimelanoma antigen scFv, and an IgG Fc-binding polypeptide. The resulting vectors have shown significantly enhanced infectivity to the cells that express the targeted receptor.

Adenovirus is a nonenveloped, linear, double-stranded DNA virus, and it attacks the cell by binding to a coxsackievirus and adenovirus receptor protein (CAR) and internalizes via the integrins $\alpha_v\beta_3$ and $\alpha_v\beta_5$.[107,108] Einfeld et al.[109] observed that compared with a native vector, gene expression in the liver and other organs is reduced significantly by the systemic administration of a vector containing mutations that ablate CAR and integrin binding, indicating that both CAR and integrin interactions play a vital role in gene delivery in vivo. Use of replication-deficient recombinant adenovirus was proven to be very efficient in transferring genes to a wide variety of target cells because it expresses the transgene not only in replicating cells but also in nonreplicating cells.[110] Adenovirus was first used for the treatment of cystic fibrosis (CF) using the cystic fibrosis transmembrane conductance regulator (*CFTR*) gene, which was identified in 1989 by Kerem et al.[111] To improve targeting efficiency, adenovirus vectors have been modified by either genetic or nongenetic means.[101] Of various approaches, two-component systems have emerged recently as promising candidates for gene delivery using adenoviruses. These approaches use a bifunctional adaptor or bridging molecule capable of binding to the vector as well as to a target receptor.[112] In contrast to genetic modifications, two-component systems are not limited by the sized or types of ligands, and thus they are useful in assessing the feasibility of attaching adenovirus vectors to various receptors such as α_v integrins, folate receptors, and endoglin. For example, Reynolds et al.[113] have demonstrated that specific targeting to the lung vasculature can be achieved by using a bispecific antibody that binds both adenovirus and the lung endothelial-specific receptor (angiotensin-converting enzyme), resulting in most of the gene expression in the lung. The most serious known side effect of adenovirus is the adverse immune reaction against the patient, and this has limited the use of adenovirus, especially after the tragedy of J. Gelsinger (ornithine transcarbamylase deficiency syndrome patient) in 1999.[114]

In general, adenoviral vectors are known to infect the target cells effectively, but it is noteworthy to keep in mind that safety always comes first because some adverse side effects could be critical to the health of patients. Other types of viral vectors, such as adeno-associated virus, also are used for gene therapy in many diseases, even though more studies on safety, as well as efficacy, still remain until successful human clinical use can be expected. As demonstrated by clinical reports, fusion of knowledge on the molecular biology of viral vectors and the diseases to be treated holds promises for the future of medicine.

10.3.3 Nonviral vector

Delivery systems based on vectors, such as liposomes or polymers, have been used for the delivery of pharmacologically active macromolecules in various applications. The reasons that these vectors are used frequently include but are not limited to the safety and efficacy. Although toxicity problems have been reported in some cases, liposome- or polymer-based vectors generally are regarded as safe for human clinical use, especially in comparison with viral vectors. Thus a novel cationic liposome or polymer, by a slight change or modification from the existing system, may be used as a delivery system in very complicated steps of exogenous gene expression.

Although liposome- or polymer-based vector systems entered the gene therapy repertoire later than viral vectors, the discovery of cationic lipids greatly stimulated interest in the development of nonviral vectors for gene therapy applications, and they became one of the most widely used vectors in human clinical trials. Furthermore, as our understanding of the mechanisms by which these vectors affect gene transfer efficiency continues to improve, novel strategies to overcome the physical and biological barriers limiting this vector system are being developed.

This section will focus primarily on the development of nonviral vectors that are based on surfactants, phospholipids, and natural or synthetic polymers for their application to the delivery of macromolecules, mainly of genetic materials such as DNA or RNA. Development of safe and efficient vectors will expedite the success of human gene therapy, which has captured the attention recently of the public and the biomedical research community. Even though more and more new delivery vectors are being introduced into the scientific milieu with enhanced efficiency, it should be emphasized that safety always goes first, especially after the tragedy of J. Gelsinger in 1999.

Lipid. Lipids or surfactants had been used widely in the delivery of pharmacologically active materials in the form of liposomes, emulsions, or micelles. Since the first description of their potential for exogenous gene transfer, much progress has been made in the development of improved cationic lipid structures and formulations with enhanced gene transfection activity.

Liposome. Liposomes were first used for the delivery of genetic materials, such as nucleic acid, in 1979 by Dimitriadis.[115] However, the liposomes used were made of negatively charged phosphatidylserine (PS). Even though phospholipids with neutral or negative charges also were used for the purpose of gene delivery earlier, encapsulation and transfection efficiency were very low, and they were rather toxic. Others have reported the idea of using cationic liposomes for gene delivery purpose[116]; however, a major breakthrough using cationic phospholipid was achieved by Felgner et al. in 1987, who showed that transfection of an exogenous gene could be achieved in vitro with high transfection efficiency by using a newly synthesized cationic phospholipid *N*-[1-(2,3-dioleoyloxy)propyl]-*N,N,N*-trimethyl ammonium chloride (or DOTMA).[117] Since then, nonviral gene delivery in vivo is based mostly on the formation of a complex of nucleic acids with polycations such as cationic liposomes or polymers.

Numerous scientific groups have been working on the development of different cationic phospholipids for gene delivery (Table 10.2), but the structure of the DNA–phospholipids complex (or lipoplex) is still poorly

TABLE 10.2 Abbreviations and Chemical Names of Some Cationic Lipids

Abbreviation	Lipid	References
DOTMA	*N*-[1-(2,3-dioleoyloxy)propyl]-*N,N,N*-trimethylammonium chloride	[115]
DOTAP	1,2-Dioleoyl 3-trimethyl ammonium propane	Boehinger Mannheim
DC-chol	3-β[*N*-(*N',N'*-Dimethylaminoethane) carbamoyl] cholesterol	[132]
DOGS	Dioctadecyl amido glycil spermine	[133]
DOSPA	2,3,Dioleoyl-*N*[sperminecaroxamino)ethyl]-*N,N*-dimeth-ly-1-propanaminium	[133]
DMRIE	Dimyristoyl oxypropyl dimethyl hydroxyethyl ammonium bromide	[134]
DOSPER	1,3-Di-oleoyloxy-2(6-carboxy spermyl) propylamide four acetate	Sigma
DORIE	Dioleoyloxypropyl dimethyl hydroxyethyl ammonium bromide	[134]
DOTIM	1-[2-(oleoyloxy)-ethyl]-2-oleoyl-3-(2-hydroxyethyl) imidazolinium chloride	[129]

understood.[118,119] In many cases, addition of colipids, such as dioleoylphosphatidyl ethanolamine (DOPE), is shown to increase transfection efficiency.[120–122] A possible explanation seemed to be the action of DOPE as a destabilizer of endosomal membrane; however, a similar increase in transfection efficiency also was achieved when a membrane stabilizer, such as cholesterols, was added. This might indicate that there are different mechanisms of cellular uptake. The presence of neutral colipid can have some other effects, including charge density and fluidity of the lipid bilayers. These effects are also important in the interaction of DNA with liposomes, which subsequently influence the transfection efficiency.

The common structure of the lipid consists of nonpolar part, the linker, and the polar head group, even though slight modifications also can exist. Lee et al.[123] had observed that spermine attached to cholesterol through the secondary amine yielded over 100-fold higher transfection efficiency as compared with attaching to the primary amine. They suggested that a certain type of structure of the lipid, such as T-shaped motif in this case, might help DNA bind to lipid and stabilize the complex.

Liposomes used for transfection are either large unilamellar vesicles (LUVs) of 100 to 200 nm in diameter or small unilamellar vesicles (SUVs) of 20 to 100 nm. Liu et al.[124] have reported that for a given liposome composition, multilamellar vesicles (MLVs) of 300 to 700 nm in diameter exhibit higher transfection efficiency than SUVs. However, more recent studies on the nature of the liposome-DNA complex (or *lipoplex*) revealed that lipoplexes from SUVs or MLVs do not differ significantly in size. On the other hand, the composition of the medium, not the type of the liposome used in the preparation of the lipoplex, plays a key role in determining the final size of the complex. And the transfection efficiency is also shown to depend on the final size of the complexes but not the type of the liposome.[125]

Complexes prepared from a lipid-DNA charge ratio of 1:1 exhibit approximately neutral zeta potential, where all the DNA molecules are neutralized by cationic lipid molecules. Neutral lipoplexes show a heterogeneous size distribution in general and usually present much lower stability in solution (or in serum) owing to a lack of electrostatic repulsive forces. This phenomenon becomes more significant when large amounts of DNA or lipid are used. Ionic strength of the solutions is also an important factor influencing the physicochemical properties of lipoplexes, including surface charge. Low ionic strength of the solution is preferred because it enhances the attractive electrostatic forces required to form a more condensed lipoplex, resulting in less aggregation formation and increased transfection efficiency.

Even though an enormous amount of work has been devoted to the development of novel cationic liposomes, the mechanisms by which the lipoplexes enter the cells are not well understood yet. Known obstacles to the success of gene delivery include but are not limited to cellular membranes, cytoplasm, endosome, or lysosome, and nuclear membrane. Fusion and endocytosis are two main routes of cell entry, and these two pathways are not exclusive to each other. Attachment of targeting ligands, such as antibodies or lectins, to liposomes is known to increase the cell internalization process. However, it also should be pointed out that promotion of the extent of cellular binding and internalization of the lipoplexes do not necessarily guarantee an increase in gene expression.[126]

Following the interaction with plasma membranes, lipoplexes arrive at the cytoplasm after endosomal (or lysosomal) release. Presence of DOPE in the liposome formulations is thought to destabilize the endosomal membranes, helping to ease escape of the lipoplexes into the cytoplasm.[127] Inclusion of other types of pH-sensitive fusogenic materials, such as GALA from influenza virus hemagglutinin, resulted in similar increases in transfection. Entry of the free DNA, but not as a lipoplex form, into the nucleus is a prerequisite for the success of gene expression becuase the lipid inhibits transcription of DNA into mRNA, as shown in a microinjection experiment.[128]

Poor correlations between the in vitro and in vivo transfection rates were observed regardless of the administration routes of the lipoplexes.[129–131] Possible explanations for this might be the differences between cell culture and the in vivo system, the dilution effect in biological fluid, and the bioavailability of the lipoplexes. Serum proteins in blood can decrease the transfection efficiency because they dissociate the complex and/or increase the size of the lipoplexes. The effect of helper lipid, DOPE or cholesterol on the transfection efficiency in vitro is different from that in vivo, where DOPE increases transfection efficiency in vitro as opposed to what cholesterol does in vivo. In this regard, use of polyethylenglycol (PEG)–modified phospholipid or stealth liposome seems to be very promising because they might circumvent the above-mentioned problems associated with the physicochemical properties of the lipoplexes as well as prolongation of circulation time in vivo.

Emulsion. For the past decade, lipid emulsions have been investigated extensively for their applications in pharmaceutical and medical fields.[135–139] In particular, oil-in-water (o/w) emulsions based on cationic lipids have received increasing attention as potent gene carriers because they can maintain the physical integrity in the presence of serum, resulting in high transfection activity.[137,139] While liposome is a closed double layer of lipids filled with water, the o/w emulsion consists of oil dispersed in the aqueous phase with suitable emulsifying agents such as a phospholipids and nonionic surfactants.[135] The o/w emulsions are

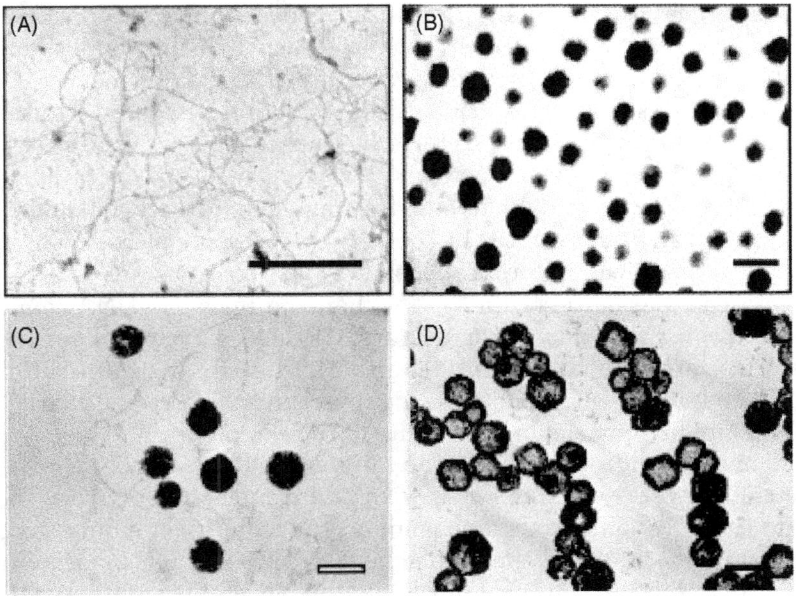

Figure 10.2 Transmission electron micrographs of (*A*) pCMV-beta and (*B*) trilaurin-cored emulsion and their complexes at different weight ratios. (*C*) DNA-emulsion (1:1). A few emulsion particles are found on the string of DNA. (*D*) DNA-emulsion (1:4). The emulsion and DNA are fully combined and form a chromatin-like structure. Scale bar = 100 nm.

known to form the complexes with DNA that has a chromatin-like structure,[139] whereas the liposome-DNA complexes exhibit a quite different structure in which DNA is aligned between lamellar lipid sheets or is arranged on a two-dimensional hexagonal lattice.[140,141] The representative morphology of the o/w emulsion-DNA complexes is shown in Fig. 10.2. Physicochemical properties of such o/w emulsions vary depending on the constituents of the emulsions, such as core oils, lipids, and surfactants.

The feasibility of o/w emulsions as gene carriers was demonstrated recently by Yi et al.[142] who formulated a cationic lipid emulsion using 1,2-dioleoyl-*sn*-glycero-3-trimethylammonium-propane (DOTAP) as a cationic emulsifier and 1-palmitoyl-2-oleoyl-*sn*-glycero-3-phosphoethanolamin-*N*-[poly (ethylene glycol) 2000] ($PEG_{2000}PE$) as an coemulsifier in the presence of soybean oil. These cationic lipid emulsions formed stable complexes with DNA and could efficiently deliver genes into COS-1 and CV-1 cells, even in the medium containing 90 percent serum; i.e., more than 60 percent of transfection efficiency was retained compared with the activity in the serum-free medium. This characteristic of the emulsions is of great interest and different from that of liposomes, which lose the ability to transfer genes in medium containing serum to as low as 5 to 10 percent.[143–145] Although the stability of the emulsions in serum is not fully understood, the PEG moiety in $PEG_{2000}PE$ used as a coemulsifier may play a significant role because PEO-containing lipids create a steric barrier to prevent the DNA-induced aggregation of lipid particles.[146, 147] Since serum contains a number of anionic materials that would strongly bind to cationic lipids, the complexes can be dissociated by competitive interactions, which are considered as one of the factors responsible for the poor stability of liposome-DNA complexes in serum.[148–150] When the liposome-DNA complexes are treated with poly(L-aspartic acid) (PLLA), a model polyanion DNA is dissociated from the complexes above a PLAA-DNA charge ratio of 1.25. The emulsion-DNA complexes, however, are stable up to a charge ratio of 200, indicating the high stability of the complexes.[139,142]

Selection of the oil component in the cationic emulsion may be a critical factor to improve its stability in vivo and transfection activity. To elucidate the effect of the oil component on the stability of emulsions, Chung et al.[151] formulated the emulsions by using egg phosphatidylcholine and 18 different natural oils, including vegetables and animal oils. Of the oils studied, small emulsions were formed by squalene, light mineral oil, and jojoba bean oil. On the contrary, cottonseed, linseed, and evening primrose oils produced large emulsions. The smaller emulsions stayed stable for a long period of time; e.g., squalene emulsion was maintained without any changes in the size distribution for 20 days, whereas the linseed oil emulsion was phase separated during the same period of time. This indicates that the size of emulsion significantly depends on the oil component and correlates with stability in an aqueous phase. In general, a more hydrophobic oil with a higher o/w interfacial tension appears to form a more stable emulsion with a small average particle size than less a hydrophobic oil. With regard to transfection activity, all the emulsions were less potent in transferring genes into COS-1 cells than the DOTAP liposome in the absence of serum; however, they showed higher transfection activity in 80 percent serum because of the higher stability of the emulsions compared with the DOTAP liposome. The highest transfection activity was found on squalene, emulsion which was the most stable in an aqueous phase. It can be suggested, therefore, that emulsion stability correlates with its function as a gene carrier.

Although the squalene emulsion was demonstrated to be the most potent carrier for gene delivery among the cationic lipid formulations, it is not yet as efficient as the use of a viral vector, thus requiring additional enhancement of its transfection activity. Since the cationic lipid formulations can be designed by various lipids to improve their in vitro and in vivo transfection activity, Kim et al.[152] prepared various squalene-based emulsions with different cationic lipids as emulsifiers and additional helper lipids as coemulsifiers to investigate their role in transfection activity. It was evident that among six different cationic lipids currently being investigated by many researchers, DOTAP produces the most stable emulsion, thus exhibiting the highest transfection activity.[152] Addition of 1,2-dioleoyl-*sn*-glycero-3-phosphoethanolamine (DOPE) into the emulsion is shown to increase the transfection activity more, since DOPE is known to facilitate endosomal escape.[148] However, it should be emphasized that emulsions with excess amounts of DOPE become unstable and have a big particle size because DOPE is not a good emulsifier, resulting in decreased transfection activity. In this regard, the highest transfection activity of the emulsions can be achieved by optimizing the balance between enhanced fusogenic ability and decreased physical stability of the emulsion.

A number of nonionic surfactants are available for use as constituents of the emulsion. The primary role of surfactants in cationic lipid formulation is to inhibit the formation of large aggregates in the process of complex formation with DNA.[146,147] In an attempt to investigate the role of surfactants in the transfection activity of the emulsions, Jeong et al.[153] prepared squalene emulsion with DOTAP and different nonionic surfactants. The surfactants containing the PEG group could generate the small and stable emulsion droplets that also form smaller complexes with DNA than the emulsions prepared in the absence of surfactants. For example, incorporation of Tween 80, which is a poly(oxyethylene sorbitan monooleate) having branched PEG chains, allowed the formation of stable emulsion-DNA complexes, leading to a significant increase in transfection activity. In addition to their stabilization effect on the complex, PEG-bearing surfactants have been proposed to possess a similar fusogenic property to DOPE.[137] On the other hand, the surfactants without the PEG group, such as Span 80, Montanide 80, and oleyl alcohol, could not provide the resistance to the size change in the presence of salts and in the process of complex formation with DNA. It should be considered, however, that the PEG groups of the emulsion may shield the positive charges of DOTAP, which interfere with the electrostatic interactions between DOTAP and DNA. This effect depends on the content and chain length of PEG groups, indicating the significant role of the PEG group in the formation of stable emulsion-DNA complexes.

Polymer. The field of nonviral gene delivery using cationic polymers is at its early stage compared to that of cationic lipid. However, this system is also known to have advantages over lipid-based systems in controlling the size, charge, and other physicochemical properties. Polycation-DNA complexes, also called as *polyplexes*, are formed by a cooperative system of ion pairs between cationic groups of polymers and anionic groups of DNA. As opposed to lipoplexes, polyplexes do not usually require interaction of the polycations molecules with each other to form a self-assembly structure. This system presents much more flexibility in forming the polyplexes by varying the molecular mass, polymer backbone structure, and side chain modification. This flexibility, however, does not necessarily guarantee enhanced transfection efficacy because the physicochemical properties of the complexes often show poor prediction of in vivo transfection efficacy. Polyplexes also have drawbacks, including low transfection efficiency and toxicity, that are similar to those of lipoplexes. Numerous natural and synthetic polycations were introduced into this field and were found to be effective in delivering exogenous gene, although the need for better understanding of the mechanisms by which polyplexes are taken up by cells and the resulting gene expression remains.

Natural polymer. Natural polymers, such as gelatin or chitosan, had been used widely as conventional pharmaceutical ingredients owing to their biocompatibility and low toxicity. Recently, they also were found to be useful in gene delivery owing to the positive charge characteristic of these polymers at certain physiological pH ranges.

Gelatin is a polyampholyte that gels below body temperature at around 35 to 40°C.[154] The major anionic side groups are aspartic acid and glutamic acid, and the major cationic groups are lysine and arginine. At pH below 5, gelatin is positively charged and therefore can form a complex with anionic materials such as DNA or antisense oligodeoxynucleotide to form a microsphres or nanosphere through a so-called coacervation process. Owing to the low transfection level of the gelatin-DNA microsphere system, attachment of targeting ligands on the surface of the complexes was shown to enhance the level by severalfold in vitro. In vivo application of the gelatin-DNA microsphere system appeared in delivery of the cystic fibrosis transmembrane conductance regulator (*CFTR*) gene to the lung airways.[155] It was shown that the *CFTR* gene persists in airway nuclei for almost a month with a high percentage of airway epithelia. Expression of the CFTR protein is highly localized to the apical membrane surface of airway cells, which is the site of function of this protein. Injection of this system into the tibialis muscle bundle of mice also expressed the protein successfully for 3 weeks.[156] Intramuscular injection of the gelatin-DNA microsphere system into mice elicits a modest antibody response with the aid of booster injections. This is a very promising result in the DNA vaccination field because the current system, such as complete Freund's adjuvant, has unavoidable side effects. Even though the in vivo biocompatibility and toxicity of gelatin as a gene carrier should be studied thoroughly before wide application to human clinical use, it has been reported that it is bioabsorbable, nontoxic, and poorly immunogenic.[154] This low toxicity may provide the opportunity for repeated administrations in vivo.

Chitin is found abundantly in the exoskeketons of crustaceans such as shrimp and crab, and chitosan is its *N*-deacetylated derivative. Since chitosan carries a positive charge at physiological pH, it has been used widely as an oral absorbent for removing fat or as a pharmaceutical ingredient for controlled release microsphere formulations.[157] The important factors for effective gene delivery are known to be the degree of chitosan deacetylation and its molecular weight, by which DNA binding and release are determined both in vitro and in vivo.[158] Oral gene delivery using chitosan is particularly interesting because the chitosan has been used as a potent absorption enhancer for many bioactive materials owing to its mucoadhesive characteristic. In vivo oral delivery of the chitosan-Arah2 antigen gene complex to mice was performed, and at a challenge test, the anaphylactic response was reduced significantly in the chitosan-antigen gene group but not in the naked DNA group.[159] Toxicity of chitosan, especially in repeated administrations, still needs to be investigated thoroughly even though low toxicity and immunogenicity have been reported.[160]

Synthetic polymer. Among the cationic synthetic polymers used for gene delivery are polyethylenimine (PEI), polyamidoamine dendrimers, and poly(2-dimethylaminoethyl methacrylate).[161–164] Depending on the flexibility (or rigidity) of the polymers, they form either a small (<100 nm) DNA polyplex or a large (>1 to10 μm) DNA polyplex.[165] More detailed physicochemical properties and their transfection efficacy are to be discussed.

PEI contains ethylamine, $-(CH_2CH_2NH)-$, as a repeating unit, thus providing very high solubility in water and a very high cationic charge density. Boussif et al.[146] showed that the in vitro transfection efficiency of PEI (MW 800,000 Da) in a variety of cell lines and primary cultures was as good as that of a representative cationic lipid such as lipofectam. However, this kind of high transfection efficiency was not seen with low-molecular-weight PEI of 2000 Da even at a very high N/P ratio. When the ligand, such as galactose, was conjugated to the amino nitrogens of PEI, the transfection efficiency to hepatocytes was shown to be enhanced, presumably owing to the asialoglycoprotein receptor (ASGP-R)–mediated endocytosis of the polyplexes. Moreover, the galactose-PEI-DNA complexes do not need to have otherwise necessary positive charges on their surfaces for desired transfection. The Arg-Gly-Asp (RGD) sequence, which is normally present in the extracellular matrix for targeting of cell surface integrin, also was exploited for receptor-mediated endocytosis of PEI-DNA complexes. Transfection efficiency of the RGD-PEI-DNA complex in integrin-expressing cells (HeLa or MRC5) was increased by 10- to 100-fold compared with simple PEI-DNA complexes. Interestingly enough, replacement of aspartic acid (D) by glutamic acid (E) significantly reversed the enhancement factor, indicating that sequence-specific interaction of RGD with integrin is necessary. In vivo delivery of luciferase plasmid complexed with linear PEI of molecular weight 22,000 Da (N/P ratio of 4) through the tail vein of adult mice revealed that 10^7 RLU/mg protein was expressed in the lung.[167] When branched PEI of similar molecular weight was used in the same experiment, cytotoxicity was observed even at lower *N/P* ratio. Boussif et al.[168] also have tried in vivo intracerebral injection of PEI-DNA complexes in mice, resulting in an equivalent gene expression level to the primary neuronal cells. Main transfected sites in the brain were neurons and glia, as revealed by immunostaining method, and the *N/P* ratio was 3 with the zeta-potential of 30 mV. PEI also was used in the delivery of oligodeoxynucleotide successfully in many types of neuronal cells, such as embryonic neurons and postnatal neurons.[168,169]

Polyamidoamine dendrimers, first introduced by Tomalia et al.,[170] are synthetic nonlinear cationic polymers and also known as *Starburst*. Owing to the ease of manipulation of synthesis, various generations of dendrimers had been used in gene delivery; especially the fifth- or sixth-generation PEI showed the highest transfection efficiency. However, nonbiodegradability and cytotoxicity of dendrimers still remain and often limit their wide applications to human clinical use. Surface modification of hydrophilic dendrimers with polyethylene oxide (PEO) was found to diminish the toxicity problems significantly. Recently, modification of the backbone of dendrimers was studied to improve the biodegradability and biocompatibility.

10.4 Future Directions

Clearly, it is necessary to improve our understanding of the mechanism by which carrier- and vector-mediated systems deliver biological molecules to specific cells, tissues, and organs. The design of the delivery system may depend on the type of biological macromolecules to be delivered, the target site, and the duration of drug action. Even though a number of approaches for the delivery of biological macromolecules have shown promising results in vitro, most are still facing barriers that need to be surmounted in vivo.

To avoid inactivation of the biological molecules in vivo by interaction with blood (or tissue) fluids and by clearance mechanisms, it may be desirable to coat the delivery system with a hydrophilic polymer such as polyethylene glycol and poly-[*N*-(2-hydroxypropyl)methacrylamide].

They may prevent carriers/vectors from interacting with harsh biological environments and may protect them from innate clearance mechanisms. Considerable effort, however, is further required to find optimal conditions such as the amount of hydrophilic polymers coated and their molecular weights, that minimize inactivation of the drugs but maximize the quantity reaching the target site. Recent advances in the ligand-guided targeting of biological macromolecules have made it possible to control the site that has defects capable of being treated by peptides, proteins, and genes. Therefore, the targeting efficiency of biological molecules would be further improved by conjugation of the ligands into coated hydrophilic polymers, carriers, or vectors. In addition to improving the targeting efficiency of active biological molecules, much effort also is needed to consider how the successful delivery systems can be scaled up readily without loss of quality for clinical trials.

References

1. Cefalu, W. T. Novel routes of insulin delivery for patients with type 1 or type 2 diabetes. *Ann. Med.* **33**:579–586, 2001.
2. Sinha, V. R., and Trehan, A. Biodegradable microspheres for protein delivery. *J. Contr. Rel.* **90**:261–280, 2003.
3. Morris, M. C., Depollier, J., Mery. J., et al. A peptide carrier for the delivery of biologically active proteins into mammalian cells. *Nature Biotechnol.* **19**:1173–1176, 2001.
4. Sakuma, S., Hayashi, M., and Akashi, M. Design of nanoparticles composed of graft copolymers for oral peptide delivery. *Adv. Drug Del. Rev.* **47**:21–37, 2001.
5. Goldberg, M., and Gomez-Orellana, I. Challenges for the oral delivery of macromolecules. *Nature Rev. Drug Disc.* **2**:289–295, 2003.
6. Yu, J., and Chien, Y. W. Pulmonary drug delivery: Physiologic and mechanistic aspects. *Crit. Rev. Ther. Drug Carrier Syst.* **14**:295–453, 1997.
7. Ishikawa, F., Katsura, M., Tamai, I., and Tsuji, A. Improved nasal bioavailability of elcatonin by insoluble powder formulation. *Int. J. Pharm.* **224**:105–114, 2001.
8. Fujita, T., Fujita, T., Morikawa, K., et al. Improvement of intestinal absorption of human calcitonin by chemical modification with fatty acids: Synergistic effects of acylation and absorption enhancers. *Int. J. Pharm.* **134**:47–57, 1996.
9. Uchiyama, T., Sugiyama, T., Quan, Y. S., et al. Enhanced permeability of insulin across the rat intestinal membrane by various absorption enhancers: Their intestinal mucosal toxicity and absorption-enhancing mechanism of *N*-lauryl-β-D-maltopyranoside. *J. Pharm. Pharmacol.* **51**:1241–1250, 1999.
10. Cox, D. S., Raje, S., Gao, H., et al. Enhanced permeability of molecular weight markers and poorly bioavailable compounds across Caco-2 cell monolayers using the absorption enhancer, zonula occludens toxin. *Pharm. Res.* **19**:1680–1688, 2002.
11. Lee, V. H. L. Protease inhibitors and penetration enhancers as approaches to modify peptide absorption. *J. Contr. Rel.* **13**:213–223, 1990.
12. Bernkop-Schnürch, A. The use of inhibitory agents to overcome the enzymatic barrier to perorally administered therapeutic peptides and proteins. *J. Contr. Rel.* **52**:1–16, 1998.
13. Allémann, E., Leroux, J., and Gurny, R. Polymeric nano- and microparticles for the oral delivery of peptides and peptidomimetics. *Adv. Drug Del. Rev.* **34**:171–189, 1998.
14. Jung, T., Kamm, W., Breitenbach, A., et al.. Biodegradable nanoparticles for oral delivery of peptides: Is there a role for polymers to affect mucosal uptake? *Eur. J. Pharm. Biopharm.* **50**:147–160, 2000.

15. Sinha, V. R., and Trehan, A. Biodegradable microspheres for protein delivery. *J. Contr. Rel.* **90**:261–280, 2003.
16. Donovan, M. D., Flynn, G. L., and Amidon, G. L. Absorption of polyethylene glycols 600 through 2000: The molecular weight dependence of gastrointestinal and nasal absorption. *Pharm. Res.* **8**:863–868, 1990.
17. Camenisch, G., Alsenz, J., van de Waterbeemd, H., and Folkers, G. Estimation of permeability by passive diffusion through Caco-2 cell monolayers using the drugs' lipophilicity and molecular weight. *Eur. J. Pharm. Sci.* **6**:317–324, 1998.
18. Pauletti, G. M., Gangwar, S., Knipp, G. T., et al. Structural requirements for intestinal absorption of peptide drugs. *J. Contr. Rel.* **41**:3–17, 1996.
19. Tomita, M., Shiga, M., Hayashi, M., and Awazu, S. Enhancement of colonic drug absorption by the paracellular permeation route. *Pharm. Res.* **5**:341–346, 1998.
20. Duizer, E., van der Wulp, C., Versantvoort, C. H., and Groten, J. P. Absorption enhancement, structural changes in tight junctions and cytotoxicity caused by palmitoyl carnitine in Caco-2 and IEC-18 cells. *J. Pharmacol. Exp. Ther.* **287**:395–402, 1998.
21. Sood, A., and Panchagnula, R. Peroral route: An opportunity for protein and peptide drug delivery. *Chem. Rev.* **101**:3275–3303, 2001.
22. Clausen, A. E., Kast, C. E., and Bernkop-Schnürch, A. The role of glutathione in the permeation enhancing effect of thiolated polymers. *Pharm. Res.* **19**:602–608, 2002.
23. Fasano, A. Innovative strategies for the oral delivery of drugs and peptides. *Trends Biotechnol.* **16**:152–157, 1998.
24. Thanou, M., Verhoef, J. C., and Junginger, H. E. Oral drug absorption enhancement by chitosan and its derivatives. *Adv. Drug Del. Rev.* **52**:117–126, 2001.
25. Schipper, N. G., Olsson, S., Hoogstraate, J. A., et al. Chitosans as absorption enhancers for poorly absorbable drugs:2. Mechanism of absorption enhancement. *Pharm. Res.* **14**:923–929, 1997.
26. Swaan, P. W. Recent advances in intestinal macromolecular drug delivery via receptor-mediated transport pathways. *Pharm. Res.* **15**:826–834, 1998.
27. Okamoto, C. T. Endocytosis and transcytosis. *Adv. Drug Del. Rev.* **29**:215–228, 1998.
28. Widera, A., Norouziyan, F., and Shen, W.-C. Mechanisms of TfR-mediated transcytosis and sorting in epithelial cells and applications toward drug delivery. *Adv. Drug Del. Rev.* **55**:1439–1466, 2003.
29. Larhed, A. W., Artursson, P., Grasjo, J., and Bjork, E. Diffusion of drugs in native and purified gastrointestinal mucus. *J. Pharm. Sci.* **86**:660–665, 1997.
30. Bernkop-Schnürch, A., Valenta, C., and Daee, S. M. Peroral polypeptide delivery. A comparative in vitro study of mucolytic agents. *Arzneimittel-Forsch.* **49**:799–803, 1999.
31. Pauletti, G. M., Gangwar, S., Siahaan, T. J., et al. Improvement of oral peptide bioavailability: Peptidomimetics and prodrug strategies. *Adv. Drug Del. Rev.* **27**:235–256, 1997.
32. Marschutz, M. K., and Bernkop-Schnürch, A. Oral peptide drug delivery: Polymer-inhibitor conjugates protecting insulin from enzymatic degradation in vitro. *Biomaterials* **21**:1499–1507, 2000.
33. Kratzel, M., Hiessbock, R., and Bernkop-Schnürch, A. Auxiliary agents for the peroral administration of peptide and protein drugs: Synthesis and evaluation of novel pepstatin analogues. *J. Med. Chem.* **41**:2339–2344, 1998.
34. Sjostrom, M., Lindfors, L., and Ungell, A. L. Inhibition of binding of an enzymatically stable thrombin inhibitor to lumenal protease as an additional mechanism of intestinal absorption enhancement. *Pharm. Res.* **16**:74–79, 1999.
35. Singla, A. K., Chawla, M., and Singh, A. Potential applications of carbomer in oral mucoadhesive controlled drug delivery system: A review. *Drug Dev. Ind. Pharm.* **26**:913–924, 2000.
36. Luessen, H. L., Verhoef, J. C., Borchard, G., et al. Mucoadhesive polymers in peroral peptide drug delivery. II. Carbomer and polycarbophil are potent inhibitors of the intestinal proteolytic enzyme trypsin. *Pharm. Res.* **12**:1293–1298, 1995.
37. Bernkop-Schnürch, A., and Göckel, N. C. Development and analysis of a polymer protecting from luminal enzymatic degradation caused by α-chymotrypsin. *Drug Dev. Ind. Pharm.* **23**:733–740, 1997.

38. Bernkop-Schnürch, A., and Krajicek, M. E. Mucoadhesive polymers as platforms for peroral peptide delivery and absorption: synthesis and evaluation of different chitosan-EDTA conjugates. *J. Contr. Rel.* **50**:215–223, 1998.
39. Rouquayrol, M., Gaucher, B., Roche, D., et al. Transepithelial transport of prodrugs of the HIV protease inhibitors saquinavir, indinavir, and nelfinair across Caco-2 cell monolayers. *Pharm. Res.* **19**:1704–1712, 2002.
40. Swaan, P. W., Hillgren, K. M., Szoka, F. C., Jr.,and Oie, S. Enhanced transepithelial transport of peptides by conjugation to cholic acid. *Bioconj. Chem.* **8**:520–525, 1997.
41. Kramer, W., Wess, G., Neckermann, G., et al. Intestinal absorption of peptides by coupling to bile acids. *J. Biol. Chem.* **269**:10621–10627, 1994.
42. Kramer, W., Wess, G., Enhsen, A., et al. Modified bile acids as carriers for peptides and drugs. *J. Contr. Rel.* **46**:17–30, 1997.
43. Enhsen, A., Kramer, W., and Wess, G. Bile acids in drug discovery. *Drug Disc. Today* **3**:409–418, 1998.
44. Russell-Jones, G. J. The potential use of receptor-mediated endocytosis for oral drug delivery. *Adv. Del. Rev.* **46**:59–73, 2001.
45. Lavelle, E. C. Targeted delivery of drugs to the gastrointestinal tract. *Crit. Rev. Ther. Drug Carrier Syst.* **18**:341–386, 2001.
46. Alsenz, J., Russell-Jones, G. J., Westwood, S., et al. Oral absorption of peptides through the cobalamin (vatamin B_{12}) pathway in the rat intestine. *Pharm. Res.* **17**:825–832, 2000.
47. Grant, G., Greer, F., Mackenzie, N. H., and Pusztai, A. Nutritional response of mature rats to kidney bean (*Phaseolus valgaris*) lectins. *J. Sci. Food Agr.* **36**:409–414, 1985.
48. Kilpatrick, D. C., Pusztai, A., Grant, G., et al. Tomato lectin resists digestion in the mammalian alimentary canal and binds to intestinal villi without deleterious effects. *FEBS Lett.* **185**:299–305, 1985.
49. Woodley, J. F. Lectins for gastrointestinal targeting—15 years on. *J. Drug Target.* **7**:325–333, 2000.
50. Naisbett, B.,and Woodley, J. The potential use of tomato lectin for oral drug delivery: 2. mechanism of uptake in vitro. *Int. J. Pharm.* **110**:127–136, 1994.
51. Lehr, C. M., Bouwstra, J. A., Kok, W., et al. Bioadhesion by means of specific binding of tomato lectin. *Pharm. Res.* **9**:547–553, 1992.
52. Clark, M. A., Hirst, B. H., and Jepson, M. A. Lectin-mediated mucosal delivery of drugs and microparticles. *Adv. Drug Del. Rev.* **43**:207–223, 2000.
53. Hussain, N., Jani, P. U., and Florence, A. T. Enhanced oral uptake of tomato lectin-conjugated nanoparticles in the rat. *Pharm. Res.* **14**:613–618, 1997.
54. Carreno-Gomez, B., Woodley, J. F., and Florence, A. T. Studies on the uptake of tomato lectin nanoparticles in everted gut sacs. *Int. J. Pharm.* **183**:7–11, 1999.
55. Irache. J. M., Durrer, C., Duchéne, D., and Ponchel, G. Bioadhesion of lectin-latex conjugates to rat intestinal mucosa. *Pharm. Res.* **13**:1716–1719, 1996.
56. Clark, M. A., Jepson, M. A., Simmons, N. L., et al. Differential expression of lectin-binding sites defines mouse intestinal M-cells. *J. Histochem. Cytochem.* **41**: 1679–1687, 1993.
57. Clark, M. A., Jepson, M. A., Simmons, N. L, and Hirst, B. H. Selective binding and transcytosis of *Ulex europaeus* 1 lectin by mouse Peyer's patch M-cells in vivo. *Cell Tissue Res.* **282**:455–461, 1995.
58. Chen, H., Torchilin, V., and Langer, R. Lectin-bearing polymerized liposomes as potential oral vaccine carriers. *Pharm. Res.* **13**:1378–1383, 1996.
59. Foster, N., Clark, M. A., Jepson, M. A., and Hirst, B. H. *Ulex europaeus* 1 lectin targets microspheres to mouse Peyer's patch M-cells in vivo. *Vaccine* **16**:536–541, 1998.
60. Clark, M. A., Blair, H., Liang L., et al. Targeting polymerized liposome vaccine carriers to intestinal M cells. *Vaccine* **20**:208–217, 2001.
61. Jepson, M. A., Mason, C. M., Clark, M. A., et al. Variations in lectin binding properties of intestinal M cells. *J. Drug Target.* **3**:75–77, 1995.
62. Russell-Jones, G. J., Arthur, L., Walker, H. Vitamin B_{12}–mediated transport of nanoparticles across Caco-2 cells. *Int. J. Pharm.* **179**:247–255, 1999.
63. Russell-Jones, G. J. Use of vitamin B_{12} conjugates to deliver protein drugs by the oral route. *Crit. Rev. Ther. Drug Carrier Syst.* **15**:557–586, 1998.

64. Panchagnula, R., and Thomas, N. S. Biopharmaceutics and pharmacokinetics in drug research. *Int. J. Pharm.* **201**:151–150, 2000.
65. Yamada, K., Murakami, M., Yamamoto, A., et al. Improvement of intestinal absorption of thyrotropin-releasing hormone by chemical modification with lauric acid. *J. Pharm. Pharmacol.* **44**:717–721, 1992.
66. Yodoya, E., Uemura, K., Tenma, T., et al. Enhanced permeability of tetragastrin across the rat intestinal membrane and its reduced degradation by acylation with various fatty acids. *J. Pharmacol. Exp. Ther.* **271**:1509–1513, 1994.
67. Uchiyama, T., Kotani, A., Tasumi, H., et al. Development of novel lipophilic derivatives of DADLE (leucine enkephalin analogue): Intestinal permeability characteristics of DADLE derivatives in rats. *Pharm. Res.* **17**:1461–1467, 2000.
68. Lee, Y., Nam, J. H., Shin, H., and Byun, Y. Conjugation of low-molecular-weight heparin and deoxycholic acid for the development of a new oral anticoagulant agent. *Circulation* **104**:3116–3120, 2001.
69. Hashizume, M., Douen, T., Murakami, M., et al. Improvement of large intestinal absorption of insulin by chemical modification with palmitic acid in rats. *J. Pharm. Pharmacol.* **44**:555–559, 1992.
70. Lee, Y., Kim, S. H., and Byun, Y. Oral delivery of new heparin derivatives in rats. *Pharm. Res.* **17**:1259–1264, 2000.
71. Arien, A., Toulme-Henry, N., and Dupuy, B. Cholate-induced disruption of calcitonin-loaded liposomes: Formation of trypsin-resistant lipid-calcitonin-cholate complexes. *Pharm. Res.* **12**:1289–1292, 1995.
72. Kisel, M. A., Kulik, L. N., Tsybovsky, I. S., et al. Liposomes with phosphatidylethanol as a carrier for oral delivery of insulin: studies in the rat. *Int. J. Pharm.* **216**:105–114, 2001.
73. Xing, L., Dawei, C., Liping, X., and Rongqing, Z. Oral colon-specific drug delivery for bee venom peptide: Development of a coated calcium alginate gel beads-entrapped liposome. *J. Contr. Rel.* **93**:293–300, 2003.
74. Li, H., Song, J. H., Park, J. S., and Han, K. Polyethylene glycol–coated liposomes for oral delivery of recombinant human epidermal growth factor. *Int. J. Pharm.* **258**:11–19, 2003.
75. Iwanaga, K., Ono, S., Narioka, K., et al. Oral delivery of insulin by using surface coating liposomes: Improvement of stability of insulin in GI tract. *Int. J. Pharm.* **157**:73–80, 1997.
76. Zeisig, R., Shimada, K., Hirota, S., and Arndt, D. Effect of sterical stabilization on macrophage uptake in vitro and on thickness of the fixed aqueous layer of liposomes made from alkylphosphocholines. *Biochim. Biophys. Acta* **1285**:237–245, 1996.
77. Iwanaga ,K., Ono, S., Narioka, K., et al. Application of surface-coated liposomes for oral delivery of peptide: effects of coating the liposome's surface on the GI transit of insulin. *J. Pharm. Sci.* **88**:248–252, 1999.
78. Takeuchi, H., Yamamoto, H., and Kawashima, Y. Mucoadhesive nanoparticulate systems for peptide drug delivery. *Adv. Drug Del. Rev.* **47**:39–54, 2001.
79. Takeuchi, H., Yamamoto, H., Niwa, T., et al. Enteral absorption of insulin in rats from mucoadhesive chitosan-coated liposomes. *Pharm. Res.* **13**:896–901, 1996.
80. Takeuchi, H., Matsui, Y., Yamamoto, H., and Kawashima, Y. Mucoadhesive properties of carbopol or chitosan-coated liposomes and their effectiveness in the oral administration of calcitonin to rats. *J. Contr. Rel.* **86**:235–242, 2003.
81. Florence, A. T. The oral absorption of micro- and nanoparticles: Neither exceptional nor unusual. *Pharm. Res.* **14**:259–266, 1997.
82. Mathiowitz, E., Jacob, J. S., Jong, Y. S., et al. Biologically erodable microspheres as potential oral drug delivery systems. *Nature* **386**:410–414, 1997.
83. Jiao, Y., Ubrich, N., Marchand-Arvier, M., et al. In vitro and in vivo evaluation of oral heparin-loaded polymeric nanoparticles in rabbits. *Circulation* **105**:230–235, 2002.
84. Jani, P. U., Halbert, G. W., Langridge, J., and Florence, A. T. Nanoparticle uptake by the rat gastrointestinal mucosa: Quantitation and particle size dependency. *J. Pharm. Pharmacol.* **42**:821–826, 1990.
85. Ebel, J. P. A method for quantifying particle absorption from the small intestine of the mouse. *Pharm. Res.* **7**:848–851, 1990.

86. Jenkins, P. G., Howard, K. A., Blackhall, N. W., et al. Microparticulate absorption from the rat intestine, *J. Contr. Rel.* **29**:339–350, 1994.
87. Carr, K. E., Hazzard, R. A., Reid, S., and Hodges, G. M. The effect of size on uptake of orally administered latex microparticles in the small intestine and transport to mesenteric lymph nodes. *Pharm. Res.* **13**:1205–1209, 1996.
88. Tabata, Y., Inoue, Y., and Ikada, Y. Size effect on systemic and mucosal immune responses induced by oral administration of biodegradable microspheres. *Vaccine* **14**:1677–1685, 1996.
89. Damgé, C., Aprahamian, M., Marchais, H., et al. Intestinal absorption of PLAGA microspheres in the rat. *J. Anat.* **189**:491–501, 1996.
90. Desai, M. P., Labhasetwar, V., Amidon, G. L., and Levy, R. J. Gastrointestinal uptake of biodegradable microparticles: Effect of particle size. *Pharm. Res.* **13**:1838–1845, 1996.
91. Desai, M. P., Labhasetwar, V., Walter, E., et al. The mechanism of uptake of biodegradable microparticles in Caco-2 cells is size dependent. *Pharm. Res.* **14**:1568–1573, 1997.
92. Hillery, A. M., Florence, A. T. The effect of adsorbed poloxamer 188 and 407 surfactants on the intestinal uptake of 60-nm polystyrene particles after oral administration in the rat, *Int. J. Pharm.* **132**:123–130, 1996.
93. Jani, P., Halbert, G. W., Langridge, J.,and Florence, T. The uptake and translocation of latex nanospheres and microspheres after oral administration to rats. *J. Pharm. Pharmacol.* **41**:809–812, 1989.
94. Kriwet, B., and Kissel, T. Poly(acrylic acid) microparticles widen the intercellular spaces of Caco-2 cell monolayers: An examination by confocal laser scanning microscopy. *Eur. J. Pharm. Biopharm.* **42**:233–240, 1996.
95. Everett, R. S., Hodges, B. L., Ding, E. Y., et al. Liver toxicities typically induced by first-generation adenoviral vectors can be reduced by use of E1, E2b-deleted adenoviral vectors. *Hum. Gene Ther.* **14**:1715–1726, 2003.
96. Medina-Kauwe, L. K. Endocytosis of adenovirus and adenovirus capsid proteins. *Adv. Drug Del. Rev.* **55**:1485–1496, 2003.
97. Dornburg, R. Reticuloendotheliosis viruses and derived vectors for human gene therapy. *Front. Biosci.* **8**:d801–d817, 2003.
98. Lundstrom, K. Latest development in viral vectors for gene therapy. *Trends Biotechnol.* **21**:117–122, 2003.
99. Couto, L. B., and Pierce, G. F. AAV-mediated gene therapy for hemophilia. *Curr. Opin. Mol. Ther.* **5**:517–523, 2003.
100. Kircheis, R., Wightman, L., and Wagner, E. Design and gene delivery activity of modified polyethylenimines. *Adv. Drug Del. Rev.* **53**:341–358, 2001.
101. Wickham, T. J. Ligand-directed targeting of genes to the site of disease. *Nature Med.* **9**:135–139, 2003.
102. Hong, J. W., Park, J. H., Huh, K. M.,et al. PEGylated polyethyleneimine for in vivo local gene delivery based on lipiodolized emulsion system. *J. Contr. Rel.* 99;167–176, 2004.
103. Blaese, R. M., Culver, K. W., Miller, A. D., et al. T-lymphocyte-directed gene therapy for ADA- SCID: Initial trial results after 4 years. *Science.* **270**:475–480, 1995.
104. Boerger, A. L., Snitkovsky, S., and Young, J. A. Retroviral vectors preloaded with a viral receptor-ligand bridge protein are targeted to specific cell types. *Proc. Natl. Acad. Sci. USA* **96**:9867–9872, 1999.
105. Snitkovsky, S., and Young, J. A. Targeting retroviral vector infection to cells that express heregulin receptors using a TVA-heregulin bridge protein. *Virology* **292**:150–155, 2002.
106. Hatziioannou, T., Delahaye, E., Martin, F., et al. Retroviral display of functional binding domains fused to the amino terminus of influenza hemagglutinin. *Hum. Gene Ther.* **10**:1533–1544, 1999.
107. Bergelson, J. M., Cunningham, J. A., Droguett, G., et al. Isolation of a common receptor for coxsackie B viruses and adenoviruses 2 and 5. *Science* **275**:1320–1323, 1997.
108. Wickham, T. J., Mathias, P., Cheresh, D. A., and Nemerow, G. R. Integrins $\alpha_v\beta_3$ and $\alpha_v\beta_5$ promote adenovirus internalization but not virus attachment. *Cell* **73**:309–319, 1993.

109. Einfeld, D. A., Schroeder, R., Roelvink, P. W., et al. Reducing the native tropism of adenovirus vectors requires removal of both CAR and integrin interactions. *J Virol* **75**:11284–11291, 2001.
110. Rosenfeld, M. A., Siegfried, W., Yoshimura, K., et al. Adenovirus-mediated transfer of a recombinant α_1-antitrypsin gene to the lung epithelium in vivo (comment). *Science* **252**:374, 1991.
111. Kerem, E., Reisman, J., Corey, M., et al. Prediction of mortality in patients with cystic fibrosis (comment) *New Engl. J. Med.* **327**:1244–1245, 1992.
112. Douglas, J. T., Rogers, B. E., Rosenfeld, M. E., et al. Targeted gene delivery by tropism-modified adenoviral vectors. *Nature Biotechnol.* **14**:1574–1578, 1996.
113. Reynolds, P. N., Nicklin, S. A., Kaliberova, L., et al. Combined transductional and transcriptional targeting improves the specificity of transgene expression in vivo. *Nature Biotechnol.* **19**:838–842, 2001.
114. NIH Report. Assessment of adenoviral vector safety and toxicity: Report of the National Institutes of Health Recombinant DNA Advisory Committee. *Hum. Gene Ther.* **13**:3–13, 2002.
115. Dimitriadis, G. J. Cellular uptake of ribonucleic acids entrapped into liposomes. *Cell Biol. Int. Rep.* **3**:543–549, 1979.
116. Fraley, R., and Papahadjopoulos, D. Liposomes: The development of a new carrier system for introducing nucleic acid into plant and animal cells. *Curr. Top. Microbiol. Immunol.* **96**:171–191, 1982.
117. Felgner, P. L., Gadek, T. R., Holm, M., et al. A highly efficient lipid-mediated DNA transfection procedure. *Proc. Natl. Acad. Sci. USA* **84**: 7413–741, 1987.
118. Ewert, K., Slack, N. L., Ahmad, A., et al. Cationic lipid-DNA complexes for gene therapy: Understanding the relationship between complex structure and gene delivery pathways at the molecular level. *Curr. Med. Chem.* **11**:133–149, 2004.
119. Wiethoff, C. M., Koe, J. G., Koe, G. S., and Middaugh, C. R. Compositional effects of cationic lipid/DNA delivery systems on transgene expression in cell culture. *J. Pharm. Sci.* **93**:108–123, 2004.
120. Elin, A. J., Slack, N. L., Ahmad, A., et al. Three-dimensional imaging of lipid gene-carriers: Membrane charge density controls universal transfection behavior in lamellar cationic liposome-DNA complexes. *Biophys. J.* **84**:3307–3316, 2003.
121. Duzgnues, N., Goldstwin, J. A., Friend, D. S., and Felgner, P. L. Fusion of liposomes containing a novel cationic lipid, *N*-[2,3-(dioleoyoxy)-propyl]-*N,N,N*-trimethylammonium: Induction by multivalent anions and asymmetric fusion with acidic phospholipid vesicles. *Biochemistry* **28**:9179–9184, 1989.
122. Zhou, X., and Huang, L. DNA transfection mediated by cationic liposomes containing lipopolylysine: Characterization and mechanism of action. *Biochim. Biophys. Acta.* **1189**:195–203, 1994.
123. Lee, E. R., Marshall, J., Siegel, C. S., et al. Detailed analysis of structures and formulations of cationic lipids for efficient gene transfer to the lung. *Hum. Gene Ther.* **7**:1701–1717, 1996.
124. Liu, Y., Mounkes, L. C., Liggitt, H. D., et al. Factors influencing the efficiency of cationic liposome-mediated intravenous gene delivery. *Nature Biotechnol.* **15**:167–173, 1997.
125. Ross, P. C., and Hui, S. W. Lipoplex size is a major determinant of in vitro lipofection efficiency. *Gene Ther.* **6**:651–659, 1999.
126. Cheng, P. W. Receptor ligand-facilitated gene transfer: Enhancement of liposome-mediated gene transfer and expression by transferrin. *Hum. Gene Ther.* **7**:275–282, 1996.
127. Noguchi, A., Furuno, T., Kawaura, C., and Nakanishi, M. Membrane fusion plays an important role in gene transfection mediated by cationic liposomes. *FEBS Lett.* **433**:169–173, 1998.
128. Zabner, J., Fasbender, A. J., Moninger, T., et al. Cellular and molecular barriers to gene transfer by a cationic lipid. *J. Biol. Chem.* **270**:18997–19007, 1995.
129. Solodin, I., Brown, C. S., Bruno, M. S., et al. A novel series of amphiphilic imidazolinium compounds for in vitro and in vivo gene delivery. *Biochemistry* **34**:13537–13544, 1995.

130. Gorman, C. M., Aikawa, M., Fox, B., et al. Efficient in vivo delivery of DNA to pulmonary cells using the novel lipid EDMPC. *Gene Ther.* **4**:983–992, 1997.
131. McLachlan, G., Davidson, D. J., Stevenson, B. J., et al. Evaluation in vitro and in vivo of cationic liposome-expression construct complexes for cystic fibrosis gene therapy. *Gene Ther.* **2**:614–622, 1995.
132. Gao, X., and Huang, L. Cationic liposome-mediated gene transfer. *Gene Ther.* **2**:710–722, 1995.
133. Behr, J. P. Synthetic gene transfer vectors. *Acc. Chem. Res.* **26**:274–278, 1993.
134. Felgner, J. H., Kumar, R., Sridhar, C. N., et al. Enhanced gene delivery and mechanism studies with a novel series of cationic lipid formulations. *J. Biol. Chem.* **269**:2550–2561, 1994.
135. Benita, S., and Levy, M. Y. Submicron emulsions as colloidal drug carriers for intravenous administration: Comprehensive physicochemical characterization. *J. Pharm. Sci.* **82**:1069–1079, 1993.
136. Wheeler, J. J., Wong, K. F., Ansell, S. M., et al. Polyethylene glycol modified phospholipids stabilize emulsions prepared from triacylglycerol. *Pharm. Res.* **83**:1558–1564, 1994.
137. Liu, F., Yang, J., Huang, L., and Liu, D. Effect of non-ionic surfactants on the formation of DNA/emulsion complexes and emulsion-mediated gene transfer. *Pharm. Res.* **13**:1642–1646, 1996.
138. Floyd, A. G. Top ten considerations in the development of parenteral emulsions. *Pharm. Sci. Technol. Today* **2**:134–143, 1999.
139. Kim, Y. J., Kim, T. W., Chung, H., et al. The effects of serum on the stability and the transfection activity of the cationic lipid emulsion with various oils. *Int. J. Pharm.* **252**:241–252, 2003.
140. Radler, J. O., Koltover, I., Salditt, T., and Safinya, C. R. Structure of DNA-cationic liposome complexes: DNA intercalation in multilamellar membranes in distinct interhelical packing regimes. *Science* **275**:810–814, 1997.
141. Koltover, I., Salditt, T., Radler, J. O., and Safinya, C. R. An inverted hexagonal phase of cationic liposome-DNA complexes related to DNA release and delivery. *Science* **281**:78–81, 1998.
142. Yi, S. W., Yune, T. Y., Kim, T. W., et al. A cationic lipid emulsion/DNA complex as a physically stable and serum-resistant gene delivery system. *Pharm. Res.* **17**:314–320, 2000.
143. Thierry, A. R., Rabinovich, P., Peng, B., et al. Characterization of liposome-mediated gene delivery: Expression, stability and pharmacokinetics of plasmid DNA. *Gene Ther.* **4**:226–237, 1997.
144. Dodds, E., Dunckley, M. G., Naujoks, K., et al. Lipofection of cultured mouse muscle cells: A direct comparison of Lipofectamine and DOSPER. *Gene Ther.* **5**:542–551, 1998.
145. Hofland, H. E., Shephard, L., and Sullivan, S. M. Formation of stable cationic lipid/DNA complexes for gene transfer. *Proc. Natl. Acad. Sci. USA* **93**:7305–7309, 1996.
146. Liu, F., Yang, J., Huang, L., and Liu, D. New cationic lipid formulations for gene transfer. *Pharm. Res.* **13**:1856–1860, 1996.
147. Hong, K., Zheng, W., Baker, A., and Papahadjopoulos, D. Stabilization of cationic liposome-plasmid DNA complexes by polyamines and poly(ethylene glycol)-phospholipid conjugates for efficient in vivo gene delivery. *FEBS Lett.* **400**:233–237, 1997.
148. Xu, Y., and Szoka, F. C. Mechanism of DNA release from cationic liposome/DNA complexes used in cell transfection. *Biochemistry* **35**:5616–5623, 1996.
149. Zabner, J. Cationic lipids used in gene transfer. *Adv. Drug Del. Rev.* **27**:17–28, 1997.
150. Escriou, V., Ciolina, C., Lacroix, F., et al. Cationic lipid-mediated gene transfer: Effect of serum on cellular uptake and intracellular fate of lipopolyamine/DNA complexes. *Biochim. Biophys. Acta* **1368**:276–288, 1998.
151. Chung, H., Kim, T. W., Kwon, M., et al. Oil components modulate physical characteristics and function of the natural oil emulsions as drug or gene delivery system. *J. Contr. Rel.* **71**:339–350, 2001.
152. Kim, T. W., Chung, H., Kwon, I. C., et al. Optimization of lipid composition in cationic emulsion as in vitro and in vivo transfection agents. *Pharm. Res.* **18**:54–60, 2001.

153. Kim, T. W., Kim, Y. J., Chung, H., et al. The role of non-ionic surfactants on cationic lipid mediated gene transfer. *J. Contr. Rel.* **82**:455–465, 2002.
154. Rose, P. I. Gelatin, in J. I. Kroschwitz (ed.), *Concise Encyclopedia of Polymer Science and Engineering*. New York: Wiley, p. 430.
155. Walsh, S. M., Flotte, T. R., Beck, S., et al. Delivery of *CFTR* gene to rabbit airways by gelatin-DNA microsphere. Int. Symp. Control. Rel. Bioact. Mater, **23**:73–74, 1996.
156. Truong-Le, V. L., August, J. T., and Leong, K. W. Controlled gene delivery by DNA-gelatin nanospheres. *Hum. Gene Ther.* **9**:1709–1717, 1998.
157. Kas, H. S. Chitosan: Properties, preparations and application to microparticulate systems. *J. Microencapsul.* **14**:689–711, 1997.
158. Kiang, T., Wen, J., Lim, H. W., and Leong, K. W. The effect of the degree of chitoan deacetylation on the efficiency of gene transfection. *Biomaterials* **25**:5293–5301, 2004.
159. Roy, K., Mao, H. Q., Lin, K. Y., et al. Oral immunization with DNA chitosan nanospheres. Int. Symp. Control. Rel. Bioact. Mater, **25:**348–349, 1998.
160. Brine, C. J., Sandford, P. A., and Zikakis, J. P. (eds.). *Advances in Chitin and Chitosan*. London: Elsevier Applied Science, 1992.
161. Haensler, J., and Szoka, F. C. Polyamidoamine cascade polymers mediate efficient transfection of cells in culture. *Bioconjug. Chem.* **4**:372–379, 1993.
162. Bielinska, A. U., Chen, C., Johnson, J., and Baker, J. R., Jr. DNA complexing with polyamidoamine dendrimers: Implications for transfection. *Bioconjug. Chem.* **10**:843–850, 1999.
163. Boussif, O., Lezoualc'h, F., Zanta, M. A., et al. A versatile vector for gene and oligonucleotide transfer into cells in culture and in vivo: Polyethylenimine. *Proc. Natl. Acad. Sci. USA.* **92**:7297–7301, 1995.
164. Cherng, J. Y., van de Wetering, P., Talsma, H., et al. Effect of size and serum proteins on transfection efficiency of poly[(2-dimethylamino)ethyl methacrylate]-plasmid nanoparticles. *Pharm. Res.* **13**:1038–1042, 1996.
165. Kabanov, A. V., Szoka, F. C., Jr., Seymour, L. W., Interpolyelectrolyte complexes for gene delivery: Polymer aspects of transfection activity. In Kabanov, A. V. Felgner, P. L., and Seymour, L. W. (eds.). *Self-Assembling Complexes for Gene Delivery: From Laboratory to Clinical Trial*, Chichester, U.K.: Wiley, 1998, pp. 197–218.
166. Boussif, O., Zanta, M. A., and Behr, J. P. Optimized galenics improve in vitro gene transfer with cationic molecules up to 1000-fold. *Gene Ther.* **3**:1074–1080, 1996.
167. Goula, D., Benoist, C., Mantero, S., et al. Polyethylenimine-based intravenous delivery of transgenes to mouse lung. *Gene Ther.* **5**:1291–1295, 1998.
168. Boussif, O., Lezoualc'h, F., Zanta, M. A., et al. A versatile vector for gene and oligonucleotide transfer into cells in culture and in vivo: Polyethylenimine. *Proc. Natl. Acad. Sci. USA* **92**:7297–7301, 1995.
169. Lambert, R. C., Maulet, Y., Dupont, J. L., et al. Polyethylenimine-mediated DNA transfection of peripheral and central neurons in primary culture: Probing Ca2+ channel structure and function with antisense oligonucleotides. *Mol. Cell. Neurosci.* **7**:239–246, 1996.
170. Tomalia, D. A., Naylor, A. M., Goddard, Q. A. III. Starburst dendrimers: Molecular level control of size, shape, surface, chemistry, topology and flexibility from atoms to macroscopic matter. *Angew Chem. Int. Ed.* **29**:138–175, 1990.

Chapter

11

Physical Targeting Approaches to Drug Delivery

Xin Guo

Thomas J. Long School of Pharmacy and Health Sciences
University of the Pacific
Stockton, California

11.1 Introduction

The beginning of the millennium has seen an urgent demand for improving the methodologies in drug delivery and drug targeting. The life-threatening diseases in the industrialized countries, such as cancer and viral infections, imposed challenging requirements on the specificity of medications. A large number of biomacromolecules of therapeutic potential have arisen from the phenomenal advances in molecular biology but are hampered in the clinic by their poor pharmacokinetic profiles. Completion of the Human Genome Project[1,2] and the heated research in genomics and proteomics are unveiling the molecular mechanisms of many diseases and hence promise more drug targets in the near future. In response, the research in advanced drug delivery is growing in an explosive manner.

In order to achieve successful administration, a parent drug chemical needs to be mixed with other ingredients into a pharmaceutical formulation. Together these accessory ingredients form the carrier of the parent drug. If the drug remains associated with the carrier after administration so that biodistribution of the drug follows that of the carrier, the carrier is then considered a delivery system for the drug. Therefore, the first step in the design of a drug delivery system is to introduce sufficient binding between the drug payload and its carrier. The drug is assembled with the delivery system either by attachment through covalent bonds or by noncovalent interactions such as encapsulation, solubilization, association, or adsorption. The delivery system thus protects the drug from degradation and elimination, and redirects distribution of the drug. Examples of the payload include cancer chemotherapy drugs,[3] cancer radiotherapy agents,[4] antibiotics,[5] proteins,[6] and nucleic acids.[7] Drug carriers can be categorized roughly into soluble polymers, polymeric micelles, liposomes, and solid nanospheres/microspheres.

This chapter focuses on the design of drug delivery systems by tailoring their physical and chemical properties such as size, surface charge, surface hydrophobicity, sensitivity to triggers, and surface activity toward the biomembranes so as to overcome the anatomical, cellular, and subcellular barriers to drug delivery and drug targeting. Technologies of active drug targeting and drug targeting by alternate routes of administration such as pulmonary drug delivery are presented in Chaps. 8 and 12 of this book and will not be covered here.

11.2 Rationale for the Design of a Drug Delivery System

In order to design a successful drug delivery system, it is critical to have a sound understanding of the barriers between the site of administration and the molecular receptor of the payload drug. Depending on the chemical natural of the parent drug, part or all of the barriers need to be overcome by its delivery system. It is also important to consider any unique features of the target tissue or target cells so that one can design a delivery system that selectively accumulates at the target site. Last but not the least, if the delivery system conceals the activity of the parent drug, it must unmask the active drug at the target site.

11.2.1 Challenges of drug delivery systems for intravenous administration

The most common administration route for targeted drug delivery is intravenous injection. Using this route, the dose is distributed rapidly throughout the vascular system. A delivery system given via this route must fulfill three requirements if it is to deliver a drug to the target cells. First, the payload needs to remain intact with the carrier before reaching the target site.[8] Premature cleavage or leakage of the drug from its carrier not only will decrease the amount of drug that reaches the target site but also will result in elevated systemic toxicity. Therefore, the delivery system must tolerate assaults from plasma such as opsonin adsorption and enzyme degradation. Second, a delivery system must remain in the circulation long enough to have time to accumulate in the target cells.[9] In order to have a long duration of circulation, the delivery system needs to avoid quick clearance by the mononuclear phagocyte system (MPS; also referred to as the *reticuloendothelial system*). If the target is the vascular endothelial cell layer, the delivery system can reach the target site readily via the blood circulation. To reach other tissues such as hepatocytes and cancer cells in solid tumors, the carrier needs to extravasate through the endothelial capillaries and diffuse to the target site. The endothelial cells that outline the capillaries enforce an upper size limit of about 100 nm if the delivery system is to reach the extravascular tissues. Third, on accumulation at the target site, the active drug must be released at a high enough level to mediate an effective therapeutic response.[10,11] For drugs that interact with receptors on the cell surface or are transported readily inside the target cells, their release at the interstitial space is sufficient for their activities. However, for many hydrophilic biomacromolecules whose receptors are inside the target cells, additional delivery steps across the biomembranes must take place.

11.2.2 Delivery across biomembranes

Many drugs function by interacting with their receptors inside the target cells, which are highly compartmentalized with multiple biomembranes such as the plasma membrane, the endoplasmic reticulum (ER), the endosomal/lysosomal membranes, and the nuclear envelope.

Therefore, these drug molecules must cross the multiple layers of cellular membranes between the site of administration (the blood circulation in the case of intravenous injection) and their intracellular receptors. Small molecules (MW < 400 Da) that are soluble in both water and octanol readily diffuse through the biomembranes. Many hydrophilic small-molecule drugs such as β-lactam antibiotics[12] and classical antifolates[13] require specific transporter proteins on the plasma membrane to enter the cytoplasm of the host cells.

In the case of biomacromolecules such as oligonucleotides, proteins, RNA, and DNA,[7] the delivery system often needs to transfer the payload across the plasma membrane either by fusion or by endocytosis.[7,14] If taken up by the cells via endocytosis, the macromolecule later must be released from the endosome into the cytoplasm to avoid degradation in the lysosomes. In the case of gene delivery, DNA must further relocate from the cytoplasm into the nucleus to direct the expression of the gene products.[15] Therefore, in order to develop biomacromolecules into a real-life medication, they are often packaged with a delivery system that possesses special mechanisms to destabilize the cellular membranes. Clearly, the delivery of high-molecular-weight hydrophilic molecules across biomembranes is one of the most challenging problems facing the pharmaceutical community.

11.2.3 Biostability and bioresponsiveness

There are three major considerations as to the stability of a delivery system. First, the system should possess enough physical and chemical stability that the drug does not decompose or dissociate from the delivery system before it reaches the target site. Second, at the target site, the carrier should release the drug at a rate that is suitable for a therapeutic response. Third, the carrier eventually should be degraded and eliminated from the body to avoid long-term toxicity or immunogenecity. The first and second considerations further imply that at the target site the delivery system should have an enhanced drug release.

Given such arguments, one may wonder why some anticancer drug delivery systems were not designed for an enhanced release at the target site but still have obtained considerable success in the clinic. In some cases (such as in liposomal formulations),[16] the drug molecules are consistently released at a slow rate from the carrier. Such a slow release does not incur intolerable toxicity or dose loss during distribution of the delivery system. After accumulation of the drug delivery system at the tumor site, such a slow release is already sufficient for a therapeutic response. In other cases, the drug is not fully concealed (as in some polymer-protein conjugates) and is at least partially active in its attached form.[17] Nonetheless, the recent literature strongly supports the hypothesis that a mechanism of triggered release at the target site should greatly improve the efficacy as well as the specificity of drug delivery. A number of polymeric delivery systems with structural features for enhanced release at the target site already have made their way to clinical trials.[18,19] A number of liposomal-triggered release systems[20–22] have obtained significantly improved anticancer efficacy in animal studies when compared with nontriggerable controls.

11.3 Design of a Physically Targeted Delivery System: Modulation of Physicochemical Parameters

The "physical targeting" approach in the design of a delivery system modulates the physicochemical parameters of a drug carrier so as to direct its disposition to the target site. Many physicochemical properties of a drug delivery system are critical for its in vivo behavior but often were undervalued. Such properties include but are not limited to molecular weight, size, surface hydrophobicity, surface charge, destabilization activity toward biomembranes, and sensitivity to triggering. Each type of delivery system also has its unique physicochemical parameters of consideration, such as the critical micelle concentration of polymeric micelles.

11.3.1 Molecular weight and size

The molecular weight or size of an optimal delivery system in vivo is imposed by the physiology of circulation and excretion. The lower limit of the molecular weight is about 30 kDa because the microtubular cells in the kidney readily excrete hydrophilic molecules whose molecular weight is 30 kDa or less.[23] In fact, this is the major route of elimination for many small-molecule drugs after they are transformed into more hydrophilic metabolites. Therefore, a drug delivery system should have a combined molecular weight of 30 kDa to avoid such a quick renal clearance.

The endothelial cells outlining the blood vascular system also impose important upper size limits on parental drug carriers. In most parts of the blood vasculature, the endothelial cells are tightly joined together on a continuous subendothelial basement membrane.[9] Particulates with a diameter greater than 10 nm cannot pass through this barrier into extravascular tissues. Under normal physiological conditions, only the liver, the spleen, and the bone marrow possess more sinusoidal capillaries, where "pores" (fenestrae) of 100 to 200 nm in diameter can be found, allowing drug delivery systems of 200 nm or smaller to diffuse into the interstitial spaces of these organs.

Another important exception to this size limitation is in solid tumors.[24,25] The endothelial cells are poorly developed, and fenestrations larger than 200 nm can be found. Furthermore, solid tumors lack a mature lymphatic system, resulting in poor drainage of the extravasated particles. This phenomenon in the pathological condition is called *enhanced permeation and retention (EPR)* effect and serves as a key rationale for passive drug targeting systems. To date, a number of polymeric[26] and particulate[27] drug delivery systems have exploited the EPR effect and have found success in treating small solid tumors in the clinic. The permeability of the capillary blood vessels is also enhanced at inflammation sites, where fenestrations up to 200 nm in diameter can be found.[28,29]

The size of a particular drug delivery system also influences its clearance kinetics by the mononuclear phagocyte system (MPS). The MPS consists of both fixed and mobile phagocytes in direct contact with the blood circulation as part of the immune system.[9] The fixed macrophages are located in the liver (Kupffer cells), spleen, bone marrow, and lymph nodes; the mobile macrophages include the blood monocytes and the tissue macrophages. Particulates in the size range of 100 nm to 7 μm are prone to be cleared by the MPS, especially by the fixed Kupffer cells in the liver. Thus any delivery systems in this size range need to be tailored to avoid such a premature clearance unless the macrophages themselves are the target cells.

11.3.2 Surface hydrophobicity[9]

The MPS is constantly alert to remove "foreign particles" such as bacteria, virus, and denatured proteins. Clearance by the MPS involves two steps. First, plasma proteins called *opsonins* adsorb onto the "foreign surface" of a particulate; second, the macrophages recognize the opsonsin-covered particles and initiate phagocytosis. Particles with hydrophobic surfaces are recognized immediately as "foreign," covered by the opsonins, and taken up by macrophages. The natural tendency of macrophages to uptake the lipidic particulates was exploited in a number of MPS targeting liposome formulations.[30] The potential therapeutic benefits of such liposomes include the treatment of macrophage-related microbial, viral, or bacterial infections; the immunopresentation of vaccines; potentiation of the immune system using a macrophage-activating agent such as interferon-γ; and treatment of lysosomal enzyme deficiencies.

However, if a delivery system is to be targeted to other cell types, its interaction with the MPS must be minimized. The standard approach is to coat the surface of the system with hydrophilic materials to reduce opsonin adsorption. A number of biological[31] and synthetic materials[32] have been grafted to the surface of liposomes for "steric stabilization," and the most popular material is the highly hydrophilic and flexible polymer poly(ethylene glycol).[20,27] Thus the PEG-coated so-called stealth liposomes (Sequus, Inc.) have a much longer circulation time (plasma half-life of up to 3 days). With a diameter (100 nm) between the size of the normal endothelial fenestrations and those of the pathological vasculature (usually from 100 to 200 nm), stealth liposomes can accumulate selectively at the tumors.

11.3.3 Charge

The surface charge of a drug carrier also plays an important role in its pharmacokinetic behavior. For liposomes, it has been shown that a neutral surface charge is optimal for a long circulation time.[33] Liposomes with a negative charge tend to be cleared more rapidly from the circulation by the Kupffer cells in the liver.[34] Positively charged particulates rapidly absorb negatively charged plasma proteins in the blood circulation and are recognized as foreign objects by the immune system.[35] If the particulates carry a large excess of positive charges on their surface that cannot be quenched immediately by plasma proteins, they will interact strongly with the highly negatively charged proteoglycans of the vascular endothelial cells and deposit predominantly at the first capillary bed they encounter after intravenous administration. For example, after tail vein injection in mice, positively charged complexes of DNA with cationic polymers or cationic lipids distribute mainly to the blood capillaries in the lung and mediate gene expression of lung endothelail cells.[36]

11.3.4 Membrane destabilization

The efficient transfer of hydrophilic macromolecules across the multiple layers of hydrophobic cellular membranes remains one of the most formidable tasks in drug delivery and drug targeting. Nevertheless, there are very instructive examples of efficient membrane destabilization in nature. The bacteriophage T_4 uses specialized proteins at the tail fiber to "pierce" the cell wall and the plasma membrane of the bacteria, followed by "injection" of the phage genome into the host cytoplasm.[37] The influenza virus enters the endosomal pathway of the host cells, and on the decrease in pH inside the endosome, the viral glycoprotein hemagglutinin[38] changes its conformation and induces fusion of the viral and the endosomal membranes. The viral genomic DNA thus escapes from the endosome into the cytoplasm. A series of subsequent targeting steps eventually result in translocation of the viral DNA into the host cell nucleus and expression of the viral proteins. A pharmaceutical scientist thus can incorporate some of the natural membrane-destabilizing proteins and peptides into a drug delivery system. Artificial membrane-destabilizing peptides, polymers, and bilayer compositions also can be designed based on the biophysical principles of membrane destabilization. Examples of such designs include the GALA peptide (a peptide sequence repetitive of EALA), the polyethyleneimine polymer, and the Gemini cationic lipids.

11.3.5 Physicochemical functionalities for triggered release

As discussed previously, a mechanism of triggered release should improve the efficacy as well as the specificity of a drug delivery system. However, the design of an optimal triggered release system is not a simple task because it should meet at least the following three criteria. First, the stimulus to trigger the drug release must be specific to the target site; second, the delivery system needs to be sensitive enough to the stimulus to yield effective release; and third, the triggered release mechanism must be compatible with all the other essential properties of the delivery system, such as stability in blood circulation and selective deposition at the target site.

In order to design a triggered release system, the drug delivery system needs to incorporate a physical or chemical functionality that is relatively stable during distribution of the delivery system but is sensitive to a certain stimulus at the target site. For example, a bilayer with a melting point a few degrees above body temperature can be used to build thermosensitive liposomes. To devise an enzyme-triggered system, a linker that is labile to the enzyme can be used to attach drug molecules to its carrier. The stimuli to induce release can be either external, such as localized heat or radiation, or supplied by a biological process, such as a drop in pH, enzymatic transformation, or change in a redox potential.

Therefore, many physicochemical properties play an important role in the vivo behavior of macromolecules and self-assembled colloids. These physicochemical parameters must be modulated carefully in the design of drug delivery and drug targeting systems. The following sections will discuss major categories of drug delivery systems and give examples of such "physical targeting" approaches in each type of the delivery system.

11.4 Current Drug Delivery Systems

11.4.1 Polymers

Polymer-based delivery systems (Fig. 11.1) include polymer-protein conjugates, polymer-drug conjugates, micelles consisting of polymeric surfactants, and complexes of cationic polymers and DNA (polyplexes).[26]

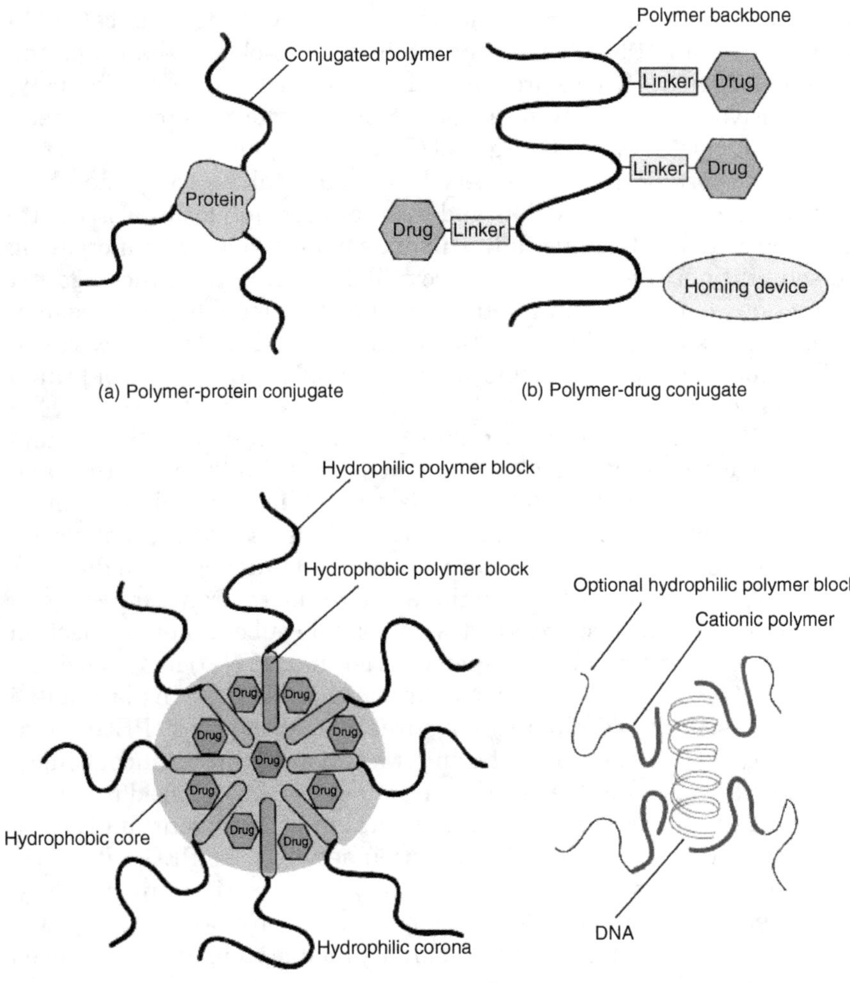

Figure 11.1 Applications of polymers in drug and gene delivery.

Since the early 1990s, a number of polymer-protein conjugates have been approved for medical use by the Food and Drug Administration (FDA), and clinical trials of the polymer conjugates of anticancer drugs have seen promising results, all of which quickly changed the conception of polymer-based therapeutics from a scientific curiosity to a clinical reality.

Polymer-protein conjugates (see Fig. 11.1a). The phenomenal advances in molecular cloning and monoclonal antibody technology have rendered numerous protein-based drug candidates.[39] However, their clinical applications are limited by their poor stability, short plasma half-life, and immunogenecity.[26] Therefore, there has been a long search for polymeric derivatives of protein drugs to improve their pharmacokinetic behavior in vivo. Polyethylene glycol (PEG)[40,41] is particularly attractive for this purpose because its safety has been established in the pharmaceutical industry. Its flexible and hydrophilic chain forms a structure with a large hydrodynamic radius in aqueous solution, thus increasing the solubility of the protein and providing a steric hindrance against interactions between the protein drug and the components of plasma. For proteins smaller than 30 kDa conjugation with PEG (now called *PEGylation*) increases the molecular weight of the drug to avoid rapid clearance via renal excretion.

The PEGylation significantly improves the stability of the protein drug in plasma, allowing less frequent parental administration and better patient compliance. For example, recombinant methionyl human granulocyte colony-stimulating factor (G-CSF) is used to prevent severe cancer chemotherapy-induced neutropenia and must be given daily for 2 weeks.[42] The PEG-G-CSF conjugate (Neulasta from Amgen) can achieve the same protective effect with a single subcutaneous injection on day 2 of each chemotherapy cycle. Interferon-α (IFN-α) has been effective in treating renal cell carcinoma, but its short half-life in plasma (2.3 hours) necessitates administration three times per week. PEGylated-α–IFN-2b (PEG-Intron from Schering) was given by subcutaneous injection once per week for 12 weeks to 44 patients with advanced renal cell carcinoma in a phase I/II study, and 6 patients (14 percent) had a positive response.[43] The hindered interaction between the PEGylated protein and the plasma components also reduces immunogenecity. PEGylated L-asparaginase (Oncaspar from Enzon) has a much lower immunogenecity compared with the unmodified protein and can be used to treat patients who are hypersensitive to the native enzyme.[44] Since the early 1990s, the clinical benefit of protein PEGylation has been well established. Other PEGylated proteins that have been approved to enter the drug market include PEG-adenosine deaminase (Adagen from Enzon) for treating the SCID syndrome,[45] PEG-human growth hormone (Somavert from Pharmacia) for treating acromegaly,[46] and PEG-α–IFN-2a (PEGASYS from Roche) for treating hepatitis C.[47]

In order to meet the quality-control standards in drug development, the synthesis of PEG-protein conjugates must be highly reproducible and the products well characterized. This means that (1) the stoichiometry of the conjugation, i.e., the number of PEG-polymer chains and protein molecules per conjugate, must be consistent from batch to batch, (2) the site of PEGylation on the protein must be specific, and (3) during the PEGylation process, most of the protein activity must remain intact.[47]

One approach to site-specific PEGylation is to first mutate one of the residues on the protein surface to cystein, followed by a cystein-specific PEGylation reaction using a methoxy-PEG-maleimide reagent.[17] Lately, Sato et al.[48] has reported a methodology for site-specific PEGylation using the enzyme transglutaminase.[48] The alkylamine derivatives of PEG were incorporated into a number of intact and chimeric proteins only at the substrate glutamine site. The resulting conjugates are highly homogeneous with intact protein activity. As in the previous approach, suitable incorporation sites can be designed by mutating part of the protein to a sequence that is recognizable by transglutaminase, the site-specific conjugation enzyme.

The only polymer-protein conjugate in the drug market that does not contain a PEG-polymer chain is the SMANCS conjugate (Yamanouchi Pharmaceuticals, marketed in Japan) used to treat hepatocellular carcinoma.[49] The protein of the SMANCS system, neocarzinostatin (NCS), is a small cytotoxic protein (MW 12 kDa) that inhibits DNA synthesis and induces DNA fragmentation. The native NCS protein is cleared rapidly by the kidney, and its cytotoxicity is not tumor-specific. In the SMANCS conjugate, two hydrophobic poly(styrene-co-maleic acid anhydride) copolymers (MW 1500 Da) are coupled to each NCS molecule. Thus the SMANCS conjugate is more hydrophobic than the parent protein and dissolves readily in organic solvents such as Lipiodol. To treat hepatocellular carcinoma, a homogeneous suspension of SMANCS with Lipiodol (SMANCS/Lipiodol) is administered via the hepatic arteries, where Lipiodol acts as a carrier of SMANCS. The SMANCS/Lipiodol retains most of the cytotoxicity of NCS, but the pharmacokinetic properties are much improved, with prolonged drug retention in the liver and increased drug uptake by hepatumor cells. The hydrophobicity of the SMANCS/Lipiodol formulation is a critical physicochemical parameter for drug targeting.

Polymer-drug conjugates. The general structure of a polymer-drug conjugate is shown in Fig. 11.1*b*. In these systems, a soluble polymeric carrier and the payload drug molecules are covalently conjugated either directly or via a designed labile spacer to optimize detachment of the drug at the target site. Active targeting devices such as a monoclonal antibody to a specific protein of the target cells also may be attached to the polymeric carrier. The emphasis of such a polymeric-drug conjugate system is that one can adjust the physicochemical properties of the soluble polymer to redirect the biodistribution of the attached drug. The conjugation of small-molecule anticancer drugs drastically changes their pharmacokinetics patterns, minimizing the toxicity from a high concentration of the free drug in circulation and promoting the accumulation of polymer-bond drug to the solid tumor owing to the above-mentioned EPR effect of the tumor vasculature. A number of the anticancer polymer drugs have entered clinical trials.[26]

Poly[*N*-(2-hydroxypropyl)methacrylamide] (pHPMA), (Fig. 11.2*a*) is the most actively investigated soluble macromolecular drug carrier owing to its excellent biocompatibility.[50] The HPMA polymer was developed initially as a plasma expander by Kopecek et al.,[51] and cumulative doses of more than 20 g/m^2 HPMA copolymer could be administered without signs of immunogenecity or polymer-related toxicity. The structure of an HPMA conjugate of doxorubicin, an anthracyclin-type anticancer drug, is shown in Fig. 11 2*b*.[52] The bulk of the conjugate (90 to 95 percent) consists of unmodified HPMA units, and the remaining units are derivatized with doxorubicin via a Gly-Phe-Leu-Gly tetrapep-

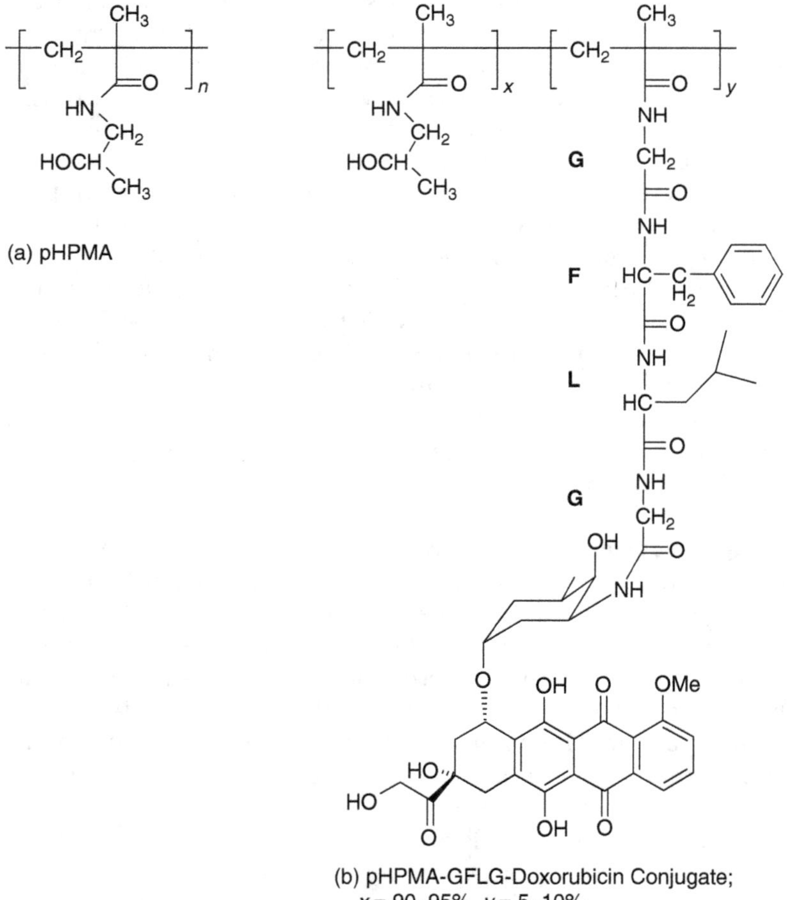

Figure 11.2 Structure of pHPMA-based polymer-drug conjugates.

tide spacer. The hydrophilic structure of the intact HPMA units on the polymer backbone is a crucial property of the conjugate, enabling its uptake by tumor cells through pinocytosis. The conjugate is processed subsequently through the endosomal pathway of the target cells to reach the lysosomal compartments, where it is exposed to a variety of decomposing enzymes. The Gly-Phe-Leu-Gly was designed to be labile in the lysosomal compartments,[50] where a thiol-dependent protease, cathepsin B, cleaves the peptide bond between the terminal glycogen and the daunosamine ring to release the free doxorubicin from the carrier. The free drug then diffuses through the intracellular membranes and exerts its activity at the nucleus of the tumor cell. Other triggerable linkers[26,53] between the doxorubicin and HPMA also were reported, including the *cis*-aconityl, hydrazone, and acetal linkages. These spacers are hydrolyzed in response to the low pH in the endosomal pathway (from 6.5 to 4.0) to liberate the free doxorubicin.

In a phase I clinical trial of the copolymer HPMA-gly-phe-leu-gly-doxorubicin (MW $\approx$ 30 kDa),[19] the maximum tolerated dose of the polymer-bound drug is four to five fold higher than the safe dose of the free drug. The cardiotoxicity of free doxorubicin was minimized in the polymer conjugate. Antitumor activities were seen in patients who were considered resistant or refractory at lower free doxorubicin doses. These observations were consistent with the targeting effect of the HPMA polymer owing to the EPR effect of solid tumors. Polyglutamate-drug conjugates and PEG-drug conjugates are in different phases of clinical investigation.[26]

There are two trends in the design of future generations of polymer-drug conjugates.[26] First, the polymer backbone tends to possess multi-branched architectures such as grafts, the stars, the dendrimers, and the linear polymers with dendron branches. These architectures offer more spherelike three-dimensional structures and more flexibility in the topological arrangement of the attached drugs and other components. Second, the conjugate tends to possess multiple types of bioactive components besides the payload drug, such as active targeting ligands (antibodies, sugars, and folic acid) and bioresponsive moieties for triggered release. As with many other types of drug carriers, more sophisticated polymer-drug conjugates are being developed in the hope of overcoming the multiple barriers of drug delivery.

Polymeric micelles. When water-soluble polymers are conjugated with lipophilic, poorly water-soluble polymers, the resulting copolymers are amphiphilic and can be used to constitute spherical micelles.[54] The sizes of the polymeric micelles range between 10 and 100 nm, which is ideal for preferential extravasation at the fenestrated capillary blood vessels. The polymeric micelles have a hydrophobic core consisting of the lipophilic polymer chains and a coating of the hydrophilic polymer chains (see Fig. 11.1*c*). Three types of amphiphilic copolymers (Fig. 11.3) can be used to design polymeric micelles: the diblock copolymers, the triblock copolymers, and the grafted copolymers.[23] The diblock copolymers have a linear structure, with the length of the hydrophilic polymer block exceeding that of the lipophilic polymer block. The triblock copolymers have two hydrophilic polymer chains covalently attached to the two ends of a lipophilic polymer. The grafted copolymers have a hydrophilic backbone grafted with lipophilic side chains. In the aqueous phase, the copolymers assemble and fold appropriately to conceal the lipophilic polymer chains from water in the core of the micelles.

In most of the reported copolymers for micelle formulations, the plasma-inert polymer, PEG of a molecular weight between 1 and 15 kDa,

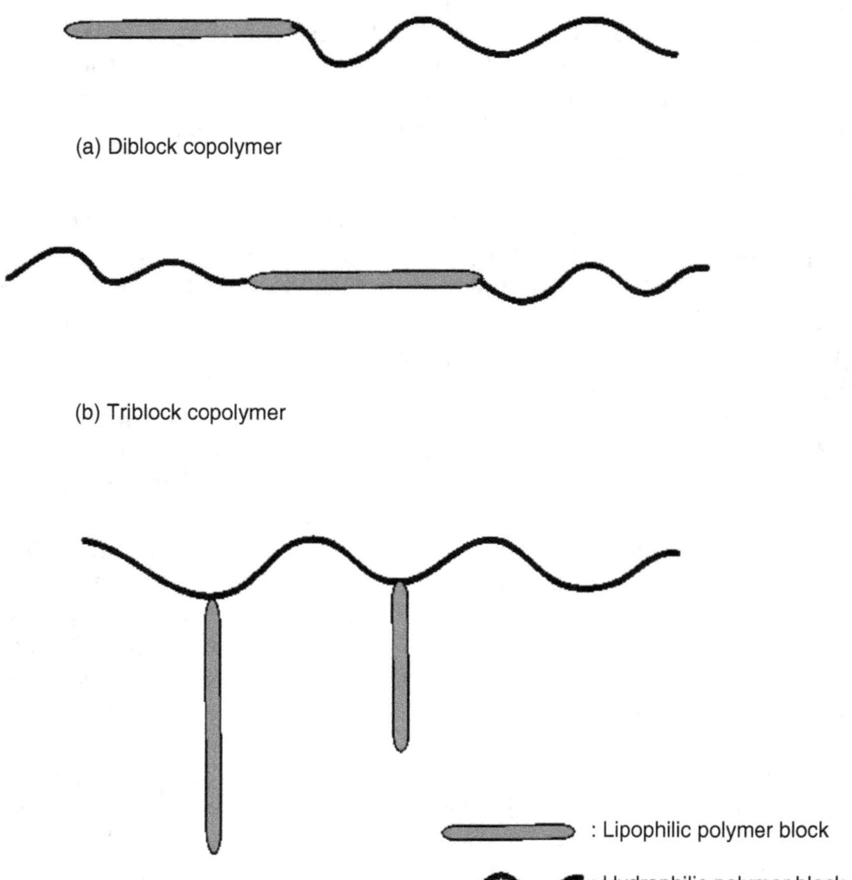

Figure 11.3 Copolymers that self-assemble into micelles.

is the leading candidate of the corona-forming hydrophilic polymer chain.[55] As in other categories of drug carriers, PEG serves to slow down the opsonization and consequently the phagocytotic clearance of the pharmaceutical colloids by the MPS. In contrast, a number of polymers have been used to build the hydrophobic core of the micelles,[23] including polymers of propylene oxide, β-benzoyl-L-aspartate, γ-benzoyl-L-glutamate, caprolactone, lysine, spermine, aspartic acid, and lactic acid. Some phospholipids of low molecular weight (~700 Da) also were selected as the core-forming block owing to the strong lipophilicity of their hydrocarbon side chains.[56] In some cases, the starting diblock polymer may consist of two hydrophilic polymer blocks. Hydrophobic drugs such as cisplatin[57] and anthracyclins[58] then can be attached covalently to one of their hydrophilic polymer blocks to form the amphiphilic self-assembling copolymer.

One physicochemical parameter that is critical to the design of polymeric micelles with sufficient in vivo stability is the *critical micelle concentration* (CMC), defined as the concentration of a monomeric amphilphile when it starts to aggregate into micelles. The low-molecular-weight surfactants that are used to solubilize hydrophobic drugs in the traditional formulation techniques[59] have a high CMC (millimolar range). They are not of great use in the construction of long-circulating micelles because the micelles so formed disintegrate quickly and release the cargo drug molecules on dilution in the blood circulation. The polymeric micelles can achieve a low CMC in the submicromolar range, and the CMC can be controlled by monitoring the structure of the monomeric copolymer.[60,61] The CMC decreases with increases in the size and hydrophobicity of the core-forming polymer as long as the hydrophilic polymer block is longer than the hydrophobic block to ensure the formation of micelles instead of other types of supermolecular structures such as rods and lamellae.[62] Increasing the size of the hydrophilic corona-forming polymer block increases the CMC, but the influence is much less compared with the change in the core-forming polymer block.[61] With similar chemical compositions and molecular weights of the polymer blocks, the micelles composed of the grafted copolymers have higher CMC values than those of the triblock and biblock copolymers.[23] This is so because the biblock and triblock copolymers are more flexible and thus can rearrange and pack into tighter hydrophobic cores than the more fixed hydrophobic blocks of the grafted copolymers. At concentrations higher than CMC, the micelles made of grafted copolymers are prone to aggregation because the hydrophobic blocks of such micelles are not fully concealed from the aqueous media and thus can interact with hydrophobic blocks of another micelle.

The core of the micelle needs to have suitable molecular interactions with the payload pharmaceuticals so as to achieve a drug loading of sufficient capacity and stability. The most stable bonding between the drug and the micellar carrier is obviously through covalent conjugation, but the lack of substantial drug release at the target site is a serious concern for this type of formulation.[20,63] Indeed, in clinical trials of a PEG-aspartic acid micelle formulation of the anticancer drug doxorubicin, the covalently bound drug was not active; it was the associated doxorubicin that slowly escaped from the micelles over 8 to 24 hours and destroyed the tumor cells.[64] The major noncovalent bonding between the drug and the micelles is the hydrophobic interaction[23] (Fig. 11.4). Water-insoluble drugs of high hydrophobicity partition deeply into the core of the micelles. Drug molecules of intermediate hydrophobicity merge partially into the core and orient their hydrophilic groups toward the corona region. Water-soluble hydrophilic drugs can adsorb only to the corona compartment and usually leak quickly from the micelles on dilution in the blood circulation. For hydrophobic drug molecules with a charged group, chemical groups of the opposite charge can be introduced to the core-forming polymer block to strengthen the drug-micelle association.[65] For example, the positively charged hydrophobic drug doxorubicin associates strongly with the negatively charged polyaspartic acid block of the PEG-aspartic acid micelle, resulting in a slow release over 24 hours in vivo.[64]

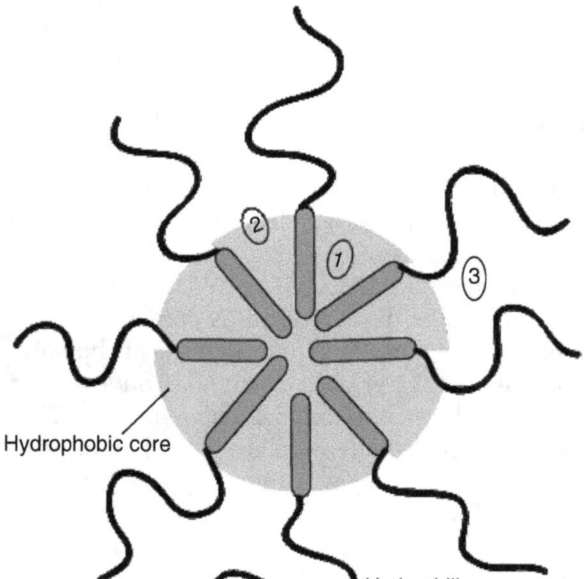

Figure 11.4 Schematic representation of drug association with polymeric micelles by hydrophobic interaction. (1 - a water-insoluble hydrophobic drug; 2 - a drug of intermediate hydrophobicity; 3 - a hydrophilic drug).

Polyplexes. When cationic polymers (polycations) are mixed with the highly negatively charged DNA, they condense into colloids called *polyplexes* (see Fig. 11.1*d*). Polyplexes serve as an important nonviral vector for gene delivery,[36,66,67] and one polyplex formulation has advanced to the clinical trials.[68]

There are two common categories of polycations. The first category is cationic polypeptides such as polylysine.[69] The second category is built from basic imines, such as polyethyleneimine[70] and fractured dendrimers.[71,72] The basic amines in these polymers are protonated at physiological pH, providing the positive charges of the polymers. If the basic amines from the polycations are in excess compared with the negatively charged phosphates in DNA, the polyplexes so formed carry positive charges on their surfaces, as reflected by a positive zeta potential. If the negatively charged phosphates are in excess compared with the basic amines, the resulting polyplexes are negatively charged. When the molar ratio of the basic amines and the DNA phosphates are close to 1, the resulting polyplexes usually are large in size and tend to precipitate out of solution. In general, positively charged polyplexes are more efficient in mediating gene transfer than negatively charged polyplexes.

Although the mechanism of polyplex-mediated gene delivery is still not fully understood, research over the past decade elucidated multiple ways[36] in which cationic polymers can facilitate gene delivery and gene transfection. First, complexation of DNA by cationic polymers hampers the enzymatic degradation of DNA,[73] resulting in more intact DNA to be transferred eventually to the cell nucleus. Second, cationic polymers probably facilitate cellular uptake in a similar way to that of lipoplex-mediated gene delivery, where the cationic complexes are absorbed to the negatively charged cell surface by ionic interaction with the negatively charged proteoglycans of the extracellular matrix.[74] This is followed by internalization via the endocytic turnover of proteoglycans. This hypothesis is consistent with the recent electron microscopic observation that the cationic polyplexes aggregated on the cell surface and were taken up by cells via a nonspecific pathway.[75] A third way of enhancing gene delivery by cationic polymers is called the *proton sponge*[70] mechanism, as in the case of polyethyleneimine (PEI) and partially fractured dendrimers.[71] PEI and partially fractured dendrimers are two of the most efficient cationic polymers reported so far. In contrast to many less successful cationic polymers, these two polymers feature titratable amines in the neutral to weakly acidic pH range, as well as flexible and expandable polymer backbones. After endocytosis of the polyplexes, the decrease in pH in the endosomes protonates the titratable amines, which, in turn, induce an expansion of the polyplex structure owing to the repulsion of their positive charges. The expansion then would destabilize the endosomal membranes and translocate DNA into the cytoplasm before they are processed to the lysosomes, where extensive degradation takes place.

The partially fractured dendrimer is now commercialized as SuperFect (*http://www.qiagen.com/transfectiontools/literature/default.asp*) and is used routinely for in vitro gene transfection in cell biology researches. Polyplexes consisting of PEI have shown transfection activity in many different tissues in animals, including the lung,[76] the brain,[77] and the kidney.[78] In all the examples of gene delivery by polyplexes, the physicochemical properties of the complexes play a definitive role in their success.

If polyplexes with excess positive charges are administered via intravenous injection, most of them will accumulate and exert their transfection activity in the lung, which is the first capillary bed that the polyplexes encounter.[36] Recently, however, significant effort has been directed toward the development of long-circulating polyplexes for the remote targeting to solid tumors.[79–81] The positive surface charges of such polyplexes are shielded by PEG so that they provoke less interaction with serum proteins and are targeted to solid-tumor tissues with leaky vasculature (the EPR effect), followed by expression of the gene product, such as a tumor necrosis factor. Shielding of the positive charges is accomplished by condensing DNA with diblock PEG-PEI copolymers, resulting in polyplexes with a PEG corona. The distal end of PEG also can be conjugated to a homing ligand to improve the targeting.[79] Such an active targeting strategy is discussed in detail in Chap. 12.

11.4.2 Lipidic colloids

Lipidic colloids are composed of amphiphilic surfactants that possess both a hydrophilic head group and a hydrophobic tail.[82] When the surfactant molecules are dispersed in aqueous medium, they self-assemble into different supermolecular scaffolds depending on the molecular shape of the surfactant. The colloids that are most relevant to drug delivery are illustrated in Fig. 11.5. If the surfactant is in a mushroom shape, meaning that it has a hydrophilic group of a significantly bigger hydrodynamic diameter than the width of the hydrophobic tail, it forms globular micelles with a hydrophobic core and a hydrophilic shell. Detergents usually fall into this type of surfactant, and they are able to dissolve greasy chemicals by incorporating them into the hydrophobic core of the micelle. Similarly, hydrophobic drugs can be complexed with lipidic micelles for targeting or improved solubulization. If the hydrophilic head group has a similar width to that of the hydrophobic tail, the surfactant is in a cylindrical shape and forms bilayers that can anneal into vesicles called *liposomes*. Drug molecules can be loaded into liposomes in a number of different ways, depending on their polarity and charge state.[83] Highly hydrophilic molecules that do not readily diffuse

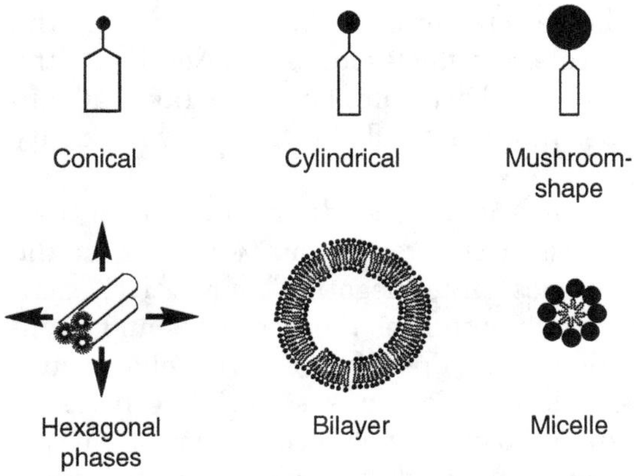

Figure 11.5 Self-assembly of surfactants in aqueous phase.

across bilayers can be encapsulated into the aqueous interior of liposomes. Highly lipophilic drugs can be incorporated into the lipid bilayers by hydrophobic interaction. Drugs with ionizable groups can be loaded by a cross-membrane pH or counterion gradient that traps the charged molecules in the aqueous liposome interior. If the hydrodynamic diameter of the head group is much smaller than the width of the hydrophobic tail, the surfactant is in a conical shape and forms hexagonal phases that tend to fuse with each other and precipitate out of solution. The hexagonal phases are not desirable for the preparation of a stable formulation but can be very useful for the delivery of hydrophilic macromolecules across biomembranes.

For example, the use of a conically shaped lipid, dioleoylphosphatidylethanolamine (DOPE), in cationic liposomes helps the destabilization of the cellular membranes, leading to a more efficient delivery of plasmid DNA in cell culture.[84] The structural diversity of the lipidic colloids offers great flexibility in their applications as drug delivery and drug targeting systems.

Conventional liposomes and lipid complexes. Liposomes were used initially as a model system for cellular membranes to study the biochemistry of membrane proteins.[85] Consequently, when liposomes were first tried as a drug delivery system, their bilayers were composed of underivatized naturally occurring lipids. Most of such "conventional liposomes" are taken up by the MPS phagocytes within a few hours of injection, mostly by liver Kupffer cells and spleen macrophages.[9] Inside the endosomes and lysosomes of those cells, liposomes are degraded. If the liposomal drugs are membrane permeable, they then can diffuse from the endosomal compartments to the cytoplasm of the macrophage cells and slowly reenter the blood circulation. Because such a clearance and redistribution is gradual, the conventional liposomes convert the liver and spleen into sustained release vehicles for drugs. Amphotec, the disklike lipid complex of the antifungal drug amphotericin, takes advantage of such a sustained release and has a substantially lower toxicity than the parent drug.[86]

The clearance of conventional liposomes depends on a number of their physicochemical properties. Excessive charges (10 mol % or more) on the surface of the conventional liposomes, either negative[34] or positive,[35] activate the complement system, which opsonizes the liposomes and facilitates recognition and elimination of the liposomes by MPS phagocytes. An increase in size also triggers the complement system.[34] In contrast, a neutral surface, a small diameter (100 nm or smaller), and a rigid-bilayer composition with cholesterol and saturated diacyl lipids all contribute to less complement activation and better stability of the conventional liposomes in circulation.[33] Senior and Gregoriadis[87] have demonstrated that small (<100 nm) neutral unilamellar liposomes with a rigid bilayer composition (cholesterol:saturated phospholipids = 1:1) have a blood half-live of up to 20 hours in rats. Optimization of the physicochemical properties of conventional liposome led to the development of DaunoXome (Gilead), an FDA-approved formulation of daunorubicin citrate encapsulated in neutral, small liposomes composed of cholesterol and the saturated phospholipid DSPC for the treatment of Kaposi's sarcoma.

Long-circulating liposomes. In 1979, Balderswieler et al.,[88] first observed that liposomes could deliver encapsulated drugs to tumors. However, the relatively rapid clearance of conventional liposomes hampered the development of effective liposomal anticancer drugs until the breakthrough discoveries in the late 1980s that surface association with certain polymers such as the ganglioside GM1[89] and polyethylene glycol (PEG)[32] increased the circulation lifetime of conventional liposomes. These observations were followed by a new generation of sterically stabilized long-circulating liposomes, whose surfaces are modified by molecules that reduce absorption of the immunoproteins and consequently slow down clearance by the MPS. In the early 1990s, the anticancer drug doxorubicin was formulated successfully by a long-circulating "stealth liposome," which contains surface-grafted PEG groups (MW ~ 2000Da) via a PEG-lipid conjugate incorporated in the lipid bilayer (Fig. 11.6). The resulting liposomal drug, Doxil (Sequus and ALZA), has a half-life of 55 hours in the bloodstream; the half-life of a conventional liposome is less than 1 hour.[86] The long plasma half-life of Doxil provides the necessary time span for the liposomes to extravasate into the tumor tissues, resulting in decreased doxorubicin to normal tissues and increased drug concentration at the tumor site. For example, the concentration of liposomal doxorubicin in Kaposi's sarcoma lesions can be more than 10

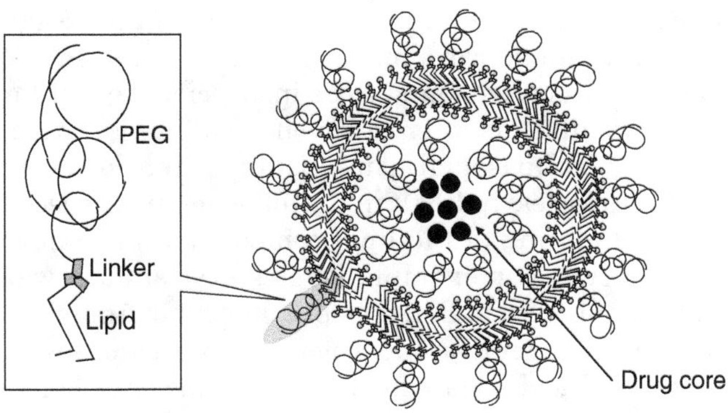

Figure 11.6 Schematic view of a sterically stabilized liposome.

times than in normal skin, leading to a tumor response rate as high as 80 percent.[90,91] Doxil was approved by the FDA in the 1990s for treating Kaposi's sarcoma and ovarian carcinoma that are refractory to other radio- or chemotherapies. At present, a number of sterically stabilized liposomal anticancer drugs have been approved for clinical use, and updated information can be found at the FDA Web site. (*http://www.fda.gov/ohrms/dockets/ac/01/slides/3763s2_08_martin/*) In all these formulations, the hydrophilic polymer PEG is grafted on the surface of the liposome in the form of a PEG-lipid conjugate.

The efficacy of steric stabilization by PEG depends on the molecular weight of the PEG group, the structure of the conjugated lipid moiety, and the mole percentage of the PEG-lipid conjugate in the lipid bilayer.[83] A molecular weight between 2000 and 5000 Da was found appropriate for the PEG group. PEG chains smaller than 2000 Da were insufficient for the steric shielding, whereas those bigger than 5000 Da tended to induce phase separation of the conjugate from the liposome lamellae owing to extensive PEG chain-chain interactions. The stable incorporation of the PEG-lipid conjugate demands long hydrocarbon side chains in the lipid moiety, and the distearoyl moiety with two 18-carbon side chains is currently the structure of choice for the liposomes on the market. The bilayers need at least 2 mol% of the PEG-lipid conjugate for effective steric stabilization. However, when the percentage is too high (> 10 percent), the liposomes are prone to problems of phase separation and micelle formation, leading to the premature leakage of their encapsulated drug.

Drug loading and retention. As discussed previously, a successful drug targeting system needs to retain the encapsulated drug until it reaches the target site. Historically, this was a serious challenge in the design of liposomal drugs because the noncovalent interactions between the drug and the lipid carrier often were found to be insufficient to last for a reasonable bench life or a dilution after intravenous administration. Significant improvements in liposomal drug loading/retention were made from the late 1980s to the early 1990s in the following two aspects.

The first aspect is the tailoring of the lipid composition to achieve retention of the loaded drug. For example, cholesteryl sulfate was selected to complex with amphotericin B in the development of Amphotec (Sequus)[86] because the lipid has multiple noncovalent interactions with the drug. The negatively charged sulfate head group forms ionic bonds with the positive charge from the amine group of amphotericin B, whereas the cholesteryl tail complements well with the hydrophobic portions of amphotericin B to offer extensive hydrophobic interactions. The 1:1 mixture of cholesteryl sulfate and amphotericin B thus forms stable dislike particles that effectively redirect the distribution of the parent drug after intravenous injection and drastically reduce the acute toxicities. Another example is optimization of the lipid composition to better retain the encapsulated anticancer drugs in sterically stabilized liposomes.[83] In order to slow down the spontaneous leakage of the drug from the liposomes through diffusion, saturated lipids (distearyl choline, hydrogenated egg PC), as well as a high percentage of cholesterol, are used in such formulations to increase the rigidity of the liposome bilayers. Such rigid "stealth liposomes" can circulate in the bloodstream for a few days with limited leakage of the free drug.[90]

The second method by which drug loading/retention can be improved involves the use of ion gradients to effect drug loading into preformed liposomes.[92,93] This technique is illustrated in Fig. 11.7 using the weakly basic drug doxorubicin. Liposomes are first prepared in a concentrated acidic buffer (citric acid buffer or ammonium sulfate buffer), followed by increasing the pH of the solution to generate a pH gradient across the liposome membranes. The drug then is applied to the solution in its free-base form. When the drug molecules are outside the liposomes, they are neutral in charge and can diffuse across the liposome bilayers. Once exposed to the acidic medium inside the liposomes, the drug molecules pick up a positive charge by protonation of the basic group and can no longer escape the liposomes through the hydrophobic bilayers. The liposome solutions usually are heated during the loading process to quicken the permeation of the drug molecules into the liposomes; once loading is completed, the temperature is decreased to reduce the permeability of the bilayers so that the drug can remain encapsulated for a longer period of time. When the concentration of the trapped drug exceeds the solubility threshold, the drug precipitates inside the liposomes in the presence of appropriate counterions. The precipitate further slows down

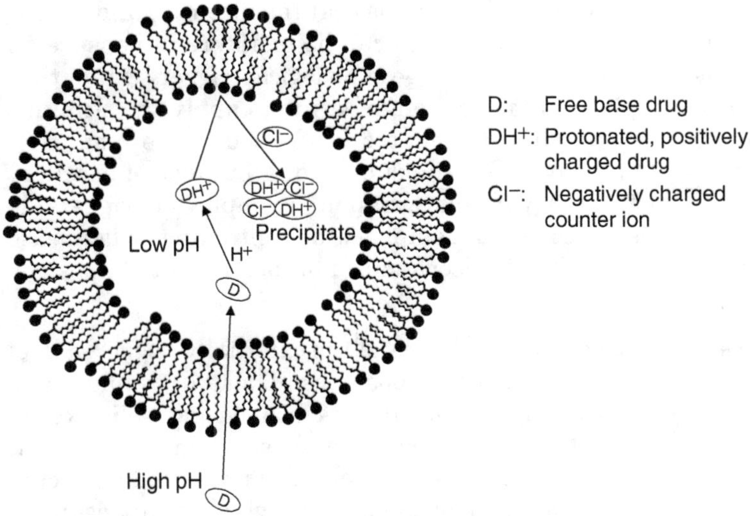

Figure 11.7 Loading of a weakly basic drug into preformed liposomes by pH gradient.

premature leakage of the liposome contents. Over the years, this procedure has proven to be particularly useful for loading high amounts of drug into preformed liposomes and can be applied to a large number of membrane-permeable drugs with a weak-acid or weak-base functional group. In the case of a weakly acidic drug, an ion gradient of a higher pH inside the liposomes is desired to achieve the loading.[94]

Triggerable liposomes. As mentioned previously, once the drug-loaded carrier has reached the target tissue, the active drug must be freed from the carrier and reach its molecular receptor. Given the requirement that the drug release before reaching the target needs to be minimized, a logical implication of such an apparent dilemma is that a mechanism for the delivery system to sense the arrival at the target followed by an elevation of drug release would greatly enhance therapeutic efficacy. Such a hypothesis is validated repeatedly in the emergence of more and more triggerable liposomes[10,20,21] and will continue to represent an important area of future research in drug delivery and drug targeting. Since the structures of the lipid colloids are versatile and very sensitive to the chemical properties of their surfactant components, numerous liposomal triggered release systems have been designed.[10,21]

The stimuli to trigger liposome release[10] can be applied either externally, such as heat or light, or naturally by the unique physiological or pathological conditions of the target site, such as the drop of pH, the elevated activity of a specific enzyme, or the change in redox potential. The advantage of externally applied stimuli is that the intensity and, to certain degree, the location of the stimuli can be monitored to ensure an effective release, whereas the advantage of biological triggers is that they do not require complicated medical engineering to apply the stimulus after the delivery system has distributed in the body. Liposomal triggered release systems have been the subject of a number of excellent reviews.[8,10,95,96] In the following paragraphs, examples of triggerable liposomes will be discussed with an emphasis on controlling drug release by carefully monitoring the physicochemical properties of the liposome components.

Thermosensitive liposomes. One category of the externally triggered liposomes consists of thermosensitive liposomes. These liposomes take advantage of the phase change of the bilayers from a rigid, solidlike gel phase to a more permeable liquid crystalline phase when the temperature is increased.[8] The temperature at which such a phase change occurs is called the *melting temperature* of the bilayer, which is analogous to the melting point of a solid crystal. In fact, during liposome loading driven by an ion gradient (see Fig. 11.7), the bilayers are converted from the solid gel phase to the liquid crystalline phase with heating so that the drug molecules can diffuse more quickly into the liposomes. Once the loading is finished, the temperature is decreased so that the bilayers are converted back to the solid gel phase.

The melting temperature is a sensitive function of the lipid composition of the bilayer.[97] Bilayers composed of diacyl lipids with saturated hydrocarbon tails have a much higher melting point (mostly higher than 20°C) than those composed of unsaturated lipids with cis double bonds (mostly lower than 0°C). The cis double bond loosens the packing of the bilayer and weakens the van der Waals interaction between the hydrophobic lipid tails. For different saturated lipids, the melting temperature of the bilayer increases as the length of the hydrocarbon tails increases. For example, the melting temperature of a bilayer composed of distearoyl phosphatidylcholine (55°C) with two 18-carbon side chains is higher than that of dipalmitoyl phosphatidylcholine (DPPC, 41°C) with two 16-carbon side chains.

In a human tumor xenograft model, Kong et al.[21] compared the antitumor effect of doxorubicin encapsulated in three types of liposome formulations, a nonthermosensitive liposome (NTSL), a traditional thermosensitive liposome (TTSL), and a low-temperature-sensitive liposome (LTSL). All three liposomes are sterically stabilized with PEG grafted on their surface. All the liposomes have a long circulation time and can accumulate selectively in tumor tissue. However, the lipid compositions of the three formulations are different. The NTSL was composed of saturated long-chain lipids such as hydrogenated soybean phosphatidylcholine and DSPE-PEG. The bilayer of the NTSL is thus very rigid, and the encapsulated doxorubicin is released at a slow rate unless the vesicles are heated above 50°C, which is not feasible under clinical settings. In the TTSL, about 50 percent of the lipids are DPPC, a saturated lipid with two shorter hydrocarbon side chains (C_{16}). Therefore, the bilayer structure of the TTSL is less compact and changes from a solidlike gel phase to the liquid crystalline phase in the temperature range of 42 to 45°C (the melting point of the bilayer). In the LTSL, another lipid (1-myristoyl-2-palmitoyl phosphatidylcholine) with an even shorter hydrocarbon side chain (a C_{14} as well as a C_{16} side chain) was introduced, and the bilayer of the liposome had a melting point of 39 to 40°C. The liposomes were administered to mice by tail vein injection, and the tumor-bearing leg of each mouse was heated immediately in a water bath. Among all the formulations and the treatments, the low-temperature-sensitive liposome (LTSL) in combination with tumor hyperthermia at 42°C released the highest amount of encapsulated doxorubicin in the tumor tissue and was the most effective in delaying tumor growth. For all treatments, the antitumor efficacy was in tight correlation with the concentration of released doxorubicin in tumor tissue.

pH-Sensitive liposomes. In 1980, Yatvin et al.[98] reported the first pH-sensitive liposome system composed of phosphatidylcholine (PC) and *N*-palmitoyl homocysteine (NPHC). At neutral pH, such liposomes carry excessive negative charges from the carboxylate group of the *N*-palmitoyl homocysteine (Fig. 11.8); when the pH is decreased, the carboxylate group is neutralized, and an elevated leakage of the liposome was observed. Since then, a number of pH-sensitive liposomes that feature a surfactant with a pH-titratable carboxylate group and a conical-shaped fusogenic lipid such as DOPE have been reported.[99] At neutral pH, the lamellar structures of these liposomes are stabilized by the negative charges of the carboxylates; when the pH is decreased, neutralization of the carboxylate group reduces the surface area of the surfactant head group and triggers the collapse of the phosphatidylethanolamine-rich bilayers into a hexagonal phase with a concomitant release of the encapsulated contents. This first generation of pH-sensitive liposomes, however, has found little application in vivo because the negative charges on their surfaces at neutral pH induce extensive interaction with complement proteins and macrophages of the MPS.[20] On intravenous administration, these liposomes are eliminated quickly from circulation.

In order to circumvent such a drawback, there has been a recent focus on the design of cleavable lipids whose hydrolysis is catalyzed by the drop in pH.[10] This approach exploits acid-labile chemical groups such as acetals, ketals, orthoesters, and vinyl ethers. Such functional groups

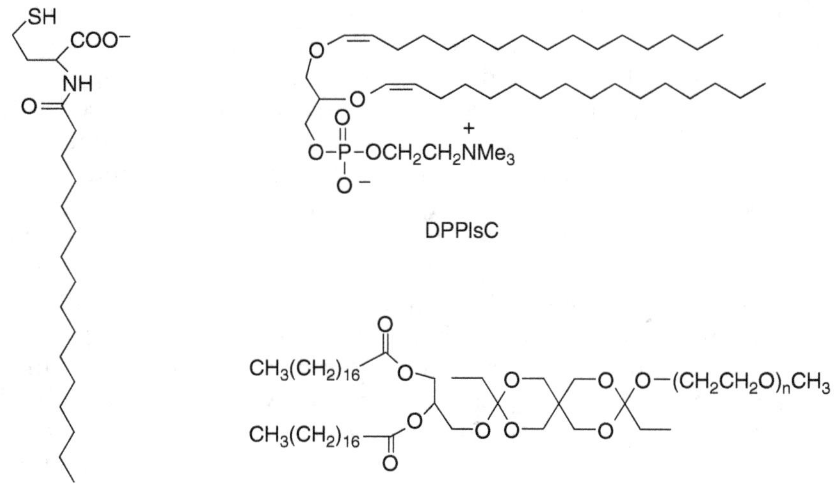

N-Palmitoyl homocysteine POD

Figure 11.8 Examples of pH-sensitive surfactants.

are used as linkers to build surfactants with a variety of head groups, lipid side chains, and linkage configurations. A pharmaceutical chemist thus enjoys considerable flexibility in tailoring the surfactant structures for specific applications such as prolonged circulation, receptor binding, or biomembrane destabilization.

Thompson et al.[100–102] designed pH-sensitive liposomes using a number of mono- and diplasmenyl lipids with an acid-labile vinyl ether linkage at the proximal end of one or both the hydrocarbon side chains. The liposomes are stable at neutral pH but release their contents more quickly when the pH is reduced. For example, liposomes composed of diplasmenyl phosphocholine (DPPlsC) (see Fig. 11.8) released 50 percent of the encapsulated calcein in 230 minutes at pH 5.3. A sterically stabilized liposome with a targeting ligand also was prepared by the DPPlsC lipid.[103] The liposome consisted of 99.5 percent DPPlsC and 0.5 percent a DSPE-PEG3350-folate conjugate, using the folate group as the targeting ligand. When KB cells (a human epithelial cell line) were incubated with such liposome loaded with propidium iodide, 83 percent of propidium iodide escaped the endosomal/lysosomal compartments within 8 hours. Liposome-encapsulated 1-β-arabinofuranosylcytosine is 6000-fold more toxic to the KB cell culture compared with the free drug. These results demonstrated that a pH-triggering mechanism can improve the efficacy of an antitumor drug significantly by a more efficient delivery of the drug to its subcellular target site.

As discussed in preceding sections, a successful triggered release system needs to be relatively stable during the circulation and yet needs to release the drug at a level sufficient for a therapeutic effect at the target site. In the case of pH triggering, this may present a serious challenge because the decrease in pH in the physiological and pathological settings is usually small. For example, the pH in inflammatory tissues[28] and solid tumors[104] is only 0.4 to 0.8 unit more acidic than that in the circulation, suggesting that the liposomes designed for triggered release at these sites need to respond to a small stimulus and release enough drug for a therapeutic effect. In an effort to develop pH-sensitive liposomes that meet such criteria, Guo and Szoka[105] designed an PEG2000-orthoester-distearoylglycerol conjugate (POD), (see Fig. 11.8). In this conjugate, the neutrally charged and hydrophilic PEG head group is selected to confer stability on the liposomes in the bloodstream; the distearoyl glycerol moiety with two long and satured 18-carbon chains serves to stably anchor the conjugate into the liposome bilayer; orthoester is one of the most acid-labile functional groups known in the literature, and the diorthoester linker derived from 3,9-diethyl-2,4,8,10-tetraoxaspiro[5,5]undecane has shown good biocompatibility based on previous research by Heller et al.[106] The conjugate is relatively stable at neutral pH but degrades completely in 1 hour when the pH is decreased to 5. Liposomes composed of 10 percent POD and 90 percent of the conical lipid DOPE are as stable as the sterically stabilized and pH-insensitive control liposomes both in vitro and in the blood circulation. However, at pH 5.5, the POD/DOPE liposomes collapsed and released most of their contents in 30 minutes. The stability in the bloodstream, as well as the quick response to small pH drops, may provide POD-based pH-sensitive liposomes significant advantages for applications in drug and gene targeting in mildly acidic bioenvironments.

Other strategies to design a pH-triggered liposome delivery system include neutralization of negatively charged polymers[96,107] or peptides,[96,108] which, in turn, absorb onto the bilayers and destabilize their structures, and protonation of neutral surfactants into their positively charged surface-active conjugate acids.[109,110] Because the decrease in pH is implicated in numerous physiological and pathological events such as endosomal processing, tumor growth, infection/inflammation, and ischemia conditions,[10,96] the design of pH-sensitive liposomes will continue to attract intensive investigations in the near future.

Cationic liposomes. Cationic liposomes represent the most investigated vector in nonviral gene delivery.[7,36] When cationic liposomes are mixed with highly negatively charged DNA, they condense spontaneously into colloidal complexes called *lipoplexes*. When the molar equivalent of positive charges from the liposomes is in excess of the negative charges from the DNA phosphates, the resulting lipoplexes carry excessive positive charges and exert a highly positive zeta potential.[111]

When cationic lipoplexes are administered via intravenous injection, most of them absorb instantly to the endothelial cells in the lung.[36] The highly positive zeta potential of the lipoplexes determines such a selective deposition. The endothelial cells in the lung outline the first capillary bed that the lipoplexes encounter after intravenous administration. When the cationic lipoplexes pass the lung, they are trapped by strong ionic interaction with the negatively charged proteoglycans composing the extracellular matrix of the endothelial cells.[74] The lipoplexes subsequently are taken up by the cells through the endocytic metabolism pathway of proteoglycans. Pretreatment of the cultured cells or the animals with the enzymes that hydrolyze the proteoglycans abolished the transfection activity of the cationic lipoplexes, demonstrating the role of proteoglycans in cationic liposome-mediated gene delivery. The destabilization activity toward the biomembranes is another important physicochemical parameter of lipoplexes. The helper lipids[84,112] and endosomotropic peptides[108] that destabilize the biomembranes enhance the transfection of complexed DNA. Liposome-mediated gene delivery is discussed further in Chap. 10.

11.4.3 Nanospheres

Since liposomes have an aqueous interior, they are also termed as *nanovesicles* or *nanocapsules*. On the other hand, nanoparticulates with a solid matrix-type interior represent another type of drug carrier called *nanospheres*.[27] Compared with liposomes, the solid interior of nanospheres generally offers a lower drug-loading capacity but at the same time provides some unique physical properties that can be advantageous in certain applications.

Nanospheres can be prepared with a variety of hydrophobic polymers, including polystyrene, poly(phosphazene), poly(methyl methacrylate), poly(butyl 2-cyanoacrylate), polylactide, and poly(DL-lactide-co-glycolide)[27]. Among them, polylactide and poly(DL-lactide-co-glycolide) attract most investigations because of their good biocompatibility.[113] These materials are biodegradable and hence eventually can be cleared from the body; the polymers and their metabolites have little known toxicities. The gradual hydrolysis of polylactide and poly(DL-lactide-co-glycolide) yields a sustained release of the complexed drugs up to weeks, which is of considerable advantage in applications including gene therapy, hormone supplement therapy, and vaccination. The physical properties of the nanospheres, such as size, charge, surface hydrophobicity, membrane-destabilization activity, and release characteristics, can be monitored by the composition of the polymers, as well as the formulation method.

As in other particulate delivery systems, the size of the nanospheres is an important physical property for drug targeting. In order to avoid blockage of the fine capillaries and screening by the spleen and liver, a nanosphere preparation for intravenous administration must have a narrow diameter distribution of around 200 nm or smaller (see Sec. 11.3.1). A submicrometer size is also required for extravasation and tissue penetration. In a rat intestinal loop model, nanospheres were able to penetrate throughout the submucosal layers, whereas microspheres accumulated on the epithelial cell layer.[114] Nanospheres can cross the blood-brain barrier following its transient disruption by hyperosmotic mannitol.[115] Nanospheres also have been used to treat restenosis,[116] the refractory narrowing of blood vessels after operational relief (e.g., balloon angioplasty or stenting) of coronary artery obstruction. In this case, a localized and sustained release of the antithrombotic and antiproliferative drugs at the injured artery is desired. Catheter-infused nanospheres with a diameter of 90 nm or smaller were able to penetrate into different layers of the artery wall, whereas larger nanospheres (120 to 500 nm in diameter) were mostly present in the exposed intima. After the infusion, the arterial pressure was relieved and the artery returned from the dilated state to the normal state. The penetrated nanospheres thus were immobilized in the artery wall, allowing a targeted exposure of the antiproliferative drug to the injured blood vessel.

Modification of the surface charge is also an important approach to targeted delivery by nanospheres. Labhasetwar et al.[117] studied the effect of surface modification of the nanospheres on their disposition at the targeted artery in the acute dog femeral artery and pig coronary artery models of restenosis. In this study, the PLGA (polylactic–polyglycolic acid copolymer) nanospheres were formulated by an oil-in-water emulsion solvent evaporation technique using 2-aminochromone (U-86983, Upjohn and Pharmacia) (U-86) as a model antiproliferative agent. The unmodified nanospheres had a zeta potential of -27.8 ± 0.5 mV (mean $\pm$ SEM, $n = 5$). When the surfaces of such nanospheres were coated with a cationic lipid, didodecyldimethylammonium bromide (DMAB), the modified nanospheres assumed a positive zeta potential of $+22.1 \pm 3.2$ mV (mean $\pm$ SEM, $n = 5$). The catheter-infused nanospheres with a positive surface charge yielded 7- to 10-fold greater arterial U-86 levels compared with the unmodified nanospheres in different ex vivo and in vivo studies. The infusion of nanospheres with higher U-86 loading reduced the arterial U-86 levels, whereas increasing the nanosphere concentration in the infusion solutions increased the arterial U-86 levels, indicating that the elevated U-86 levels are caused by better retention of the cationic nanospheres at the targeted artery owing to their better absorption to the negatively charged cell surface.

The surface charge of the PLGA nanospheres also was thought to be responsible for their quick escape from the endosomes after endocytosis.[118] The PLGA nanospheres possess a negative zeta potential at neutral pH. However, as the pH decreases to 4 to 5, the zeta potential turns from negative to positive and rises up to +20 mV. Fluorescent microscopy and transmission electron microscopic (TEM) studies of human arterial smooth muscle cells treated with the PLGA nanospheres showed that the nanospheres translocated from the endosomes to the cytoplasm within 10 minutes following the endocytosis. The authors proposed that as the pH inside the endosomes decreased, the surface charge of PLGA nanospheres turned from negative to positive, leading to a strong interaction with the negatively charged endosomal membrane and escape of the nanospheres. This hypothesis is consistent with the observation that polystyrene nanospheres, which retain a negative zeta potential at pH 4 to 5, remained trapped in the endosomal and lysosomal compartments following the endocytosis. The PLGA nanospheres encapsulating plasmid DNA mediated a high level of gene expression (up to 1 ng luciferase per milligram of cellular protein), indicating that a portion of the encapsulated DNA eventually transferred from the cytoplasm to the nucleus. In the 3-day experimental period, gene expression was the highest on the third day, in contrast to gene expressions mediated by cationic lipoplexes or polyplexes, which usually diminish after 48 hours. This demonstrates the ability of the PLGA particles to deliver DNA in sustained release kinetics. Gene delivery may be further enhanced by attaching a nuclear localization sequence to the nanosphere surfaces or to the DNA molecules.[15]

Similar to sterically stabilized liposomes, long-circulating nanospheres can be devised by coating their surfaces with neutral and amphiphilic polymers.[27] Two types of PEG-polyoxypropylene copolymers, namely, poloxamines and poloxamers, are able to adsorb to the hydrophobic surface of numerous nanospheres. The copolymer adsorption and the consequent increase in the nanosphere half-life in circulation depend on the physicochemical properties of the nanospheres, as well as on the structure of the copolymers. Polystyrene nanospheres coated with poloxamine-908 had a half-life of up to 2 days in mice and rats, whereas the coating of albumin, poly(phosphazene), and PLGA nanospheres does not generate long blood circulation (half-lives shorter than 2 to 3 hours). This is probably because the polystyrene nanospheres have a more hydrophobic surface and hence a more stable coating of the triblock copolymers via hydrophobic interactions. For a given nanosphere and a given coating copolymer, nanospheres with a diameter below 100 nm adsorb fewer copolymer molecules per unit area than larger ones. As to the structure of the PEG-polyoxypropylene copolymer, the size of the hydrophobic polyoxypropylene center block determines the coating density, whereas the size of the hydrophilic PEG blocks determines the thickness and the dynamics of the coating. The PEG blocks of the coating copolymers need a molecular weight between 2000 and 5000 Da to suppress opsonization of the nanosphere surface. The sterically stabilized nanospheres also could be prepared by covalent attachment of the PEG chains onto the nanosphere surface or by incorporation of diblock PEG-R copolymers during nanosphere preparation, where R is a hydrophobic polymer. However, only limited experimental data are currently available to confirm their success in prolonging circulation half-life in vivo.

11.5 Future Outlook for Physically Targeted Drug Delivery Systems

A drug delivery system needs to pass a series of anatomical, cellular, and subcellular barriers to reach the molecular target of the parent drug. Furthermore, the active drug must be released selectively at the target site at the right level for the right duration. To meet such challenging goals, the physicochemical properties of a drug carrier need to be optimized for a targeted drug therapy. Such physicochemical properties include size, surface charge, surface hydrophobicity, membrane destabilization activity, and sensitivity to triggering at the target site.

Given the complicated requirements for an efficient and specific delivery and the multiplicity of physicochemical parameters in consideration, chimeric delivery systems with multiple functionalities would become a new trend in physical targeting. For example, tumor-targeting liposomes may have a PEG coating for steric stabilization, cleavable lipids for triggered release or destabilization of cancer cell membranes, and ligands for active targeting. Another trend for physical targeting is that delivery systems will be more tailored for specific applications. For example, the optimal physicochemical properties of locally administered nanospheres to treat arterial restenosis are very different from those of long-circulating particulates for remote targeting to solid tumors. Therefore, a pharmaceutical scientist in the field of drug delivery will enjoy significant advantages if he or she can acquire and use information from multiple disciplines including physiology, pathology, pharmacokinetics, pharmacodynamics as well as the physical chemistry, for drug formulation.

References

1. Myers, E. W., Sutton, G. G., Smith, H. O., et al. On the sequencing and assembly of the human genome. *Proc. Natl. Acad. Sci. USA* **99**(7):4145–4146. 2002.
2. Venter, J. C., Adams, M. D., Myers, E. W., et al. The sequence of the human genome. *Science* **291**(5507):1304–1351. 2001.

3. Thanou, M., and Duncan, R. Polymer-protein and polymer-drug conjugates in cancer therapy. *Curr. Opin. Invest. Drugs* **4**(6):701–709. 2003.
4. Fisher, D. R. Assessments for high dose radionuclide therapy treatment planning. *Radiat. Prot. Dosimetry* **105**(1–4):581–586. 2003.
5. Herbrecht, R., Natarajan-Ame, S., Nivoix, Y., and Letscher-Bru, V. The lipid formulations of amphotericin B. *Exp. Opin. Pharmacother.* **4**(8):1277–1287. 2003.
6. Molineux, G. PEGylation: Engineering improved biopharmaceuticals for oncology. *Pharmacotherapy* **23**(8 Pt 2):3S–8S. 2003.
7. Huang, L., Hung. M. C., and Wagner, E. *Nonviral Vectors for Gene Therapy*. San Diego: Academic Press, 1999.
8. Needham, D. Materials engineering of lipid bilayers for drug carrier performances. *MRS Bull.* **24**:32–40. 1999.
9. Crommelin, D. J. A., Hennink, W. E., and Storm, G. Drug targeting systems: Fundamentals and applications to parental drug delivery, in Hillery, A. M., Lloyd, A. W., and Swarbrick, J., (eds.), *Drug Delivery and Targeting for Pharmacists and Pharmaceutical Scientists*. London: Taylor and Francis, 2001.
10. Guo, X., and Szoka, F. C. Chemical approaches to triggerable lipid vesicles for drug and gene delivery. *Acc. Chem. Res.* **36**(5):335–341. 2003.
11. Lim, H. J., Masin, D., McIntosh, N. L., et al. Role of drug release and liposome-mediated drug delivery in governing the therapeutic activity of liposomal mitoxantrone used to treat human A431 and LS180 solid tumors. *J. Pharmacol. Exp. Ther.* **292**(1):337–345. 2000.
12. Adibi, S. A. Regulation of expression of the intestinal oligopeptide transporter (Pept-1) in health and disease. *Am. J. Physiol. Gastrointest. Liver Physiol.* **285**(5):G779–788. 2003.
13. Matherly, L. H., and Goldman, D. I. Membrane transport of folates. *Vitam. Horm.* **66**:403–456. 2003.
14. Xu, Y., and Szoka, F. C., Jr. Mechanism of DNA release from cationic liposome/DNA complexes used in cell transfection. *Biochemistry* **35**(18):5616–5623. 1996.
15. Zanta, M. A., Belguise-Valladier, P., and Behr, J. P. Gene delivery: A single nuclear localization signal peptide is sufficient to carry DNA to the cell nucleus. *Proc. Natl. Acad. Sci. USA* **96**(1):91–96. 1999.
16. Tejada-Berges, T., Granai, C. O., Gordinier, M., and Gajewski, W. Caelyx/Doxil for the treatment of metastatic ovarian and breast cancer. *Exp. Rev. Anticancer Ther.* **2**(2):143–150. 2002.
17. Goodson, R. J., and Katre, N.V. Site-directed PEGylation of recombinant interleukin-2 at its glycosylation site. *Biotechnology* **8**(4):343–346. 1990.
18. Seymour, L. W., Ferry, D. R., Anderson, D., et al. Hepatic drug targeting: Phase I evaluation of polymer-bound doxorubicin. *J. Clin. Oncol.* **20**(6):1668–1676. 2002.
19. Vasey, P. A., Kaye, S. B., Morrison, R., et al. Phase I clinical and pharmacokinetic study of PK1 [*N*-(2-hydroxypropyl)methacrylamide copolymer doxorubicin]: First member of a new class of chemotherapeutic agents-drug-polymer conjugates. Cancer Research Campaign Phase I/II Committee. *Clin. Cancer Res.* **5**(1):83–94. 1999.
20. Drummond, D. C., Meyer, O., Hong, K., et al. Optimizing liposomes for delivery of chemotherapeutic agents to solid tumors. *Pharmacol. Rev.* **51**(4):691–743. 1999.
21. Kong, G., Anyarambhatla, G., Petros, W.P., et al. Efficacy of liposomes and hyperthermia in a human tumor xenograft model: Importance of triggered drug release. *Cancer Res.* **60**(24):6950–6957. 2000.
22. Needham, D., Anyarambhatla, G., Kong, G., and Dewhirst, M.W. A new temperature-sensitive liposome for use with mild hyperthermia: Characterization and testing in a human tumor xenograft model. *Cancer Res.* **60**(5):1197–1201. 2000.
23. Torchilin, V.P. Structure and design of polymeric surfactant-based drug delivery systems. *J Cont. Rel.* **73**(2–3):137–172. 2001.
24. Fang, J., Sawa, T., and Maeda, H. Factors and mechanism of "EPR" effect and the enhanced antitumor effects of macromolecular drugs including SMANCS. *Adv. Exp. Med. Biol.* **519**:29–49. 2003.
25. Maeda, H. The enhanced permeability and retention (EPR) effect in tumor vasculature: The key role of tumor-selective macromolecular drug targeting. *Adv. Enzyme Regul.* **41**:189–207. 2001.

26. Duncan, R. The dawning era of polymer therapeutics. *Nature Rev. Drug Disc.* **2**(5):347–360. 2003.
27. Moghimi, S. M., Hunter, A. C., and Murray, J. C. Long-circulating and target-specific nanoparticles: Theory to practice. *Pharmacol. Rev.* **53**(2):283–318. 2001.
28. Gallin, J. I., Goldstein, I. M., and Snyderman, R. *Inflammation: Basic Principles and Clinical Correlates*, 2d, ed. New York: Raven Press, 1992.
29. Lindbom, L. Regulation of vascular permeability by neutrophils in acute inflammation. *Chem. Immunol. Allergy* **83**:146–166. 2003.
30. Arikan, S., and Rex, J. H. Lipid-based antifungal agents: Current status. *Curr. Pharm. Des.* **7**(5):393–415. 2001.
31. Gabizon, A., and Papahadjopoulos, D. Liposome formulations with prolonged circulation time in blood and enhanced uptake by tumors. *Proc. Natl. Acad. Sci. USA* **85**(18):6949–6953. 1988.
32. Papahadjopoulos, D., Allen, T. M., Gabizon, A., et al. Sterically stabilized liposomes: Improvements in pharmacokinetics and antitumor therapeutic efficacy. *Proc. Natl. Acad. Sci. USA* **88**(24):11460–11464. 1991.
33. Bradley, A. J., and Devine, D. V. The complement system in liposome clearance: Can complement deposition be inhibited? *Adv. Drug Del. Rev.* **32**(1–2):19–29. 1998.
34. Devine, D. V., Wong, K., Serrano, K., et al. Liposome-complement interactions in rat serum: Implications for liposome survival studies. *Biochim. Biophys. Acta* **1191**(1):43–51. 1994.
35. Plank, C., Mechtler, K., Szoka, F. C., Jr., and Wagner, E. Activation of the complement system by synthetic DNA complexes: A potential barrier for intravenous gene delivery. *Hum. Gene Ther.* **7**(12):1437–1446. 1996.
36. Brown, M. D., Schatzlein, A. G., and Uchegbu, I. F. Gene delivery with synthetic (nonviral) carriers. *Int. J. Pharm.* **229**(1–2):1–21. 2001.
37. Miller, E. S., Kutter, E., Mosig, G., et al. Bacteriophage T4 genome. *Microbiol. Mol. Biol. Rev.* **67**(1):86–156. 2003.
38. Huang, Q., Sivaramakrishna, R. P., Ludwig, K., et al. Early steps of the conformational change of influenza virus hemagglutinin to a fusion active state: Stability and energetics of the hemagglutinin. *Biochim. Biophys. Acta* **1614**(1):3–13. 2003.
39. Nagle, T., Berg, C., Nassr, R., and Pang, K. The further evolution of biotech. *Nature Rev. Drug Disc.* **2**(1):75–79. 2003.
40. Harris, J. M., and Chess, R. B. Effect of PEGylation on pharmaceuticals." *Nature Rev. Drug Disc.* **2**(3):214–221. 2003.
41. Chapman, A. P., Antoniw, P., Spitali, M. et al. Therapeutic antibody fragments with prolonged in vivo half-lives. *Nature Biotechnol.* **17**(8):780–783. 1999.
42. Kinstler, O., Molineux, G., Treuheit, M., et al. Mono-*N*-terminal poly(ethylene glycol)-protein conjugates. *Adv. Drug Deliv. Rev.* **54**(4):477–485. 2002.
43. Bukowski, R., Ernstoff, M. S., Gore, M.E., et al. PEGylated interferon alfa-2b treatment for patients with solid tumors: A phase I/II study. *J. Clin. Oncol.* **20**(18):3841–3849. 2002.
44. Graham, M. L. PEGaspargase: A review of clinical studies. *Adv. Drug Deliv. Rev.* **55**(10):1293–1302. 2003.
45. Levy, Y., Hershfield, M. S., Fernandez-Mejia, C., et al. Adenosine deaminase deficiency with late onset of recurrent infections: Response to treatment with polyethylene glycol-modified adenosine deaminase. *J. Pediatr.* **113**(2):312–317. 1988.
46. Parkinson, C., Kassem, M., Heickendorff, L., et al. PEGvisomant-induced serum insulin-like growth factor-I normalization in patients with acromegaly returns elevated markers of bone turnover to normal. *J. Clin. Endocrinol. Metab.* **88**(12): 5650–5655. 2003.
47. Pedder, S. C. PEGylation of interferon alfa: Structural and pharmacokinetic properties. *Semin. Liver Dis.* **23** (suppl 1):19–22. 2003.
48. Sato, H. Enzymatic procedure for site-specific pegylation of proteins. *Adv. Drug Deliv. Rev.* **54**(4):487–504. 2002.
49. Abe, S., and Otsuki, M. Styrene maleic acid neocarzinostatin treatment for hepatocellular carcinoma. *Curr. Med. Chem. AntiCancer Agents* **2**(6):715–726. 2002.

50. Kopecek, J., Kopeckova, P., Minko, T., and Lu, Z. HPMA copolymer-anticancer drug conjugates: Design, activity, and mechanism of action. *Eur. J. Pharm. Biopharm.* **50**(1):61–81. 2000.
51. Sprincl, L., Exner, J., Sterba O., and Kopecek, J. New types of synthetic infusion solutions: III. Elimination and retention of poly-[*N*-(2-hydroxypropyl)methacrylamide] in a test organism. *J. Biomed. Mater. Res.* **10**(6):953–963. 1976.
52. Duncan, R. Preclinical evaluation of polymer-bound doxorubicin. *J. Contr. Rel.* **19**:331–346. 1992.
53. Etrych, T., Jelinkova, M., Rihova, B., and Ulbrich, K. New HPMA copolymers containing doxorubicin bound via pH-sensitive linkage: Synthesis and preliminary in vitro and in vivo biological properties. *J. Contr. Rel.* **73**(1):89–102. 2001.
54. Mittal, K. L., Lindman, B., eds. *Surfactants in Solution.* New York: Plenum Press, 1991.
55. Torchilin, V. P., and Trubetskoy, V. S. Which polymers can make nanoparticulate drug carriers long-circulating? *Adv. Drug Deliv. Rev.* **16**(2–3):141–155. 1995.
56. Bedu-Addo, F. K., and Huang, L. Interaction of PEG-phospholipid conjugates with phospholipid: Implications in liposomal drug delivery. *Adv. Drug Del. Rev.* **16**(2–3):235–247. 1995.
57. Yokoyama, M., Okano, T., Skurai, Y., et al. Introduction of cisplatin into polymeric micelles. *J. Contr. Rel.* **39**:351–356. 1996.
58. Yokoyama, M., Satoh, A., Sakurai, Y., et al. Incorporation of water-insoluble anticancer drug into polymeric micelles and control of their particle size. *J. Contr. Rel.* **55**(2–3):219–229. 1998.
59. Florence, A. T., Attwood, D. *Physicochemical Principles of Pharmacy*, 3d ed. New York: Palgrave, 1998.
60. Kwon, G. S., Yokoyama, M., Okano, T., et al. Biodistribution of micelle-forming polymer-drug conjugates. *Pharm. Res.* **10**:970–974. 1993.
61. Kwon, G., Naito, M., Yokoyama, M., et al. Micelles based on AB block copolymers of poly(ethylene oxide) and poly(benzyl-aspartat). *Langmuir* **9**:945–949. 1993.
62. Zhang, L., and Eisenberg, A. Multiple morphologies of "crew-cut" aggregates of polystyrene-*b*-poly(acrylic acid) block copolymers. *Science* **268**:1728–1731. 1995.
63. Lim, H. J., Masin, D., Madden, T. D., and Bally, M. B. Influence of drug release characteristics on the therapeutic activity of liposomal mitoxantrone. *J. Pharmacol. Exp. Ther.* **281**(1):566–573. 1997.
64. Nakanishi, T., Fukushima, S., Okamoto, K., et al. Development of the polymer micelle carrier system for doxorubicin. *J. Contr. Rel.* **74**(1–3):295–302. 2001.
65. Kataoka, K., Harada, A., and Nagasaki, Y. Block copolymer micelles for drug delivery: Design, characterization and biological significance. *Adv. Drug Deliv. Rev.* **47**(1):113–131. 2001.
66. Oupicky, D., Ogris, M., and Seymour. L. W. Development of long-circulating polyelectrolyte complexes for systemic delivery of genes. *J. Drug Target.* **10**(2):93–98. 2002.
67. Howard, K. A., and Alpar, H. O. The development of polyplex-based DNA vaccines. *J. Drug Target.* **10**(2):143–151. 2002.
68. Morse, M. A. Technology evaluation: VEGF165 gene therapy, Valentis Inc. *Curr. Opin. Mol. Ther.* **3**(1):97–101. 2001.
69. Kim, H. H., Lee, W. S., Yang, J. M., and Shin, S. Basic peptide system for efficient delivery of foreign genes. *Biochim. Biophys. Acta* **1640**(2–3):129–136. 2003.
70. Boussif, O., Lezoualc'h, F., Zanta, M. A., et al. A versatile vector for gene and oligonucleotide transfer into cells in culture and in vivo: polyethylenimine. *Proc. Natl. Acad. Sci. USA* **92**(16):7297–7301. 1995.
71. Tang, M. X., Redemann, C. T., and Szoka, F. C., Jr. In vitro gene delivery by degraded polyamidoamine dendrimers. *Bioconjug. Chem.* **7**(6):703–714. 1996.
72. Haensler, J., and Szoka, F. C., Jr. Polyamidoamine cascade polymers mediate efficient transfection of cells in culture. *Bioconjug. Chem.* **4**(5):372–379. 1993.
73. Mullen, P. M., Lollo, C. P., Phan, Q. C., et al. Strength of conjugate binding to plasmid DNA affects degradation rate and expression level in vivo. *Biochim. Biophys. Acta* **1523**(1):103–110. 2000.

74. Mounkes, L. C., Zhong, W., Cipres-Palacin, G., et al. Proteoglycans mediate cationic liposome-DNA complex–based gene delivery in vitro and in vivo. *J. Biol. Chem.* **273**(40):26164–26170. 1998.
75. Cartier, R., Velinova, M., Lehman, C., et al. Ultrastructural analysis of DNA complexes during transfection and intracellular transport. *J. Histochem. Cytochem.* **51**(9):1237–1240. 2003.
76. Goula, D., Benoist, C., Mantero, S., et al. Polyethylenimine-based intravenous delivery of transgenes to mouse lung. *Gene Ther.* **5**(9):1291–1295. 1998.
77. Abdallah, B., Hassan, A., Benoist, C., et al. A powerful nonviral vector for in vivo gene transfer into the adult mammalian brain: Polyethylenimine. *Hum. Gene Ther.* **7**(16):1947–1954. 1996.
78. Boletta, A., Benigni, A., Lutz, J., et al. Nonviral gene delivery to the rat kidney with polyethylenimine. *Hum. Gene Ther.* **8**(10):1243–1251. 1997.
79. Kursa, M., Walker, G. F., Roessler, V., et al. Novel shielded transferrin–polyethylene glycol–polyethylenimine/DNA complexes for systemic tumor-targeted gene transfer. *Bioconjug. Chem.* **14**(1):222–231. 2003.
80. Ogris, M., and Wagner, E. Tumor-targeted gene transfer with DNA polyplexes. *Somat. Cell. Mol. Genet.* **27**(1–6):85–95. 2002.
81. Ogris, M., Walker, G., Blessing, T., et al. Tumor-targeted gene therapy: Strategies for the preparation of ligand–polyethylene glycol–polyethylenimine/DNA complexes. *J. Contr. Rel.* **91**(1–2):173–181. 2003.
82. Gennis, R. B. *Biomembranes: Molecular Structure and Function.* New York: Springer-Verlag, 1989.
83. Janoff, A. S., ed. *Liposomes: Rational Design.* New York: Marcel Dekker, 1999.
84. Mok, K. W., and Cullis, P. R. Structural and fusogenic properties of cationic liposomes in the presence of plasmid DNA. *Biophys. J.* **73**(5):2534–2545. 1997.
85. Bangham, A. D., Standish, M. M., and Watkins, J. C. Diffusion of univalent ions across the lamellae of swollen phospholipids. *J. Mol. Biol.* **13**(1):238–252. 1965.
86. Kling, J. The liposome maker's art: Wrapping toxic drugs in lipid disguises. *Mod. Drug Disc.* **2**(4):41–49. 1999.
87. Senior, J., and Gregoriadis, G. Is half-life of circulating liposomes determined by changes in their permeability? *FEBS Lett.* **145**(1):109–114. 1982.
88. Proffitt, R. T., Williams, L. E., Presant, C. A., et al. Tumor-imaging potential of liposomes loaded with In-111-NTA: Biodistribution in mice. *J. Nucl. Med.* **24**(1):45–51. 1983.
89. Allen, T. M., and Chonn, A. Large unilamellar liposomes with low uptake into the reticuloendothelial system. *FEBS Lett.* **223**(1):42–46. 1987.
90. Northfelt, D. W., Martin, F. J., Working, P., et al. Doxorubicin encapsulated in liposomes containing surface-bound polyethylene glycol: Pharmacokinetics, tumor localization, and safety in patients with AIDS-related Kaposi's sarcoma. *J. Clin. Pharmacol.* **36**(1):55–63. 1996.
91. Northfelt, D. W., Dezube, B. J., Thommes, J. A., et al. Efficacy of PEGylated-liposomal doxorubicin in the treatment of AIDS-related Kaposi's sarcoma after failure of standard chemotherapy. *J. Clin. Oncol.* **15**(2):653–659. 1997.
92. Mayer, L. D., Tai, L. C., Bally, M. B., et al. Characterization of liposomal systems containing doxorubicin entrapped in response to pH gradients. *Biochim. Biophys. Acta* **1025**(2):143–151. 1990.
93. Haran, G., Cohen, R., Bar, L.K., and Barenholz, Y. Transmembrane ammonium sulfate gradients in liposomes produce efficient and stable entrapment of amphipathic weak bases. *Biochim. Biophys. Acta* **1151**(2):201–215. 1993.
94. Clerc, S., and Barenholz, Y. Loading of amphipathic weak acids into liposomes in response to transmembrane calcium acetate gradients. *Biochim. Biophys. Acta* **1240**(2):257–265. 1995.
95. Meers, P. Enzyme-activated targeting of liposomes. *Adv. Drug Deliv. Rev.* **53**(3):265–272. 2001.
96. Drummond, D. C., Zignani, M., and Leroux, J. Current status of pH-sensitive liposomes in drug delivery. *Prog. Lipid Res.* **39**(5):409–460. 2000.
97. Silvius, J. R. *Lipid-Protein Interactions.* New York: Wiley, 1982.

98. Yatvin, M. B., Kreutz, W., Horwitz, B. A., and Shinitzky, M. pH-sensitive liposomes: Possible clinical implications. *Science* **210**(4475):1253–1255. 1980.
99. Chu, C.-J., and Szoka, F. C. pH-Sensitive liposomes. *J. Liposome Res.* 4:361–395. 1994.
100. Boomer, J. A., and Thompson, D. H. Synthesis of acid-labile diplasmenyl lipids for drug and gene delivery applications. *Chem. Phys. Lipids* **99**(2):145–153. 1999.
101. Boomer, J. A., Thompson, D. H., and Sullivan, S. M. Formation of plasmid-based transfection complexes with an acid-labile cationic lipid: Characterization of in vitro and in vivo gene transfer. *Pharm Res* **19**(9):1292–1301. 2002.
102. Thompson, D. H., Gerasimov, O. V., Wheeler, J. J., et al. Triggerable plasmalogen liposomes: Improvement of system efficiency. *Biochim. Biophys. Acta* **1279**(1):25–34. 1996.
103. Rui, Y., Wang, S., Low, P. S., and Thompson, D.H. Diplasmenyl-choline-folate liposomes: An efficient vehicle for intracellular drug delivery. *J. Am. Chem. Soc.* **120**:11213–11218. 1998.
104. Gerweck, L. E. Tumor pH: Implications for treatment and novel drug design. *Semin. Radiat. Oncol.* **8**(3):176–182. 1998.
105. Guo, X., and Szoka, F. C., Jr. Steric stabilization of fusogenic liposomes by a low-pH sensitive PEG–diortho ester–lipid conjugate. *Bioconjug. Chem.* **12**(2):291–300. 2001.
106. Heller, J. Controlled drug release from poly(ortho esters): A surface eroding polymer. *J. Contr. Rel.* **2**:167–177. 1985.
107. Thomas, J., and Tirrell, D. Polyelectrolyte-sensitized phospholipid vesicles. *Acc. Chem. Res.* **25**:336–342. 1992.
108. Nir, S., Nicol, F., and Szoka, F. C., Jr. Surface aggregation and membrane penetration by peptides: relation to pore formation and fusion. *Mol. Membr. Biol.* **16**(1):95–101. 1999.
109. Liang, E., and Hughes, J. Characterization of a pH-sensitive surfactant, dodecyl-2-(1′-imidazolyl)propionate (DIP), and preliminary studies in liposome mediated gene transfer. *Biochim. Biophys. Acta* **1369**:39–50. 1998.
110. Liang, E., and Hughes, J. Membrane fusion and rupture in liposomes: effect of biodegradable pH-sensitive surfactants. *J. Membr. Biol.* **166**:37–49. 1998.
111. Xu, Y., Hui, S. W., Frederik, P., and Szoka, F. C., Jr. Physicochemical characterization and purification of cationic lipoplexes. *Biophys. J.* **77**(1):341–353. 1999.
112. Bennett, M. J., Nantz, M. H., Balasubramaniam, R. P., et al. Cholesterol enhances cationic liposome-mediated DNA transfection of human respiratory epithelial cells. *Biosci. Rep.* **15**(1):47–53. 1995.
113. Panyam, J., and Labhasetwar, V. Biodegradable nanoparticles for drug and gene delivery to cells and tissue. *Adv. Drug Deliv. Rev.* **55**(3):329–347. 2003.
114. Desai, M. P., Labhasetwar, V., Amidon, G. L., and Levy R. J. Gastrointestinal uptake of biodegradable microparticles: Effect of particle size. *Pharm. Res.* **13**(12):1838–1845. 1996.
115. Kroll, R. A., Pagel, M. A., Muldoon, L. L., et al. Improving drug delivery to intracerebral tumor and surrounding brain in a rodent model: A comparison of osmotic versus bradykinin modification of the blood-brain and/or blood-tumor barriers. *Neurosurgery* **43**(4):879–886; discussion 886–879. 1998.
116. Labhasetwar, V., Song, C., and Levy, R. J. Nanoparticle drug delivery for restenosis. *Adv. Drug Deliv. Rev.* **24**:63–85. 1997.
117. Labhasetwar, V., Song, C., Humphrey, W., et al. Arterial uptake of biodegradable nanoparticles: Effect of surface modifications. *J. Pharm. Sci.* **87**(10):1229–1234. 1998.
118. Panyam, J., Zhou, W.Z., Prabha, S., et al. Rapid endolysosomal escape of poly(DL-lactide-co-glycolide) nanoparticles: Implications for drug and gene delivery. *FASEB. J.* **16**(10):1217–1226. 2002.

Chapter

12

Ligand-Based Targeting Approaches to Drug Delivery

Andrea Wamsley

Thomas J. Long School of Pharmacy and Health Sciences
University of the Pacific
Stockton, California

12.1 Introduction

From the herbs of antiquity to the blockbuster drugs of our time, drug discovery and development largely have been based on the idea that all chemicals can be used safely simply by reducing dose, whence the universally accepted belief that beneficial effect outweighs toxicity. Hence the oldest and most venerated axiom of toxicology is: "All substances are poisons; there is none which is not a poison. The right dose differentiates a poison and a remedy" (Paracelsus, 1493–1541). With such mentality, it is not surprising that current drug therapies for the treatment of a variety of disease states, such as cancer, lack the capability to specifically target the disease. A complex disease such as cancer, where the cancerous cells are as diverse as and apparently indistinguishable from normal cells, requires greater selectivity of action by the drug. For many drugs, the dose responses for beneficial and adverse effects overlap. This usually results in unwanted side effects because both healthy and diseased tissues are affected,[1] which induces as much complaint as the disease itself. Therefore, there is a great global need and challenge for the pharmaceutical industry to develop therapies with site-specific functioning drug modalities. As scientific advances are made in the fields of molecular and cellular biology, unique molecular ligands for each disease will become available. New therapies that exploit such specific ligands have the potential to minimize side effects on normal cells.

The ultimate goal of targeted drug delivery is to increase control of drug dosing at specific physiological sites, such as cells, tissues, or organs, thereby reducing the unwanted side effects at nontarget sites.[2] The concept of drug targeting began at the turn of the twentieth century when Paul Ehrlich coined the phrase *magic bullet*, which emphasized that drug delivery should be aimed to a specific site.[3] Since then, many drug delivery systems that are biodegradable, nontoxic, nonimmunogenic, site-specific, and capable of undergoing many physicochemical manipulations have been developed. However, a great majority of these drug delivery systems are based on passive targeting. For greater control of targeting, drug delivery systems must be rendered smarter by incorporating unique ligands that are specifically recognized by target disease cells, thereby converting passive into ligand-targeting drug delivery systems.

The long-term potential benefits of targeted drug delivery systems in the treatment of pathological disease are immeasurable. The primary reason for developing drug targeting is to decrease toxicity by reducing the side effects that are prevalent in traditional forms of drug therapy. In addition, targeting would allow therapeutic drug levels to be maintained specifically at the site of action, thereby decreasing the necessary amount of drug as well as the number of doses to treat disease. Drug targeting has the potential to be less invasive, lead to improved patient compliance, and reduce health care costs.[4]

12.2 Rationale for Targeting Drug Delivery Systems

12.2.1 Concepts of active and passive targeting

Drug targeting can be divided into two general categories: passive and active targeting. The objective of passive targeting with a drug delivery carrier is to increase the concentration of drug at target tissue by reducing nonspecific interactions, which is accomplished by exploiting physicochemical interactions and physical properties of the carrier system. Physiochemical interactions such as hydrophobic effect and physical features such as size and mass of the drug targeting system have been used to passively target sites of interest.[2,5] Examples of drug delivery carriers that take advantage of passive targeting include polymeric conjugates,[6] micelles,[5] liposomes,[7] and micro- and nanoparticles,[8] as discussed in the previous chapter. In contrast, active targeting with drug delivery carriers exploits specific biological processes, such as specific ligand-receptor recognition and interaction, to increase the concentration of drug at a particular site. Active targeting systems use antibodies, peptides, sugars, vitamins, and other ligands as a homing device that specifically interacts with the receptor of the target cell.[2] These targeted drug delivery systems are basically passive carriers that have been rendered more specific by the addition of a homing device; consequently, important physical attributes such as prolonged circulation and accumulation at target tissue are preserved. However, the effectiveness of a targeting system depends on its specificity and capacity to deliver drug at the required dosage to target cells.

12.2.2 Organ/cellular/subcellular targeting

Targeting can be further classified into three different levels.[9,10] The first, also known as *organ targeting*, delivers the drug to a specific organ or tissue. An example of organ targeting is when microparticle drug delivery carriers are used to target the liver. Owing to the "leaky" nature of liver tissue, the carriers can accumulate within the tissue and slowly release the drug. Other tissues are unaffected because they form tight junctions and prevent the microparticles/macromolecules from being deposited.[11,12] The second level is *cellular targeting* where drug is delivered to a particular cell within an organ or tissue. Cellular targeting can be achieved by conjugating targeting moieties such as antibodies onto macromolecular drug carriers that specifically recognize and bind to complementary antigens and receptors on the cell surface.[4] The third level of targeting is *subcellular* which involves delivery at specific intracellular sites. The most common example of subcellular targeting is the use of gene targeting, where delivery of a gene into specific cells is essential to the success of the therapy.[13]

12.3 Design of Ligand-Based Targeting Drug Delivery Systems

Drug targeting embodies many areas of research and therefore requires a multidisciplinary approach. Several fundamental factors must be considered when developing a targeting drug delivery system. These include developing systems that are biodegradable and biocompatible, that show no inherent toxicity, and that do not accumulate in the body. Also, ideal drug delivery carriers must contain adequate functional groups or components for attachment or enclosure of drugs and/or targeting moiety without compromising their original attributes. In addition to these basic criteria, drug delivery carriers must be capable of targeting specific sites. This entails identification of unique ligands and/or receptors that can function as targeting motifs and specifically interact with target receptors/ligands, selection of an appropriate targeting motif, and construction of the carrier system based on the specific interaction of the targeting motif with the target site. The effectiveness of a targeted drug carrier to specifically interact and deliver drug at the target site must be evaluated, followed by selection of a compatible drug candidate and the type of linkage between the drug and the carrier. Figure 12.1 illustrates the basic design of a targeting polymer-drug conjugated system. For colloidal systems (Fig. 12.2), a drug candidate must be able to be loaded either by encapsulation or entrapment within their core or boundary of the carriers. Ultimately, the success of therapy with targeting drug delivery carriers rests on the specificity of the targeting moiety and the

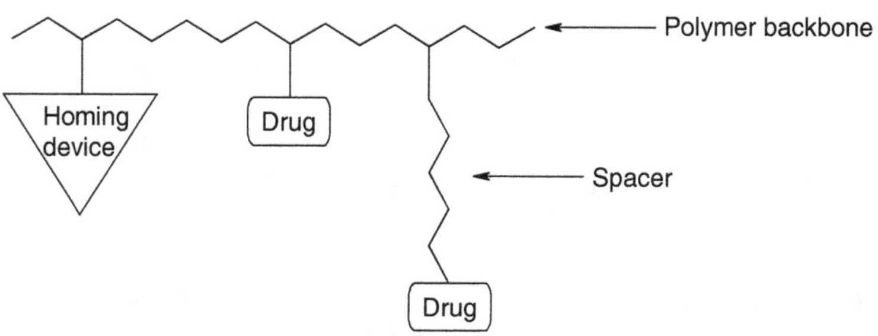

Figure 12.1 Schematic design of polymeric drug targeting carrier.

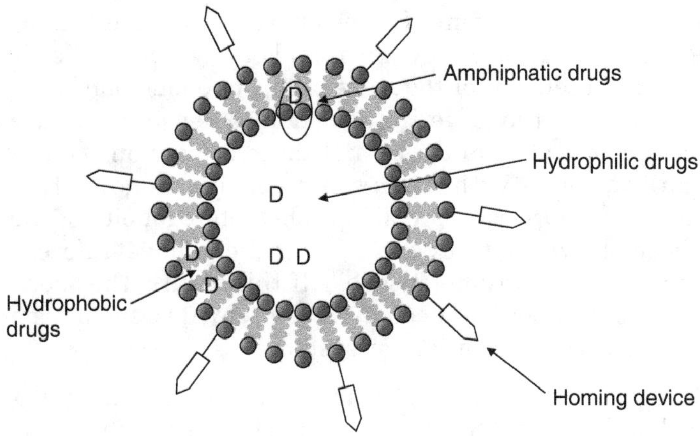

Figure 12.2 Schematic design of colloidal drug targeting carrier.

ability of the carrier to adequately deliver drug to the site of interest for treatment of the disease.[14]

12.3.1 Ligand-receptor-based interaction

Ligand, as described in this chapter, is a general term that characterizes a molecule that specifically interacts with the receptor of another molecule on the surface of a cell, tissue, or organ. Ligands may bind to a specific receptor site, where they then can be internalized into the cell. As a result, ligands represent a diverse class of molecules that can be exploited for targeted drug delivery because the ligand-receptor complex is the result of a specific molecular interaction that requires structural complementarily. The following subsections discuss some of the most commonly studied ligands and their complementary receptors used in active targeting systems.

Receptors/ligands for targeting

Antigens. The use of tumor-associated antigens for the targeting of antibodies has been the most widely exploited form of anticancer targeting drug delivery.[15] The basis for this is that certain antigens usually are expressed in lesser degree in normal tissues than in tumor tissues. Several such antigens have been identified.[15] For example, the carcinoembryonic antigen, prevalent in gastrointestinal (GI), lung, and breast tumors, was the first to be identified, and it has been used extensively as a target.[16] Even though antibodies and their interactions with antigens have been used as homing devices for specific drug delivery, there are generally two problems associated with this type of system.

The first is shedding of the antigen.[17] Studies have shown that tumor-associated antigens shed because they are devoid of any control that is typically seen in normal cells. A consequence of this phenomenon is the decrease in effectiveness of the targeting drug delivery system because the actual targets are the shed antigens and those located on the cell surface. When antibody binds with the shed antigen, it is removed rapidly from circulation.[18] The second problem is the heterogeneity of the antigen. It has been shown that any given antigen can have different patterns of expression within tumor tissue.[19] If this occurs, the specificity of the targeting antibody is decreased, thereby decreasing the effectiveness of the targeting drug carrier system.

Cadherins. Cadherins are a group of glycoproteins that facilitate Ca^{2+}-dependent cell-cell adhesive interactions.[20] When cadherin function is disrupted, the release of a tumor cell can result. It was shown that the aggressive metastasis of undifferentiated epithelial carcinoma cells that had lost cell-cell adhesion could be stopped by transfection with E-cadherin cDNA.[21] Therefore, it was suggested that E-cadherin suppresses metastasis,[22] which was further supported by more recent studies that showed the loss of adhesion of human gastric, prostatic, and lung cancer cells was due to the gene mutation of a protein associated with the proper function of cadherin.[23]

Selectins. Selectins are another type of cell adhesion molecule that is responsible for carbohydrate binding.[24] Selectins mediate cell adhesion by recognizing specific carbohydrate ligands arranged on the surfaces of cells. In addition, it has been suggested that cell-cell adhesion may not be a result of only a single selectin-carbohydrate ligand binding but rather the cumulative effect of multiple interactions of many sugar moieties, the so-called polyvalency or cluster effect.[25] Two examples of this phenomenon were demonstrated by the binding affinity and selectivity of the tetrasaccharide glycolipids sialyl-Lewis X (sLe^x) and sialyl-Lewis A (sLe^a), which play important roles in inflammation, reperfusion, and metastases. As monomers, these carbohydrate moieties are bound with low affinity and selectivity by specific selectins.[26–28] When presented in a cluster, as in a liposome, not only could they significantly increase solubility and bioavailability of the complex, but they also yield higher affinity and selectivity than their respective monomers.

Integrins. The integrin family has been identified as excellent candidates for the development of target specific cancer therapies.[29] Integrins are heterodimeric glycoproteins that consist of α and β subunits, which combine to form the various types of integrins. Currently, 18 α subunits, 8 β subunits, and approximately 22 different integrins have been identified.[29,30]

Adhesion between cell-cell and cell–extracellular matrix occur when integrin recognizes and specifically interacts with minimal peptide core sequences within the extracellular matrix. Minimal peptide core sequences such as arginine–glycine–aspartic acid (RGD), tyrosine–isoleucine–glycine–serine–arginine (YIGSR), and leucine–aspartic acid–valine (LDV) have been identified.[31–33] The binding to minimal core sequences found in the extracellular matrix depends on the degree of integrin expression on the cell surface. Integrin expression is upregulated in cancer cells compared with normal cells. Integrin function could be disrupted by blocking with monoclonal antibodies, peptide antagonists, and small molecules.[34] In addition, the type of integrin expressed varies for different cell lines.[29]

Vitamins. Vitamins are essential for normal cellular function and growth. In pathological conditions, these vitamins also play crucial roles. As such, cell surface receptors for vitamins have been considered as drug targets because vitamins generally are internalized into the cell by receptor-mediated endocytosis. Such vitamins as folic acid, riboflavin, biotin, and vitamin B_6 all have been evaluated as potential ligands for targeted delivery of therapeutic agents to specific cells.[35] However, more research is needed to elucidate the specific function of these and other vitamins and what potential roles they might play in targeted drug delivery.

Folic acid, shown in Fig. 12.3, is a common component of food and vitamins.[36] The receptor for the vitamin folic acid is known as the folate receptor or the high-affinity membrane folate-binding protein.[37] These folate receptors typically are not present in normal cells, but in a variety of human tumors they are greatly overexpressed.[36–38] In addition, folate receptors mediate internalization by endocytosis. For these reasons, folic acid has been examined as a homing device. In general, studies have shown that by simple conjugation of folic acid to macromolecules, internalization into cells can occur similarly to that of free folic acid.[39]

In an effort to determine if other vitamins are capable of similar specific receptor-mediated endocytosis, a BSA-riboflavin conjugate was studied[35] (Fig. 12.3). Riboflavin (vitamin B_2) is an essential water-soluble vitamin that is necessary for cellular functions. Indeed, the BSA-riboflavin conjugate was determined to be internalized by receptor- mediated endocytosis. However, the bovine serum albumin (BSA) conjugation with riboflavin enters through a different pathway than free riboflavin. Although it was shown that the BSA-riboflavin conjugate could be internalized into endosomal compartments of cultured human cells, the transport mechanism and the degree of specificity have not been determined. It is still a subject of much controversy. A clearer understanding of the riboflavin transport mechanism is needed before its potential for targeted drug delivery can be assessed.

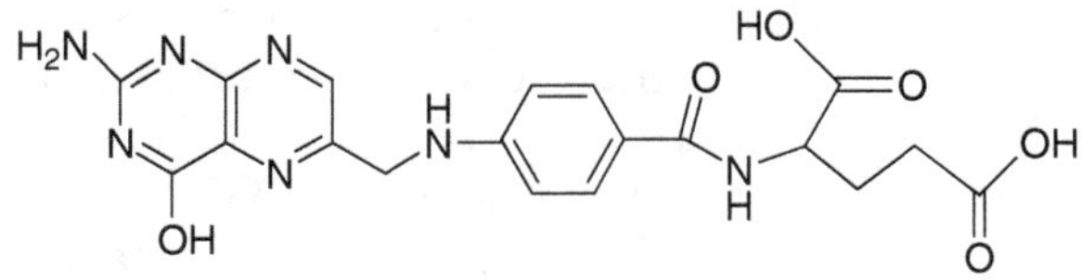

Folic acid

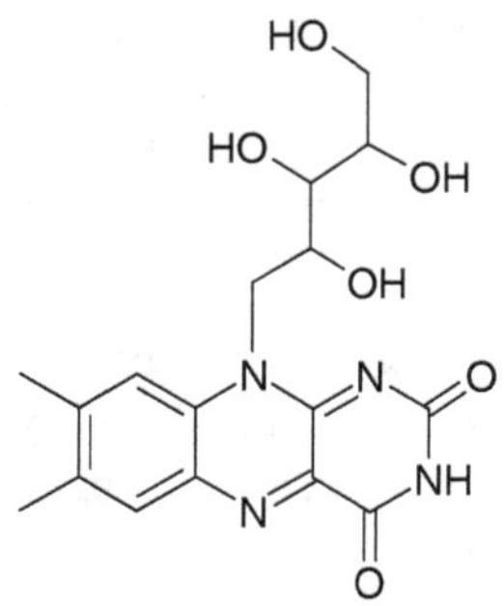

Riboflavin

Figure 12.3 Structures of common vitamins used in targeting drug delivery systems.

Transferrin. Transferrin is a glycoprotein responsible for transporting iron into cells.[13] Iron binds to transferrin and enters the cell through a highly specific receptor-mediated endocytosis via the transferrin receptor. Rapid recycling of the transferrin receptor allows for approximately 2×10^4 molecules of transferrin to be internalized into each cell per minute.[40] The transferrin receptor is expressed on the surfaces of cells in both proliferating and nonproliferating normal tissue, but it is highly upregulated in tumor cells, as evident by reduced transferrin levels in patients with cancer.[13,41] The use of transferrin to target the transferrin receptor has been investigated for targeted drug delivery.[42]

Hormone. The presence of hormone receptors in hormone-sensitive cancers presents potential applications of hormone-targeted drug delivery of traditional drugs. In fact, studies have demonstrated that when doxorubicin is conjugated to an analogue of the luteinizing hormone–releasing hormone (LH-RH), cancer cell death was observed at subnanomolar concentrations.[43] In addition, since LH-RH receptors are found mostly in the pituitary gland, toxicity was found to be exclusively localized to the LH-RH gonadotroph cells. This approach is potentially applicable to ovarian, endometrial, and breast cancers because the onset of tumor growth in each of these tissues is accompanied by an increase in hormone receptors.[44]

Low-density lipoprotein (LDL). Lipoproteins are naturally occurring spherical macromolecular particles that transport lipids such as cholesterols and triacylglycerols through the blood–cell membrane barrier of various cells.[15] The function of these LDL particles is twofold; the first is to solubilize hydrophobic lipids, and the second is to transport lipids to specific cells and tissues throughout the body. As such, they promise to be very good candidates for the targeted drug delivery of traditional drugs to various tissues. In particular, various tumor cells aggressively overexpress the LDL receptors that recognize lipoproteins such as apolipoprotein E (apoE) and apolipoprotein B-100. Since they are of natural origin, these molecules are biodegradable and biocompatible, nonimmunogenic, and freed from being recycled by the reticuloendothelial system (RES).[45]

Targeted drug delivery systems. In order to improve selectivity and reduce toxicity, a number of drugs have been linked or encapsulated into different types of macromolecules.[4] Based on the current understanding of different receptor sites, targeting drug delivery systems can be designed to carry drugs specifically to the site of interest. In general, targeting drug delivery systems have been created to carry low-molecular-weight drugs to a specific site of action in the body.[46] The net effect of conjugating drugs to macromolecules allows for modification of the toxicity of the drug, the rate at which the drug is excreted from the body, and immunogenicity. Since macromolecular drug carriers predominantly enter cells via endocytosis, they can be designed to deliver drugs to specific areas where activity is required.[47]

The use of polymers in targeted drug delivery has led to the development of a wide variety of systems. Depending on the application, polymeric systems can be either natural or synthetic. Typical examples of natural polymers include polysaccharides (dextran, chitosan, and agarose) and proteins (albumin and collagen).[48] Synthetic polymers can be divided into two classes biodegradable and nonbiodegradable. The biodegradable polymers consist of polyamides, polyesters, polyanhydrides, and phosphorous-based polymers, whereas the nonbiodegradable polymers include acrylic polymers and silicones.[48]

Targeted drug delivery with polymeric systems generally can be divided into three main categories: (1) The targeting moiety and drug are chemically conjugated to the polymer carrier, (2) the drug is chemically conjugated to a polymer that functions both as the drug carrier and the targeting moiety, and (3) the drug is physically entrapped within a carrier that has been modified with a targeting moiety. Several examples of these targeting drug delivery systems are described in the following subsections.

Immunoglobulin-directed targeting. The application of antibodies and antibody fragments (Fig. 12.4) as targeting moieties has increased significantly over the past couple of decades, especially for the treatment of cancer.[49] The approach has many advantages over other types of targeting system, such as higher affinity and specificity while maintaining native structure, common coupling and interchangeability of other antibodies and antibody fragments, biocompatibility, and well-established chemistry.[24] In general, there are four different approaches to targeted drug delivery with antibodies: (1) The antibody acts as an inhibitor that targets cells that are overexpressing a particular antigen, (2) the drug could be conjugated to the antibody in such a way that the drug will be released by nonspecific processes such as hydrolysis or dissociation,

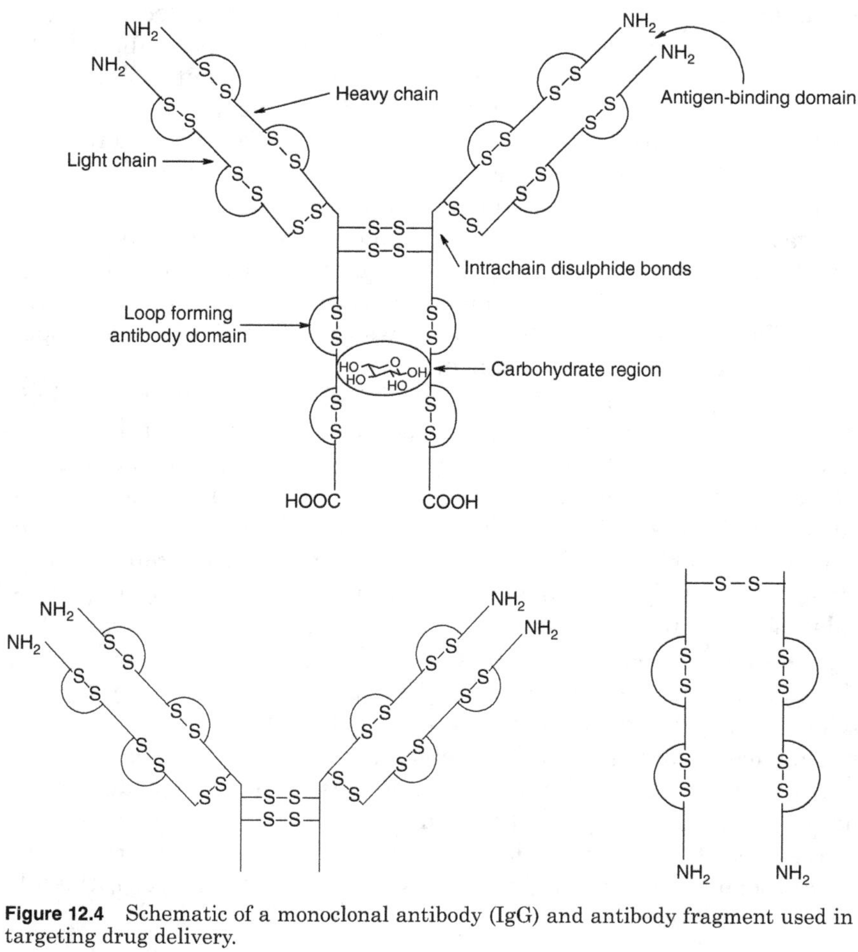

Figure 12.4 Schematic of a monoclonal antibody (IgG) and antibody fragment used in targeting drug delivery.

(3) the drug could be targeted to a specific site and released by mechanisms in the cellular environment, and (4) the antibody-macromolecule complex is internalized into the cells, and the drug is released by lysosomal degradation.[1]

A substantial amount of research has been carried out with antibodies to target drugs to a specific site of action. Initially, antibodies were used to specifically block overexpressed antigens on the surfaces of cells. Currently, such drugs as Herceptin, which targets and blocks the overexpression of HER2 antigens on metastatic breast cancer cells, are clinically approved for use in humans. Next, drugs were chemically conjugated to antibodies, which allowed the drug to be internalized directly into cells that expressed a particular antigen.[1] However, this approach initially met with only limited success. By conjugating a hydrophobic drug to the antibody, the solubility was greatly altered, thereby limiting the number of drug molecules that could be conjugated to the antibody. Since insufficient amounts of drug were conjugated to the antibody, potency sometimes was not sufficient to affect cancer cells in vivo. In addition, the chemical conjugation of drugs to antibodies could result in changes in the activity of the antibody. However, drugs such as Mylotarg are being approved for clinical use for a new class of anticancer therapy called *antibody-targeted chemotherapy*. Mylotarg is an antibody that recognizes the CD33 molecule on the surfaces of acute myeloid leukemia (AML) cells that is conjugated to a toxic chemotherapeutic agent called *calicheamicin*. In addition to Mylotarg, Bexxar and Zevalin are drugs that consist of an antibody conjugated to a radiopharmaceutical, also known as *radioimmunotherapeutics*, for the treatment of B-cell lymphomas. These drugs and others to follow will offer new hope in the fight against cancer.

Antibody-targeted therapy could be further improved by conjugation of the antibodies to water-soluble polymers, thereby maintaining solubility of the attached hydrophobic drugs while delivering the appropriate dosage required for activity. One such polymer is *N*-(2-hydroxypropyl) methacrylamide (HPMA). HPMA (Table 12.1) is a highly water-soluble polymer, and its conjugate with doxorubicin already has been shown to accumulate in tumors.[50] It has oligopeptide side chains that can be activated as a *p*-nitrophenyl ester derivative and coupled with drugs and/or targeting moieties. Drugs and targeting moieties that contain an amino group can be conjugated easily via the activated ester group. HPMA polymers have been shown to be valuable for targeted drug delivery with antibodies because they are biocompatible, retain solubility when conjugated with antibody, and still can undergo receptor-mediated endocytosis.[3] For example, targeting of human and mouse T-lymphocytes has been achieved by monoclonal antibody–HPMA–doxorubicin conjugates directed at the T-cell surface

TABLE 12.1 Typical Examples of Targeting Drug Delivery Carriers

Targeting drug delivery carriers	Targeting moiety	Reference
HPMA copolymer	Antibody	51
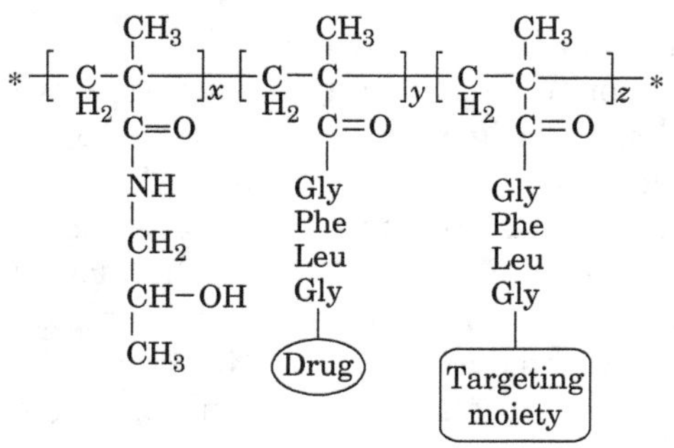	Carbohydrates Galactose Mannose Hyaluronic acid	52, 53
Colloidal systems Liposome Micelle	Antibody fragment	54, 55
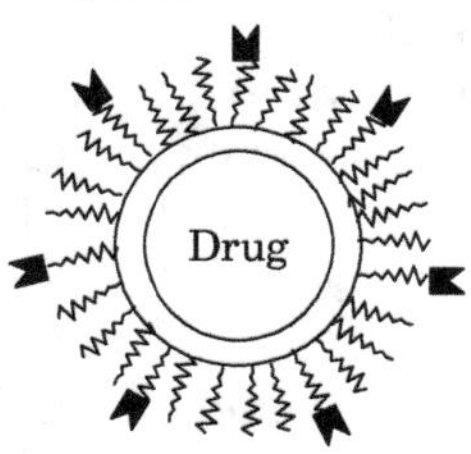	Folic acid	56
	Transferrin	41, 42
	Carbohydrates	57
	Peptides RGD	58, 59
Polysaccharide		60
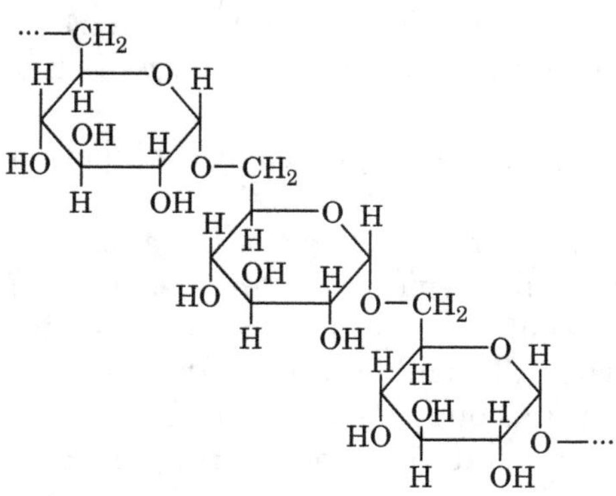	Polymer backbone Antibodies	15
Poly(amino acids) Poly-L-glutamic acid Poly-L-lysine	Antibodies	61
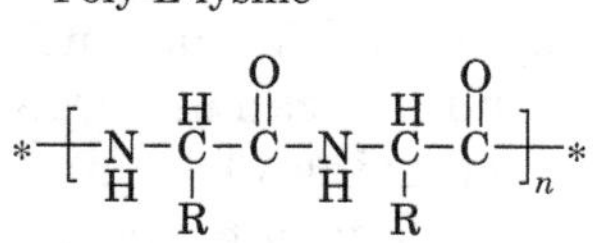	Carbohydrates Galactose Mannose Fucose Xylosyl	62, 63
	Folic acid	36
	Transferrin	13

antigens.[64] Other polymers that have been modified with immunoglobulins are poly(amino acids). For example, the in vitro and in vivo activities of targeted poly(L-glutamic acid)–drug conjugates have been evaluated extensively.[65] The results showed that increased activity and specificity were observed with these immunoconjugates.

Another approach that has employed antibodies extensively as the targeting moiety is delivery with colloidal systems, in particular, liposomes. As described in Chap. 11, liposomes are microparticulate systems constructed with one or more phospholipid bilayers that have been specifically designed to carry therapeutic agents by entrapment within its phospholipid bilayers, the bilayer interface, or the entrapped aqueous volume.[66] A diverse array of lipophilic drugs can be entrapped within liposomes, with the release profile of each drug found to depend on the stability and/or half life of the liposome-drug conjugate.[67] Stability of liposomes could be increased with attachment of hydrophilic long-chain polymers such as polyethylene glycol (PEG).[54] In addition, the incorporation of PEG on the liposome surface prevents absorption and elimination from the circulation by the reticuloendothelial system (RES).[55] To achieve targeting, antibodies or antibody fragment could be attached either directly on the liposome or at the terminal end of the PEG chain.[54] Conjugation with the antibody Fab fragments appeared to have increased circulation and accumulation at tumor tissues than the respective whole antibody.[68] The reason for this dramatic difference was attributed to the enhanced clearance rate of whole antibody.

Selectin-directed targeting. Natural polysaccharides have been used extensively in the formulation of solid dosage forms of drugs. Recently, these polysaccharides have been garnering great interest for use as targeted delivery systems of drugs to the colon because of the presence of large amounts of polysaccharidases in the human colon.[60] The rationale is that the polysaccharide-drug conjugates will release the drugs only after degradation in the colon, where a large number and variety of bacterial colonies that secrete enzymes such as β-D-glucosidase, β-D-galactosidase, amylase, pectinase, and dextranase are located. These conjugates exemplify the use of a polymer backbone to function as both the carrier and the targeting moiety.

Polysaccharide conjugates have not been applied extensively to the targeting of other tissues owing to the limitation of specific enzymes necessary for degradation of the polysaccharide conjugates and release of the drugs. Nonetheless, carbohydrates as targeting moieties (Fig. 12.5) offer a versatile approach for the targeting of selectins. Specifically, D-galactose and mannose have been used in monosaccharide–poly(amino acid) conjugates as targeted delivery systems to specific tissues such as the liver.[62–63,69] For instance, poly-L-glutamic acid and poly-L-lysine have

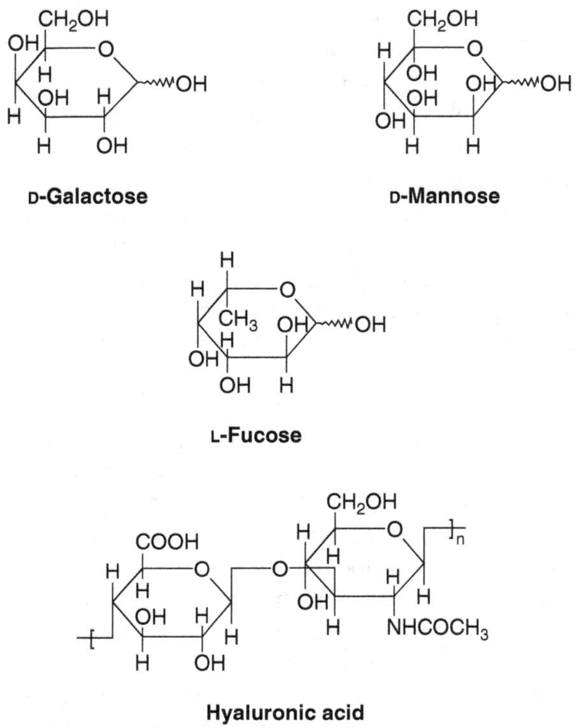

Figure 12.5 Structure of common sugars used in targeting.

been used widely as carrier systems because they are biocompatible and biodegradable, and the degradation products have low or no toxicity.[48] Some interesting examples of glycosylated poly(amino acids) involve the modification of poly-L-lysine by δ-gluconolactone and D-mannose[51] and *N*-acetylmuramyldipeptide[69] for the targeting of macrophages.

Another popular polymer that has been modified by glycosylation is HPMA. For example, an HPMA–*N*-acetyl galactosamine (GalN)–adriamycin conjugate was used to target adriamycin to ovarian carcinoma cell lines, which substantially increased the intracellular concentration of adriamycin when compared with incubation with free adriamycin.[51] When doxorubicin was attached to HPMA conjugations of hyaluronic acid, which is a naturally occurring linear polysaccharide of alternating D-glucuronic acid and *N*-acetyl-D-glucosamine units, enhanced uptake of doxorubicin into cancer cells was observed.[52] In addition, intracellular uptake of the targeted conjugates appeared to be receptor mediated. This is expected because the levels of hyaluronic acid are elevated in the surrounding environment of tumor cells, and most metastatic cells overexpress the hyaluronic acid receptors.[70,71]

Since the surfaces of cells usually contains a high density of polysaccharides, it also was reasonable to coat the surfaces of liposomes with specific carbohydrate moieties for targeted drug delivery. There are a number of advantages to the modification of liposomes with carbohydrate moieties: (1) The carbohydrates reduce the permeability of water-soluble material from blood plasma or serum, (2) degradation by enzymes such as pullulanase is restricted to the exposed parts of the carbohydrate coat, and (3) the liposomal phospholipids are also protected against the action of lipases.[72] A number of carbohydrate moieties have been used to coat liposomes, but the effectiveness of these targeted drug delivery systems ultimately depends on the specificity of the carbohydrate moiety and its interaction with the target selectin.[57]

Integrin-directed targeting. Based on the interactions between integrins and peptide sequences within the extracellular matrix, monomers, oligomers, and polymers containing these minimal core peptide sequences, RGD and YIGSR, have been developed for use as drugs that act as target-specific inhibitors of cancer metastasis. Studies have shown that oligomers and polymers that contain repeating sequences of RGD, RGD analogues, and YIGSR inhibit metastasis more effectively than the monomer peptide sequence.[33,73] In addition, RGD has been used extensively to increase target specificity by attaching or incorporating it into drug delivery carriers such as liposomes and polyethylene glycol.[74–76] LDV terpolymers that contain randomly distributed leucine–aspartate–valine sequences within the polymer backbone also have been shown to be capable of targeted delivery of doxorubicin to human melanoma cells that overexpress the $\alpha_4\beta_1$.[77,78]

Folic acid receptor–directed targeting. It has been shown that simple conjugation of folic acid to macromolecules can enhance internalization into cells.[39] The first clinically relevant folic acid conjugate was the application of folate-deferoxamine for the targeting of tumors with gallium 67 (^{67}Ga), a gamma-emitting radionuclide that is fatal to cells and has a half life of 78 hours.[79] The folate-deferoxamine complex chelates and delivers ^{67}Ga to tumor sites with high tumor–nontarget-tissue ratios because of the increased uptake of folic acid and folic acid conjugates by tumor cells. Improvement of this approach was made with other folate conjugates of radionuclide.[36]

Folate conjugation with polymers such as polylysine and polyethylenimine have allowed the delivery of transgenes and oligonucleotides into cancerous cells, which have been shown to significantly enhance expression of the respective transgenes and antisene effects.[80] More recently, this enhancement could be further increased if a polyethyleneglycol spacer was used along with the polycationic polymers.[81] In addition, folate-liposome conjugates made possible the targeted delivery of doxorubicin to epithelial cancer cells, which overexpresses receptors for folic acid.[19]

Transferrin-directed targeting. The use of transferrin to target transferrin receptors in cancer therapy, as targeted drug delivery to tumor cells, also has garnered considerable interest over the past few years.[42] This application of transferrin conjugation to specifically deliver drugs, toxins, and proteins to cells has been well investigated.[13] Studies have shown that when small drugs, such as doxorubicin or methotrexate, are chemically conjugated to transferrin, transport can be facilitated without disrupting the mechanism of transferrin receptor–mediated endocytosis.[82,83] In addition, compared with free doxorubicin, a decrease in the toxicity and development of drug resistance was observed.[84] For example, when a diphtheria toxin was conjugated to transferrin, it was found that the transferrin-toxin system could selectively affect cells that overexpressed the transferrin receptor.[85]

The versatility of transferrin conjugation also has led to the development of transferrin-polymer conjugates for the delivery of DNA.[13] DNA was condensed with poly-L-lysine–transferrin complex to form a transferrin-polylysine-DNA structure that exposed the transferrin on the surface,[86] which could interact with the transferrin receptor to internalize the transferrin-polylysine-DNA complex.[87] This system was further improved when DNA-polylysine complexes were coated with HPMA and conjugated with transferrin.[88] An alternative to this conjugation with polylysine and HPMA was to encapsulate a drug or DNA into a PEGylated (polyethylene glycol) cationic liposome that had transferrin conjugated to the tail of the polyethylene glycol.[41,42] The overall effect, once transferrin had been attached to these systems, was that both cellular uptake and transfection of DNA were enhanced. Clearly, these polymer-transferrin conjugates offer significant progress in the development of targeted gene delivery.

LDL-directed targeting. LDL has been used as a targeting drug delivery system primarily in photodynamic therapy.[89] Photodynamic therapy uses photosensitizers that once selectively taken up by disease tissues are activated with visible light at a certain wavelength to generate toxic compounds and results in elimination of the targeted tissues. This mode of delivery could selectively eliminate diseased tissue without damaging the surrounding healthy tissue if the photosensitizers could be selectively delivered to the target tissue. Studies have demonstrated that when photosensitizers were incorporated with LDL and administered to tumor-bearing mice, there was an increase in therapeutic efficacy when compared with free photosensitizers. In cells that are rapidly proliferating, as in the case of tumor cells, LDL receptors are upregulated because of the need for more cholesterol.[27] It is this need that drives the increased uptake of LDL and which provides the basis for targeting with LDL.

12.3.2 Targeted enzyme prodrug therapy

In order to achieve successful delivery of therapeutic agents using macromolecule-drug conjugates, as discussed earlier, there are several barriers that must be surmounted. An alternative approach to macromolecular active targeting is to use a prodrug strategy. The use of prodrugs, which are covered separately in this book, traditionally has not been considered to be a targeting drug delivery system. However, in disease states such as cancer, elevated levels of specific enzymes can be associated with the cancer cells relative to normal cells. The higher levels of enzymes associated with the tumor may lead to selective activation of the prodrug, through metabolism, at the desired site of action. In the case of prolidase, which has high levels of expression on breast cancer cells, drugs such as melphalan that are conjugated to a C-terminal proline group via an imido bond are selectively cleaved in the presence of this enzyme.[90,91] This is followed by a nonselective mechanism of drug uptake by the tumor. Ultimately, the effectiveness of this type of system would be based on the identification of unique exploitable enzymes associated with a disease state. To further improve on this drug delivery system, monoclonal antibodies are used as a ligand to actively target an enzyme to the cancer cell surface, which, in turn, activates the prodrug at the tumor site.[15]

Antibody-directed enzyme prodrug therapy (ADEPT) systems use a two-step process for treating the disease state. In the first step, enzymes are conjugated to an antibody and delivered to specific target cells that express the antibody's complementary antigen. Second, the prodrug is administered after the antibody-enzyme conjugate has sufficiently localized at the site of action, and the prodrug is activated once it is metabolized by the enzyme at the site of action. However, it is necessary that the antibody-enzyme conjugates remain bound to the cell surface rather than being internalized for antibody targeted prodrug therapy to be effective. This characteristic allows for adequate prodrug interaction with the enzyme for delivery of the parent drug to treat the disease.

Several ADEPT systems that are beyond the scope of this chapter have been studied extensively and are the subject of many reviews.[15,92–93] One of the most widely cited ADEPT systems employs a glucouronidase (GUS)–monoclonal antibody conjugate that activates a doxorubicin-glucuronide prodrug. Studies have shown that there is a substantial therapeutic benefit, a higher level of drug in tumor tissue when compared with nontumor tissue, using the ADEPT system compared with systemic administration of doxorubicin alone,[92] thus showing the potential therapeutic effectiveness of this type of treatment.

However, ADEPT systems have some potentially negative drawbacks. First, if any circulating unbound antibody-enzyme conjugate remains, it potentially could react systemically with the prodrug and produce a toxic response in an unwanted location. Therefore, it is necessary to optimize tumor uptake of the antibody-enzyme conjugate while decreasing the amount remaining in circulation to ensure maximum effectiveness of this therapy. In addition, the use of monoclonal antibodies for ADEPT systems is susceptible to the same limitations as described with macromolecular-drug carriers, such as immunogenecity and poor tumor permeability.

12.4 Factors Affecting Design of Ligand-Based Targeting Drug Delivery Systems

The effectiveness of ligand-based targeting drug delivery systems to deliver drugs at the site of action does not rely exclusively on the preferential binding of the carrier to the target. It is also determined by the distribution, internalization, and release of an active form of the drug from the drug delivery system for treatment of the disease state. In addition, whenever a foreign substance, such as a targeting drug delivery carrier, is injected into a biological system, an immune response may be triggered by the biological system, which may decrease the effectiveness of the carrier. Therefore, in designing targeting drug delivery systems, these factors should be considered because they may affect the overall performance of the drug delivery system.

12.4.1 Kinetics of active targeting systems

The in vivo distribution of drug targeting macromolecule systems is distinctly different from that observed with traditional small molecules. Low-molecular-weight chemotherapeutic drugs used for the treatment of cancer usually have little selectivity in their pharmacological action. These highly toxic agents diffuse rapidly throughout the body, which makes it difficult to control the in vivo distribution of such drugs. Thus they exhibit severe toxicity toward both normal and diseased cells. Passive targeting by macromolecular carriers allows a degree of selective distribution from capillaries to tissues, which depends largely on the anatomical and physiological characteristics of the body.[94] Consequently, uptake and clearance of drugs loaded on or into macromolecular carriers are restricted to tissues having leaky capillaries, such as the liver, spleen, bone marrow, and kidney and most solid tumors. Active targeting gives an additional level of targeting, which directs drugs specifically to a desired site of action. Once the drug is released, uptake is localized to the target site, thereby reducing toxicity to nontarget tissue. For ligand-based targeting drug carriers, the loaded macromolecules could be internalized into target cells via receptor-mediated endocytosis to deliver their content.[95] Such targeted drug delivery carriers are removed from the circulation only when they bind to specific receptors, which essentially could eliminate toxicity owing to drug interaction with normal cells. In addition, ligand-based targeted drug delivery carriers should have long circulation times and be maintained at high enough levels to provide a continuous source of uptake. To maximize the area under the curve (high amounts of pharmacologically active agents that reach the desired site), a sufficiently high number of receptors should be expressed and accessible on the surfaces of target cells to overcome the relatively slow rate of uptake via receptor-mediated endocytosis (RME) compared with diffusion of low-molecular-weight chemotherapeutic drugs across cell membrane. Ideally, the receptor should be overexpressed on target cells to further differentiate diseased cells from normal healthy cells.

12.4.2 Internalization of ligand-based targeting drug delivery systems

Given that the drug alone cannot specifically accumulate at the desired site and simply diffuse through the cellular membrane, it can be chemically conjugated to a targeted polymeric drug carrier to deliver it to specific target cells. As discussed previously, there are three levels of targeting. The first level can be seen with most polymer-drug conjugates, which are able to passively accumulate in target tissues such as a tumor. It was shown that tumor tissues can uptake macromolecules in the range of 15 to 70 kDa.[96] This affinity for macromolecules is known as the *enhanced permeation and retention (EPR) effect*. This effect is due to enhanced angiogenesis or vascularization of the tissue associated with the cancerous cells, which results in increased vascular permeability. Concurrently, the functioning of the lymphatic system also decreases, which leads to limited clearance of macromolecules by venules of the tumorous tissue.[97]

Once a drug is conjugated to a macromolecule, it cannot enter the cell by simple diffusion but rather requires endocytosis, which could be enhanced in cancerous cells by overexpression of receptors. Therefore, to achieve cellular targeting, macromolecules require access to the interior of a cell through endocytosis, a process in which a portion of the plasma membrane invaginates to create a new intracellular vesicle (Fig. 12.6). Internalization via endocytosis can occur in one of two ways:

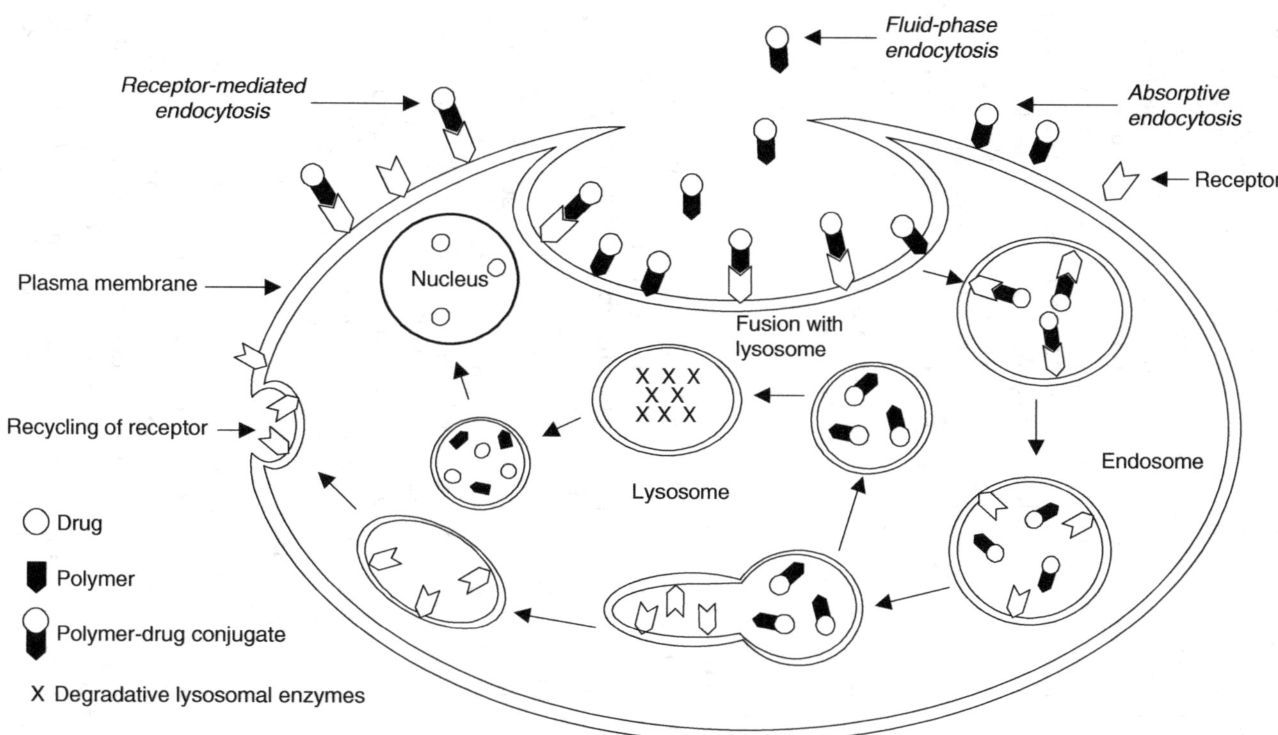

Figure 12.6 Schematic of endocytosis.

by phagocytosis or by pinocytosis. The process of phagocytosis entails the engulfing of particulate material (10 to 20 μm), and it is a specialized process usually associated with such cells as macrophages.[98] This mode of entry generally is more advantageous for colloidal targeting drug delivery systems. On the other hand, pinocytosis is an event that occurs in all mammalian cells that involves the continual uptake of extracellular fluid and any material that happens to be in it or on the cell surface.[99] There are three types of pinocytosis, which include absorptive, fluid-phase, and receptor-mediated endocytosis. Absorptive endocytosis involves the absorption of a polymer-drug conjugate onto the cell surface and internalization into the cell. In fluid-phase endocytosis, macromolecules enter the cell as liquid beads without physically interacting with specific receptors. The most specific form is receptor-mediated endocytosis. Receptor-mediated endocytosis is activated when a macromolecule conjugated to a targeting device specifically binds to a cell surface receptor. The advantage of receptor-mediated endocytosis is that it potentially can deliver compounds specifically to the site of action where drug activity is required.[100] After a polymer-drug conjugate binds to a receptor with clathrin-coated pits, the receptor invaginates to form an intracellular vesicle or endosome that contains the conjugate and transports it into the cellular environment. Once the polymer-drug conjugates dissociate from the receptors and are released into the cytosol or lysosomal compartments, the receptors are recycled to the cell surface for additional binding and internalization. The polymer-drug conjugates then undergo degradation to release drug inside the cell.

To prepare a targeting drug delivery system that can exploit the specific ligand-receptor interaction and consequent receptor-mediated endocytosis, a distinctive process intrinsic to the disease state must be identified, from which key receptors, antigens, and/or binding domains associated with the disease can be used as homing devices. Receptors such as the cell adhesion molecules include immunoglobulins, cadherins, selectins, and integrins.[22,101] An example of a targeting polymeric drug delivery carrier that uses cell adhesion molecules to target integrin receptors is the novel LDV terpolymer that contains the targeting peptide sequence leucine–aspartate–valine distributed within the polymer backbone, which was shown to specifically interact with human melanoma cells that overexpress the $\alpha_4\beta_1$ integrin.[77,78]

Similar incorporation of ligand-based targeting motifs is applicable to colloidal systems. In the final analysis, the effectiveness of ligand-based targeting drug delivery systems depends on the density of the target receptor on the cell surface, the rate in which the systems are internalized into the cell, and recycling and reexpression of the receptor on the cell surface after internalization.

12.4.3 Drug release from delivery systems

Generally, there are three different methods to incorporate drug with a carrier. These include (1) directly conjugating the drug to the carrier, (2) chemical conjugation through a spacer, and (3) encapsulation. Each method affects the rate at which the drug is released from the delivery carrier. The bond between the carrier and the drug or encapsulation of the drug by carrier must be stable in the bloodstream during transport. For chemical conjugation, the drug must have appropriate functional groups that allow for binding with carrier or spacer.

In the case of chemical conjugation, drugs could be linked directly or through a spacer group to the polymer backbone.[102] Traditionally, direct conjugation has offered a more facile procedure, but one limitation is slow release of the drug because the bond between the carrier and the drug could be sterically hindered toward cleavage.[46] Over the past decade, systems that make use of a spacer group have been shown to internalize more efficiently via receptor-mediated endocytosis.[103] In addition, the site and rate at which the drug is released can be controlled by using different spacers. A number of spacers have been used to conjugate drugs to macromolecules to release at specific sites by passive hydrolytic cleavage, pH, or specific enzymes.[46]

Another determining factor of the release profile of a drug is the type of chemical conjugation used whether the linkage is an ester, amide, urethane, or carbonate bond. The ease of hydrolysis is ester > carbonate > urethane > amide, with the ester linkage being the easiest and the amide the hardest to cleave. For example, peptidyl spacers have been employed to conjugate drugs to a polymer.[103] These spacers are predominantly cleaved intracellularly in the lysosomal compartment of the cell. Other linkers, such as the pH-sensitive *N-cis*-aconityl and *N*-maleyl, once internalized and localized within the endosome and lysosomal compartments, will induce hydrolytic cleavage and release the drug because of the relatively acidic pH conditions of these environments. An example of this pH-controlled release is illustrated by the poly(D-lysine) delivery of daunomycin to the lysosomal compartment with the help of *cis*-aconityl spacer.[104] When a disulfide spacer was used, as reported for the delivery of methotrexate with poly(D-lysine),[105] reductive cleavage and release of the drug occurred in the cytosol compartments, whereas the acidic pH conditions of the endosomes or lysosomal proteases had no effect. If delivery at the lysosomes is desired, there are more than 40 enzymes that can break down all major classes of biologically relevant materials such as fats, proteins, and carbohydrates.[106] Even though targeting the lysosomal compartment of cells with spacers can add another element to specific targeting, bear in mind that lysosomes are present in all cells, both normal and tumor. Ultimately, specific delivery to the desired tissue or site must be achieved by the primary ligand-based targeting at the cell surface, with control of intracellular release as secondary targeting.

12.4.4 Immunogenicity

Over the past few decades, therapeutic application of antibodies has been limited by induced immune responses of the host to administered antibodies, which were predominantly comprised of foreign proteins. It is generally known that conjugation of proteins to soluble polymers can reduce the proteins' immunogenicity. Until recently, conjugations with PEG were investigated to reduce immunogenicity.[49] However, the need to do this has disappeared with the advent of technological advances in the production of humanized and human antibodies.[107] Another potential solution to the elicitation of immune responses has been to express fragments of antibodies such as Fab' and scFv.[49]

Considerable research has been devoted to assessment of the immunogenicity of HPMA copolymers.[108–110] By themselves, HPMA copolymers do not elicit any immune response. When HPMA copolymers were modified with oligopeptides, only weak immunogenic responses were observed, and their intensity depended on the structure of the oligopeptide sequence and the host.[111]

In the development of any drug delivery system, especially ones that are protein based, difficulty with the elicitation of an immune response from the host must be considered. As such, the application of poly(amino acids) as drug carriers requires a complete assessment of the potential antigenicity of the system. This also entails the incorporation of possible strategies to resolve anticipated difficulties associated with immunogenicity into the design of the targeted drug delivery system.

The ability of synthetic poly(amino acids) to induce an immune response has been studied extensively, specifically, to determine which chemical structures and what composition render them antigenic and how the antigenicity could be reduced by chemical modification.[112] A review of the current literature indicates that the antigenicity of a poly(amino acid) depends on the polymer's size, composition, shape, and accessibility of particular amino acids to the biosynthesis site where the antibody is generated.[112] In addition, the stereochemistry of the amino acids determines whether a poly(amino acid) is capable of inducing an immune response. Maurer[113] found that terpolymers of three L-amino acids are antigenic, wherears terpolymers that contain one or more D-amino acids are not. In general, polymers constructed from three or more amino acids are more likely to be good immunogens than homopolymers.[114] Terpolymers and copolymers are variable in their ability to generate an immune response.[114] In brief, each homopolymer, copolymer, or terpolymer must be tested individually in a specific species to determine its ability to elicit an immune response, and the conclusion thus drawn regarding its immunogenicity does not necessarily apply to other species.

12.5 Current Status and Future of Actively Targeted Drug Delivery Systems

Despite decades of experimentation, ligand-based targeting for the treatment of disease has yet to lead to widespread commercial therapy. Over the past few years, a select number of antibodies and antibody-drug conjugates have been clinically approved to target the overexpression of antigens on cells for the treatment of specific cancers, thus indicating that the promise of actively targeted drug delivery is real and that its potential is immeasurable. Numerous new targeting motifs have been identified, which when coupled with a carrier loaded with a generic drug, whether it be a small molecule, peptide/protein, or gene, will allow for a diverse array of ligand-based targeted drug delivery systems to be constructed and tailored for specific disease. These systems have the potential to revolutionize drug development, with the focus shifting from creating the drugs to identifying targets unique to specific diseases. As such, the search is not for a "magic bullet" but rather for a "smart bomb" loaded with a conventional explosive that is guided by a homing device.

References

1. Garnett, M. Targeted drug conjugates: Principles and progress. *Adv. Drug Del. Rev.* **53**:171–216, 2001.
2. Yokoyama, M. and Okano, T. Targetable drug carriers: Present status and a future perspective. *Adv. Drug Del. Rev* **21**:77–80, 1996.
3. Rihova, B. Targeting of drugs to cell surface receptors. *Crit. Rev. Biotechnol.* **17**:149–169, 1997.
4. Torchilin, V. P. Drug targeting. *Eur. J.* of *Pharm. Sci. 11*:S81–S91, 2000.
5. Kwon, G. S., and Okano, T. Polymeric micelles as new drug carriers. *Adv. Drug Del. Rev.* **21**:107–116, 1996.
6. Seymour, L. W., Ulbrich, K., Steyger, P. S., et al. Tumor tropism and anticancer efficacy of polymer-based doxorubicin prodrugs in the treatment of subcutaneous murine B16F10 melanoma. *Br. J. Cancer* **70**:636–641, 1994.
7. Gabizon, A. A. Selective tumor localization and improved therapeutic index of anthracyclines encapsulated in long-circulating liposomes. *Cancer Res.* **52**:891–896, 1992.
8. Couvreur, P., and Puisieux, F. Nano-, microparticles for the delivery of polypeptides and proteins. *Adv. Drug Del. Syst.* **10**:141–162, 1993.
9. Takakura, Y., Mahato, R. I., Nishikawa, M., and Hashida, M. Control of pharmacokinetic profiles of drug-macromolecule conjugates. *Adv. Drug Del. Rev.* **19**:377–399, 1996.
10. Widder, K. J., Senyei, A. E., and Ranney, D. F. Magnetically responsive microspheres and other carriers for the biophysical targeting of antitumor agents. *Adv. Pharmacol. Chemother.* **16**:213–271, 1997.
11. Maruyama, K., Ishida, O., Takizawa, T., and Moribe, K. Possibility of active targeting to tumor tissues with liposomes. *Adv. Drug Del. Rev.* **40**:89–102, 1999.

12. Davis, S. S. Biomedical applications of nanotechnology: Implications for drug targeting and gene therapy. *Trends in Biology* **15**:217–224, 1997.
13. Wagner, E., Curiel, D., and Cotten, M. Delivery of drugs, proteins and genes into cells using transferrin as a ligand for receptor-mediated endocytosis. *Adv. Drug Del. Rev.* **14**:113–135, 1994.
14. Hashida, M., and Takakura, Y. Pharmacokinetics in design of polymeric drug delivery systems. *J. Contr. Rel.* **31**:163–171, 1994.
15. Dubowchik, G. M., and Walker, M. A. Receptor-mediated and enzyme-dependent targeting of cytotoxic anticancer drugs. *Pharmacol. Ther.* **83**:67-123, 1999.
16. Johnson, J. P. Cell adhesion molecules in neoplastic disease. *Int. J. Clin. Lab. Res.* **22**:69–72, 1992.
17. Kiessling, L. L., and Gordon, E. J. Transforming the cell surface through proteolysis. *Chem. Biol.* **5**:R49–62, 1998.
18. Pimm, M. V., Durrant, L. G., and Baldwin, R. W. Influence of circulating antigen on the biodistribution and tumor localization of radiolabelled monoclonal antibody in a human tumor: Nude mouse xenograft model. *Eur. J. Cancer. Clin. Oncol.* **25**:1325–1332, 1989.
19. Hernando, J. J., von Kleist, S., and Grunert, F. A repertoire of monoclonal antibodies reveals extensive epitope heterogeneity in CEA purified from neoplasms originating from different organs. *Int. J. Cancer* **56**:655–661, 1994.
20. Kadler, K. Extracellular matrix: I. Fibril-forming collagens. *Protein Profile* **1**:519–638, 1994.
21. Shimoyama, Y., and Hirohashi, S. Cadherin intercellular adhesion molecule in hepatocellular carcinoma: Loss of E-cadherin expression in an undifferentiated carcinoma. *Cancer Lett.* **57**:131–135, 1991.
22. Price, J. T., Bonovich, M. T., and Kohn, E. C. The biochemistry of cancer dissemination. *Crit. Rev. Biochem. and Mol. Biol.* **32**:175–253, 1997.
23. Shimoyama, Y., Nagafuchi, A., Fujita, S., et al. Cadherin dysfunction in a human cancer cell line: Possible involvement of loss of alpha-catenin expression in reduced cell-cell adhesiveness. *Cancer Res.* **52**:5770–5774, 1992.
24. Forssen, E., and Willis, M. Ligand-targeted liposomes. *Adv. Drug Del. Rev.* **29**:249–271, 1998.
25. Palomino, E. "Carbohydrate handles" as natural resources in drug delivery. *Adv. Drug Del. Rev.* **13**:311–323, 1994.
26. Berg, E. L., Robinson, M. K., Mansson, O., et al. A carbohydrate domain common to both sialyl Le(a) and sialyl Le(X) is recognized by the endothelial cell leukocyte adhesion molecule ELAM-1. *J. Biol. Chem.* **266**:14869–14872, 1991.
27. Foxall, C., Watson, S. R., Dowbenko, D., et al. The three members of the selectin receptor family recognize a common carbohydrate epitope, the sialyl Lewis(x) oligosaccharide. *J. Cell. Biol.* **117**:895–902, 1992.
28. Lefer, A. M., Weyrich, A. S., and Buerke, M. Role of selectins, a new family of adhesion molecules, in ischaemia-reperfusion injury. *Cardiovasc. Res.* **28**:289–294, 1994.
29. Humphries, M. J. Integrin cell adhesion receptors and the concept of agonism. *Trends Pharmacol. Sci.* **21**:29–32, 2000.
30. Spring, F. A., Parsons, S. F., Ortlepp, S., et al. Intercellular adhesion molecule-4 binds alpha 4 beta 1 and alpha V-family integrins through novel integrin-binding mechanisms. *Blood* **98**:458–466, 2001.
31. Pierschbacher, M. D., and Ruoslahti, E. Cell attachment activity of fibronectin can be duplicated by small synthetic fragments of the molecule. *Nature* **309**:30–33, 1984.
32. Komoriya, A., Green, L. J., Mervic, M., et al. The minimal essential sequence for a major cell type-specific adhesion site (CS1) within the alternatively spliced type III connecting segment domain of fibronectin is leucine-aspartic acid-valine. *J. Biol. Chem.* **266**:15075–15079, 1991.
33. Saiki, I., Murata, J., Iida, J., et al. Antimetastatic effects of synthetic polypeptides containing repeated structures of the cell adhesive Arg-Gly-Asp (RGD) and Tyr-Ile-Gly-Ser-Arg (YIGSR) sequences. *Br. J. Cancer* **60**:722–728, 1989.
34. Kerr, J. S., Slee, A. M., and Mousa, S. A. Small molecule alpha(v) integrin antagonists: Novel anticancer agents. *Exp. Opin. Invest. Drugs* **9**:1271–1279, 2000.

35. Holladay, S. R., Yang, Zhen-fan, Kennedy, M. D., et al. Riboflavin-mediated delivery of a macromolecule into cultured human cells. *Biochim. Biophys. Acta* **1426**:195–204, 1999.
36. Leamon, C. P., and Low, P. S. Folate-mediated targeting: From diagnostics to drug and gene delivery. *Drug Disc. Today* **6**:44–51, 2001.
37. Sudimack, J., and Lee, R. J. Targeted drug delivery via the folate receptor. *Adv. Drug Del. Rev.* **41**:147–162, 2000.
38. Lee, R. J., and Low, P. S. Folate-mediated tumor cell targeting of liposome-entrapped doxorubicin in vitro. *Biochim. Biophys. Acta* **1233**:134–44, 1995.
39. Leamon, C. P., and Low, P. S. Delivery of macromolecules into living cells: A method that exploits folate receptor endocytosis. *Proc. Natl. Acad. Sci. USA* **88**:5572–5576, 1991.
40. Ciechanover, A., Schwartz, A. L., Dautry-Varsat, A., and Lodish, H. F. Kinetics of internalization and recycling of transferrin and the transferrin receptor in a human hepatoma cell line. Effect of lysosomotropic agents. *J. Biol. Chem.* **258**:9681–969, 1983.
41. Xu, L., Pirollo, K. F., and Chang, E. H. Tumor-targeted p53-gene therapy enhances the efficacy of conventional chemo/radiotherapy. *J. Contr. Rel.* **74**:115–128, 2001.
42. Li, H., Sun, H., and Qian, Z. M. The role of the transferrin-transferrin-receptor system in drug delivery and targeting. *Trends Pharmacol. Sci.* **23**:206–209, 2002.
43. Nagy, A., Schally, A. V., Armatis, P., et al. Cytotoxic analogs of luteinizing hormone–releasing hormone containing doxorubicin or 2-pyrrolinodoxorubicin, a derivative 500–1000 times more potent. *Proc. Natl. Acad. Sci. USA* **94**:652–656, 1996.
44. Bulun, S. E., Noble, L. S., Takayama, K., et al. Endocrine disorders associated with inappropriately high aromatase expression. *J. Steroid Biochem. Mol. Biol.* **61**:133–139, 1997.
45. Rensen, P. C., de Vrueh, R. L., Kuiper, J., et al. Recombinant lipoproteins: Lipoprotein-like lipid particles for drug targeting. *Adv. Drug Deliv. Rev.* **47**:251–276, 2001.
46. Kratz, F., Beyer, U., and Schutte, M. T. Drug-polymer conjugates containing acid-cleavable bonds. *Crit. Rev. Ther. Drug Carrier Syst.* **16**:245–288, 1999.
47. Ringsdorf, H. Structures and properties of pharmacologically active polymers. *J. Polymer Sci.* **51**:135–153, 1975.
48. Pillai, O., and Panchagnula, R. Polymers in drug delivery. *Curr. Opin. Chem. Biol.* **5**:447–451, 2001.
49. Chapman, A. P. PEGylated antibodies and antibody fragments for improved therapy: A review. *Adv. Drug Del. Rev.* **54**:531–545, 2002.
50. Vasey, P. A., Kaye, S. B., Morrison, R., et al. Phase I clinical and pharmacokinetic study of PK1 [*N*-(2-hydroxypropyl)methacrylamide copolymer doxorubicin]: First member of a new class of chemotherapeutic agents-drug-polymer conjugates. Cancer Research Campaign Phase I/II Committee. *Clin. Cancer Res.* **5**:83–94, 1999.
51. Derrien, D., Midoux, P., Petit, C., et al. Muramyl dipeptide bound to poly-L-lysine substituted with mannose and gluconoyl residues as macrophage activators. *Glycoconj. J.* **6**:241–255, 1989.
52. Monsigny, M., Roche, A. C., and Bailly, P. Tumoricidal activation of murine alveolar macrophages by muramyl dipeptide substituted mannosylated serum albumin. *Biochem. Biophys. Res. Commun.* **121**:579–584, 1984.
53. Seymour, L. Soluble polymers for lectin-mediated drug targeting. *Adv. Drug Del. Rev.* **14**:89–111, 1994.
54. Mastrobattista, E., Koning, G. A., and Storm, G. Immunoliposomes for the targeted delivery of antitumor drugs. *Adv. Drug Del. Rev.* **40**:103–127, 1999.
55. Bendas, G., Krause, A., Bakowsky, U., et al. Targetability of novel immunoliposomes prepared by a new antibody conjugation technique. *Int. J. Pharm.* **181**: 79–93,1999.
56. Wang, S., and Low, P. S. Folate-mediated targeting of antineoplastic drugs, imaging agents, and nucleic acids to cancer cells. *J. Cont. Rel.* **53**:39–48, 1998.
57. Jones, M. N. Carbohydrate-mediated liposomal targeting and drug delivery. *Adv. Drug Del. Rev.* **13**:215–250, 1994.

58. Oku, N., Tokudome, Y., Koike, C., et al. Liposomal Arg-Gly-Asp analogs effectively inhibit metastatic B16 melanoma colonization in murine lungs. *Life Sci.* **58**:2263–2270, 1996.
59. Gyongyossy-Issa, M. I., Muller, W., and Devine, D. V. The covalent coupling of Arg-Gly-Asp-containing peptides to liposomes: Purification and biochemical function of the lipopeptide. *Arch. Biochem. Biophys.* **353**:101–108, 1998.
60. Sinha, V. R., and Kumria, R. Polysaccharides in colon-specific drug delivery. *Int. J. Pharm.* **224**:19–38, 2001.
61. Trubetskoy, V. S., Torchilin, V. P., Kennel, S. J., and Huang, L. Use of *N*-terminal modified poly(L-lysine)-antibody conjugate as a carrier for targeted gene delivery in mouse lung endothelial cells. *Bioconjug. Chem.* **3**:323–327, 1992.
62. Hirabayashi, H., Nishikawa, M., Takakura, Y., and Hashida, M. Development of pharmacokinetics of galactosylated poly-L-glutamic acid as a biodegradable carrier for liver-specific drug delivery. *Pharm. Res.* **13**:880–884, 1996.
63. Gonsho, A., Irie, K., Susaki, H., et al. Tissue-targeting ability of saccharide-poly (L-lysine) conjugates. *Biol. Pharm. Bull.* **17**:275–282, 1994.
64. Jelinkova, M., Strohalm, J., Plocova, D., et al. Targeting of human and mouse T-lymphocytes by monoclonal antibody-HPMA copolymer-doxorubicin conjugates directed against different T-cell surface antigens. *J. Cont. Rel.* **52**:253–270, 1998.
65. Li, C. Poly(L-glutamic acid)-anticancer drug conjugates. *Adv. Drug Del. Rev.* **54**:695, 2002.
66. Sharma, A., and Sharma, U. S. Liposomes in drug delivery: Progress and limitations. *Int. J. Pharm.* **154**:123–140, 1997.
67. Lasic, D. D., and Papahadjopoulos, D. Liposomes revisited. *Science* **267**:1275–1276, 1995.
68. Maruyama, K., Takahashi, N., Tagawa, T., et al. Immunoliposomes bearing polyethyleneglycol-coupled Fab' fragment show prolonged circulation time and high extravasation into targeted solid tumors in vivo. *FEBS Lett.* **413**:177–180, 1997.
69. Akamatsu, K., Imai, M., Yamasaki, Y., et al. Disposition characteristics of glycosylated poly (amino acids) as liver cell-specific drug carrier. *J. Drug Target.* **6**:229–239, 1998.
70. Lokeshwar, V. B., Iida, N., and Bourguignon, L. Y. The cell adhesion molecule, GP116, is a new CD44 variant (ex14/v10) involved in hyaluronic acid binding and endothelial cell proliferation. *J. Biol. Chem.* **271**:23853–23864, 1996.
71. Lokeshwar, V. B., Lokeshwar, B. L., Pham, H. T., and Block, N. L. Association of elevated levels of hyaluronidase, a matrix-degrading enzyme, with prostate cancer progression. *Cancer Res.* **56**:651–657, 1996.
72. Sunamoto, J., and Iwamoto, K. Protein-coated and polysaccharide-coated liposomes as drug carriers. *CRC Crit. Rev. Ther. Drug Carrier Sys.* **2**:117–136, 1987.
73. Saiki, I., Murata, J., Matsuno, K., et al. Anti-metastatic and anti-invasive effects of polymeric Arg-Gly-Asp (RGD) peptide, poly(RGD), and its analogues. *Jpn. J. Cancer Res.* **81**:660–667, 1990.
74. Pakalns, T., Haverstick, K. L., Fields, G. B., et al. Cellular recognition of synthetic peptide amphiphiles in self-assembled monolayer films. *Biomaterials* **20**:2265–2275, 1999.
75. Dillow, A. K., Ochsenhirt, S. E., McCarthy, J. B., et al. Adhesion of alpha 5 beta 1 receptors to biomimetic substrates constructed from peptide amphiphiles. *Biomaterials* **22**:1493–1505, 2001.
76. Maeda, M., Izuno, Y., Kawasaki, K., et al. Amino acids and peptides. XXXII. A bifunctional poly(ethylene glycol) hybrid of fibronectin-related peptides. *Biochem. Biophys. Res. Commun.* **241**:595–598, 1997.
77. Wamsley, A., Jasti, B., Phiasivongsa, P., and Li, X. Synthesis of random terpolymers and determination of reactivity ratios of *N*-carboxyanhydrides of leucine, β-benzyl aspartate, and valine. *J. Polymer Sci. [A] Polymer Chem.* **42**:317–325, 2004.
78. Wamsley, A., Shen, S., Jasti, B., and Li, X. Unpublished results.
79. Wang, S., Lee, R. J., Mathias, C. J., et al. Synthesis, purification, and tumor cell uptake of 67Ga-deferoxamine-folate, a potential radiopharmaceutical for tumor imaging. *Bioconjug. Chem.* **7**:56–62, 1996.

80. Mislick, K. A., Baldeschwieler, J. D., Kayyem, J. F., and Meade, T. J. Transfection of folate-polylysine DNA complexes: Evidence for lysosomal delivery. *Bioconjug. Chem.* **6**:512–515, 1995.
81. Leamon, C. P., Weigl, D., and Hendren, R. W. Folate copolymer–mediated transfection of cultured cells. *Bioconjug. Chem.* **10**:947–957, 1999.
82. Kanellos, J., Pietersz, G. A., and McKenzie, I. F. Studies of methotrexate-monoclonal antibody conjugates for immunotherapy. *J. Natl. Cancer Inst.* **75**:319–332, 1985.
83. Yeh, C. J., and Faulk, W. P. Killing of human tumor cells in culture with adriamycin conjugates of human transferrin. *Clin. Immunol. Immunopathol.* **32**:1–11, 1984.
84. Singh, M. Transferrin as a targeting ligand for liposomes and anticancer drugs. *Curr. Pharm. Des.* **5**:443–451, 1999.
85. Laske, D. W., Youle, R. J., and Oldfield, E. H. Tumor regression with regional distribution of the targeted toxin TF-CRM107 in patients with malignant brain tumors. *Nature Med.* **3**:1362–1368, 1997.
86. Wagner, E., Cotton, M., Mechtler, K., et al. DNA-binding transferrin conjugates as functional gene-delivery agents: Synthesis by linkage of polylysine or ethidium homodimer to the transferrin carbohydrate moiety. *Bioconjug. Chem.* **2**:226–231, 1991.
87. Wagner, E., Cotton, M., Foisner, R., and Birnstiel, M. L. Transferrin-polycation-DNA complexes: The effect of polycations on the structure of the complex and DNA delivery to cells. *Proc. Natl. Acad. Sci. USA* **88**:4255–4259, 1991.
88. Dash, P. D., Read, M. L., Fisher, K. D., et al. Decreased binding to proteins and cells of polymeric gene delivery vectors surface modified with a multivalent hydrophilic polymer and retargeting through attachment of transferrin. *J. Biol. Chem.* **275**:3793–3802, 2000.
89. Konan, Y. N., Gurny, R., and Allemann, E. State of the art in the delivery of photosensitizers for photodynamic therapy. *J. Photochem. Photobiol. [B]* **66**:89–106, 2002.
90. Chrzanowski, K., Bielawska, A., and Palka, J. Proline analogue of melphalan as prodrug susceptible to the action of prolidase in breast cancer MDA-MB 231 cells. *Il Farmaco* **58**:1113–1119, 2003.
91. Bielawska, A., Bielawska, K., Chrzanowski, K., and Wolczynski, S. Prolidase-activated prodrug for cancer chemotherapy cytotoxic activity of praline analogue of chlorambucil in breast cancer MCF-7 cells. *Il Farmaco* **55**: 736–741, 2000.
92. Senter, P. D. and Springer, C. J. Selective activation of anticancer prodrugs by monoclonal antibody-enzyme conjugates. *Adv. Drug Del. Rev.* **53**: 247–264, 2001.
93. Kearney, A. Prodrugs and targeted drug delivery. *Adv. Drug Del. Rev.* **19**:225–239, 1996.
94. Nishikawa, M., Takakura, Y., and Hashida, M. Pharmacokinetic evaluation of polymeric carriers. *Adv. Drug Del. Rev.* **21**:135–155, 1996.
95. Kato, Y., Seita, T., Kuwabara, T., and Sugiyama, Y. Kinetic analysis of receptor-mediated endocytosis (RME) of proteins and peptides: Use of RME as a drug delivery system. *J. Contr. Rel.* **39**:191–200, 1996.
96. Matthews, S. E., Pouton, C. W., and Threadgill, M. D. Macromolecular systems for chemotherapy and magnetic resonance imaging. *Adv. Drug Del. Rev.* **18**:219–267, 1996.
97. Matsumura, Y., Maeda, H. A new concept for macromolecular therapeutics in cancer chemotherapy: Mechanism of tumoritropic accumulation of proteins and the antitumor agent smancs. *Cancer Res.* **46**:6387–6392, 1986.
98. Pratten, M. K., and Lloyd, J. B. Pinocytosis and phagocytosis: The effect of size of a particulate substrate on its mode of capture by rat peritoneal macrophages cultured in vitro. *Biochim. Biophys. Acta* **881**:307–313, 1986.
99. Mellman, I. Endocytosis and molecular sorting. *Annu. Rev. Cell Dev. Biol.* **12**:575–625, 1996.
100. Tartakoff, A. M. *The Secretory and Endocytic Paths*. New York: Wiley, 1987.
101. Price, J. T., Bonovich, M. T., and Kohn, E. C. The biochemistry of cancer dissemination. *Crit. Rev. Biochem. Mol. Biol.* **32**:175–253, 1997.
102. Soyez, H., Schacht, E., and Vanderkerken, S. The crucial role of spacer groups in macromolecular prodrug design. *Adv. Drug Del. Rev.* **21**:81–106, 1996.

103. Robinson, J. R., and Lee, V. H. *Controlled Drug Delivery: Fundamentals and Applications*. New York: Marcel Dekker, 1987.
104. Hudecz, F. Design of synthetic branched-chain polypeptides as carriers for bioactive molecules. *Anticancer Drugs* **6**:171–193, 1995.
105. Shen, W. C., Ryser, H. J., and LaManna, L. Disulfide spacer between methotrexate and poly(D-lysine). A probe for exploring the reductive process in endocytosis. *J. Biol. Chem.* **260**:10905–10908, 1985.
106. Dingle, J. T. *Lysosomes: A laboratory Handbook*. Amsterdam: North-Holland Publishing Co., 1972.
107. Vaughan, T. J., Osbourn, J. K., and Tempest, P. R. Human antibodies by design. *Nature Biotechnol.* **16**:535–539, 1998.
108. Rihova, B., Ulbrich, K., Kopecek, J., and Mancal, P. Immunogenicity of *N*-(2-hydroxypropyl)-methacrylamide copolymers: Potential hapten or drug carriers. *Folia Microbiol. (Praha)* **28**:217–227, 1983.
109. Rihova, B., Kopecek, J., Ulbrich, K., et al. Effect of the chemical structure of *N*-(2-hydroxypropyl)methacrylamide copolymers on their ability to induce antibody formation in inbred strains of mice. *Biomaterials* **5**:143–148, 1984.
110. Rihova, B., and Riha, I. Immunological problems of polymer-bound drugs. *Crit. Rev. Ther. Drug Carrier Syst.* **1**:311–374, 1985.
111. Rihova, B., Bilej, M., Vetvicka, V., et al. Biocompatibility of *N*-(2-hydroxypropyl) methacrylamide copolymers containing adriamycin: Immunogenicity, and effect on haematopoietic stem cells in bone marrow in vivo and mouse splenocytes and human peripheral blood lymphocytes in vitro. *Biomaterials* **10**:335–342, 1989.
112. Stahmann, M. A. *Polyamino Acids, Polypeptides, and Proteins*. Madison: The University of Wisconsin Press, 1962, p. 394.
113. Maurer, P. H. Antigenicity of polypeptides (poly-alpha-amino acids). XVII. Immunologic studies in humans with polymers containing L or D and L-alpha-amino acids. *J. Immunol.* **95**:1095–1099, 1965.
114. Anderson, J. M. In vitro and in vivo studies of drug-releasing poly(amino acids). *Ann. N. Y. Acad. Sci.* **446**:67–75, 1985.

Chapter

13

Programmable Drug Delivery Systems

Shiladitya Bhattacharya, Appala Raju Sagi,* Manjusha Gutta

Thomas J. Long School of Pharmacy and Health Sciences
University of the Pacific
Stockton, California

Rajasekhar Chiruvella

College of Pharmaceutical Sciences
Kakatiya University
Warangal, India

Ramesh R. Boinpally

OSI Pharmaceuticals
Boulder, Colorado

*Present affiliation: Corium International, Inc., Redwood City, California.

13.1 Introduction

Design of drug delivery systems tailored to the needs of a patient is a formidable challenge that pharmaceutical scientists are trying to overcome. As a result, the design of drug delivery systems has gone through evolutionary changes. In the first-generation drug delivery systems, such as tablets and capsules, there was no control over the drug release/delivery rate. In the second generation of drug delivery systems, zero-order or constant release was achieved. However, none of these systems, including the novel drug delivery systems, such as osmotic pumps, implants, transdermal drug delivery systems, liposomes, and other particulate carriers, offers efficacy (including targeting ability) and safety to the patients. This could be due to a lack of correlation between delivered drug and biological markers that assess the disease activity.

The current level of understanding of different diseases, especially the disease correlation with biological markers, together with the revolution in the field of microelectronics, provides an opportunity to optimize drug delivery using programmable drug delivery systems. These delivery systems include magnetically stimulated systems, ultrasonically stimulated systems, electrically stimulated systems, photo-stimulated systems, and self-triggered/regulated systems.

13.2 Rationale for Programmed Drug Delivery Systems

As discussed in Chap. 1, the logic behind the design of second-generation drug delivery systems such as zero-order release systems (similar to IV infusion) was to attain and maintain therapeutic concentrations. However, a constant-rate intravenous (IV) infusion does not always guarantee constant plasma concentrations because of chronopharmacokinetic mechanisms (Table 13.1). For example, the plasma concentration-time profile following constant IV infusion of ketoprofen exhibits peaks and troughs during a 24-hour circadian cycle (Fig. 13.1).

Further, in the case of certain chronic diseases, plasma drug concentrations are required to be maintained based on the severity of the disease condition. For example, asthma as a disease state does not occur constantly, but asthma attacks show a rhythm at certain times of a day. In the early morning hours (3 A.M. to 5 A.M.), the disease is more severe, as manifested by least expiratory flow rate, lowest plasma cortisol levels, etc., than at any other time. It is clear that a patient requires more therapeutic agents at such time than at other times. Even if a person takes a zero-order drug delivery system before going to bed, the concentration will be maintained throughout the night. It may prevent the attack in the early-morning hours, but the unwanted (unnecessary) concentration maintained throughout the night may lead to adverse effects.

TABLE 13.1 Chronokinetics after Intravenous Infusion at a Constant Rate

Drug	Administration regimen	Significant findings	Reference
Bupivacaine	Continuous peridural infusion over 36 h, 0.25 mg/kg/h	Maximum clearance at 0630 h (~60% change)	1
Fluorouracil	Continuous IV infusion over 5 days, 450–966 mg/m^2/day	~58% change in mean highest plasma concentrations at 0100 h	2
Ketoprofen	Continuous IV infusion over 24 h, 5 mg/kg/day	~51% variation in plasma concentration with a peak at 2100 h.	3
Ranitidine	Continuous IV infusion 6.25 mg/h (or) Sinusoidal infusion from 3.0–9.4 mg/h over 24 h	Highest plasma concentrations at 2200 h (~41% change) No circadian rhythm after sinusoidal rate infusion	4
Terbutaline	Constant-rate infusion 0.033 mg/kg over 24 h preceded by a 2.94 µg/kg bolus	Highest plasma concentrations at 2300 h (~43% change)	5
Vindesine	Continuous IV infusion 3 mg/m^2/day over 48 h	Highest serum concentration at 1200 h (~30% change)	6

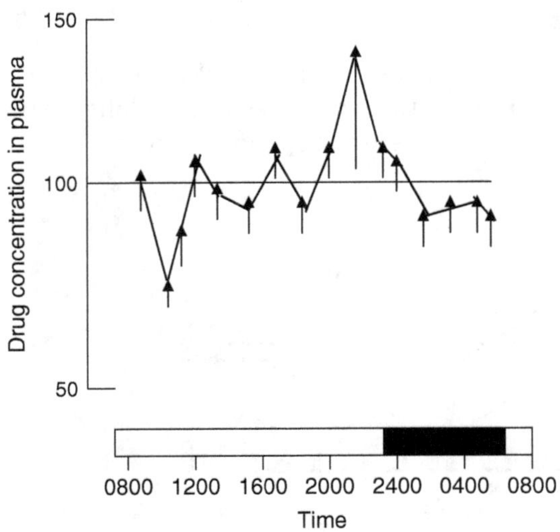

Figure 13.1 The continuous intravenous infusion of ketoprofen does not lead to a constant plasma concentration. Changes in concentration are expressed as an averaged percentage of each individual 24 hour mean.[7] Vertical bar represent SEM. ■ Sleep, ❑ Wakefulness.

The same is true with cytotoxic (anticancer) drugs, which have many adverse effects. Many neoplastic diseases (leukemia, breast cancer, etc.) show definite rhythms in presenting their symptoms. Thus, if the therapy is tuned according to these rhythms, not only is the disease better treated, but the patient also is protected from side effects. *Thus the drug delivery should be programmed according to the disease condition.*

The body functions by biological feedback control systems; secretion of hormones and other active substances follows such a pattern and responds to rising levels of certain marker molecules in the blood. Endogenous insulin is secreted depending on the feedback from the blood glucose level. When insulin is delivered as a therapeutic agent by a drug delivery system to diabetic patients, a constant rate of insulin delivery irrespective of blood glucose level may lead to hypoglycemia. Therefore, it is essential to mimic the physiological feedback mechanism of insulin to maintain the glucose at a normal level. Hence a *feedback-controlled drug delivery system is needed to deliver the drug in a rational way.*

The relationship between the concentration of drug and the disease state is illustrated schematically in Fig. 13.2. Because of the feedback mechanism, the programmable delivery systems for drug delivery can be thought of as an artificial imitation of healthy functioning.

13.3 Classification, Design, and Functioning of Programmed Drug Delivery Systems Based on Design Principles

The programmable drug delivery systems can be classified as open- or closed-loop systems. Although both systems are intended to deliver the drug as it is needed, the open-loop system depends on an external

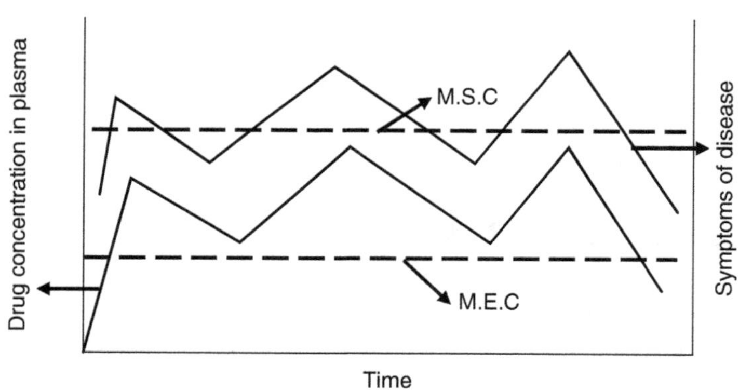

Figure 13.2 Plasma concentration-time profile with a programmable drug delivery system.

stimulating signal to deliver the drug. Thus, in an open-loop system, drug delivery is preprogrammed or needs the intervention of the patient or a health care provider. Since the drug needs of a patient are not always preprogrammable (as in diabetes, the insulin needs of a patient depend on diet, exercise, etc.), these systems are far from ideal for drug delivery. On the contrary, the closed-loop systems are self-regulated. The drug release is governed by the feedback information collected from the body by the system. The rate-controlling mechanism in closed-loop systems is accomplished by using pH-sensitive polymers, binding of antibodies, hydrolytic reactions, pH-sensitive drug solubilization, enzyme substrate reactions, and microelectronics.

13.3.1 Open-looped (pulsatile) systems

Open-looped systems operate on trigger signals sent from an external device that thereby activates the implanted system and causes drug release within the body. Pulse delivery occurs by magnetic, ultrasonic, or electronic means in open-looped systems.

Microfabricated systems. Based on the characteristics, microfabricated systems include microchips and micropumps.

Microchips. Pulsatile drug delivery is advantageous when the continuous presence of drugs leads to the development of tolerance. The conventional pulsatile drug delivery systems are large in size, and when there is a need for multiple drug delivery, the situation turns more complex, and controlled drug delivery through conventional routes gets extremely complicated.

Microchips, being miniature in size, can closely mimic many endogenous components through a pulsatile pattern of drug release. While a computer microchip breaks down any of the complex operations into a series of binary numbers, 0s and 1s, and makes logical decisions, a biochemical microchip interprets 0 and 1 signals as commands for closing and opening the reservoir, respectively. On the implementation side, the signal for opening a reservoir dissolves the small anodic membrane covering that reservoir, thereby releasing the drug in the reservoir.[8]

The microchip design can be classified into passive and active microchips. Passive microchips are made up of polymeric materials without any electronic components and control endocrine function by temporal release of therapeutic agents.[9] Active microchips are made of silicon and are the electronic version of passive microchips. They supplement or manipulate endocrine function through release controlled by sensor activation. The microchips can be designed to store single or multiple chemicals in the reservoirs, and a complex release pattern can be achieved by opening different reservoirs at different times with the help of a preprogrammed clock, biosensor feedback, or a remote control device used by the patient. When the active microchips operate based on biosensor feedback, they fall under the category of closed-loop systems.

Passive microchips are made up of polymers such as polyesters. The polymer is molded into the required shape with a small opening, drug is placed within it using microinjection, and the opening is sealed with a degradable or nondegradable membrane coating. Biodegradable membrane has an advantage because the implanted microchip need not be removed after drug release. Release of the drug from the reservoir depends on the degradation rate of the membrane or diffusion of the drug through a nondegradable membrane.

Active microchips consist of drug reservoirs covered with a 0.2 to 0.3-μm-thick anodic membrane and are wired with final circuitry controlled by a microprocessor.[10] After the drug reservoirs are filled, they are sealed with a waterproof material. Three cathodes made of gold are placed at various intervals[11] in the microchip for the electrochemical reaction to occur. When the system is triggered with an electrical potential, the anodic gold membrane forms a soluble complex with chloride that is present in the saline, forming aurium chloride complex, and releases the drug from the reservoir.

Some microchips control drug release by valves made of polymeric materials and are therefore called *artificial muscle valves*.[12] Artificial muscle is a chemomechanical actuator consisting of a blend of a hydrogel and an electronically conducting redox polymer.[13] The redox polymer is sensitive to the pH, electrical potential, and chemical potential of its microenvironment, whereas the hydrogel exhibits dramatic swelling and shrinking on changes in pH, solvent, temperature, electrical field, or ambient light conditions. By electropolarizing these polymers onto electrodes, reservoirs can be opened or closed, and the drug compound released or retained, via the swelling and shrinking processes of the polymer system in response to electrochemical actuation.

The absence of moving parts makes microchips more sturdy and reliable, providing the drugs with an inert environment that results in improved drug stability. Multiple drugs can be loaded and released accurately. Thus complex patterns of multiple drug delivery can be achieved with these devices. Furthermore, they can be implanted locally for onsite drug activity and can be coupled with or controlled by devices generating electrical signals for triggering drug release. The rate of release from any reservoir can be controlled by proper selection of the materials (e.g., pure drugs or drugs with polymers) placed inside the reservoir. A material that can dissolve quickly when the reservoir is opened can give pulsatile release, whereas a material that dissolves slowly after the reservoir is opened can be used to achieve sustained release.

Micropumps. The invention of microcomputers, microsensors, and micropower sources revolutionized many fields. The field of drug delivery systems is not an exception to this. Microprocessor-controlled drug delivery systems that can be implanted into patients are the recent advances in programmable implantable medication systems (PIMS). The first programmable implantable medication system was developed at the Johns Hopkins University Applied Physics Laboratory. Implantable medication systems consist of a group of devices that are surgically placed in the subcutaneous or intraperitoneal region of the patient's body. They may function independently, as in polymer-based pulsatile delivery systems, or may be controlled by the patient or feedback-regulated, as in implanted devices working with microelectronics and micromachines. Programmable implantable medication systems usually are composed of four components: a command system, a telemetry system, a power system, and microminiature drug delivery devices.

Command system. Implanted electronic devices use command systems that operate from a radiosignal originating in physician's console, i.e., exterior to the patient. The PIMS command system is used to change the basal delivery rate, to turn the device on and off, to set the limits on medication usage. A command system for an implant allows it to adapt to the patient's changing needs.

Telemetry system. Telemetry involves the transmission of data from a remote location. The typical implant telemetered data might be the confirmation of the parameters that have been commanded into it, the battery voltage, and the rate of infusing medication both in real time and as stored data.

Power system. Previously, rechargeable nickel-cadmium cells were used in implant systems. More recently, the power systems of implantable medical electronic devices have become so small that a single AA size lithium primary cell can be used without recharging for more than 5 years.

Microminiature drug delivery device. This is the heart of the implanted medication systems. The drug delivery device often consists of a diaphragm-operated infusion pump that supplies drug at a constant predetermined rate to the body. The pump is connected to the drug reservoir, which, in some cases, may be recharged when exhausted. Today, the available electronic components are so miniature that they can be implanted comfortably even in newborn babies. The pump can be programmed for a constant or variable basal infusion of medication with a repetitive period of from 1 hour to 60 days. By far the most frequently used basal period is 24 hours. A period of 28 days is available, particularly for the infusion of sex hormones to mimic the female menstrual cycle.

As shown in Fig. 13.3, the patient is provided with a handheld device called the *patient's programming unit*, by which the patient can initiate

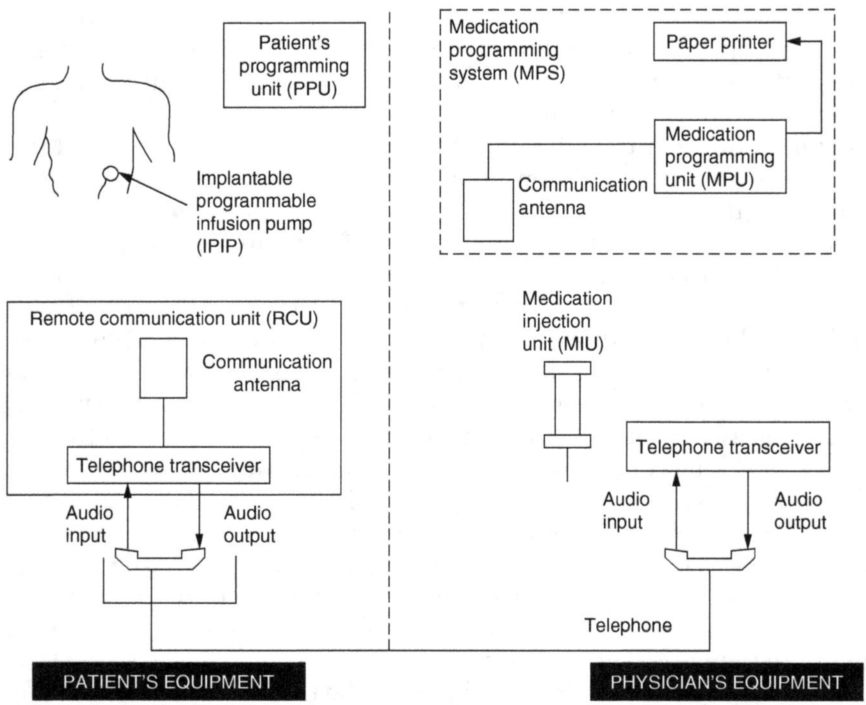

Figure 13.3 Programmable implantable medication system (PIMS).

self-medication from the implantable programmable infusion pump (IPIP) within the constraints programmed into the IPIP by the physician. The IPIP can be refilled through the medication injection unit.

The last major portion of the PIMS is a telephone communication system by means of which the patient at a location remote from the medication programming system can have the IPIP reprogrammed with a new prescription or can have stored telemetry data in the IPIP read out and displayed by the medication programming system.

Advantages of PIMS include precise medication rate, patient compliance, ease of delivery, and accurate periodicity, which leads to a requirement for lesser amounts of medication, and accidental overdose is prevented.

Programmable implantable medication systems find use in the treatment of diabetes, conditions of chronic pain, and Parkinson's and Alzheimer's diseases. Infusions of morphine can be programmed to deliver the drug when the intensity of pain is high. This reduces the chance of the patient developing tolerance to the opioid. Daily injection of insulin subcutaneously causes pain and sometimes infection in diabetic patients. A delivery system that has high accuracy of pumping insulin at micro- to nanoliter levels over a wide range of external conditions (pressure, temperature, viscosity) and also has a lifetime ranging from several days to several months depending on the application will circumvent the problems associated with daily administration.

Silicon micropumps offer major advantages in terms of system miniaturization and control over low flow rates with a stroke volume 160 nL.[14] The micropump has the characteristics of very small in size, implantability in the human body, low flow rates (in the range of 10 μL/min), moderate pressure generation from the microactuator to move the drug, biocompatibility, and most important, a reliable design for safe operation. The implantable device is particularly suitable (over the injectable drug delivery systems) for patients with Parkinson's disease, Alzhiemer's disease, diabetes, and cancer, as well as chronically ill patients, because the catheter that is attached to the device can transport drug to the required site.

Micropumps can be classified into two groups: mechanical pumps with moving parts and nonmechanical pumps without moving parts. In mechanical micropumps, movement is obtained by reciprocating movement or peristaltic movement. Implantable micropumps deliver drugs by using peristaltic pump technology to give a steady flow rate.

Micropumps based on piezoelectrics are made of pumping chambers that are actuated by three piezoelectric lead zirconate titanate disks (PZT). The pump consists of an inlet, pump chambers, three silicon membranes, three normally closed active valves, three bulk PZT actuators, three actuation reservoirs, flow microchannels, and outlet. The actuator is controlled by the peristaltic motion that drives the liquid in the pump. The inlet and outlet of the micropump are made of a Pyrex glass, which makes it biocompatible. Gold is deposited between the actuators and the silicon membrane to act as an upper electrode. Silver functions as a lower electrode and is deposited on the sidewalls of the actuation reservoirs. In this design, three different pump chambers can be actuated separately by each bulk PZT actuator in a peristaltic motion.

Van Lintel et al.[15] designed a volumetric pump with a pumping membrane that compresses the pumping chamber, and a pair of inlet and outlet check valves that direct the liquid flow. The pump can be operated by a remote control device to provide precise dosing. It has improved safety, resolution, programmability, and autonomy, which make it ideal in peptide delivery.

Micropumps have been designed using polymers such as polymethyl methacrylate (PMMA) as the base material. The basic design of this pump consists of six layers. The first is made of a PMMA plate that has an opening at the center for fixing a peizodisk, which, in turn, works as both the actuator and the pump membrane. The peizodisk and a second PMMA plate act as the pump chamber. The second PMMA plate has two access holes for inlet and outlet. The fourth and fifth layers are made up of another polymer, and the check valves are incorporated in these two layers. All these layers are then fixed with four screws. These polymers have low Young's modulus, and thus the check valves require less pressure for operation when compared with silicon micropumps. The polymeric material attains flow rates of up to 1 mL/min and also offers better sealing characteristics.

Rizk and Stevens[16] devised an implantable electrophoretic pump for the delivery of ionic drugs. The device consists of a drug reservoir with a filling opening and a discharge opening. The discharge port has a diffusion membrane with a pair of electrodes lining it. A battery is included in the device that generates the potential difference between the electrodes, facilitating the ionic transport of drug into the patient's body. This device has been found to be beneficial in the delivery of drugs such as insulin, blood thinners, and certain antibiotics. The pump has an exterior housing, on the walls of which the exit port is located, and a resealable refilling port is located on the face of the device perpendicular to the exit port. Thus, if the device is implanted into the patient's peritoneum, the refilling port forms the base, and when in need, it can be recharged easily by an injection through a hypodermic needle. The housing is made of an inert material such as titanium, medical-grade stainless steel, or reinforced polymers, and the inner walls are coated with silicone rubber to prevent reaction, precipitation, and/or adhesion of the drug slurry. Inside the pump houses, two lithium iodide batteries are connected to a capacitor, which, in turn, is connected to the electrodes attached to the membrane at the exit port. These capacitors are electrical energy reservoirs that provide the required current during periods of bolus dosing. A set of wires wound around the reservoir acts as an antenna to transmit and receive information from an external programmer. Usually in such systems a basal program is set on timing principles such that the batteries will energize the electrodes adjacent to the membrane at a certain current level for a predetermined period of time at predetermined times during the day. At other times of the day, the system will be at an "off state," wherein transport of ions continues by diffusion only. In an emergency, the patient may activate the system so as to provide a bolus dose or additional quantities of the ionic drugs, such as the administration of insulin during mealtimes. This can be done by having the patient hold a magnet in close proximity to the unit for a predetermined time. The magnetic field from the handheld magnet turns on a magnetic read switch in the electric circuit of the pump, thereby putting the pump in the "on" state as desired by the patient.

Portner and Jassawalla[17] designed an implantable infusion pump that includes a reservoir for containing the drug, a catheter for delivering drug to the body, and a solenoid-driven miniature pump with a power supply that is responsive to a signal applied externally for initiating delivery of a precisely regulated dosage. The implanted controller provides a basal dosage rate that itself may be variable or constant and may be altered by telemetry signals delivered from outside the body. The internal device is also operable by an external unit. The reservoir is maintained at zero gauge or slightly negative gauge pressure to prevent loss of the contents of the reservoir into the body in which the device is implanted in the event of a failure. The interior of the device houses a drug reservoir, a pumping chamber, a microelectronics chamber, and an array of valves arranged for communication with the catheter, which is disposed outside the housing. The device is capable of a basal mode delivery, as well as bolus dosing. The drug is dispensed by a positive-displacement pump of fixed stroke volume. An external programmer controls both delivery rates and the duration of pump action when operating in the basal and bolus-dose modes. The implanted control circuit and battery automatically provide a programmed delivery of insulin appropriate for the basal requirement. During mealtime, the external (palm-sized) programmer/power unit is placed over the subcutaneously implanted system, and a predetermined insulin dose is entered by means of a suitable key pad. The external unit is inductively coupled to the implanted module and overrides the internal circuit, instructing the pump to deliver the selected amount of insulin with a series of pump pulses. The energy required for the delivery of these bolus doses is also transmitted transcutaneously, thus saving the implanted battery life for basal delivery. The drug reservoir of the device includes a variable-volume bellows-type expansion chamber. The expansion chamber is filled with a saturated vapor mixture of Freon 113 and Freon 11, which maintains a constant pressure as long as a constant temperature is maintained. Thus the expansion chamber pressure variation is essentially constant throughout the volume change of the reservoir. An inlet tube extends into the drug reservoir through an inlet port and is looped at its end so as to prevent the entry of air bubbles if formed in the reservoir. The tube forms the outlet port of the device and is connected to a series of check valves before it communicates with the outside. When the pump is activated, insulin in the reservoir is pressurized by an elastomeric diaphragm, which, in turn, is driven by a small solenoid via a plunger. The fluid pressure opens the check valves, and the drug exits through the catheter to the delivery site. When the solenoid completes its stroke, the current pulse is turned off, and a spring returns the armature to its initial position. During the return stroke, insulin from the reservoir flows through the inlet port to fill the pump chamber. The pump is then ready to be activated again. The housing has an opening through which the device can be refilled. An injection port is fitted with a resealable septum and a check valve to control the flow of fluid to the reservoir chamber. A hypodermic needle may be inserted through the skin and the septum for injecting a fresh supply of drug.

As described earlier, the basic design of the micropump involves a drug reservoir attached to a pumping device with or without a sensor. The inclusion of a sensor with the device makes it a closed-loop system, where the device can check the levels of marker molecules such as glucose and deliver the therapeutic agent. In many systems the devices are implanted such that the reservoir can be charged when it is exhausted. The device may pump at a basal rate or may be controlled by a circuit connected in a closed loop with a sensor or by a handheld remote control device by the patient.

In polymeric membrane and matrix-based micropumps, the membrane or the matrix makes the essential component of the delivery device that controls the rate of release. In matrix controlled delivery, the rate of the hydrolytic breakdown of the matrix is the governing process. In polymeric membrane-controlled release, the rate of hydration of the membrane and the subsequent diffusion of drug are the rate-controlling steps.

Electrical responsive systems. Drug delivery system can be designed based on polyelectrolyte hydrogels that respond to an electrical field. With this kind of response, pulsatile drug delivery can be achieved by alternating application of current mimicking the endogenous components.[18] An electroconducting patch could be placed on the skin over the gel. When release of drug is required, electrodes are placed, followed by application of an electrical field onto the skin.[19] The applied electrical field stimulates the gel that is placed on the skin, which causes deswelling of the gel and release of the drug. The drug is pushed out of the gel when the fluid in the gel exudates owing the influence of the electrical field applied above a threshold value. The patch can be removed when a calculated amount of drug is released. This patch could be designed as a wristwatch for patients to wear.

Polymer-based systems. Pulsatile release of medicaments can be obtained from polymer-based delivery systems. Based on the mechanism of drug release from the polymer, these systems can be divided into various classes and subclasses. Broadly, they can be classified into three classes: delivery by hydrolysis of polymers, delivery by osmotic pressure, and delivery by both hydrolytic degradation and osmotic effects.

Hydrolysis of the polymers can be affected by spontaneous degradation of the material from the surface or the bulk or by enzymatic hydrolysis. Copolymers of lactic acid and glycolic acid (PLGA) are by far the most common biocompatible polymers that undergo bulk erosion by ester linkage hydrolysis. PLGA-based implants consisting of a core made of antigen adsorbed on dibasic calcium phosphate and coated with a blend of PLGA, ethyl cellulose, and Eudragit S were prepared by Khoo and Thiel.[20] The outer layer gets hydrated, followed by degradation of the PLGA to form pores. The Eudragit layer then erodes, and the antigen is released after the core is hydrated. Coadministration of the uncoated implant together with a similar coated implant resulted in the release of antigen after 1 and 75 days, respectively, from the dosage forms. The Eudragit layer is used to delay release of the antigen from the core, which results in a pulsatile release profile. Dextran-hydroxyethylmethacrylate (dex-HEMA) microspheres for pulsatile delivery of proteins were formulated by Franssen et al.[21] The carbonate esters hydrolyze under physiological conditions to release the entrapped proteins in a controlled manner. The authors altered the size of the dextran and the cross-link density to block the protein molecules from leaving the system by diffusion. In vitro release of IgG from such a system showed an initial release that was lower than 10 percent of the total amount. An increase in cross-link density could delay the release from 5 to 15 days. In another variation of the system, the microspheres were loaded with the liposomes carrying the model drug. The initial delay could be lengthened further, with pulsed release occurring for over a month. A similar study was reported by Kikuchi et al.,[22,23] where physical cross-linking was used to cross-link aqueous alginate solution. Calcium alginate beads were prepared, and exchange of sodium from the medium caused degradation of the beads to release the model compound dextran loaded into them. The lag time was proportional to the amount of calcium alginate used and the size of the bead. A pulsatile release was obtained from a mixture of beads of three different diameters.

Matrices made from polyanhydrides and polyorthoesters exhibit surface erosion, an often-sought-after property in designing pulsatile delivery. This is due to the rate of hydrolytic breakdown of the polymer at the surface in contact with water being several folds greater than the flux of water molecules into the matrix. Gîpferich et al.[24] designed a layered implant composed of surface and bulk eroding polymers with pulsatile release characteristics. A core of poly[1,3-bis(carboxy phenoxypropane)-co-sebacic acid] (p(CPP-SA)) was loaded with drug and coated with a drug-free layer of p(CPP-SA), followed by another coat of polylactic acid (PLA) to further delay drug release from the core. A final coat of p(CPP-SA) with drug was applied for initial release of the active agent. An initial pulse of the active ingredient from the outermost layer lasted a week, followed by release from the core after 2 weeks. In these systems, the pulse duration of the initial release from the outermost layer depends largely on the hydrophilicity and molecular weight of the drug molecule. Jiang et al. described a protein delivery system where the protein drug is complexed with poly(methacrylic acid)/polyethoxazoline and is layered alternately with polyanhydride in a cylindrical polycarbonate case open at one end.[25] Dissolution of the drug-free polyanhydride layer provides the delay in between pulses. Qiu et al.[26] formulated a similar device consisting of pH-sensitive degradable layers of polyphosphazene containing the drug sandwiched between layers of polyanhydrides. In this case, the release profiles and the lag time between the pulses were found to depend on the hydrophilicity of the drug.

Hydrolysis of polymers also can be affected by the inclusion of polymer-degrading enzymes in the formulation or from the physiological environment. Kibat et al.[27] formulated drug-loaded liposomes coated by phosholipase A2, encapsulated them in alginate microgels, and further coated them with polylysine. Drug release from this system followed hydration and subsequent degradation of the liposomes by the enzyme, after which the free drug diffused out of the alginate microgel. The liposomes acted as a drug depot, which was held intact in the alginate gel and which was further protected by the insoluble polylysine coat. The lag time was found to be proportional to the amount of enzyme loaded into the system. Moriyama et al.[23] designed a delivery system on similar principles consisting of alternate layers of dextran and PEG grafted dextran layers containing insulin cased in a nondegradable silicone tube with an open end. The inclusion of dextranase in the dextran layers degraded the dextran and was responsible for the 10- and 50-hour delays in the pulses of insulin release.

Pulsatile release also may be achieved by the use of an osmotically active pumping device, where the rate of drug release from the device is controlled by water permeation through the semipermeable membrane to activate the osmotic agent. Concerta, used for the treatment of attention-deficit hyperactivity disorder (ADHD) in school-aged children, is a pulse delivery system for methylphenidate hydrochloride. The capsule consists of three layers and an outer coating containing drug for initial release. Of the three layers, the first two are the drug-containing layers, and the third is an osmotically driven push layer. After the formulation is exposed to gastric contents, water permeates through the semipermeable coat of the capsule, hydrating the push layer and the two drug layers. The push layer expands to release the drug from a laser-drilled precision hole at the drug-layer end of the capsule. Within the first 2 hours of ingestion, a therapeutic level of the agent is reached by drug dissolution from the outer coat. After 5 or 6 hours, the agent is again released in a pulsed manner.

Some osmotic systems have been designed based on the principle that the semipermeable membrane will break open owing to osmotic pressure created by swelling of the osmagent to release the medicament. Time-controlled exploding systems (TES) were designed by Ueda et al.[28]

The delivery system is spherical and has a four-layer structure. The core is made up of polystyrene, on which the drug is loaded as a concentric coating. This is further coated by a swelling agent, followed by a coat of water-insoluble ethyl cellulose. On exposure to water, the swelling agent grows in volume to disrupt the ethyl cellulose coat, and the drug is released all at once. The thickness of the ethyl cellulose coat determines the rate of hydration of the swelling agent, which, in turn, determines the lag time for drug release. Administration of TES particles with varying lag times resulted in pulsatile release of the medicaments. Thiel et al.[24] formulated an osmotic bursting implant for the delivery of a model protein. The protein, an antigen, was mixed and compressed with Ex-plotab (sodium starch glycolate), which formed the core of the formulation. The core was then coated with a pH-sensitive Eudragit S100 film, and further coated with an insoluble Eudragit NE30D film containing hydroxypropylmethylcellulose (HPMC) as a pore former. When exposed to water, the HPMC in the outer coat dissolves, making pores in the Eudragit NE30D film, and the physiological fluids hydrate the inner Eudragit film and the core. At physiological pH, Eudragit S100 film will dissolve, and the core gets hydrated, expanding and disrupting the outer insoluble coat and leading to antigen release. Cardamone et al.[30] designed a pulsed release formulation of antigen by packing active tablets and dummy spacer tablets in a water-impermeable tube. One end of the tube had a water-swellable agent covered with a porous membrane, and the other end had a removable plug. When the delivery system was exposed to water, the hydrophilic agent at one end swelled, causing the plug at the other end to drop off and exposing the first tablet to the dissolution medium.

In another design, the same material was used to serve as both the osmagent and the drug-dispersal matrix in the delivery device. Thus the lag time is governed not only by rupture of the barrier membrane but also by diffusion of the drug from the bulk erodible matrix. Iskakov et al.[31] prepared calcium alginate beads coated with a polyacrylamide layer. On hydration, drug release takes place by sodium ions diffusing from the medium into the gel through the polyacrylamide coat, causing polymer relaxation. At the same time, the sodium alginate solution formed absorbs water owing osmotic effect and finally breaks open the polyacrylamide coat to release the entire drug at once. Applying thicker coats retarded drug release, and the initial release could be lowered to less than 10 percent by using drugs with molecular weights higher than 145 kD.

A pulsatile release diltiazem hydrochloride dosage form with a blend of fast, medium, and slow release fractions of a multilayered diltiazem bead was designed.[32] Polymeric membrane coating was applied to modulate the time of release. The fast, medium, and slow release fractions are composed of diltiazem beads coated with weight gain of 14 to 18 percent, 39 to 43 percent, and 63 to 67 percent polymeric membrane coating, respectively. Diltiazem has site-specific absorption in the gastrointestinal tract, with the highest absorption in the duodenum, followed by the ileum and the colon. The multilayered drug delivery system delivers diltiazem hydrochloride in a site-specific and time-controlled manner at gastrointestinal sites. Thus the time-controlled delivery of drug to the duodenal, ileal, and colonic sites achieves a pulsatile release kinetics.

Polymer matrices for implants are made of ethyl vinyl acetate containing a uniform dispersion of drug and magnetic beads. Drug is released into the body fluids by diffusion. On excitation by an oscillating magnetic field, however, the drug release can be increased when required. Both the magnetic field strength and the mechanical properties of the polymer govern the release characteristics from such dosage forms.

Drug release can be modulated with the help of ultrasound from bioerodable and nonbioerodable matrices implanted inside the body.[33] Copolymers of sebacic acid with polyglycolide, polylactide, and poly[bis(*p*-carboxyphenoxy)] alkane anhydrides release entrapped drugs when triggered with ultrasound, and the effect is reversible in the absence of the stimulus. Release of zinc bovine insulin from nonbioerodible matrices made of ethyl vinyl acetate can be increased 15 times with the application of ultrasound, thereby imparting a pulsatile control over the delivery system. The use of ultrasound is not limited to implants; it also can be applied to the transdermal route of delivery. Reports of the permeation enhancement of various small molecules such as salicylic acid and cortisone and larger molecules such as insulin, γ-interferon, and erythropoietin across human skin also have been published.[33] Cavitation, mixing, thermal effects, generation of convective velocities, mechanical properties of polymers, and acoustic streaming are proposed to be the factors for higher release rates from implants on being triggered with ultrasound.

13.3.2 Closed-looped (feedback-controlled) systems

In the programmable devices, the delivery rate of drug is preprogrammed, and it can be altered. In the feedback-controlled systems, continuous monitoring of biological markers and feedback of this information to the delivery system control the release of drug to meet the therapeutic needs of a patient. Such a system could be considered as an artificial gland and represents the ideal therapeutic system. A general design scheme of feedback-controlled delivery systems includes a sensor, a processor, and a delivery system. A feedback-controlled drug delivery system, the Glucose Controlled Insulin Infusion System (GSIIS), known by the trade name Biostator, was available.[34–36] Table 13.2 summarizes some aspects of closed-loop delivery systems.

As shown in Fig. 13.4, the blood is drawn from the patient at a continuous slow rate of 50 mL per 24 hour. Before the blood enters the sensor, it is diluted with anticoagulant buffer mixture (heparin) to prevent coagulation in the transport tubing and analyzer. The sensor contains immobilized glucose oxidase as the primary reagent. The diluted blood flows past a membrane (30-nm pores), through which the glucose passes selectively by diffusion to react with the reagent. One of the reaction products, hydrogen peroxide, is measured polarographically. The computer/process controller receives the input from the sensor and activates a four-channel infusion pump system. The total response time, including blood transport from the patient, is less than 90 seconds.

This Biostator was a large instrument that could be used only in hospitals. But advancements in the fields of electronics and computers make miniature components available. By using microprocessors and microsensors, it is possible to prepare a device, similar to the size of PIMS, that can be implanted without any difficulty.

Blanco and Samadani[37] obtained a patent for the construction of a microprosessor-based insulin pump that works in a similar fashion to the Biostator. The implantable infusion device consists of a catheter, an information-transmitting sensor located in the catheter, a microprocessor, a pump, the drug reservoir, and a power source. The pump, the sensor, and the valves are connected by appropriate leads to the microprocessor. The device is implanted in the subcutaneous tissue in the chest area, and the infusion catheter is tethered intravenously to a central location, such as the right atrium. The device is inserted with the inlet port facing outward so that it may be refilled periodically by a physician.

Palti[38] designed a system that consists of living cell–based sensors that can be used to sense blood constituents. These sensors can be used in combination with pumps and controllers that make up the delivery systems. The sensors are made of living cells that generate a varying electrical signal in the presence of constituents to which the cells are sensitive and are encapsulated in a biocompatible semipermeable membrane. Sensors are implanted in the body of a patient that generate electrical or other detectable signals in response to the presence of particular constituents.

Changes in pH cause phase changes in certain random copolymers of acrylic acid and methacrylamidopropyl trimethylammonium chloride, which make them ideal candidates for pH-responsive systems. At different pH values, the positively and negatively charged groups interact

TABLE 13.2 Closed-Loop Delivery Systems and Their Trigger Mechanisms

Type of system	Design principle	Reservoir drugs	Drug release triggering agent	Reference
Electromechanical (Biostator)	Enzyme reactions	Insulin	Blood glucose level	34–36
Electromechanical (implantable microprocessor-based pump)	Enzyme reactions	Insulin	Blood glucose level	37
Living cells	Electrical responses from living cells	Not specific to any drug	Electrical responses	38
Polymer-based (random copolymers of acrylic acid and methacrylamidopropyl trimethylammonium chloride)	Changes in polymer relaxation	Not specific to any drug	pH change	33
Polymer-based (ethyl vinyl acetate)	Changes in solubility of the drug	Insulin	pH change	33
Polymer-based [ethylidene-2,4,8,10-tetraoxaspirol(5,5)undecane and *N*-methyldiethanolamine]	Bioerosion of drug polymer complex	Insulin	pH change	39
Polymer-based (copolymer of 2-hydroxyethylacrylate, *N,N*-dimethylaminoethyl methacrylate and 4-trimethylsilyl styrene)	Polymer relaxation	Insulin	pH change	33
Immobilized enzyme-based	Enzyme reactions	Naltrexone	Plasma levels of morphine	33
Copolymer of *m*-acrylamidophenylboronic acid with polyvinylpyrrolidone	Sol-gel transformation	Insulin	Plasma glucose levels	40
Nylon microcapsules with concavalin-bound insulin	Competitive binding	Insulin	Plasma glucose levels	41,42
Polymer-based (*N*-isopropylpolyacrylamide)	Phase change	Not specific to any drug	Temperature	43–46
Polymer-based (hydrogel cross-linked with hyaluronic acid)	Oxidative damage of crosslinker	Antiinflammatory drugs	Free radicals	47,48

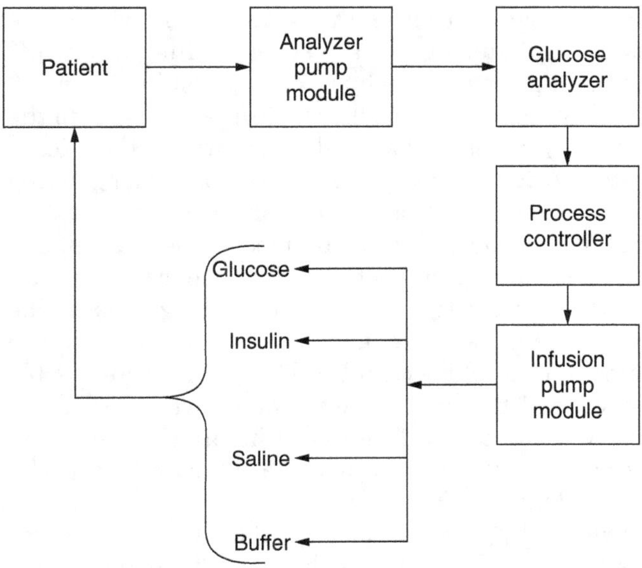

Figure 13.4 Block diagram of Biostator operation.

through attractive and repulsive forces and hydrogen bonding to yield different phases. The interchange between these phases is marked by abrupt swelling or shrinking of the matrix. Thus these polymers can respond to changes in environmental pH and act accordingly with desired release characteristics.[33]

Alterations in pH also can be responsible for the increase in solubility of loaded active agents. Glucose oxidase immobilized on sepharose beads were incorporated into ethyl vinyl acetate matrices along with insulin in the solid form. Glucose from blood enters these matrices, gets oxidized to glucuronic acid, and the decrease in pH increases the solubility of insulin, which diffuses out. The insulin in this case was modified by the addition of three extra lysine residues that ensured an isoelectric point of pH 7.4 for the molecule.[33]

In a similar design, the decrease in pH was linked to polymer degradation, which causes release of the drug. Insulin was immobilized in a pH-sensitive bioerodible polymer made from ethylidene-2,4,8,10-tetraoxaspirol(5,5)undecane and *N*-methyldiethanolamine, which was further coated with a hydrogel containing immobilized glucose oxidase. In this case, however, it was later observed that degradation of the polymer was not hydronium ion–specific but depended on the constitution of the buffering system.[39]

pH-dependent polymer relaxation also has been explored for pulsatile insulin delivery. Hydrogels made of 2-hydroxyethylacrylate,

N,*N*-dimethylaminoethyl methacrylate, and 4-trimethylsilyl styrene and loaded with insulin and glucose oxidase show swelling at low pH caused by the oxidation of glucose, thereby releasing insulin.[33]

Binding and competitive binding of marker molecules, present in the environment of the delivery system, act as feedback control to the release of drugs from the system. A delivery system for naltrexone, a long-acting opioid antagonist, was designed to release the drug from a matrix system triggered by the presence of morphine in the blood. The drug is dispersed in a bioerodible polymer matrix coated with a lipid layer that prevents water entry into the matrix, thereby preventing drug release. The delivery system is placed in a dialysis bag which also contains a lipase conjugated to morphine-antimorphine complex. The enzyme is reversibly activated by the presence of free morphine in the dialysis bag as the antibody competitively binds to the free morphine and releases the active enzyme. The active enzyme degrades the lipid layer, thereby "turning on" naltrexone delivery.[33]

Insulin can be derivatized by glycosylation, and the glycosylated product is bound to concavalin A and coated with hydrophilic nylon to form microcapsules. When free glucose levels in the blood increases, free glucose displaces glycosylated insulin from the concavalin complex. The release rate depends on the polymer coatings or the matrix materials and the relative affinity of the glycosylation units on insulin toward concavalin A compared with glucose.[41, 42]

In temperature-sensing systems, phase changes owing to changes in temperature cause polymers such as *N*-isopropylpolyacrylamide to swell and shrink, restricting or releasing the entrapped components.[43–46]

Biodegradable hydrogels cross-linked with hyaluronic acid serve as inflammation-sensing systems. At sites of inflammation leukocytes and macrophages produce hydroxyl radicals, which specifically break down the hyaluronic acid cross-linking, thus aiding in drug release at the site of inflammation.[47, 48]

13.4 Current Systems and Future Potential

Biomicroelectronic and microfabricated systems are actively targeted by many investigators and companies, as evident by efforts from MicroCHIPS to commercialize an electronically activated drug delivery microchip, ChipRx, to introduce systems that integrate silicon and electroactive polymer technologies for controlled delivery and iMEDD to develop nanoporous membranes and micromachined particles for a variety of drug delivery applications. An implantable biosensor, which is in the early stages of development and is being developed by ChipR$_X$, can be used in on-demand therapy. The system is the size of a matchstick.

Products that are currently under development for commercialization are external and implantable microchips for the delivery of proteins, hormones, pain medications, and other pharmaceutical compounds. Microchips can be developed as a "pharmacy on a chip" because different drugs can be placed in different reservoirs of the same microchip, and the release could be achieved by applying the electrical potential to a specific reservoir.

Two major impediments to successful commercialization of bio-MEMS devices have been evaluation of these systems and their packaging. It is estimated that packaging constitutes of up to 80 percent of the device final cost. Packaging is a particularly challenging task if the device is to be implanted, where hermeticity and biocompatibility are of paramount importance. Careful considerations in the design and engineering of the systems at several different levels (material, device, system, testing, and packaging) are critical to the success of the final product.

It is reasonable to recognize that bio-MEMS are a promising field in drug delivery research. However, it encompasses a tremendously enabling array of techniques that provide new approaches for solving outstanding problems in drug delivery.

Some of the current programmable drug delivery systems that employ polymer-based programmable release include Covera-HS, Verelan PM, Cardizem LA, Innopran XL, Uniphyl, and naproxen sodium extended release formulation from Andrx Pharmaceuticals. Covera-HS releases verapamil 4 to 5 hours after ingestion. This lag time is introduced and controlled by a soluble membrane coating between the drug core and the outer semipermeable membrane. Water from the gastrointestinal tract enters the tablet, dissolves the soluble coating causing the osmotic layer to expand, which pushes the drug layer, releasing the drug through a precision laser-drilled orifice in the outer membrane at a constant rate. The biologically inert components of the delivery system remain intact during GI transit and are eliminated in the feces as an insoluble shell.

The sodium naproxen extended release formulation from Andrx Pharmaceuticals contains a portion of the drug for initial burst release and another portion in the form of a sustained release matrix. Plasma levels of the drug are detected within 30 minutes of dosing, with peak plasma levels occurring at about 5 hours after dosing.

The major drawbacks in current polymer-based programmable drug delivery systems arise from biological variations among individuals, and hence these systems may exhibit variations in in vitro models. Food effects in the gastrointestinal tract often lead to altered release patterns and result in a deviation from predicted plasma drug concentrations. The implantable systems may suffer from biofouling, i.e., growth of cells and tissues over the surface of the implant. This often hampers their release kinetics and performance. Some closed-loop systems have been successfully applied and the sensors give predictable results when incorporated in ex-vivo systems (e.g., Biostator).

13.5 Conclusions

To conclude, it is time for all pharmaceutical technologists to reevaluate the potential of programmable systems for drug delivery, which are still in the developmental stage. The key considerations in the design of these systems are their biocompatibility and the toxicity of the polymer-based devices, response time and sensitivity to the stimulus and markers, the ability to maintain desired levels of the drugs, and the routine formulation issues regarding dosage design, sterility, shelf life, and reproducibility. The immense pharmacological benefit from these systems should make this an important and rewarding area for future research.

References

1. Reinberg, A., and Smolensky, M. H. Circadian changes of drug disposition in man. *Clin. Pharmacokinet.* **7**:401–420, 1982.
2. Petit, E., Milano, G., Levi, F., et al. Circadian rhythm-varying plasma concentration of 5-fluorouracil during a five-day continuous venous infusion at a constant rate in cancer patients. *Cancer Res.* **48**:1676–1679, 1988.
3. Queneau, P., Decousus, H., Ollagnier, M., et al. [Chronokinetics of ketoprofen administered orally and by continuous venous infusion.] *Rev. Rhum. Mal. Osteoartic.* **52**:403–408, 1985.
4. Sanders, A. B., and Moore, J. G. Intragastric pH and pharmacokinetics of intravenous ranitidine during sinusoidal and constant-rate infusions. *Chronobiol. Int.* **8**:267–276, 1991.
5. Philip-Joet, F., Bruguerolle, B., Lagier, F., et al. Effects of a constant dose rate of terbutaline on circadian peak expiratory flow, heart rate and systolic arterial pressure in patients with asthma exacerbation. *Respiration* 59:197–200, 1992.
6. Focan, C., Doalto, L., Mazy, V., et al. [48-hour continuous infusion of vindesine (followed by cisplatin) in advanced lung cancer: Chronopharmacokinetic data and clinical efficacy.] *Bull. Cancer* **76**:909–912, 1989.
7. Decousus, H., Ollagnier, M., Cherrah, Y., et al. Chronokinetics of ketoprofen infused intravenously at a constant rate. *Annu. Rev. Chronopharmacol.* **3**:321–322, 1986.
8. Santini, J. T., Jr., Richards, A. C., Scheidt, R., et al. Microchips as controlled drug-delivery devices. *Angew Chem. Int. Ed. Engl.* **39**:2396–2407, 2000.
9. Santini, J. T., Jr., Richards, A. C., Scheidt, R. A., et al. Microchip technology in drug delivery. *Ann. Med.* **32**:377–379, 2000.
10. Santini, J. T., Jr., Cima, M. J., and Langer, R. A controlled-release microchip. *Nature* **397**:335–338, 1999.
11. LaVan, D. A., McGuire, T., and Langer, R. Small-scale systems for in vivo drug delivery. *Nature Biotechnol.* **21**:1184–1191, 2003.
12. Low, S. S., He, K., and Madou, M. J. Microactuators toward microvalves for responsive controlled drug delivery. *Sensors Actuators [B] Chem.* **67**:149–160, 2000.
13. Madou, M., and Florkey, J. From batch to continuous manufacturing of microbiomedical devices. *Chem. Rev.* **100**:2679–2692, 2000.
14. Maillefer, D., Rey-Mermet, G., Hirschi, R. A high-performance silicon micropump for an implantable drug delivery system. 12th IEEE MEMS 1999, *Technical Digest*, Orlando, FL, 1999.

15. van Lintel, H., and Bouwstra, S. A piezoelectric micropump based on micromachining of silicon. *Sensors Actuators [A]* **15**:153–168, 1988.
16. Rizk, N. I., and Stevens, C. L. Implantable electrophoretic pump for ionic drugs and associated methods, U.S. Patent 4639244, 1987.
17. Portner, P. M., and Jassawalla, J. S. Implantable infusion device, U.S. Patent 4360019, 1982.
18. Murdan, S. Electro-responsive drug delivery from hydrogels. *J. Contr. Rel.* **92**:1–17, 2003.
19. Ramanathan, S., and Block, L. H. The use of chitosan gels as matrices for electrically modulated drug delivery. *J. Contr. Rel.* **70**:109–123, 2001.
20. Khoo, W. J. T. Veterinary implants for delayed pulsed release of antigen, in *Proceed. Int. Symp. Control. Rel. Bioact. Mater.*, San Diego, CA, 1994, pp. 116–117.
21. Franssen, O., Vandervennet, L., Roders, P., and Hennink, W. E. Degradable dextran hydrogels: Controlled release of a model protein from cylinders and microspheres. *J. Contr. Rel.* **60**:211–221, 1999.
22. Kikuchi, A., Kawabuchi, M., Watanabe, A., et al. Effect of Ca^{2+}-alginate gel dissolution on release of dextran with different molecular weights. *J. Contr. Rel.* **58**:21–28, 1999.
23. Stubbe, B. G., De Smedt, S. C., and Demeester, J. "Programmed polymeric devices" for pulsed drug delivery. *Pharm. Res.* **21**:1732–1740, 2004.
24. Gopferich, A. Bioerodible implants with programmable drug release. *J. Contr. Rel.* **44**:271–281, 1997.
25. Jiang, H. L., and Zhu, K. J. Pulsatile protein release from a laminated device comprising of polyanhydrides and pH-sensitive complexes. *Int. J. Pharm.* **194**:51–60, 2000.
26. Qiu, L. Y., and Zhu, K. J. Design of a core-shelled polymer cylinder for potential programmable drug delivery. *Int. J. Pharm.* **219**:151–160, 2001.
27. Kibat, P. G., Igari, Y., Wheatley, M. A., et al. Enzymatically activated microencapsulated liposomes can provide pulsatile drug release. *Faseb J.* **4**:2533–2539, 1990.
28. Ueda, S., Hata, T., Asakura, S., et al. Development of a novel drug release system, time-controlled explosion system (TES): I. Concept and design. *J. Drug Target.* **2**:35–44, 1994.
29. Thiel, W. J., Barr, I., and Kleinig, M. In vitro and in vivo testing of a single dose vaccination system. *Int. Symp. Control. Rel. Bioact. Mater.*, 1994.
30. Cardamone, M., Lofthouse, S. A., Lucas, J. C., et al. In vitro testing of a pulsatile delivery system and its in vivo application for immunisation against tetanus toxoid. *J. Contr. Rel.* **47**:205–219, 1997.
31. Iskakov, R. M., Kikuchi, A., and Okano, T. Time-programmed pulsatile release of dextran from calcium-alginate gel beads coated with carboxy-*n*-propylacrylamide copolymers. *J. Contr. Rel.* **80**:57–68, 2002.
32. Sharma, V. K. Process for the pulsatile delivery of diltiazem HCL and product produced thereby, U.S. Patent 5,914,134, 1999.
33. Wise, D. L. *Handbook of Pharmaceutical Controlled Release Technology*. New York: Marcel Dekker, 2000, pp. 65–89.
34. Clemens, A. H., Chang, P. H., and Myers, R. W. The development of Biostator, a glucose controlled insulin infusion system (GCIIS). *Horm. Metab. Res.* **7**:23–33, 1977.
35. Fogt, E. J., Dodd, L. M., Jenning, E. M., and Clemens, A. H. Development and evaluation of a glucose analyzer for a glucose controlled insulin infusion system (Biostator). *Clin. Chem.* **24**:1366–1372, 1978.
36. Lysy, Y., Globus, M., and Chowers, I. Control of hypoglycemia by the Biostator. *Isr. J. Med. Sci.* **20**:727–728, 1984.
37. Blanco, E. E., and Samadani, R. Automatic blood monitoring for medication delivery method and apparatus, U.S. Patent 5109850, 1992.
38. Palti, Y. Implantable drug delivery pump, U.S. Patent 5,474,552.
39. Ng, S. Y., and Heller, J. Block copolymers based on poly(ortho esters) containing amine groups, U.S. Patent 6667371, 2003.
40. Kitano, S., Koyama, Y., Kataoka, K., et al. A novel drug delivery system utilizing a glucose responsive polymer complex between poly (vinyl alcohol) and poly(*N*-vinyl-2-pyrrolidone) with a phenylboronic acid moiety. *J. Contr. Rel.* **19**:161–170, 1992.

41. Shiino, D., Murata, Y., Kubo, A., et al. Amine containing phenylboronic acid gel for glucose-responsive insulin release under physiological pH. *J. Contr. Rel.* **37**:269–276, 1995.
42. Sung Wan Kim, Chaul Min Paii, Kimiko Makino, et al. Self-regulated glycosylated insulin delivery. *J. Contr. Rel.* **11**:193–201, 1990.
43. Liu, X.-M., Wang, L. S., Wang, L., et al. The effect of salt and pH on the phase-transition behaviors of temperature-sensitive copolymers based on *N*-isopropylacrylamide. *Biomaterials* **25**:5659–5666, 2004.
44. Motonaga, T., and Shibayama, M. Studies on pH and temperature dependence of the dynamics and heterogeneities in poly(*N*-isopropylacrylamide-co-sodium acrylate) gels. *Polymer* **42**:8925–8934, 2001.
45. Mi, K. Y., Yong, K. S., Young, M. L., and Chong, S. C. Effect of polymer complex formation on the cloud-point of poly(*N*-isopropylacrylamide) (PNIPAAm) in the poly(NIPAAm-co-acrylic acid): polyelectrolyte complex between poly(acrylic acid) and poly(l-lysine). *Polymer* **39**:3703–3708, 1998.
46. Akerman, S., Viinikka, P., Svarfvar, B., et al. Drug permeation through a temperature-sensitive poly(*N*-isopropylacrylamide) grafted poly(vinylidene fluoride) membrane. *Int. J. Pharm.* **164**:29–36, 1998.
47. Nobuhiko Yui, Teruo Okano, and Yasuhisa Sakurai. Inflammation responsive degradation of cross-linked hyaluronic acid gels. *J. Contr. Rel.* **22**:105–116, 1992.
48. Nobuhiko Yui, Jun Nihira, Teruo Okano, and Yasuhisa Sakurai. Regulated release of drug microspheres from inflammation responsive degradable matrices of crosslinked hyaluronic acid. *J. Contr. Rel.* **25**:133–143, 1993.

Index

ABOUT THE EDITORS

XIAOLING LI, PH.D., is a professor and chair of the Department of Pharmaceutics and Medicinal Chemistry, Thomas J. Long School of Pharmacy & Health Sciences, University of the Pacific, Stockton, California. Professor Li received his Ph.D. degree from the University of Utah and had his postdoctoral training at Ciba-Geigy (now Novartis). His research interest areas are design and synthesis of novel polymers for pharmaceutical and biomedical applications, targeted drug delivery, and transport of drugs across biological barriers. He holds two patents, has published more than 30 papers, and had more than 70 presentations at national and international conferences. He serves as a consultant for various pharmaceutical and biotechnology companies.

BHASKARA R. JASTI, PH.D. is an associate professor in the Department of Pharmaceutics and Medicinal Chemistry, Thomas J. Long School of Pharmacy & Health Sciences, University of the Pacific, Stockton, California. Prior to joining the University of the Pacific, he worked as a staff scientist at Cygnus Therapeutics Systems and as an assistant professor at Wayne State University in the Departments of Pharmacy and Internal Medicine, where he also acted as an assistant director of pharmacology core. His current research interests are identifying the barriers for drug delivery and the design of targeted and mucosal drug delivery systems. Dr. Jasti has published more than 30 papers and presented more than 60 papers at various national and international meetings.